Advances in poultry nutrition

It is widely recognised that agriculture is a significant contributor to global warming and climate change. Agriculture needs to reduce its environmental impact and adapt to current climate change whilst still feeding a growing population, i.e. become more 'climate-smart'. Burleigh Dodds Science Publishing is playing its part in achieving this by bringing together key research on making the production of the world's most important crops and livestock products more sustainable.

Based on extensive research, our publications specifically target the challenge of climate-smart agriculture. In this way we are using 'smart publishing' to help achieve climate-smart agriculture.

Burleigh Dodds Science Publishing is an independent and innovative publisher delivering high quality customer-focused agricultural science content in both print and online formats for the academic and research communities. Our aim is to build a foundation of knowledge on which researchers can build to meet the challenge of climate-smart agriculture.

For more information about Burleigh Dodds Science Publishing simply call us on +44 (0) 1223 839365, email info@bdspublishing.com or alternatively please visit our website at www.bdspublishing.com.

Related titles:

Achieving sustainable production of poultry meat Volume 1
Print (ISBN 978-1-78676-064-7); Online (ISBN 978-1-78676-066-1, 978-1-78676-067-8)

Achieving sustainable production of poultry meat Volume 2
Print (ISBN 978-1-78676-068-5); Online (ISBN 978-1-78676-070-8, 978-1-78676-071-5)

Developing animal feed products
Print (ISBN 978-1-78676-463-8); Online (ISBN 978-1-78676-465-2, 978-1-78676-466-9)

Improving gut health in poultry
Print (ISBN 978-1-78676-304-4); Online (ISBN 978-1-78676-306-8, 978-1-78676-307-5)

Improving poultry meat quality
Print (ISBN 978-1-80146-103-0); Online (ISBN 978-1-80146-105-4, 978-1-80146-106-1)

Optimising poultry flock health
Print (ISBN 978-1-78676-887-2); Online (ISBN 978-1-78676-889-6, 978-1-78676-890-2)

Seaweed and microalgae as alternative sources of protein
Print (ISBN 978-1-78676-620-5); Online (ISBN 978-1-78676-622-9, 978-1-78676-623-6)

Chapters are available individually from our online bookshop:
https://shop.bdspublishing.com

climate•SMART•publishing
IN AGRICULTURAL SCIENCE

BURLEIGH DODDS SERIES IN AGRICULTURAL SCIENCE
NUMBER 159

Advances in poultry nutrition

Edited by: Professor Todd J. Applegate, University of Georgia, USA

burleigh dodds
SCIENCE PUBLISHING

Published by Burleigh Dodds Science Publishing Limited
82 High Street, Sawston, Cambridge CB22 3HJ, UK
www.bdspublishing.com

Burleigh Dodds Science Publishing, 1518 Walnut Street, Suite 900, Philadelphia, PA 19102-3406, USA

First published 2024 by Burleigh Dodds Science Publishing Limited

Library of Congress Control Number: 2024933618

British Library Cataloguing in Publication Data
A catalogue record for this book is available from the British Library

ISBN 978-1-80146-731-5 (print)
ISBN 978-1-80146-733-9 (PDF)
ISBN 978-1-80146-732-2 (ePub)
ISSN 2059-6936 (print)
ISSN 2059-6944 (online)

DOI: 10.19103/AS.2024.0143

Typeset by Deanta Global Publishing Services, Dublin, Ireland

Contents

Part 4 Novel feed sources

Series list

Title	Series number
Achieving sustainable cultivation of maize - Vol 1 From improved varieties to local applications Edited by: Dr Dave Watson, CGIAR Maize Research Program Manager, CIMMYT, Mexico	001
Achieving sustainable cultivation of maize - Vol 2 Cultivation techniques, pest and disease control Edited by: Dr Dave Watson, CGIAR Maize Research Program Manager, CIMMYT, Mexico	002
Achieving sustainable cultivation of rice - Vol 1 Breeding for higher yield and quality Edited by: Prof. Takuji Sasaki, Tokyo University of Agriculture, Japan	003
Achieving sustainable cultivation of rice - Vol 2 Cultivation, pest and disease management Edited by: Prof. Takuji Sasaki, Tokyo University of Agriculture, Japan	004
Achieving sustainable cultivation of wheat - Vol 1 Breeding, quality traits, pests and diseases Edited by: Prof. Peter Langridge, The University of Adelaide, Australia	005
Achieving sustainable cultivation of wheat - Vol 2 Cultivation techniques Edited by: Prof. Peter Langridge, The University of Adelaide, Australia	006
Achieving sustainable cultivation of tomatoes Edited by: Dr Autar Mattoo, USDA-ARS, USA and Prof. Avtar Handa, Purdue University, USA	007
Achieving sustainable production of milk - Vol 1 Milk composition, genetics and breeding Edited by: Dr Nico van Belzen, International Dairy Federation (IDF), Belgium	008
Achieving sustainable production of milk - Vol 2 Safety, quality and sustainability Edited by: Dr Nico van Belzen, International Dairy Federation (IDF), Belgium	009
Achieving sustainable production of milk - Vol 3 Dairy herd management and welfare Edited by: Prof. John Webster, University of Bristol, UK	010
Ensuring safety and quality in the production of beef - Vol 1 Safety Edited by: Prof. Gary Acuff, Texas A&M University, USA and Prof. James Dickson, Iowa State University, USA	011
Ensuring safety and quality in the production of beef - Vol 2 Quality Edited by: Prof. Michael Dikeman, Kansas State University, USA	012
Achieving sustainable production of poultry meat - Vol 1 Safety, quality and sustainability Edited by: Prof. Steven C. Ricke, University of Arkansas, USA	013
Achieving sustainable production of poultry meat - Vol 2 Breeding and nutrition Edited by: Prof. Todd Applegate, University of Georgia, USA	014

Introduction

Poultry nutrition faces many challenges, including the need to meet the changing requirements of growing birds whilst also avoiding over-nutrition and nutrient losses which can be both environmentally and economically damaging. With the sector also facing increasing pressure to reduce its reliance on antibiotics, a greater understanding of the poultry gut, its function and its role in optimising overall bird health is required.

This volume provides a detailed overview of these issues. Discussion is split into four parts: Part 1 chapters cover the nutrient requirements of broilers, poultry digestive anatomy and physiology, the digestive kinetics of feedstuffs in broilers, poultry feed intake regulation, the cellular basis of feed efficiency in poultry muscle, and genetic factors affecting feed efficiency in meat-type chickens. Chapters in Part 2 explore the safety of poultry feed, feedstuff management and feed formulation utilizing big data, and broiler breeder nutrition. Part 3 chapters describe the use of amino acids in poultry nutrition, the use of enzymes in broiler nutrition, the use of phosphorus, and the use of probiotics and prebiotics to improve gut function. Finally, chapters in Part 4 review emerging protein sources for poultry feed, the use of phytogenics in optimising gut function, the use of macro- and microalgae in poultry feed, synthetic biology to improve poultry production, and high protein corn fermentation products for poultry derived from corn ethanol production.

Part 1 Physiology of feed intake

The first chapter of Part 1 provides an overview of the nutrient requirements of broilers. Chapter 1 begins by outlining the factors affecting broiler feed formulation, including: feed intake, cost, and societal needs. Next, trends in feed formulation are discussed, covering energy content reduction, crude protein, amino acids, dietary fibre and macro-minerals. This is followed by a section describing current feeding programmes. Finally, the chapter highlights ways to improve feeding programmes, such as phase feeding, adapting feed form and particle size, and precision nutrition.

Chapter 2 focuses on advances in understanding poultry digestive anatomy and physiology. It opens with an exploration of the factors that affect the attractiveness of feed, before turning attention to the role of the crop in the digestive system. This is followed by sections on the role of the gizzard, small intestine, and caeca and colon.

Next, Chapter 3 discusses the digestive kinetics of feedstuffs in broiler chickens. It begins by reviewing digestion and intestinal uptakes of nutrients, in particular the transition across the gut mucosa. Following this, the chapter examines rapidly digestible protein, comparing intact protein to non-bound amino acids and oligopeptides, and discussing ammonia overload. A section on slowly digestible starch is also included. The chapter concludes with an exploration of starch-protein digestive dynamics.

The following chapter offers a review of the current understanding on feed intake regulation, highlighting the potential molecular mechanisms by which environmental and nutritional stressors modulate energy intake. Chapter 4 begins by outlining the factors influencing feed intake, including flock size and stocking density and lighting. It then covers the ways in which stress and stressors, such as noise and sounds, and heat and cold stress, can affect flocks. Sections describing hypoxia, and nutritional stress and nutrient sensing systems are also included.

Chapter 5 concerns the advances in understanding the cellular basis of feed efficiency in poultry muscle. It opens with an overview of the mitochondria, before delving into mitochondrial physiology, covering: oxidative phosphorylation and stress, proton leak, the phosphocreatine mitochondrial-cytosol energy transfer system, and mitochondrial biogenesis. Next, the chapter looks at understanding feed efficiency in muscle at the cellular level, including: mitochondrial function, biochemistry, and oxidative stress in feed efficiency. Sections investigating enhanced energy production, transfer systems, nucleotide metabolism, ribosome assembly, and protein translation in feed efficiency, as well as muscle development and progesterone signalling follow.

The final chapter of Part 1 explores the genetic factors affecting feed efficiency in meat-type chickens. Chapter 6 starts with an overview of the concepts of feed efficiency, in particular feed conversion ratio and residual feed intake. Next, genetic factors affecting feed efficiency are described: environmental feed factors, bid factors, and feed efficiency and output factors.

Part 2 Feed formulation

Chapter 7 highlights those contaminants that are considered to pose the most significant risks to poultry and human health, namely mycotoxins, dioxins, and bacterial pathogens. Each section discusses the negative effects of these factors as well as possible control measures that may be implemented to reduce feed contamination and secure feed and food safety.

Next, Chapter 8 addresses current challenges facing feed formulation programmes, including the need to account for sustainability issues as well as the increasing range and volume of high-quality data on feedstuff composition and effects on poultry growth. The chapter discusses solutions, such as tools for importing new data into the formulation database, developing linear and non-linear optimizers to handle a greater range and volume of data and the role of artificial intelligence (AI) in future feed formulation.

Chapter 9 describes advances in broiler breeder nutrition. It beings with an exploration of the diet composition of hens, covering diluted diets, body composition, low dietary protein during the laying phase, transgenerational effects, and vitamins and minerals. The chapter then discusses feeding strategies for hens, including feeding every day and feeding on the litter during rear, and twice-a-day feeding during lay. Water management for breeders is also examined. The chapter ends with a section on the diets of male chickens.

Part 3 Individual nutrients and additives

The first chapter of Part 3 looks at the use of amino acids in poultry nutrition. Chapter 10 reviews four amino acid minimums: arginine, glycine and serine, histidine, and phenylalanine and tyrosine.

Chapter 11 provides a review of the use of enzymes in broiler nutrition. It begins with an exploration of phytases, covering factors affecting phytase efficacy, such as the concentration of calcium and inorganic phosphorus in the diet, and the inositol-related effects of phytase. Next, the chapter looks at proteases and their effects on growth performance and nutrient digestibility, as well as their practical applications. The chapter then brings into focus the carbohydrases xylanases and amylase and their individual potentials in enzyme combinations.

Following this, Chapter 12 discusses the use of the macro-mineral phosphorus (P) in poultry nutrition. The chapter begins by outlining the challenges to optimising mineral nutrition in poultry, covering the characterisation of the P value in feed ingredients and factors that affect P availability. Next, a section on case studies is explored, including a meta-analysis for assessing the P value of a diet and a model where the P and calcium (Ca) quantities required for maintenance and growth are aggregated to determine the overall needs of the animals. Finally, future trends are discussed, covering precision feeding for sustainability, the potential for compensatory mechanisms and genetic determinism, the balance between Ca and P in laying hens, and a digestible Ca system.

The last chapter of Part 3 focuses on the use of probiotics and prebiotics to improve gut function and immunity in poultry. Chapter 13 describes probiotic and prebiotic interactions with avian gut flora, as well as their impacts on poultry intestine bacteria balance, architecture, nutrient digestion, and immune control. The chapter also discusses issues in using probiotics and prebiotics.

Part 4 Novel feed sources

Part 4 begins with Chapter 14 which examines emerging protein sources for poultry feed. The challenges of using conventional protein sources in poultry diets are described before the chapter brings into focus potential alternative sources of protein. Both animal and plant sources are discussed, as well as the chemical, biological, and physical treatments that may be used to improve the nutritional value of these alternative protein sources. A section on the environmental sustainability of using alternative protein sources is also included. Finally, the chapter looks at consumer perceptions of feeding animals with alternative feed sources.

Chapter 15 covers the use of phytogenics in optimising gut function and health in poultry. It begins by evaluating the main types of phytogenic compound (terpenoids, flavonoids and alkaloids), as well as general issues relating to phytogenic variability and bioavailability. Following sections review relevant aspects of poultry gut function, particularly critical signalling pathways and gene components that are directly linked to optimal gut function and health. The chapter then discusses the beneficial effects of phytogenics in the maintenance and protection of gut function and health, focussing particularly on gut barrier integrity, cytoprotection and inflammatory response.

Chapter 16 describes the main nutritional characteristics of algae and the current knowledge on their effects in poultry production with consideration to their impacts on animal health, growth performance, and product quality (eggs and meat). The chemical and nutritional value of algae is explored, along with extraction methods and the benefits of algae to animals. The chapter then describes algae's impacts on laying hen production, covering egg productivity, quality and composition, as well as hens' health. Finally, algae's effect on broiler chicken production is reviewed, covering growth performance, meat production and quality, and broiler health.

Next, Chapter 17 looks at advances in synthetic biology to improve poultry production. The chapter explores the syntheses of proteins including enzymes by bacterial, fungal, and plant systems. It also covers application of synthesis biology to produce alternative proteins, nutrient supplements, antimicrobial peptides, and prebiotics for poultry feed. Finally, the use of synthetic biology in waste valorisation is reviewed.

The final chapter of the book explores high protein corn fermentation products for poultry derived from corn ethanol production. Chapter 18 starts by describing the ethanol production by-product, distillers dried grains with solubles, and its use as poultry feed. Next, the production process and value of corn fermented protein (CFP) are covered, followed by the challenges to its production. Focus then turns to two case studies. These are: the use of corn-fermented protein as a feed for broilers and for turkey poults.

Part 1

Physiology of feed intake

Chapter 1

Nutrient requirements of broilers: an overview of factors affecting feed formulation

Gonzalo G. Mateos, Guillermo Talegón, Lewis Aguirre, Nereida L. Corrales, and Rosa Lázaro, Universidad Politécnica de Madrid, Spain

1 Introduction

The demand for poultry meat has increased substantially over the past 40 years because of a growing global population with increasing incomes and the move of populations from rural to urban areas. The reasons for this increased demand for poultry meat include properties such as high protein and amino acid (AA) content, affordable cost, acceptance by all cultures, reduced greenhouse emissions and carbon footprint from broiler production, and an image among consumers as a healthy food (Magdelaine et al., 2008; Zuidhof et al., 2014; Moss et al., 2021). As a result, poultry meat has become the main source of high-quality animal protein for the world population and is expected to account for over 40% of all meat produced by 2030 (Poultry Trends, 2023).

The increase in broiler growth, especially in Asian Pacific and Latin American countries, has been a consequence of breeding programs, followed by improvements in flock management (e.g. increases in farm size, better equipment, and adequate vaccination programs) and nutrition practices. From hatch, broilers might now reach a bodyweight (BW) of 3.3 kg, with a feed efficiency (g of feed per kg of BW gain) of 1.500 at 40 days of age (Cobb, 2022).

http://dx.doi.org/10.19103/AS.2024.0143.01

Table 1 shows comparative data on the growth performance of Cobb 500 and Ross 308 broilers according to 2022 guidelines, including results from a recent experimental trial at the Universidad Politécnica de Madrid. The data show that the potential for growth of current broiler strains is even greater than that suggested by the industry.

2 Factors affecting broiler feed formulation

Main factors affecting nutrient requirements of broilers and nutrient percentages in feed formulas are feed intake (FI), growth potential, lean tissue to fat ratio, and feed efficiency. These, in turn, are affected by factors such as:

- Bird characteristics (e.g. genetics (e.g. fast vs. low growth rate), age, sex, feathering);
- Environment (e.g. barn design, ambient temperature, ventilation system, bedding material, litter quality);
- Management (e.g. stocking density and thinning according to age, litter management, biosecurity, disease prevention, vaccination programs);
- Feeding programs (e.g. number of phases);
- Feed composition (e.g. use of traditional and new ingredients, additives, antibiotics) and amounts of different nutrients;
- Feed quality (e.g. consistent nutrient content, safety, ingredient availability); and
- Feed form (e.g. mash vs. crumbles vs. pellets) and particle size.

This section looks at a number of factors.

Table 1 Growth potential of straight run broilers from hatch to 42 d of age.

	Ross 308		Cobb 500		UP Madrid[1]
Year	(2019)	(2022)	(2018)	(2022)	(2022)
Initial body weight (g)	43	44	42	42	46
Final body weight (g)	2,918	2,998	2,952	3,278	3,426
Body weight gain (g/d)	69.5	70.0	70.0	77.1	84.5
Feed intake (g/d)	111.9	109.2	113.4	121.4	122.8
Feed conversion ratio (g/g)	1.611	1.531	1.610	1.555	1.453

[1] Universidad Politécnica de Madrid (Cobb 500 broilers fed corn/soybean meal diets) Kadardar et al. (2022).

2.1 Feed intake

Feeding strategies and nutritional practices need to take account of constant changes in feed intake. Originally, genetic programs were based on increasing the capacity of the birds to eat which in turn resulted in an increase in growth rate and feed efficiency. Consequently, poultry nutritionists must continuously evaluate the adequacy of their diets for the type of bird and the phase of the feeding program to provide optimal levels of nutrients at an economical cost.

Leeson and Summers (2001) have defined the requirement for a given nutrient as 'the minimum amount of the nutrient needed to produce the best weight gain and feed efficiency, resulting at the same time in a lack of any sign of nutritional deficiency'. In general, it is accepted that birds eat to satisfy their energy requirements, provided that the ratio between energy and key nutrients is maintained constant and that the capacity of the gastrointestinal tract (GIT) does not limit FI, and the birds are physically able to eat enough feed to satisfy their requirements. Consequently, the accuracy in estimating voluntary FI is the starting point in formulating diets. Once FI is estimated, the requirements for energy, AA, and other nutrients are calculated. In most cases, the factorial method is used, in which the needs for maintenance (e.g. metabolic body weight (BW), environmental temperature, and physical activity) and production (BW gain and composition of the gains) are estimated (FEDNA, 2018).

2.2 Cost

Cost is the underpinning of any feed formulation program. Feed represents 60–65% of the total cost of chicken-meat production (Table 2). In addition to managing the cost of feed ingredients, improvements in feed efficiency help the sector to maintain a leading position as the main supply of animal protein in many parts of the world. This requires improving knowledge of broiler requirements, and more precisely adjusting the composition and nutritive value of diets to broilers' needs.

Table 2 Cost of production of broiler meat (% of live cost)

	2000	2005	2010	2015	2020	2023[1]
Chick	16.2	14.6	12.7	13.0	15.9	14.6
Housing	19.6	19.9	15.8	16.3	18.6	14.7
Feed	52.0	53.5	62.1	61.4	55.1	61.9
Catch	5.0	5.1	4.0	3.9	4.4	3.7
Others	7.2	6.9	5.4	5.4	6.0	5.1

[1] First trimester 2023, USA data Agri Stats, 2023.

2.3 Societal needs

Broader societal needs affecting poultry feed refer to those areas beyond the immediate focus of the poultry industry on targets such as ensuring a sufficient supply and quality of meat at a competitive price. These broader needs include:

- Nutritional quality and acceptability;
- Food safety;
- Animal welfare; and
- Environmental sustainability.

Measures to address these areas will increase feed formulation cost in different ways, because the nutritionist will be obliged to eliminate the use of certain ingredients and additives, increase quality control testing, and add new restrictions into the feed formulation process. This and the following sections look at these issues.

2.3.1 Societal needs: nutritional quality and acceptability

An important aspect of poultry production is the carcass lean to fat ratios of modern broilers, since the proportion of lean meat in the carcass determines meat yield and the commercial value of the final product. Leanness is the most important criterion by which consumers judge quality and value for money. Broiler carcasses should have a maximum lean content, an optimum level of fatness, and a minimum of bone tissues. Depending on these factors, the proportion of energy and nutrients consumed by the bird for maintenance or meat production will change. Feed programs will also be affected by strategies to enhance nutritional content or sensory quality via feed ingredients and additives (e.g. high W-3 FA, vitamin E-enriched meat, pigmented carcasses using xanthophylls, etc.)

2.3.2 Societal needs: food safety

Food safety issues which affect feed formulation include:

- The ban on 'in feed' antibiotics as growth promoters or to prevent diseases such as clostridiosis and coccidiosis;
- Increased quality control of the feed production process and the ingredients used because of potential contamination risks from pathogenic bacteria (e.g. *Salmonella* spp.), fungi and mycotoxins or from residues of herbicides, insecticides or heavy metals; and

- The ban on use of certain meat and bone products from ruminant origin (e.g. UK and European Union-27 (EU-27)).

Any improvement in food safety has a cost. It also needs to be made clear to customers, especially in the developed countries, that 'zero' risk is not practically possible.

One of the main challenges facing the broiler industry is the ban on antibiotics at sub-therapeutic doses as growth promoters, as is the case in the EU-27. Many countries have also introduced bans on selected antibiotic growth promoters due to concerns over the development of antibiotic resistance in the microbiome of the GIT, with a consequent risk of transferring antibiotic-resistant pathogens to humans. The experience of the EU-27 after 18 years of the ban, has been positive, with a general reduction of antibiotics for growth promotion without showing any negative effect on broiler growth. The removal of antibiotics as growth promoters or as coccidiostats in feed facilitates the use of feed additives to promote GIT function and defense mechanisms. The withdrawal of in-feed antibiotics, including ionophores coccidiostats, has led to various changes in feed:

- A reduction in the level of crude protein (CP);
- An increase in particle size in feeds (despite its potential negative effects on pellet quality) through coarser grinding of the cereal; and
- An increase in the insoluble lignified, fiber content of the diets to promote anti-peristaltic movements, greater water reabsorption in the large intestine, and an increase in hydrochloric acid (HCl) production by the proventriculus.

These changes have aimed to increase the retention time of the digesta in the upper part of the digestive tract (González-Alvarado et al., 2008; Kalmendal et al., 2011; Mateos et al., 2012), resulting in:

- An improvement in GIT and gizzard function, with better development of the digestive tract with an increase in HCl production;
- A reduction in the multiplication of *Clostridium perfringens*, *Enterobacteriaceae*, and other intestinal pathogens; and
- Better quality of the excreta, with a reduction in wet litter problems.

Improved HCl production favors pepsin activity and phytate solubilization, facilitating protein digestibility, phytase activity, and phosphorus (P) and calcium (Ca) absorption (González-Alvarado et al., 2007, 2008; Jiménez-Moreno et al., 2019), with a reduction in the incidence of wet litter problems (González-Alvarado et al., 2010; Mateos et al., 2012; Jiménez-Moreno et al., 2016, 2019).

In addition, a reduction of the pH of the excreta reduces the growth of 'alkaline' bacteria and ammonia concentration in the litter, which in turn, benefits broiler production and animal welfare (Ferguson et al., 1998). More indirectly, this change has required broader work in biosecurity and management practices which have improved overall productivity in the poultry sector (Rafiq et al., 2022; Rahman et al., 2022).

The ban (or voluntary withdrawal) of antibiotics in feeds creates potential problems in feeding practices that have to be addressed by the nutritionist. Firstly, the reduction in CP content of the diet often requires extra control of certain AA, such as Gly + Ser, Val, Arg, Ile, and others that did not previously limit broiler performance. In addition, new studies on the minimum level of protein required by broilers at different ages might be needed (Wu and Lin, 2022).

The change has also put additional pressure to ensure feeds are free of potential bacterial contamination. This requires the use of adequate temperatures to eliminate any potential contamination post-mixing, and the inclusion of organic acids (e.g. formic acid) to inhibit bacterial activity. The need to control and reduce microbial counts in feed has increased the temperatures applied during the pelleting processes to values close to 90°C which, if exceeded, will reduce nutrient digestibility, especially in diets based on wheat or broken rice (González-Alvarado et al., 2007). It is also important to evaluate the effects of an excess of heat on the digestibility of nutrients, especially Lys, Cys, and Arg.

2.3.3 Societal needs: animal welfare

The increases in growth potential and feed efficiency of the birds observed for the last 40 years is not without cost and has often resulted in increases in metabolic problems such as ascites, cellulitis, sudden death syndrome, and skeletal disorders (Julian, 1998, 2005; Wideman, 2016; Biswas, 2019; Korver, 2023). More recently, the emphasis on broiler growth has affected the muscular tissues of rapidly growing broilers slaughtered at heavy weights with an increase in myopathies such as the Oregon disease (green psoas minor) wooden breast and spaghetti type meat (Rauw et al., 1998; Huang and Ahn, 2018; Petracci et al., 2019; Baldi et al., 2021; Soglia et al., 2021). This could be addressed by reducing daily growth and the BW of the birds at slaughter but this would have huge implications for the poultry industry (Lilburn et al., 2019; Korver, 2023). Biological limits will inevitably limit genetic improvements in the bird, which, based on recent trends, a BW of 3.0 kg, could theoretically be reached at one d of age in broilers in 2070 (Korver, 2023).

There are now additional need to estimate feed requirements for slow growth and organic chickens. The production of meat from broiler with medium or low rate of growth has increased in recent decades in the EU-27 with the objective to promote animal welfare. Other areas which affect feed requirements include reduction in rearing densities on welfare grounds with more space per bird, or allowing the access of the birds to open areas which increases energy needs without any change in the AA and mineral requirements.

2.3.4 Societal needs: environmental impact

Although poultry systems are more environmentally friendly than many other livestock systems, their environmental impact is still of concern. Improving the efficiency of chicken meat production will reduce environmental impacts in various ways. Reducing the amount of feed required per bird will require fewer crops to provide raw materials for feed, reducing pressure to clear land, reducing climate change impacts and promoting biodiversity. Higher feed efficiency also reduces the potential amount of poultry manure and thus its impact as an environmental pollutant. More broadly, an improvement in growth rate will reduce the time needed to reach slaughter weight. A reduction in the fattening period suggests that a higher proportion of the energy supplied to the bird is used for growth rather than for maintenance. Data provided by the industry show that the energy needed by the bird to reach commercial BW at slaughter decreased by 12% from 1995 to 2009. On the other hand, new models of feeding poultry, meeting at the same time the society needs, include the reduction in CP while maintaining AA content, the reduction in phosphorus by the use of increasing levels of novel phytases, and the use of organic chelates in substitution of inorganic sources of the trace minerals, such as Zn, Cu, and Mn.

3 Trends in feed formulation: energy, crude protein and amino acids, dietary fiber, and macro-minerals

There are numerous studies dedicated to pinpointing the requirements in key nutrients of broilers at different ages (FEDNA, 2018; Cobb, 2022; CVB, 2022; Ross, 2022). More recently, the use of models and prediction equations has allowed more accurate estimation of the energy and ileal digestibility of nutrients. However, an accurate determination of broiler requirements based on models is not available. The use of models helps to optimize the dose-response relation at each age and production state (e.g. EFG, 1995; Hauschild et al., 2015; Oviedo-Rondón, 2015; Sakomura et al., 2015; Da Costa et al., 2017). All models attempt to integrate information about the bird, feed characteristics (including cost), and the environmental conditions in which it is kept, with a final objective of simulating its potential performance (Emmans, 1981; EFG Software, 2006; Gous, 2007). However, despite the progress made in the last three decades, there are still many challenges to improve the accuracy of model predictions. All models contain some assumptions and thus, none can claim to be absolutely accurate.

Energy, AA, phosphorus (P), and additives are the most expensive components in poultry diets. Broiler diets are usually high in energy because of the high growth rate and fast rate of passage of digesta through the GIT. Most nutritionists formulate diets with a minimum of CP and a maximum of crude fiber (CF) in order to meet the requirements for AA and nitrogen-corrected apparent metabolizable energy (AME_n). Similarly, diets are formulated with high levels of calcium (Ca) and P sources ($CaCO_3$, monoCa, and diCa phosphate) to avoid leg problems on the farm and bone breakage at the slaughterhouse, and low levels of sodium chloride (NaCl) to avoid wet litter problems. Nutritionists have also taken advantage of other developments such as new macro-mineral sources (e.g. mono-calcium phosphate) and a growing range of additives (e.g. AA, organic acids, and phytase and other enzymes) to improve the sustainability and economy of the broiler production system.

Current feeding programs rely more than two decades ago on the use of synthetic AA, exogenous enzymes (especially phytases, carbohydrases, and proteases), and chelates minerals to reduce the levels of phosphorus (P) and calcium (Ca), protein (Nx6.25), and inorganic minerals (e.g. Cu and Zn) which, in turn, improves sustainability. Better design of poultry farms and improved poultry management practices, have also affected poultry nutritional requirements, resulting in modifying energy and nutrient content of diets, in part to control feed cost. Particular trends are discussed below.

3.1 Reducing energy content

Cost of production, per kg of carcasses weight, depends on the cost of the feed (approximately 60–65% of the total production cost). Energy and AA levels constitute up to 80% of the cost per unit of feed. Consequently, selecting the ideal level of energy and standardized ileal digestible (SID) Lys (and consequently of all other AA) in poultry diets depends not only on nutritional considerations but also on economic considerations.

Energy content of a feed formula will vary, depending on factors such as:

- Cost of essential AA and energy sources;
- Objectives of the production model (BW gain, feed efficiency, carcass and breast yield, etc.);
- Health status of birds;
- Uniformity of the flock (average BW at slaughter within an individual farm); and
- Safety margins applied to diet formulations.

In recent years, the trend toward reducing the energy and protein content of broiler diets, while maintaining and even increasing essential AA levels, has increased worldwide. The main reasons for the trend to decrease energy content are:

- The change in relative cost of lipid vs. cereal sources of feed and in the relative cost of using food by-products compared to both lipid and cereal sources;
- The improvement in broiler growth and management (e.g. bedding material and use of environmental-controlled buildings (often with a tunnel design); and
- The increase in FI and BW gain of the broilers which results in extra production of heat by the bird.

In the last few years, because of the development of the biofuel industry, the price of plant oils, cereals, and other ingredients high in starch has increased significantly. In addition, new ideas about developing a more circular economy have resulted in a relative increase in the availability of high-energy ingredients for feed manufacturing.

Current dietary formulation programs for broilers thus show a clear reduction in energy in favor of AA content. This is partly because broilers grow and eat twice as much as they did a few decades ago. Growth results from oxidation of feed to form meat and other tissues, a process that results in a high production of heat which, in turn, reduces the need to use food for heating purposes. Modern farms are also better insulated which reduces energy requirements without any effect on AA requirements. This has resulted in reducing the nitrogen-corrected apparent metabolizable energy:digestible lysine (AME_n:DLys) ratio of diets for modern broiler production compared with the recent past.

3.2 Crude protein

The inclusion of digestible AA in the diet reduces the need for CP. However, the inclusion of a minimum and a maximum of CP might be convenient in feed formulation:

- A minimum to offset any deficiency in AA becoming a limiting factor in bird growth; and
- A maximum to improve protein metabolism and bird physiology while, at the same time, reducing environmental contamination from the excreta.

Diets are now formulated with 2–4% less CP than few decades ago. The use of crystalline AA has increased to meet the requirements for increasing breast meat production. A reduction in CP benefits flock health and decreases the burden of excess N on the environment. In addition, animals cannot store any excess of protein in their bodies and therefore, the excess has to be eliminated via urine which is costly and with potential negative effects on the bird, because of the presence of toxic substances such as NH_3, indole, and other metabolites. Moreover, undigested protein favors the multiplication of *Clostridium perfringens* and other microbes present in the GIT that use N substances as the main substrate for growth, resulting in a higher incidence of GIT and wet litter problems. A reduction of the protein content of a broiler diet of 1–2%, while maintaining levels of AA content, will reduce litter N content by 10–20% (Ferguson et al., 1998).

3.3 Amino acids

The requirement for AA is based on the ideal protein concept. This concept refers to a protein source that provides the precise balance of AA required for optimal performance and to maximize growth, taking account of factors such as size, carcass weight, yield, and body composition (Glista et al., 1951; Fisher, 1980; Baker et al., 1996; Emmert et al., 1997). The ideal protein concept assumes that the amount of digestible AA supplied to the bird exactly meets its requirements and thus, all AA are limiting at the same time, growth performance.

The SID of Lys is used as a control. Lys has been widely adopted as the reference AA in poultry because, together with Met + Cys, it is typically the most limiting AA in practical broiler diets. Lys is also abundant in poultry ingredients, easy to analyze in the lab, and does not interact significantly with other AAs. Lys is primarily utilized for synthesis of body protein. Lys requirements in broilers change with age. Requirements have increased in fast growth-rate chickens because of continuous improvements in both FI and BW gain of birds and changes in the lipid to lean tissue ratio of carcasses (Rostagno et al., 2017; FEDNA, 2018).

3.4 Dietary fiber

Originally, dietary fiber was considered as an 'antinutritional factor' in poultry diets because of its negative effects on feed palatability and nutrient digestibility. The development of high nutrient density diets in the 1970s meant undervaluing the importance of the fiber fraction. Numerous studies emphasized the negatives effects of non-starch polysaccharides (NSP), present in small white grains (e.g. barley and wheat), on digesta viscosity and energy, AA, and nutrient digestibility. However, there is a renewed interest in fiber nutrition given potentially positive effects of inert fiber on gut health and microbiota profile.

A key factor is the type of fiber (soluble vs. insoluble or high vs. low in lignin) (de Lange, 2000; Leeson, 2012; Mateos et al., 2019). Soluble, non-lignified fiber affects digesta viscosity and has negative effects on litter quality, FI, and feed efficiency. High levels of certain soluble fiber sources, such as NSP (xylans and b-glucans) in cereals, and α-galactosidades (such as stachyose and raffinose), in soybeans and in most legumes, affect gut microflora, especially in diets without antibiotics (de Lange, 2000; Mateos et al., 2017a; Teague et al., 2023). Fermentation of NSP to volatile fatty acids, such as butyrate, can control pathogen growth and improve gut health (Brons et al., 2002). Non-lignified, insoluble fibers, such as oat hulls, will improve GIT function and health, although it can also reduce carcass yield because of the increase in size and weight of the gizzard and of the whole GIT (Jiménez-Moreno et al., 2013, 2019).

3.5 Macro-minerals

Five macro-minerals are usually considered in tables on nutrient requirements for poultry because in many cases the needs are over the amounts supplied by typical ingredients: Ca, P, and the electrolytes: Na, Cl, and K. The requirements in Ca and P are covered at present by the use of decreasing amounts of mineral sources such as calcium carbonate and dicalcium and monocalcium phosphate in most cases. The needs of both minerals decrease with age because most of these two minerals are used for bone mineralization. In addition, the use of phytase has reduced further the need of including extra amounts of Ca and P mineral sources in the diets. Moreover, the better knowledge on Ca and P requirements of the birds and on the availability of the sources, have permitted to reduce to a minimum the amount supplied, especially with the use of the new phytases at extra doses in the finisher phases of the broiler diets.

Feed phosphates are costly because of the limited reserves available worldwide of rock phosphates from which they are derived. Moreover, P-rich animal proteins are not approved in several regions worldwide. On the other hand, an excess of P in animal diets will harm the environment, specifically surface water bodies. Consequently, broilers must receive adequate amounts of dietary P to support their health and productivity, but a maximum amount should be set to avoid any oversupply. In this respect, new studies conducted worldwide (Angel et al., 2006; Mutucumarana et al., 2014) indicated the benefits of increasing the amount of digestible P in young broilers over current recommendations while reducing by over 30% the amounts used after 21–27 days of age.

4 Current feeding programs

There are numerous feeding programs available for broilers, with nutritional recommendations that differ widely according to factors such as:

- The genetic background of the bird (e.g. Cobb vs. Ross vs. Hubbard);
- The main objective of the program (e.g. maximize growth vs. improve carcass and meat quality); and
- The marketing channel of the final products (e.g. whole carcass vs. cutting parts vs. further processing).

Most of the information on broiler nutrition presented in this chapter will refer, unless otherwise indicated, to traditional broiler chickens with a high feed intake capacity and a fast rate of growth. The energy, AA, minerals, and vitamins needs for maintenance, or per gram of fat or lean tissue deposited, have not changed much in recent decades with no critical changes likely in the near future (Leeson, 2008, 2012).

Traditionally, rations are formulated based on the nutritional recommendations of breeding companies, universities or research centers, and feed or additives manufacturers. However, the values based on these recommendations (typically presented in tables) do not take into account the high variability existing among broiler farms, including those related to the birds (genetics, age, sex), the building (equipment, bedding, temperature, and cooling and ventilation systems), and management practices (rearing density, thinning according to age, health status, vaccination program, reduction of stress). Sources of variability affect both, AA and energy requirements of the birds (CVB, 2022).

The recommendations on nutrient requirements of broilers from hatch to 48 days of age, under a four-phase feeding program, as proposed by Mateos (personal communication), are shown in Table 3. A comparative study of current recommended requirements for pre-starter and finisher diets together with the evolution with time of these requirements are shown in Tables 4 and 5, and Tables 6 and 7, respectively. Recommended levels of vitamins and trace minerals for starter and finisher diets are presented in Table 8. The information in these tables shows the high variability in recommendation for key nutrients in meeting the main objective of broiler production: provide energy and nutrients to the bird to maximize benefits at a minimum cost.

Information on the energy and nutrient content of the pre-starter and finisher diets recommended according to the growth potential of different broiler lines is provided in Tables 9 and 10, respectively. The main difference among broiler lines is the energy and nutrient concentration of the diets which are reduced as the growth rate potential of the birds decreases.

In practical feeding, all nutritionists add 'safety margins' to nutrient requirements in order to protect chicken growth from contingencies, which are inherent to the manufacturing, processing itself and to factors such as (a) individual responses to particular environments, (b) variation in nutrient delivery to the birds vs, specifications in the initial formulation, (c) variation in ingredient

Table 3 Nutrient recommendations (%) of broilers from hatch to slaughter (Mateos, 2024)

Age (days)	0-10	11-24	25-37	>38
AMEn (kcal/kg)	2,930	2,990	3,070	3,110
Crude protein	21.2	19.8	18.2	17.2
Digestible amino acids				
Lys	1.26	1.15	1.03	0.95
Met + Cys	0.93	0.85	0.78	0.72
Thr	0.83	0.76	0.68	0.63
Trp	0.20	0.18	0.19	0.17
Ile	0.82	0.75	0.70	0.65
Val	0.95	0.86	0.81	0.75
Arg	1.31	1.20	1.09	1.01
Crude fiber[1]	2.9	3.2	3.2	3.2
Neutral detergent fiber[1]	9.0	9.5	10.0	10.0
Calcium	1.00	0.85	0.73	0.69
Digestible phosphorus	0.48	0.39	0.34	0.30
Potassium	0.51	0.50	0.46	0.40
Sodium[2]	0.21	0.18	0.17	0.16
Chlorine	0.17	0.17	0.16	0.15
Linoleic acid	0.65	0.60	0.55	0.50

[1] Levels could be increased in winter or in the presence of wet litter problems for greater gastrointestinal tract size and development.

nutrient composition, and (d) feed management and equipment systems that modify voluntary FI by the birds (Applegate and Angel, 2014).

5 Improving feeding programs

This section reviews a number of options for improving feeding programs.

5.1 Phase feeding

Modern broilers grow rapidly, with nutrient requirements changing daily throughout their production cycle. Ideally, this should mean a continuous change in the composition and nutritive value of a diet but, in practice, this is not feasible. A compromise is to supply a flock with 3–5 different diets to match different stages in their lives. However, there is a potential to increase the number of phases (and diets), optimizing the opportunity to blend two diets prior to feeding: a protein-dense and an energy-dense diet (Moss et al., 2020, 2021).

Table 4 Nutrient recommendations of broilers as hatched (%): Pre-starter diets

Institution	NRC[1] (1994)	Rostagno[2] (2017)	CVB (2018)	FEDNA (2018)	Cobb (2022)	Ross (2022)	Mateos (2024)
Age (days)	<21	<7	<7	<14	<10	<10	<10
AMEn (kcal/kg)	3,200	3,000	2,850	2,950	2,900	2,975	2,930
Crude protein	23.0	25.3	–	21.2	21.0	23.0	21.2
Digestible Lys	1.10	1.36	1.20	1.22	1.26	1.32	1.26
Ideal protein[3]							
Met + Cys	82	73	73	74	75	76	74
Thr	73	65	64	65	68	67	66
Trp	18	18	15	17	16	16	16
Ile	73	66	60	67	64	67	65
Val	82	75	77	79	76	76	75
Arg	114	105	107	105	108	106	104
Crude fiber	–	–	–	2.85	–	–	2.90
Calcium	1.00	1.01	0.88	0.98	0.96	0.95	1.00
Available phosphorus	0.45	0.48	–	0.48	0.58	0.50	0.52
Digestible phosphorus	–	0.42	0.40	0.45	–	–	0.48
Potassium	0.30	0.60	–	0.51	0.60	0.60	0.51
Sodium	0.20	0.23	–	0.19	0.16	0.18	0.21
Chlorine	0.20	0.20	–	0.17	0.16	0.18	0.17
Linoleic acid	1.00	1.10	–	0.80	1.20	1.25	0.65

[1] The values provided correspond to total amino acids.
[2] Nutritional requirements for high-performing male broiler chickens.
[3] Digestible AA: Digestible Lys.

Table 5 Nutrient recommendations (%) of broilers as hatched: finisher diets

Institution	NRC (1994)	Rostagno (2017)	CVB (2018)	FEDNA (2018)	Cobb (2022)	Ross (2022)	Mateos (2024)
Age (days)	>42	>43	>40	>37	>40	>40	>38
AMEn	3,200	3,300	3,000	3,120	3,100	3,125	3,110
Crude protein	18.0	17.7	–	17.5	17.0	18.0	17.2
Digestible Lys	0.85	0.97	0.97	0.92	0.96	1.02	0.95
Ideal protein[1]							
Met + Cys	71	74	73	74	77	76	76
Thr	80	66	64	65	65	67	66
Trp	19	18	15	17	18	16	18
Ile	73	68	60	67	66	67	68
Val	82	77	77	79	77	76	79
Arg	118	107	107	105	109	106	106
Crude fiber	–	–	–	3.05	–	–	3.2
Calcium	0.80	0.58	0.59	0.70	0.72	0.60	0.69
Available phosphorus	0.30	0.27	–	0.35	0.36	0.34	0.32
Digestible phosphorus	–	0.25	0.27	0.32	–	–	0.30
Potassium	0.30	0.60	–	0.40	0.60	0.60	0.40
Sodium	0.12	0.20	–	0.15	0.16	0.18	0.16
Chlorine	0.12	0.17	–	0.15	0.16	0.18	0.15
Linoleic acid	1.00	1.03	–	0.50	1.00	1.00	0.50

[1] Digestible AA: Digestible Lys, except for NRC (1998).

Table 6 Evolution with time of the nutrient recommendations of broilers as hatched (%): pre-starter diets

Institution	FEDNA (2008)	Cobb (2013)	Ross (2014)	FEDNA (2018)	Cobb (2018)	Ross (2020)	Cobb (2022)	Ross (2022)	Mateos (2024)
Age (days)	<7	<10	<10	<14	<8	<10	<10	<10	<10
AMEn (kcal/kg)	3,000	3,035	3,000	2,950	2,975	3,000	2,900	2,975	2,930
Crude protein	21.8	21.0	23,0	21.2	21.0	23.0	21.0	23.0	21.2
Digestible Lys	1.27	1.18	1,28	1.22	1.22	1.28	1.26	1.32	1.26
Ideal protein[1]									
Met + Cys	73	74	74	74	75	74	75	76	74
Thr	63	65	67	65	68	67	68	67	66
Trp	17	16	16	17	16	16	16	16	16
Ile	–	–	67	67	63	67	64	67	65
Val	–	75	75	79	73	75	76	76	75
Arg	–	105	107	105	105	107	108	106	104
Crude fiber	2.30			2.85	–		–	–	2.90
Calcium	1.00	0.90	0.96	0.98	0.90	0.96	0.96	0.95	1.00
Available phosphorus	0.45	0.45	0.48	0.48	0.45	0.48	0.58	0.50	0.52
Digestible phosphorus	0.40	–	–	0.45	–	–	–	–	0.48
Potassium	0.51	0.60	0.40	0.51	0.60	0.40	0.60	0.60	0.51
Sodium	0.22	0.16	0.16	0.19	0.16	0.16	0.16	0.18	0.21
Chlorine	0.17	0.17	0.16	0.17	0.16	0.16	0.16	0.18	0.17
Linoleic acid	1.50	1.00	1.25	0.80	1.00	1.25	1.20	1.25	0.65

[1] Digestible AA: Digestible Lys.

Table 7 Evolution with time of the nutrient recommendations (%) of broilers as hatched: finisher diets

Institution	FEDNA (2008)	Cobb (2013)	Ross (2014)	FEDNA (2018)	Cobb (2018)	Ross (2020)	Cobb (2022)	Ross (2022)	Mateos (2024)
Age (days)	>38	>43	>40	>37	>29	>40	>40	>40	>38
AMEn	3,170	3,203	3,225	3,120	3,150	3,200	3,100	3,125	3,110
Crude protein	18.2	17.0	18,0	17.5	17.0	18.3	17.0	18.0	17.2
Digestible Lys	0.91	0.90	0.96	0.92	0.97	0.96	0.96	1.02	0.95
Ideal protein[1]									
Met + Cys	76	78	78	74	78	78	77	76	76
Thr	65	68	67	65	65	67	65	67	66
Trp	18	18	16	17	18	16	18	16	18
Ile	–	–	69	67	66	69	66	67	68
Val	–	77	78	79	75	76	77	76	79
Arg	–	108	108	105	105	107	109	106	106
Crude fiber	3.0	–	–	3.05	–		–	–	3.2
Calcium	0.86	0.76	0.74	0.70	0.76	0.75	0.72	0.60	0.69
Available phosphorus	0.38	0.38	0.38	0.35	0.38	0.38	0.36	0.34	0.32
Digestible phosphorus	0.33	–	–	0.32	–	–	–	–	0.30
Potassium	0.40	0.60	0.40	0.40	0.60	0.40	0.60	0.60	0.40
Sodium	0.14	0.15	0.16	0.15	0.16	0.16	0.16	0.18	0.16
Chlorine	0.15	0.15	0.16	0.15	0.16	0.16	0.16	0.18	0.15
Linoleic acid	0.40	1.00	1.00	0.50	1.00	1.00	1.00	1.00	0.50

[1] Digestible AA: Digestible Lys.

Table 8 Vitamins and trace minerals recommendations for broilers (per kg of diet)

Institution	NRC		FEDNA		Cobb		Ross		Mateos[1]	
	(1994)		(2018)		(2022)		(2022)		(2024)	
Period	Starter	Finisher	Starter	Finisher	Starter	Finisher	Starter	Finisher	Starter	Finisher
Vitamins										
Vitamin A (IU × 10^3)	1.5	1.5	10.0	7.0	10.0	10.0	13.0	10.0	10.5	8.0
Vitamin D3 (IU × 10^3)	2.0	2.0	3.5	2.5	5.0	5.0	5.0	4.0	4.1	3.0
Vitamin E (IU)	10	10	25	20	80	50	80	55	30	20
Vitamin B1 (mg)	1.8	1.8	2.0	0.6	3.0	2.0	5.0	3.0	2.3	0.7
Vitamin B2 (mg)	3.6	3.0	6.5	3.1	9.0	6.0	9.0	7.0	6.0	3.3
Vitamin B6 (mg)	3.5	3.0	3.0	1.0	4.0	3.0	5.0	3.0	3.2	2.0
Vitamin B12 (µm)	10	7	16	11	20	15	20	15	17	12
Biotin[2] (µm)	150	120	130	30	150	120	350	220	150	50
Choline[3] (mg)	1300	750	1250	1100	500	350	1700	1500	1300	1050
Trace minerals (mg)										
Copper	8	8	8	5	15	15	16	16	10	8
Manganese	60	60	90	65	100	100	120	120	90	80
Zinc	40	40	75	55	100	100	120	120	80	60
Selenium	0.15	0.15	0.34	0.34	0.35	0.35	0.30	0.30	0.35	0.35
Iron	80	80	33	25	40	40	20	20	30	20
Iodine	0.35	0.35	1.10	0.55	1.00	1.00	1.25	1.25	1.10	0.70

[1] Unpublished data.
[2] Increased in diets based on wheat.
[3] Added choline for Cobb (2022).

Table 9 Nutrient recommendations for broilers according to type of the bird and production system (%): pre-starter diets

	Industrial	Medium growth rate		
Production system		Certified[1]	Free-range[2]	Label[3]
Age (days)	(<14)	(<21)	(<28)	(<28)
AMEn (kcal/kg)	2,950	2,900	2,850	2,850
Crude protein	21.2	20.5	19.9	19.6
Digestible lysine	1.22	1.09	1.03	1.00
Crude fiber	2.85	3.00	3.00	3.00
Calcium	0.98	1.00	0.97	0.95
Digestible phosphorus	0.45	0.45	0.44	0.43
Potassium	0.51	0.65	0.65	0.65
Sodium	0.19	0.18	0.18	0.17
Chlorine	0.17	0.16	0.16	0.16
Linoleic acid	0.80	0.80	0.80	0.80

[1] Slaughter: >56 days. FEDNA, 2018.
[2] Slaughter: >65 days. Management requirements on legislation or local agreements.
[3] Slaughter: >81 days. Mandatory access to pasture (free range).

Table 10 Nutrient recommendations for broilers according to type of the bird and production system (%): finisher diets

	Industrial	Medium growth rate		
Production system		Certified[1]	Free-range[2]	Label[3]
Age (days)	(>37)	(>43)	(>57)	(>57)
AMEn (kcal/kg)	3,120	3,050	3,000	2,950
Crude protein	17.5	16.4	15.6	14.3
Digestible lysine	0.92	0.84	0.78	0.70
Crude fiber	3.05	3.50	3.75	4.00
Calcium	0.70	0.70	0.65	0.63
Digestible phosphorus	0.32	0.29	0.27	0.26
Potassium	0.40	0.60	0.60	0.55
Sodium	0.15	0.15	0.15	0.15
Chlorine	0.15	0.16	0.16	0.16
Linoleic acid	0.50	0.60	0.60	0.60

[1] Slaughter: >56 days. FEDNA, 2018
[2] Slaughter: >65 days. Management requirements on legislation or local agreements.
[3] Slaughter: >81 days. Mandatory access to pasture (free range).

5.2 Feed form and particle size

Pelleting is the thermal processing most used in broiler diets. Pelleting reduces ingredient segregation, facilitates the handling and flowability of the feed and facilitates the use of alternative ingredients, reducing feed cost. Technologies based on reducing particle size, together with the application of heat, can improve pellet quality, improve FI, and reduce feed wastage (Mateos et al., 2012; Moran, 1987; Zelenka, 2003). On the other hand, an excess of fine particles might reduce growth performance in broiler (Sharma et al., 2014).

Research comparing the benefits of pellets vs. mash in poultry diets highlights an increase in nutrient digestibility through pelleting (Jiménez-Moreno et al., 2016; Massuquetto et al., 2019). The pelleting process reduces the particle size of ingredients, increasing the rate of passage through the upper part of the GIT, reducing HCl production by the proventriculus and increasing pH (Mateos et al., 2012). Pelleting increases, in relative terms, the amount of energy and nutrients available for growth vs. maintenance. However, research has clearly shown that the effects of pelleting on protein or starch digestibility are very limited and that most of the benefits are explained by the reduction in feed wastage and the increase in FI (Svihus et al., 2005; Jiménez-Moreno et al., 2016, 2019; Abdollahi et al., 2013, 2019). By reducing feed wastage, pelleting improves feed efficiency and reduces environmental contamination.

However, conventional pelleting to achieve good pellet quality is usually obtained at the expense of physiological and nutritional changes, which can then result in changes in microbiota profile, reduced motility of the walls of the GIT, and an increase in water intake which, in turn, can result in an increase in the incidence of wet litter problems (Engberg et al., 2002; Abdollahi et al., 2013; Jiménez-Moreno et al., 2016).

5.3 Precision nutrition

Precision nutrition involves the adjustment of diets offered to birds to ensure that dietary nutrient supply precisely matches daily nutrient requirements. Successful implementation of a precision nutrition requires meeting key requirements:

- Accurate characterization of ingredients to ensure that the final feed represent the intended formulated diet;
- Accurate determination of broiler nutrient requirements according to age; and
- Combining these two requirements in a way that avoids either over-supply or under-supply of nutrient requirements for each bird.

In theory, models could predict precise feed formulations for groups of broilers (e.g. classified by age) but, as noted earlier, current models still struggle to reflect actual conditions in the production environment. The main reason is that voluntary FI of broilers is very variable and depends on a complex range of factors such as genetic background, age, environmental conditions (e.g. temperature) management, feed phases, and feed form. If FI is not uniform and differs among birds in the flock, estimated nutrient content of the feed might not match bird requirements.

Accurate ingredient characterization is a must to ensure that finished feed represent the intended feed formulation. However, this is a quite difficult task when we used tabulated values because the chemical composition and nutritive value of the ingredients vary due to breed, year of the crop, area of production environmental condition during crop, harvesting, storage, and transport of the ingredient. These difficulties particularly affect ingredients such as soybean meal and DDGS that needs processing before use in feed production (García-Rebollar et al., 2016; Ibáñez et al., 2020; Aguirre et al., 2022; Fondevila et al., 2023).

Information on the chemical composition and energy content of high protein soybean meal (SBM) is provided in Tables 11 and 12. In spite of the similar CP and EE content, the AME_n of the SBM varies from 2,180 kcal/kg (CVB, 2022) to 2,550 kcal/kg (RPRI, 2014). For CP, the SID varies from 85% (CVB, 2022) to 91% (Kadardar et al., 2022) and for Lys from 88.0% to 93.3%. Recently published data on key chemical composition (CP, Lys, and dietary fiber), protein quality indicators of the SBM are presented in Table 13, whereas data on SID of CP and AA are shown in Table 14. Finally, Table 15 offers data of the CP, NDF, EE, and AMEn content of corn DDGS from different sources. Again, all the information provided indicates the need to improve our knowledge on the nutrient requirements of broilers, depending on type and management of the birds as well as the chemical composition and nutritive value of the ingredients.

6 Conclusion

Nutrient requirements per unit of production will not change significantly in the near future. However, feeding programs and dietary specifications will continue to evolve according to new objectives and market needs. New niche markets, especially for added value products, will develop to satisfy new consumer demands.

Key challenges include:

- The potential increase in metabolic disorders, especially those related with skin and meat quality and Ca metabolism, which will likely pose the limit to further increases in productivity of the birds,

Table 11 Proximal analyses and energy content of high protein soybean meal (%, as fed)

Institution	Year	Country	Species	CP (%)	NDF[1] (%)	EE[2] (%)	AMEn (kcal/kg)
WPSA[3]	1989	Europe	Poultry	47.5	–	2.5	2,412
NARO	2009	Japan	Poultry	45.0	13.7	1.9	2,400
NRC[4]	2012	USA	Poultry	47.7	8.2	1.52	2,474
RPRI[3]	2014	Russia	Poultry	47.0	–	1.4	2,550
Rostagno et al.	2017	Brazil	Poultry	46.5	13.6	2.9	2,396
Premier Atlas	2019	UK	Broilers	46.5	8.6	2.7	2,364
Feedpedia	2020	France	Broilers	46.3	12.5	1.6	2,249
FEDNA	2021	Spain	Poultry	47.0	9.2	1.8	2,345
Evonik	2021	Germany	Poultry	46.5	11.3	2.1	2,341
INRAE	2021	France	Broilers	46.2	12.5	1.5	2,260
CVB	2022	Netherlands	Broilers	46.8	8.4	2.4	2,180
Aguirre et al.	2023	Spain	Broilers	46.3	11.4	2.3	2,298
Mateos[5]	2024	Spain	Broilers	46.4	10.5	1.8	2,300

[1] Neutral detergent fiber.
[2] Ether extract with previous HCl hydrolysis, except for WPSA (1989) and Rostagno et al. (2017).
[3] Russian Poultry Research Institute.
[4] AMEn value according to NRC (1994).
[5] Personal communication. Average value for soybeans meals from Argentina, Brazil, and USA.

Table 12 Crude protein and standardized ileal digestibility (SID) of crude protein, Lys and total sulfur amino acids (%) contents of high protein soybean meal, according to Institutions (88% dry matter)

Institution	Year	Country	CP	SID CP	SID Lys	SID TSAA[1]
WPSA[1]	1989	Europe	47.5	87.0	–	–
Rostagno et al.	2017	Brazil	46.5	91.0	92.4	89.1
Premier Atlas	2019	UK	46.5	89.0	88.0	85.4
INRAE	2021	France	46.2	–	88.0	82.0
Evonik	2021	Germany	46.5	–	89.0	84.0
FEDNA[2]	2021	Spain	47.0	88.0	89.0	86.0
CVB	2022	Netherlands	46.9	–	88.0	82.4
Aguirre et al.	2023	Spain	46.3	91.5	2.66	93.3

[1] Data presented are for apparent ileal digestible values.
[2] Data presented are for apparent fecal digestible values.

- The ban on the use of antibiotics and other pharmaceutical products as growth promoters or as prevention of problems, which will conduct to specific modifications on the use of ingredients and nutrients to better control the dynamics and the profile of the bacterial population of the GIT,
- Extra efforts to meet consumer and legislation demands in relation to food safety, environment sustainability, and animal welfare.

Future research should aim to improve: (a) current knowledge on energy and nutrient requirements of broilers, which vary with the genetic background, age, management conditions, and voluntary feed intake of the birds, (b) improve the methodology used to determine the energy (e.g. AME_n vs. AME vs. NE energy systems) and nutritional value (e.g. standardized ileal amino acid digestibility and digestible P and Ca content) of the ingredients, (c) increase the use of new ingredients, usually high in fiber content, and (d) reduce, as much as possible, the safety margins used by the nutritionists in the final diets. However, these objectives are not easy to reach because of the numerous factors that affect the nutritional requirements of broilers and the energy and nutrient content of the ingredients, which include not only the type of bird but also factors affecting the voluntary feed intake, such as health status, flock management (e.g. vaccination programs, health status, sanitary conditions, housing, stocking density, light program, and temperature), and physicochemical characteristics of the diet (e.g. feed form, particle size, distribution, ingredient composition, energy, and amino acid and mineral profile).

Table 13 Chemical composition and protein quality traits of soybeans meals according to the country of origin of the soybeans[1,2]

	García-Rebollar et al. (2016)[3]			Ibáñez et al. (2020)[4]			Aguirre et al. (2022)[5]		
	Argentina	Brazil	USA	Argentina	Brazil	USA	Argentina	Brazil	USA
Number of samples	(170)	(165)	(180)	(481)	(531)	(819)	(40)	(40)	(40)
Crude protein (%)	45.5	46.8	46.8	45.5	47.0	46.4	46.2	46.9	46.2
Lys (% crude protein)	6.10	6.07	6.17	6.19	6.12	6.23	6.11	6.08	6.13
Carbohydrates (%)									
Crude fiber	5.2	6.1	4.3	4.3	5.0	3.9	4.2	4.6	4.2
Neutral detergent fiber	10.2	11.8	9.0	9.6	10.8	8.7	8.8	9.3	8.7
Sucrose	7.8	6.4	8.4	6.4	5.2	7.0	6.9	5.8	7.3
Stachyose	5.7	5.3	6.4	4.2	3.8	4.8	4.7	4.5	5.1
Raffinose	1.4	1.6	1.1	1.2	1.3	1.0	1.3	1.4	1.1
Protein quality									
PDI[6] (%)	16.0	15.0	19.5	14.0	13.5	17.8	11.0	12.9	14.6
KOH solubility (%)	81.2	82.0	86.1	76.6	78.2	82.1	74.6	77.6	80.4
TIA[7] (mg/g)	2.6	2.7	3.5	2.5	2.8	3.2	1.9	2.4	2.6

[1] Urease activity <0.03 mg N/g for all origins ($P < 0.01$).
[2] Within each trial, the differences among origin were significant ($P < 0.001$) for all variables presented.
[3] Samples collected in 2015 in the country of origin.
[4] Meta analytical study (17 sets of data included).
[5] Samples of soybean meals collected in Europe in 2021.
[6] Protein dispersibility index.
[7] Trypsin inhibitor activity.

Table 14 Standardized ileal digestibility (SID) of crude protein and selected amino acids of the soybean meal (%). Average data

Reference	Year	Samples[1]	CP	Arg	Lys	Met	Thr	Trp	Val	Cys
Adedokun et al.[2]	2008	1	–	91.5	89.5	91.5	87.3	–	87.0	81.8
Frikha et al.	2012	21	92.2	94.0	91.6	94.8	91.7	–	91.3	83.7
Ravindran et al.[2,3]	2014	42	83.3	88.7	86.3	87.0	79.7	–	83.0	68.3
Perryman and Dozier[3]	2012	2	–	93.0	90.1	91.1	83.6	89.8	85.5	79.9
Leung and Kiarie	2020	1	–	89.7	90.6	92.9	91.4	–	92.3	87.3
Sung et al.	2023	6	92.1	95.2	92.6	94.5	89.2	93.7	91.4	87.3
Siegert et al.	2023	18	–	91.5	85.9	89.8	77.8	–	85.4	77.8
Ali et al.	2023	1	89.0	94.1	91.9	94.7	87.9	92.4	88.8	84.5
Barau et al.	2023	1	87.3	91.6	90.3	90.5	83.4	86.4	87.3	76.3
Aguirre et al.	2023	27	91.3	94.1	93.0	96.0	90.4	92.6	92.9	80.3

[1] Number of samples used.
[2] Mean value does not include data of the Indian soybean meals.
[3] Apparent ileal digestible values.

Table 15 Chemical composition of selected traits (%, as fed) and energy content (kcal/kg) of corn DDGS

Institution	Year	Country	Species	CP (%)	Lys (%)	NDF (%)	EE (%)	AMEn (kcal/kg)
WPSA	1989	Europe	Cock	28.0	–	–	7.5	2.699
NARO	2009	Japan	Poultry	30.8	0.82	35.4	11.0	2,900
NRC[2]	2012	USA	Poultry	27.4	0.90	30.5	8.9	2,503
Premier Atlas	2017	UK	Broiler	30.3	0.85	33.0	4.0	2,462
Feedpedia	2017	France	Broiler	26.3	0.79	30.4	9.79	2.615
CVB	2022	Netherlands	Broiler	26.8	0.64	28.8	12.5	–
Evonik	2022	Germany	Poultry	27.4	0.79	38.5	10.6	2.565
INRA	2023	France	Broiler	24.6	0.73	31.9	4.5	2.210
FEDNA	2021	Spain	Poultry	27.5	0.81	33.2	7.5	2,055
FEDNA[3]	2023	Spain	Poultry	30.3	0.93	33.0	10.0	2.286

[1] Values presented correspond to average. The number of samples analyzed for each ingredient varied widely.
[2] AMEn was calculated using the NRC (1994) equation.
[3] Corresponding to samples processed in Spain, exclusively.

7 References

Abdollahi, M. R., Ravindran, V. and Svihus, B. (2013). Pelleting of broiler diets: an overview with emphasis on pellet quality and nutritional value. *Anim. Feed Sci. Technol.* 179(1–4), 1–23.

Abdollahi, M. R., Zaefarian, F. and Ravindarn, V. (2019). Maximising the benefits of pelleting diets for modern broilers. *Anim. Prod. Sci.* 59(11), 2023–2028.

Agri Stats (2023). Available at: Poultry production cost. Available at: https://www.agristats.com/.

Aguirre, L., Cámara, L., Smith, A., Arroyo, J. J., de Juan, A. F., Fondevila, G. and Mateos, G. G. (2022). Chemical composition, protein quality indicators and in vitro protein digestibility of commercial soybean meals from four different origins. *Anim. Feed Sci. Technol.* 293, 115473.

Angel, R., Saylor, W. W., Mitchell, A. D., Powers, W. and Applegate, T. J. (2006). Effect of dietary phosphorus, phytase, and 25-hydroxycholecalciferol on broiler chicken bone mineralization, litter phosphorus and processing yields. *Poult. Sci.* 85(7), 1200–1211.

Applegate, T. J. and Angel, R. (2014). Nutrients requirements of poultry publication: history and need for an update. *J. Appl. Poult. Res.* 23(3), 567–575.

Baldi, G., Soglia, F. and Petracci, M. (2021). Spaghetti meat abnormality in broilers: current understanding and future research directions. *Front. Physiol.* 12, 684497.

Baker, D. H., Fernadez, S. R., Webel, D. M. and Parsons, C. M. (1996). Sulfur amino acid requirement and cystine replacement value of broiler chicks during the period three to six weeks posthatching. *Poult. Sci.* 75(6), 737–742.

Biswas, A. (2019). Pulmonary hypertension syndrome in broiler chickens: a review. *Vet. Arhiv* 89(5), 723–734.

Brons, F., Kettlitz, B. and Arrigoni, E. (2022). Resistant starch and the butyrate revolution. *Trends Food Sci. Technol.* 13, 251–261.

Cobb (2022). *Performance and Nutrition Supplement*. Cobb Vantress Inc., Siloam Springs, AR.

CVB (Centraal Veevoeder Bureau) (2022). *Feed Table: Chemical Composition and Nutritional Values of Feedstuffs. Veevoedertabel (Livestock Feed Table).* Central Veevoeder Bureau, The Netherlands.

Da Costa, M. J., Colson, G., Frost, T. J., Halley, J. and Pesti, G. M. (2017). Evaluation of starter dietary digestible lysine level on broilers raised under a sex-separated or straight-run housing regime, part 2: economics of sex separation and digestible lysine level for maximum returns. *Poult. Sci.* 96(9), 3282–3290.

de Lange, D. F. M. (2000). Characterization of the non-starch polysaccharides in feeds. In: Moughan, P. J., Verstegen, M. W. A. and Visser-Reyneveld (Eds). *Feed Evaluation-Principles and Practice*. Wageningen Pers, Wageningen, pp. 77–92.

EFG (1995). *EFG Broiler Growth Model*. Available at: http://www.efgsoftware.net.

EFG Software (2006). Broiler growth model (version 6), Broiler nutrition optimizer (version 1), Pig growth model (version 2) and Pig nutrition optimizer (version 1). Retrieved 2006. Available at: www.efgsoftware.com.

Emmans, G. C. (1981). A model of the growth and feed intake of ad libitum fed animals, particularly poultry. In: Hillyer, G. M., Whittemore, C. T. and Gunn, R. G. (Eds). *Computers in Animal Production*. British Society of Animal Production Occasional Publication, BSAP, Edinburgh, pp. 103–110.

Emmert, J. L. and Baker, D. H. (1997). Use of the ideal protein concept for precision formulation of amino acid levels in broiler diets. *J. Appl. Poult. Res.* 6(4), 462–470.

Engberg, R. M., Hedemann, M. S. and Jensen, B. B. (2002). The influence of grinding and pelleting of feed on the microbial composition and activity in the digestive tract of broiler chickens. *Br. Poult. Sci.* 43(4), 569–579.

FEDNA (Fundación Española Desarrollo Nutrición Animal) (2018). *Necesidades Nutricionales para Avicultura: Normas FEDNA*. Fundación Española Desarrollo Nutrición Animal, Madrid.

Feedipedia (2017). *Animal Feed Resources Information System*. Accessed April 2017. Available at: http://www.feedipedia.org/.

Ferguson, N. S., Gates, R. S., Taraba, J. L., Cantor, A. H., Pescatore, A. J., Straw, M. L., Ford, M. J. and Burnham, D. J. (1998). The effect of dietary protein and phosphorus on ammonia concentration and litter composition in broilers. *Poult. Sci.* 77(8), 1085–1093.

Fisher, C. (1980). Protein deposition in poultry. In: Buttery, E. and Lindsay (Eds). Proceedings of the 29th Easter School of Agriculture Science, Butterworts, London, UK.

Fondevila, G., Corrales, N. L., Talegón, G. and Mateos, G. G. (2023). Actualización tablas FEDNA (nuevos ingredientes): DDGs de maíz parcialmente desgrasados. In: FEDNA (Ed). *Avances en Nutrición y Alimentación Animal*. FEDNA, Madrid, pp. 331–338. ISBN: 978-84-09-56569-6.

García-Rebollar, P., Cámara, L., Lázaro, R. P., Dapoza, C., Pérez-Maldonado, R. and Mateos, G. G. (2016). Influence of the origin of the beans on the chemical composition and nutritive value of commercial soybean meals. *Anim. Feed Sci. Technol.* 221, 245–261.

Glista, W. H., Mitchell, H. H. and Scott, H. M. (1951). The amino acid requirements of the chick. *Poult. Sci.* 30, 915.

González-Alvarado, J. M., Jiménez-Moreno, E., Lázaro, R. and Mateos, G. G. (2007). Effect of type of cereal, heat processing of the cereal, and inclusion of fiber in the diet on productive performance and digestive traits of broilers. *Poult. Sci.* 86(8), 1705–1715.

González-Alvarado, J. M., Jiménez-Moreno, E., Valencia, D. G., Lázaro, R. and Mateos, G. G. (2008). Effects of fiber source and heat processing of the cereal on the development and pH of the gastrointestinal tract of broilers fed diets based on corn or rice. *Poult. Sci.* 87(9), 1779–1795.

González-Alvarado, J. M., Jiménez-Moreno, E., González-Sánchez, D., Lázaro, R. and Mateos, G. G. (2010). Effect of inclusion of oat hulls and sugar beet pulp in the diet on productive performance and digestive traits of broilers from 1 to 42 days of age. *Anim. Feed Sci. Technol.* 162(1–2), 37–46.

Gous, R. M. (2007). Predicting nutrient responses in poultry: future challenges. *Animal* 1(1), 57–65.

Hauschild, L., Sakomura, N. K. and Silva, E. P. (2015). AvinespModel: predicting poultry growth, energy and amino acid requirements. In: Sakomura, N. K., Gous, R. M., Kyriazakis, I. and Huschild, L. (Eds). *Nutritional Modelling for Pigs and Poultry*. CABI Digital Library, Wallingford, UK, pp. 188–208.

Huang, X. and Ahn, D. U. (2018). The incidence of muscle abnormalities in broiler breast meat: a review. *Korean J. Food Sci. Anim. Resour.* 38(5), 835–850.

Ibáñez, M. A., de Blas, C., Cámara, L. and Mateos, G. G. (2020). Chemical composition, protein quality and nutritive value of commercial soybean meals produced from beans from different countries: a meta-analytical study. *Anim. Feed Sci. Technol.* 267, 114531.

INRAE (Institut National de la Recherche Agronomique) (2021). Tables *de Composition et de Valeur Nutritive des Matieres Premières Destinées aux Animaux d'élevage*. INRA (Ed.), Paris, Cedex, France. Available at: https://feedtables.com/fr.

Jiménez-Moreno, E., Frikha, M., de Coca-Sinova, A., Lázaro, R. P. and Mateos, G. G. (2013). Oat hulls and sugar beet pulp in diets for broilers. 2. Effects on the development of the gastrointestinal tract and on the structure of the jejunal mucosa. *Anim. Feed Sci. Technol.* 182(1–4), 44–52.

Jiménez-Moreno, E., de Coca-Sinova, A., González-Alvarado, J. M. and Mateos, G. G. (2016). Insoluble fiber sources in mash or pellets diets for young broilers. 1. Effects on growth performance and water intake. *Poult. Sci.* 95(1), 41–52.

Jiménez-Moreno, E., González-Alvarado, J. M., de Coca-Sinova, A., Lázaro, R. P., Cámara, L. and Mateos, G. G. (2019). Insoluble fiber sources in mash or pellets diets for young broilers. 2. Effects on gastrointestinal tract development and nutrient digestibility. *Poult. Sci.* 98(6), 2531–2547.

Julian, R. J. (1998). Rapid growth problems: ascites and skeletal deformities in broilers. *Poult. Sci.* 77(12), 1773–1780.

Julian, R. J. (2005). Production and growth related disorders and other metabolic diseases of poultry – a review. *Vet. J.* 169(3), 350–369.

Kadardar, H. , Aguirre, L., Talegón, G., Lázaro, R., Cámara, L. and Mateos, G. (2022). Effects of the nutritional density of the diet and whole wheat supplementation on growth performance and gizzard development of broilers from 0 to 40 days of age. *Poult. Sci.* 102(E-Suppl 1; Abstract), 136.

Kalmendal, R., Elwinger, K., Holm, L. and Tauson, R. (2011). High-fibre sunflower cake affects small intestinal digestion and health in broiler chickens. *Br. Poult. Sci.* 52(1), 86–96.

Korver, D. R. (2023). Review: current challenges in poultry nutrition, health, and welfare. *Animal* 17(2), 100755.

Leeson, S. (2008). Predictions for commercial poultry nutrition. *J. Appl. Poult. Res.* 17(2), 315–322.

Leeson, S. (2012). Future considerations in poultry nutrition. *Poult. Sci.* 91(6), 1281–1285.

Leeson, S. and Summers, J. D. (2001). *Nutrition of the Chicken* (4th edn.). University Books, Guelph, Ontario, Canada.

Lilburn, M. S., Griffin, J. R. and Wick, M. (2019). From muscle to food: oxidative challenges and developmental anomalies in poultry breast muscle. *Poult. Sci.* 98(10), 4255–4260.

Magdelaine, P., Spiess, M. P. and Valceschini, E. (2008). Poultry meat consumption trends in Europe. *J. World's Poult. Sci.* 64(1), 53–64.

Massuquetto, A., Panisson, J. C., Marx, F. O., Surek, D., Krabbe, E. L. and Maiorka, A. (2019). Effect of pelleting and different feeding programs on growth performance, carcass yield, and nutrient digestibility in broiler chickens. *Poult. Sci.* 98(11), 5497–5503.

Mateos, G. G., Jiménez-Moreno, E., Serrano, M. P. and Lázaro, R. P. (2012). Poultry response to high levels of dietary fiber sources varying in physical and chemical characteristics. *J. Appl. Poult. Res.* 21(1), 156–174.

Mateos, G. G., Aguirre, L., Kadardar, H. and Lázaro, R. P. (2017a). *New Insights on the Use of Fiber Sources in Diets for Poultry, 21st BOKU-Symposium, Tierernährung*, Boku TTE, Viena, Austria, pp. 1–7.

Mateos, G. G., Cámara, L., Fondevila, G., Mandalawi, H., Arch, J. and Lázaro, R. (2017b). Insoluble fiber for gut development, microbial management and performance. In: Abdollahi, R. and Ravindran, V. (Eds). 6th International Broiler Nutrition Conference. Poultry Beyond 2023, New Zeland, Queenstown, pp. 189–216.
Mateos, G. G., Fondevila, G. and Cámara, L. (2019). The importance of the fiber fraction of the feed in non-ruminant diets. In: González-Ortiz, Bedford, M. R., Bach Knudsen, K. E., Courtin, C. M. and Classen, H. L. (Eds). *The Value of Fibre. Engaging the Second Brain for Animal Nutrition*. Wageningen Publishers, The Netherlands, pp. 61–83.
Moran, E. T. (1987). Pelleting affects feed and its consumption. *World's Poult.* 5, 30–31.
Moss, A. F., Chrystal, P. V. and Cadogan, D. J. (2020). Precision feeding enhances feed efficiency and carcass yield compared to broilers offered standard feeding programs. International Poultry Scientific Forum, Atlanta, GA. North American Meat Institute, Washington, DC, p. 32.
Moss, A. F., Chrystal, P. V., Cadogan, D. J., Wilkinson, S.J., Crowley, T. M. and Choct, M. (2021). Precision feeding and precision nutrition: a paradigm shift in broiler feed formulation? *Anim. Bio Sci.* 34(3), 354–362.
Mutucumarana, R. K., Ravindran, V., Ravindran, G., et al. (2014). Measurement of true ileal digestibility of phosphorus in some feed ingredients for broiler chickens. *J. Anim. Sci.* 92(12), 5520–5529.
NARO (National Agriculture and Food Research Organization) (2009). *Standard Tables of Feed Composition in Japan Livestock Industry Association*. NARO, Tokyo.
Oviedo-Rondón, E. O. (2015). Model applications in poultry production and nutrition. In: Sakomura, N. K., Gous, R. M., Kyriazakis, I. and Hauschild, L. (Eds). *Nutritional Modelling for Pigs and Poultry*. CABI Digital Librery, Wallingfor, UK, pp. 120–140.
Petracci, M., Soglia, F., Madruga, M., Carvalho, L., Ida, E. and Estévez, M. (2019). Wooden-breast, white striping, and spaghetti meat: causes, consequences and consumer perception of emerging broiler meat abnormalities. *Rev. Food Sci, Compr. Rev. Food Sci. Food Saf.* 18(2), 565–583.
Poultry Trends (2023). The statistical reference for poultry executives. In: *WATT Executive Guide to World*(https://img.wattagnet.com/files/base/wattglobalmedia/all/document/2015/04/an.2009_2010_Poultry_Trends.pdf); (accessed January 15, 2024).
Premier Atlas (2014). *Ingredients Matrix*. Premier Nutrition Products Ltd, AB Agri Ltd, Rugeley Staffs.
Rafiq, K., Tofazzal Hossain, M., Ahmed, R., Hasan, M. M., Islam, R., Hossen, M. I., Shaha, S. N. and Islam, M. R. (2022). Role of different growth enhancers as alternative to in feed antibiotics in poultry industry. *Front. Vet. Sci.* 8, 794588.
Rahman, M. R. T., Fliss, I. and Biron, E. (2022). Insights in the development and uses of alternatives to antibiotic growth promoters in poultry and swine production. *Antibiotics* 11(6), 766.
Rauw, W. M., Kanis, E., Noordhuizen-Stassen, E. N. and Grommers, F. J. (1998). Undesirable side effects of selection for high production efficiency in farm animals: a review. *Livest. Prod. Sci.* 56(1), 15–33.
RPRI (Russian Poultry Research Institute) (2014). *Manual for Optimization of Formulation of Compound Feeds for Poultry*. Russian Poultry Research Institute, Sergiev Posade, Russian Federation.
Rochell, S. J., Applegate, T. J., Kim, E. J. and Dozier, W. A. (2012). Effects of diet type and ingredient composition on rate of passage and apparent ileal amino acid digestibility in broiler chicks. *Poult. Sci.* 91(7), 1647–1653.

Ross (2022). *Broiler: Performance Objectives*. Ross Aviagen, Huntsville, AL.

Rostagno, H. S., Albino, J. L. L. F. T., Hannas, M. I., Donzele, J. L., Sakomura, N. K., Perazzo, F. G., Saraiva, A., Teixeira, M. L., Rodrigues, P. B., Oliveira, R. F., Barreto, S. L. and de Oliveira Brito, C. (2017). *Tabelas Brasileiras para Aves e Suínos. Composiçao de Alimentos e Exigencias Nutricionais* (4th edn.). Universidade Federal de Viçosa, Viçosa.

Sakomura, N. K., Silva, E. P., Dorigam, J. C. P., Gous, R. M. and St-Pierre, N. (2015). Modeling amino acid requirements of Poultry, *J. Appl. Poult. Res.* 24(2), 267–282.

Sharma, N. K., Creswell, D. and Swick, R. A. (2014). Effect of feeding whole wheat and cracked corn on performance and carcass yield of broilers. In: Proceedings of the 14th European Poultry Conference, June 23–27, Stavanger, Norway.

Soglia, F., Petracci, M., Davoli, R. and Zappaterra, M. (2021). A critical review of the mechanisms involved in the occurrence of growth-related abnormalities affecting broiler chicken breast muscles. *Poult. Sci.* 100(6), 101180.

Svihus, B., Uhlen, A. K. and Harstad, O. M. (2005). Effect of starch granule structure, associated components and processing on nutritive value of cereal starch: a review. *Anim. Feed Sci. Technol.* 122(3–4), 303–320.

Teague, K. D., Tellez-Isaias, G., Chai, J., Petrone-Garcia, V., Vuong, C. N., Blanch, A., Rasmussen, S. H., Brown, K., Zhao, J. and Rochell, S. J. (2023). Dietary soy galactooligosaccharides affects the performance, intestinal function, and gut microbiota composition of growing chicks. *Poult. Sci.* 102(4), 102440.

Wideman Jr, R. F. (2016). Bacterial chondronecrosis with osteomyelitis and lameness in broilers: a review. *Poult. Sci.* 95(2), 325–344.

Wu, G. and Lin, P. (2022). The "ideal protein" concept is not ideal in animal nutrition. *Exp. Bio. Med.* 247(13), 1191–1201.

WPSA (World's Poultry Science Association) (1989). Subcommittee of Energy of the Working Group no 2. *European Table of Energy Values for Poultry Feedstuffs* (3rd edn.). Spelderholt Centre for Poultry Research, Nutrition of the European Federation of Branches of the World's Poultry Science Association, Beekbergen.

Zelenka, J. (2003). Effect of pelleting on digestibility and metabolizable energy values of poultry diets. *Czech J. Anim. Sci.* 48(6), 239–242.

Zuidhof, M. J., Schneider, B. L., Carney, V. L., Korver, D. R. and Robinson, F. E. (2014). Growth, efficiency, and yield of commercial broilers from 1957, 1978, and 2005. *Poult. Sci.* 93(12), 2970–2982.

Chapter 2

Advances in understanding poultry digestive anatomy and physiology

Birger Svihus, Norwegian University of Life Sciences, Norway

1 Introduction

The digestive tract is faced with a massive task in modern poultry. A fast-growing broiler may consume dry feed equivalent to 15% of its body weight per day. To put this in perspective, a 75 kg person would have had to consume almost one loaf of bread (750 g) per hour around the clock to reach the same intake. With such an overwhelming task, it is reasonable to assume that the digestive system is pushed to its limit, and with very little room for suboptimal functionality of any part.

Speculation has abounded that, in fact, the digestive tract is not able to complete the digestion and absorption process to its full in for example broilers (Svihus and Hetland, 2001; Carré et al., 2008), leading to a heightened level of undigested nutrients reaching the lower digestive tract. In addition to the loss of nutrients such an effect may represent (with impacts on nutrient use efficiency and the overall sustainability of poultry production), it may also lead to vulnerability to microbiota-induced gut disorders (Apajalahti and Vienola, 2016). A thorough understanding of the functionality and physiology of the digestive tract is thus crucial for a successful poultry industry.

http://dx.doi.org/10.19103/AS.2024.0143.02

It will not be possible to cover all advances in the understanding of poultry digestive anatomy and physiology in this chapter. Specifically, the large and evolving field of gut microbiota will not be covered. Several good reviews have been written recently, for example the critical appraisal of microbiome science in animal production in general by Bajagai et al. (2022), or the review specifically on microbiota and poultry by Yadav and Jha (2019). As always, the description of the digestion process will rest on a solid foundation of previous summaries of the topic. In particular, the classic texts in *Sturkie's Avian Physiology* have been formative (Sturkie, 1986), not the least the excellent descriptions of the digestive tract. In addition, more recent work has been emphasized, not least review articles covering more recent advances in the understanding of the functionality of the different segments of the digestive tract, with a particular emphasis on digestive functionalities peculiar to poultry.

2 Factors affecting the attractiveness of feed

Consumption of feed clearly relies on the fact that the bird has a sufficiently high desire to eat. With high-performing poultry species, not least broilers, this is a critical issue. Maximizing feed intake is necessary for maximized performance which, again, is an important driving factor for the efficiency of poultry production. A high performance reduces the relative contribution of maintenance requirements to total nutrient use, thus improving feed efficiency.

Although the perception of domestic poultry as being rather insensitive to taste and olfactory cues has been challenged (Roura et al., 2013), such factors do not seem to be of great importance to poultry. However, visual and tactile cues play an important role in feed intake. It has been shown that when given a choice, poultry will tend to select some particle sizes before others. Generally, particles smaller than 0.5 to 1 mm are avoided, and birds prefer larger particles which they can grasp with their beak. Size preference increases with age (Portella et al., 1988) although the exact optimal particle preference at different ages does not seem to have been carefully studied. However, pelleting to create pellets with a diameter of 3–5 mm results in a considerable increase in feed intake, at least in broilers.

As opposed to other farm animals, the oral cavity of poultry does not play a large role in the digestion process. The feed is not chewed, and only small amounts of saliva are secreted. A feed particle simply passes through and is swallowed involuntarily upon contact with the surface of the tongue. Some birds do use the beak and tongue to get rid of the hull fraction of seeds, but such behaviour is not commonly observed with commercial poultry species for which swallowing is a reflex.

3 The role of the crop in the digestive system

As the dry feed passes down the esophagus, it may pass directly to the gastric region. This will happen if the bird has been starved and there is therefore space available in the gizzard. However, if the gizzard is full due to a short interval since last feeding or if the feeding bout results in a higher feed intake than available space in the gastric region, the feed will enter the crop. This extended portion of the esophagus serves as an intermediary store of food, allowing birds to eat enough feed to keep the digestive system fully occupied over a long period. In other animal species, the stomach can serve this role but, due to the grinding function of the stomach region in birds, the amount of feed in the stomach must be much more limited. The crop therefore serves this role in birds. Broiler chickens trained to intermittent feeding have been shown to be able to portion out feed for as much as 5–6 h before the crop is empty (Sacranie et al., 2017). For broiler breeders fed only once or twice per day, the crop can hold contents for an even longer period.

With the dry feed generally given to poultry, the feed in the crop will initially also be dry. However, birds will usually drink quickly after feeding, and thus the crop contents will be moisturized, usually reaching 50% moisture content within 30 min. Although no enzymes are secreted into the crop, the moisturization process itself may be beneficial to the digestion process simply because the digestion process depends on water to take place. In addition, with the increasing use of exogenous enzymes added to poultry diets, the crop can serve as an incubation chamber for the degradation process carried out by these enzymes. It has been demonstrated that a considerable proportion of phytate will be degraded in the crop when a phytase is added (Svihus et al., 2010; Kristoffersen et al., 2022).

Conflicting results have been found on the effect of crop retention on the digestion process. Several studies have been carried out where the crop retention time in broiler chickens has been substantially increased by switching from ad libitum feeding to intermittent feeding. Typically, feed access has been restricted to 4–6 feeding bouts lasting 1–2 h and spread throughout the day. When this is done, the birds quickly learn to crop up when feed is available, in many cases managing to maintain normal weight gain. In some of these experiments, a significant improvement in feed conversion efficiency has been observed, while in others no effect has been seen. One possibility is that the benefits of the moisturization and stimulation of enzyme activity in some cases are balanced out by the considerable nutrient requirement of lactic acid bacteria. For example, the lactic acid bacteria may metabolize essential amino acids.

As discussed extensively by Classen et al. (2016), the crop serves as a very suitable fermentation chamber, being quickly dominated by lactobacilli.

As competitive exclusion is very important in the gut microbiome ecosystem, the lactobacilli-dominated crop environment can serve as a barrier to the proliferation of harmful microbiota such as *Salmonella*. The mechanisms involved can both be competition for adherence sites and through immune system stimulation. In addition, the pH will quickly drop as lactic acid levels increase with increasing lactic acid bacteria numbers, further making the environment hostile to pathogenic bacteria.

Apart from a better understanding of the complex microbiological ecosystem residing in the crop, a few fundamental new insights into crop function have emerged lately. However, there has been an interest in stimulating crop retention in order to facilitate the types of potential benefits crop retention may have on nutrient digestion. While the crop is extensively used in some production systems, such as layers not fed overnight and in restrictively fed broiler breeders, it has been demonstrated that the crop is not extensively used in other poultry production systems with differing feeding patterns. Kristoffersen et al. (2022), for example, demonstrated that one-third to half the broiler chickens in commercial farms with ad libitum feeding had insignificant amounts of feed in the crop when they were assessed at 10, 20 and 30 days of age. While broilers trained to intermittent feeding can hold more than 50 g feed dry matter in the crop, only 17% of the broilers had more than 15 g feed dry matter in the crop at 30 days of age.

To facilitate an increased retention time in the crop, birds must be exposed to periods without feed. They will then learn to crop up when feed is available. In broiler chickens, 3 h without feed will be enough to stimulate such intermittent feeding behaviour. This is an example of new practices which has emerged as a consequence of heightened awareness of the role the digestive system can play in a successful poultry production.

4 The gastric compartments: the role of the gizzard

The stomach region of poultry differs from the stomach of other farm animals in profound ways. As described in more detail in Svihus (2011), the stomach is divided into two compartments:

- The proventriculus; and
- The gizzard.

The proventriculus is the glandular part of the stomach where parietal and chief cells produce hydrochloric acid and pepsinogen, respectively. The small volume of this compartment results in a very short retention time. A majority of the digestive action of these secretions therefore takes place in the gizzard, which is separated from the proventriculus by a constricted junction. Since the material will be refluxed into the proventriculus during contractions of the pair of large gizzard muscles, this will allow for additional secretions to be added to the contents (Duke, 1992). The proventriculus and gizzard can thus be considered as working as an integrated unit in gastric digestion.

The main body of the gizzard comprises two thick, opposed lateral muscles and two thin anterior and posterior muscles. The reason for this is that the gizzard fulfills the role of mechanical particle size reduction which is carried out in the oral cavity in other animals. Relative to the long axis of the gizzard, thick and thin muscles are asymmetrically arranged, resulting in a rotary as well as crushing movement when the gizzard contracts. A thick layer of glycoproteins that is hardened by the low pH covers the inside of the gizzard. The koilin layer is continuously renewed by secretions from glands, crypt and surface epithelial cells as it is worn, with the hardened rods formed from 5 to 6 glands and the softer koilin layer between the rods formed by epithelial cells. Some birds, however, are known to shed the whole koilin layer at intervals (McLelland, 1989). The rod-like hardened glycoprotein will stand out from the remaining glycoprotein matrix and will give the surface its characteristic sand-paper-like appearance.

Grinding activity and the regulation of this activity in the gizzard has been described in detail by Duke (1992). The grinding cycle begins with contraction of the thin muscles, followed by the opening of the pylorus and a powerful peristaltic contraction in the duodenum. The pair of thick muscles contracts immediately after the commencement of the duodenal contraction, which means that some gastric material is pushed into the duodenum and some material is pushed into the proventriculus. As the thick muscles begin to relax, the proventriculus contracts and returns content to the gizzard. This contraction cycle takes place up to four times per minute. Material is ground down due to rubbing against the koilin layer on the inside of the gizzard and against other particles in the gizzard during contraction of the large muscles, while the small muscles move material towards the grinding zones between contractions of the large muscles.

The grinding that takes place in the gizzard is very effective and produces very small particles, thus enhancing the digestion process. Analysis of the resulting particle size is not easy since digestion will rapidly reduce particle size, but measurements of duodenal content have revealed an average particle size varying between 0.1 and 0.2 mm, and with around 40% particles smaller than 0.04 mm (Hetland et al., 2002). To a large extent, the gizzard will selectively retain particles until they have reached a sufficiently small size before passing them out of the gizzard. Large particles are retained due to the simultaneous contraction of the duodenum and the thick gizzard muscles, resulting in the contents of the gizzard being forced through a sieve-like narrow pylorus. The many folds in the pylorus opening also contribute to this sieving process.

Large particles such as grit stones and whole cereal particles are known to be occasionally passed on despite not being degraded. In fact, it is still not confirmed whether large particles will be retained under all circumstances. It is for example possible that, when the load of coarse particles exceeds the grinding capacity of the gizzard, the gizzard will start passing large particles through. Experiments have demonstrated a significant passage through the gizzard of whole wheat and grit particles (Svihus et al., 2017) as well as oat hulls (Ghimire et al., personal comm.) when provided in large quantities. This is a logical response to diets with a high level of large fibre particles and will allow the bird to compensate for a low nutrient content with an increased intake.

Assessments based on the use of a marker indicate a retention time of around 1 h in the gizzard. Taking selective retention into consideration, the retention time in the gizzard will obviously vary with particle size and toughness of the feed material. In addition, it will vary with how well the gizzard has developed due to the inclusion of coarse particles in the diet, as will be discussed below. As coarse particles tend to be retained until sufficiently broken down, the retention time can be very long. Even after 16 h of feed withdrawal, broiler chickens that did not have access to litter had large amounts of feed in the gizzard (Sacranie et al., 2017). Even when correcting for contents in the crop, this indicates that feed material can be retained in the gizzard for many hours. Most birds will consume small stones if available, presumably to aid the digestion of seeds and other coarse particles. Although no beneficial effect of grit stones on nutrient availability was observed in a series of experiments, grit stones were found in the gizzard as long as 9 days after being fed (Svihus et al., 2017). On the other hand, material seem to be able to pass through almost with no retention time. When refeeding birds after a long feed withdrawal period, content will reach the small intestine within minutes.

The pH of the contents of the gizzard will vary with retention time and gizzard functionality. It will also to a large extent be affected by the buffering capacity of the feed. Due to the high calcium content in the diet, pH may be higher than 4 in the gizzard content of layers. A pH below 2 has often been recorded, and indicates the large pH-reducing capacity of the gastric secretions. Kristoffersen et al. (2022) measured an average pH of 1.71 in broiler chickens that were starved for 8 h. Once fed, the pH increased to around 3, and then slowly decreased to 2 or lower within 2 h.

It is now firmly established that a well-functioning gizzard is a prerequisite for a high digestibility and a healthy digestive tract. In order for this to happen, the material that the bird eats must contain coarse materials. When the diet contains whole or coarsely ground seeds such as cereals, the gizzard will be stimulated to contract, resulting in a growth in the size of the gizzard and the surrounding gizzard muscles. This will facilitate the functionality of the gizzard. The junction between the gizzard and the proventriculus will become more conspicuous, as well as the pyloric junction between the gizzard and the small intestine. Experiments comparing the flow of material through the gizzard have shown clearly how coarse ingredients have resulted in a gizzard that produces more finely ground material entering the small intestine, and resulting in a significantly lower pH of the gizzard contents.

A well-functioning gizzard is also able to regulate flow at a more steady state into the small intestine. Just like the stomach of other animals, the gizzard has an important role as a feed-flow regulator, passing an appropriate amount of material into the small intestine to maximize digestion and absorption. It has been demonstrated that broilers with a well-developed gizzard passed material into the small intestine at a much more steady rate than birds where the gizzard was poorly developed due to a lack of consumption of large coarse materials (Sacranie et al., 2017). Recent work has shown that broilers with a well-developed gizzard had significantly less starch in the excreta than broilers with a poorly developed gizzard due to lack of coarse materials (Itani and Svihus, 2019). The feed flow role of the gizzard is therefore important for optimized digestibility.

As pellets disintegrate quickly when moisturized, they will not stimulate gizzard development, independent of size. Surprisingly, adding grit stones to broiler diets neither stimulated gizzard development, presumably partly due to the fact that they were not retained in the gizzard (Svihus et al., 2017). Exactly what amounts and nature of structural material is needed to stimulate gizzard development will vary with species and life stage. In broilers, where this has been studied to the greatest extent, it has been shown that, when structural material is in the form of tough fibres such as oat hulls, less material is needed to stimulate the gizzard. In a recent experiment, the weight of the gizzard increased as levels of coarse oat hulls increased to 11% of the diet, but even as little as 3% resulted in a 27% increase in size (Itani et al., 2024). When the oat hulls were finely ground to a size smaller than 1.6 mm, the gizzard responded much less to increasing levels, demonstrating that particle size is of great importance, as already discussed.

When whole or coarsely ground cereals are used, higher levels must be used to significantly affect gizzard development. When 12.5% whole wheat was used from 10 days of age and 30% from 24 days of age, the gizzard grew in size as expected (Hetland et al., 2002). However, increasing the level of whole wheat to 30% and 44%, respectively, did not result in a further increase in gizzard size. The broilers were able to process these large quantities of whole wheat with only a small increase in coarse particles in the small intestine. Just as for the diet with a more moderate level of whole wheat, starch digestibility increased compared to the diet without whole wheat. When as much as 50–60% of whole wheat was added to the diet, the performance of broilers was compromised, indicating that the level exceeded the capacity of the gizzard to grind while, at the same time, being able to keep feed intake high.

As the previous discussion demonstrates, the capacity of the gizzard to handle large quantities of coarse material is very large indeed. But it seems that the beneficial effects of coarse particles maximizes at a much lower level. When coarsely ground seeds are used, the gizzard-stimulating effect will depend on the size and particle distribution of the ground material. When wheat-based diets containing different levels of coarse particles were fed to broilers, increasing the amount of coarse material to 40% larger than 1 mm, of which 10% was larger than 2 mm, significantly increased gizzard weight. However, increasing the level of coarse particles further did not increase gizzard size (Svihus et al., 2004b). Although more systematic studies are needed, this indicates that, when coarse cereal particles are the only source, the diet should contain a large proportion of particles larger than 1 mm, of which perhaps as much as 20% of the particles in the diet should be larger than 2 mm.

5 The small intestine

The small intestine is a core segment of the digestive tract. This is because it is the site for most of the digestion and practically all absorption of nutrients. The anatomy and general structure of the small intestine is not significantly different from the small intestine of other farm animals. However, the digestive tract is rather small, growing rapidly until reaching approx. 5–10% of live weight at 10 days of age, followed by a relative reduction in size to 3–4% of live weight by 20–30 days of age (Wijtten et al., 2012). The villi height, however, continues to increase, probably as a consequence of increased life span of the enterocytes, thus compensating for a relative reduction in small intestine mass. It is thought that this is an adaptation to flight, which limits the potential for an increased intestinal weight. In contrast, the relative size of the small intestine continues to increase with age in pigs, while villous height decreases (Wijtten et al., 2012). In fact, cross-sections of the upper small intestine of broilers often reveal a very small and sometimes almost non-existent lumen, due to villi and/or folds protruding all the way to the centre of the small intestinal tube.

The small intestine has been described in detail before (Svihus, 2014) and will only be briefly outlined here. The first part of this segment of the digestive tract is the duodenal loop. Although this segment ends at the point where the pancreatic and bile ducts empty the digestive juices into the small intestine, the acidic contents from the gizzard are mixed with bile and pancreatic juices through gastroduodenal refluxes during the very short retention time here of less than 5 min. Consequently, pH quickly rises to a level above six, and the process of digestion starts. Sklan et al. (1975) reported that 95% of fat was broken down in the duodenum.

Although there are no histologically distinct segments posterior to the duodenum, the adjacent segment that ends at the yolk sack residue (Meckel's diverticulum) is usually referred to as the jejunum. This segment has a key role, as all the major nutrients are to a large extent digested and absorbed here. The prominent role is reflected in the fact that the empty weight of this segment is considerably higher than the ileum. Despite the large size, retention time in this segment is usually reported to be only around an hour, which is approximately half the retention time of material in the ileum. The shorter retention time, despite a larger holding capacity, is a logical consequence of a larger amount of material entering this segment compared to the ileum. It has been demonstrated that absorption of digestion products from major nutrients is to a large extent completed by the end of the jejunum.

The ileum is the last segment of the small intestine and ends at the ileo-ceco-colic junction. Despite the fact that the length of this segment is approximately the same as the jejunum, weight is much lower as discussed above. Although some digestion and absorption of fat, protein and starch may take place, this segment is mainly thought to play a role as a site for water and mineral absorption. As will be discussed further below, it has been shown that the ileum may play a significant role for digestion and absorption of starch in fast-growing broiler chickens. Svihus et al. (2004a) observed that starch digestibility increased from 91% to 99% from the anterior third to the posterior third of the ileum. Likewise, some fat absorption has been shown to take place in the ileum. Due to the fact that a majority of the feed dry matter has been absorbed, the passage through this segment is much slower than through the jejunum, as discussed earlier. As there are villi below Meckel's diverticulum, this at least partly can explain the significant starch digestion taking place in the ileum.

A large internal surface of the small intestine of poultry will compensate for a small size, as discussed above, and a short length, which reaches 1.5–2 m within 4 weeks of age (Moradi et al., 2021). However, reflux may also contribute to a high digestive capacity. It has been demonstrated that reflux is particularly prominent in birds. Through his classical work with turkeys, Gary Duke at the University of Minnesota demonstrated a very poultry-specific regular reflux of material from the upper digestive tract into the gizzard and proventriculus. Further work demonstrated that this reflux originated in the ileum as much as 88% of the time and that birds are therefore able to reflux material along the whole small intestine (Basha and Duke, 1999). It has been shown that this reflux increases in starved birds, demonstrating the likely function of this movement in maximizing digestibility in a short digestive tract. Reflux may explain the very rapid flow of material into the whole length of the small intestine when starved birds are refed. In fact, it has been demonstrated that starved birds with an empty small intestine will have filled the whole length of the small intestine within 1 h after refeeding (Sacranie et al., 2017). The material will then probably be refluxed to assure maximized digestion. Another possible explanation for a high digestion capacity of birds despite a short digestive tract, is an increased capacity to absorb nutrients. While no difference in transporter capacity has been shown, there are indications that birds have a higher capacity for passive transport (McWhorter et al., 2009).

The passage rate through the digestive tract is very fast in poultry. When measured as the first appearance in the excreta of a marker, Svihus and Itani (2019) found an average across a number of studies of 177 minutes, which varied across studies from approximately 2–4 h. However, mean retention time is a more accurate and relevant measure of passage rate. On average, 50% of a marker in an eaten bout of feed can be found in the excreta within 5–6 h. Retention time specifically in the small intestine has been summarized from a number of studies to be an average of around 3 h, of which one-third of the time is spent in the jejunum (Svihus and Itani, 2019). Due to the much longer retention time, the small intestine posterior to Meckel's diverticulum could contribute significantly to the digestion and absorption process, although the extent to which this segment of the digestive tract is able to digest and absorb nutrients has been questioned.

There is significant starch digestion taking place posterior to Meckel's diverticulum. Gutierrez del Alamo et al. (2009) demonstrated that although half the starch was digested by the proximal jejunum, considerable amounts of starch were digested in the distal jejunum and the proximal ileum. However, little further digestion took place at the distal ileum. A similar pattern was observed for protein, although only a small part of the protein was digested by the proximal jejunum. These observations indicate that little digestion takes place in the distal ileum. However, the cause for this could simply be that the remaining part of the diet reaching the distal ileum is not digestible. It is interesting that Yamauchi (2007), in his review of his and other work on the functionality of the small intestine, noted that, when the jejunum was resected, the ileum resumed a considerable digestive and absorptive capacity, resulting in normal digestion in the resected birds. It is therefore possible that a large part of the ileum is able to take part in digestion and absorption if needed.

As the above demonstrates, the small intestine is able to carry out the process of digesting and absorbing nutrients in a remarkably short time and despite a rather small intestinal tract. However, due to the very high feed intake, particularly in rapidly growing birds, it is important to be aware of how sensitive these birds are to disturbances that will reduce small intestinal functionality. In addition to reduced sustainability due to a poorer feed efficiency, birds will also be more vulnerable to microbiological disturbances as they proliferate when more nutrients become available in the lower small intestine and caeca. As has been discussed, stimulating gizzard development through the use of coarse feed materials will have a significant impact on small intestinal functionality, both through the grinding action and the crucial role of the gizzard as a feed flow regulator. To what extent it is possible to facilitate small intestinal functionality is less clear, not the least since more research is needed to elucidate what limits small intestinal digestion in birds.

6 Caeca and colon

Just like most species, the last part of the digestive tract of poultry contains a segment characterized by a large concentration of microbiota. Functionally, the segment thus serves as a source of nutrients from feed materials which remain undigested after passing through the small intestine, since some nutrients such as volatile fatty acids and vitamins produced by the microbiota can be absorbed and utilized by the bird. In addition, this segment serves as a major site for water absorption. Although fermentation and water reabsorption to a large extent is confined to the two blind sacs representing the caeca, with a short colon characterized by a limited retention time, these segments work in concert and will thus be considered together.

Material passes through the sphincter between the ileum and colon due to peristalsis of the small intestine, probably combined with the relaxation of the sphincter muscle. Some of this material then enters the caeca, due to simultaneous anti-peristaltic movements in the colon. Due to the narrow opening into the caeca which contains numerous folds, villi and microvilli, only small particles and non-viscous material will pass into the caeca (Svihus et al., 2013). In addition, material such as uric acid originating from the cloaca and lower colon will be refluxed into the caeca due to regular anti-peristaltic contractions in the colon. Son et al. (2002) found that 18% of the dry matter and 17% of the water entered the caeca of cockerels, and thus concluded that material entering the caeca is of significant quantity. For starved birds, the proportion of material entering the caeca was even higher.

The dynamics of frequent aborad movement are still not fully understood. Duke (1989) hypothesized that urine and other fluid materials are refluxed along the periphery of the colon due to low-amplitude antiperistaltic movements while the core, consisting of less soluble material, is moved towards the cloaca by higher-amplitude peristaltic movements. Such selective retention allows the bird to combine a rapid passage of coarse un-fermentable material with an extended retention time of fine, fermentable material, since retention time in the caeca is much longer than in the rest of the digestive tract.

Due to the abovementioned mechanisms, only easily fermentable material enters the caeca, while coarse particles will pass through rather quickly. In fact, birds with a very high feed intake such as broilers, will defecate approximately twice per hour, demonstrating the rapid passage of material not entering the caeca. This mechanism will facilitate a high fermentation rate with minimized weight, suitable for flight animals. In addition, the reflux of uric acid into the caeca will potentially boost fermentation taking place there. It has also been speculated that some of the nitrogen (N) lost in uric acid may be recycled through being resynthesized into amino acids by the microbiota. However, although both amino acid transporters and amino acid absorption has been demonstrated, it is uncertain whether amino acids absorbed in the caeca are of significant quantity.

Volatile fatty acids (VFAs) produced by the microbiota will be easily absorbed and thus contribute with energy to the bird. While Annison et al. (1968) indicated that VFAs could provide up to 11% of metabolisable energy for mature chickens, subsequent work has estimated the amount of energy that can be attained from caecal fermentation to be around 3-5% of the total energy needs of the chicken (Choct et al., 1992; Jorgensen et al., 1996; Jamroz et al., 2002). Although the mechanisms have not been studied in detail, it appears that at least some vitamins such as Vitamin K and folate are absorbed in the caeca.

One important function of the caeca is electrolyte and water absorption, where the caeca have been described as the quantitatively most important segment of the gut. Thomas (1982), in his comprehensive review of water and electrolyte absorption in the fowl, stated that net water absorption in the gut does not occur until after the ileum, and is mainly due to reabsorption of electrolytes and water of intestinal and renal origin in the caeca. Thomas (1982) estimated that 36% of the water and 75% of the sodium of renal origin were absorbed from the lower digestive tract, with the caeca being the most important organ. In fact, the whole lower digestive tract may be said to be 'an integrator of renal and intestinal function' (Thomas, 1982), where electrolytes, water and nitrogen balance are secured through reabsorption and recycling, and where the caeca play a major role.

Although far more research is needed to understand the intricate relationship between diet, feeding regime and function of the caeca and colon, it is clear that this digestive compartment plays a crucial role in finalizing the digestion process, not the least in relation to water reabsorption and some energy provision through fermentation. However, a more precise understanding of the selective passage of materials into the caeca is still needed, including the extent and role of reflux of urinary content into the caeca. Likewise, the positive role of caeca fermentation as a source of nutrients and through other roles such as immune function also needs to be elucidated.

7 Conclusion and future trends

Maximizing gut functionality is a key factor for a sustainable poultry production. This is because gut functionality is necessary for maximized nutrient utilization and for an a ability to handle a large feed intake and thus a maximized bird performance which will assure an effective conversion of feed to human food. In addition, bird health and welfare is, to a large extent, dependent on a well-functioning digestive tract. As this review shows, the digestive tract of poultry is remarkably efficient, despite the tremendous increase in digestive load due to a high feed intake. However, some changes to feed and feeding management may still improve functionality. The importance of stimulating gizzard development through structural components is now well established, although implementation in commercial feed production is still in progress and varies considerably throughout the world.

In the future, more emphasis on crop retention may result in changes to feeding practices which will also have an effect on the use of functionality-enhancing additives such as exogenous enzymes and probiotics, provided that more research is diverted into this area. The same holds true for the caeca, where more research into nutritional roles of the caeca may allow for more targeted feed composition to stimulate the fermentation activity taking place there. Potential changes to facilitate urine reflux into the caeca, given that research confirms a beneficial effect, may also be undertaken. Lastly, as more research emerges to provide more insight into the mechanism allowing for the high digestive capacity of the small intestine, targeted additives or feeding practices may allow for a further increase in nutrient uptake, further improving sustainability in an already very efficient animal food production system.

8 Where to look for further information

As already mentioned, *Sturkie's Avian Physiology* and similar detailed basic textbook texts are very valuable sources of fundamental information about the anatomy and physiology of birds in general, and the digestive tract specifically. In fact, the digestive tract closely interacts with many other organs and tissues in the body. Thus, reading textbooks on the anatomy and physiology of the bird in general will be very useful for an in-depth understanding of the complex role of the digestive tract and its interaction with the birds as a whole. For those particularly interested in specific digestive functions, systematic searches in renowned databases such as Web of Science is key to staying updated on the latest progress. Journals such as *Poultry Science*, *Animal Feed Science and Technology*, *Journal of Applied Poultry Research*, and *British Poultry Science* are particularly well regarded within poultry science, although very good papers can also be found in many other journals. For reviews, *World's Poultry Science Journal* may be a particularly suitable source. In fact, good recent reviews will in many cases be the best source to getting an overview of the current state of knowledge within digestive physiology in general or the functionality of specific organs. In the reference list, several recommendable reviews can be found.

9 References

Annison, E. F., Hill, K. J. and Kenworthy, R. (1968). Volatile fatty acids in the digestive tract of the fowl. *Br. J. Nutr.* 22(2), 207–216.

Apajalahti, J. and Vienola, K. (2016). Interaction between chicken intestinal microbiota and protein digestion. *Anim. Feed Sci. Technol.* 221, 323–330.

Bajagai, Y. S., Trotter, M., Williams, T. M., Costa, D. F. A., Whitton, M. M., Ren, X., Wilson, C. S. and Stanley, D. (2022). The role of microbiota in animal health and productivity: misinterpretations and limitations. *Anim. Prod. Sci.* 62(11), 944–956.

Basha, M. E. and Duke, G. E. (1999). Effect of fasting on small intestinal antiperistalsis in the Nicholas Turkey (*Meleagris gallopavo*). *J. Exp. Zool.* 283(4–5), 469–477.

Carré, B., Mignon-Grasteau, S. and Juin, H. (2008). Breeding for feed efficiency and adaptation to feed in poultry. *World's Poult. Sci. J.* 64(3), 377–390.

Choct, M., Annison, G. and Trimble, R. P. (1992). Soluble wheat pentosans exhibit different antinutritive activities in intact and cecectomized broiler chickens. *J. Nutr.* 122(12), 2457–2465.

Classen, H. L., Apajalahti, J., Svihus, B. and Choct, M. (2016). The role of the crop in poultry production. *Worlds Poult. Sci. J.* 72(3), 459–472.

del Alamo, A. G., Verstegen, M. W. A., Den Hartog, L. A., de Ayala, P. P. and Villamide, M. J. (2009). Wheat starch digestion rate affects broiler performance. *Poult. Sci.* 88(8), 1666-1675.

Duke, G. E. (1989). Relationship of cecal and colonic motility to diet, habitat, and cecal anatomy in several avian species. *J. Exp. Zool. Suppl.* 3, 38–47.

Duke, G. E. (1992). Recent studies on regulation of gastric motility in turkeys. *Poult. Sci.* 71(1), 1–8.

Hetland, H., Svihus, B. and Olaisen, V. (2002). Effect of feeding whole cereals on performance, starch digestibility and duodenal particle size distribution in broiler chickens. *Br. Poult. Sci.* 43(3), 416–423.

Itani, K. and Svihus, B. (2019). Feed processing and structural components affect starch digestion dynamics in broiler chickens'. *Br. Poult. Sci.* 60(3), 246–255.

Itani, K., Apajalahti, J., Smith, A., Ghimire, S. and Svihus, B. (2024). The effect of increasing levels of oat hulls, extent of grinding and their interaction on the performance, gizzard characteristics and gut health of broilers fed oat-based diets. *Anim. Feed Sci. Technol.* 308, 115858.

Jamroz, D., Jakobsen, K., Knudsen, K. E. B., Wiliczkiewicz, A. and Orda, J. (2002). Digestibility and energy value of non-starch polysaccharides in young chickens, ducks and geese, fed diets containing high amounts of barley. *Comp, Biochem. Physiol.* 131A, 657–668.

Jorgensen, H., Zhao, X. Q., Knudsen, K. E. B. and Eggum, B. O. (1996). The influence of dietary fibre source and level on the development of the gastrointestinal tract, digestibility and energy metabolism in broiler chickens. *Br. J. Nutr.* 75(3), 379–395.

Kristoffersen, S., Wisniewska, Z., Kazcmarek, S., Gjefsen, T., Kjos, N. P., Cowieson, A. J. and Svihus, B. (2022). Assessment of crop usage in ad libitum fed birds and short-term phytase efficiency as affected by acid addition. *Br. Poult. Sci.* 63(3), 414–420.

McLelland, J. (1989). Anatomy of the avian cecum. *J. Exp. Zool. Suppl.* 3, 2–9.

McWhorter, T. J., Caviedes-Vidal, E. and Karasov, W. H. (2009). The integration of digestion and osmoregulation in the avian gut. *Biol. Rev.* 84(4), 533–565.

Moradi, S., Moradi, A., Elmi, V. A. and Abdollahi, M. R. (2021). Interactive effect of corn particle size and insoluble fiber source on performance, nutrient utilization and intestine morphology in broilers fed pelleted diets. *J. Anim. Physiol. Anim. Nutr.* 105(6), 1113–1126.

Portella, F. J., Caston, L. J. and Leeson, S. (1988). Apparent feed particle size preference by broilers. *Can. J. Anim. Sci.* 68(3), 923–930.

Roura, E., Baldwin, M. W. and Klasing, K. C. (2013). The avian taste system: potential implications in poultry nutrition. *Anim. Feed Sci. Technol.* 180(1–4), 1–9.

Sacranie, A., Adiya, X., Mydland, L. T. and Svihus, B. (2017). Effect of intermittent feeding and oat hulls to improve phytase efficacy and digestive function in broiler chickens. *Br. Poult. Sci.* 58(4), 442–451.

Sklan, D., Hurwitz, S., Budowski, P. and Ascarelli, I. (1975). Fat digestion and absorption in chicks fed raw or heated soybean meal. *J. Nutr.* 105(1), 57–63.

Son, J. H., Ragland, D. and Adeola, O. (2002). Quantification of digesta flow into the caeca. *Br. Poult. Sci.* 43(2), 322–324.

Sturkie, P. D. (1986). *Avian Physiology*. Springer-Verlag, New York.

Svihus, B. (2011). The gizzard: function, influence of diet structure and effects on nutrient availability. *World's Poult. Sci. J.* 67(2), 207–224.

Svihus, B. (2014). Function of the digestive system. *J. Appl. Poult. Sci.* 23(2), 306–314.

Svihus, B. and Hetland, H. (2001). Ileal starch digestibility in growing broiler chickens fed on a wheat-based diet is improved by mash feeding, dilution with cellulose or whole wheat inclusion. *Br. Poult. Sci.* 42(5), 633–637.

Svihus, B. and Itani, K. (2019). Intestinal passage and its relation to digestive processes'. *J. Appl. Poult. Res.* 28(3), 546–555.

Svihus, B., Choct, M. and Classen, H. L. (2013). Function and nutritional roles of the avian caeca: a review. *Worlds Poult. Sci. J.* 69(2), 249–264.

Svihus, B., Itani, K., Borg, K., Larsson, E. C., Ao, R., Sudubilige, A., Fuerjiafu, B., Liu, H., Hetland, H., Sanson, G., Kierończyk, B., Rawski, M. and Józefiak, D. (2017). Performance and digestive function of broiler chickens given grit in the diet. *Br. Poult. Sci.* 58(5), 530–535.

Svihus, B., Juvik, E., Hetland, H. and Krogdahl, Å. (2004a). Causes for improvement in nutritive value of broiler chicken diets with whole wheat instead of ground wheat. *Br. Poult. Sci.* 45(1), 55–60.

Svihus, B., Kløvstad, K. H., Perez, V., Zimonja, O., Sahlström, S., Schüller, R. B., Jeksrud, W. K. and Prestløkken, E. (2004b). Physical and nutritional effects of pelleting of broiler chicken diets made from wheat ground to different coarsenesses by the use of roller mill and hammer mill. *Anim. Feed Sci. Technol.* 117(3–4), 281–293.

Svihus, B., Sacranie, A., Denstadli, V. and Choct, M. (2010). Nutrient utilization and functionality of the anterior digestive tract caused by intermittent feeding and inclusion of whole wheat in diets for broiler chickens. *Poult. Sci.* 89(12), 2617–2625.

Thomas, D. H. (1982). Salt and water excretion by birds: the lower intestine as an integrator of renal and intestinal excretion. *Comp, Biochem. Physiol.* 71A, 527–535.

Wijtten, P. J. A., Langhout, D. J. and Verstegen, M. W. A. (2012). Small intestine development in chicks after htch and in pigs around the time of weaning and its relation with nutrition: a review. *Acta Agric. Scand.* 62, 1–12.

Yadav, S. and Jha, R. (2019). Strategies to modulate the intestinal microbiota and their effects on nutrient utilization, performance, and health of poultry. *J. Anim. Sci. Biotechnol.* 10(1), 2.

Yamauchi, K. (2007). Review of a histological intestinal approach to assessing the intestinal function in chickens and pigs. *Anim. Sci. J.* 78(4), 356–370.

Chapter 3

Digestive kinetics of feedstuffs in broiler chickens

Sonia Yun Liu[1], Shemil P. Macelline and Peter H. Selle[1], The University of Sydney, Australia

1 Introduction

As Rérat (1993) has commented: Study of kinetics associated with differences in the relative absorption rate of the various nutrients is necessary for a fuller understanding of variations in the nutritive value of feeds. The growth performance of modern genotype broiler chickens is dynamic. From just 44 g at hatch, straight-run Ross 308 chickens achieve body weights of 2998 g at 42 days post hatch based on 2022 Aviagen performance objectives. This rapid growth rate is not adequately reflected in static apparent digestibility coefficients of protein, amino acids and starch determined at the terminal ileum. While ileal digestibilities eliminate the confounding influence of hind-gut fermentation, amino acids and glucose are mainly absorbed in the jejunum. Moreover, digesta is retained for approximately only 49 min in the duodenum and jejunum and for 109 min in the ileum (Del Alamo et al., 2009; Gutierrez del Alamo et al., 2009). Total tract retention times are in the order of 340 min in broiler chickens (Svihus and Itani, 2019). The synchronicity of intestinal uptakes of amino acids and glucose and their extents, rates and sites of digestion and absorption are pivotal to the digestive kinetics of feedstuffs. Careful attention needs to be paid to the inherently different digestive kinetics of protein-bound

1 Corresponding authors: peter.selle@sydney.edu.au; sonia.liu@sydney.edu.au.

http://dx.doi.org/10.19103/AS.2024.0143.03

versus non-bound (synthetic, crystalline) amino acids and their post-enteral availability. This is mandated by the likelihood that the dietary incorporation of non-bound amino acids will escalate with the increasing adoption of reduced crude protein (CP) diets and that the bioavailabilities of non-bound versus protein-bound amino acids are not identical (Selle et al., 2022b).

Instructively, static apparent ileal digestibility coefficients of protein (N) and starch were not related to weight gain or feed conversion ratio (FCR) in a jointly reported study (Selle et al., 2012; Liu et al., 2013b). Conversely, kinetic protein (N) digestion rates were related to both weight gain (r = 0.353; P = 0.009) and FCR (r = −0.341; P = 0.012) as were starch digestion rates, albeit to lesser extents, related to weight gain (r = 0.311; P = 0.022) and FCR (r = −0.238; P = 0.083). This comparison illustrates the need to consider the digestive kinetics of feedstuffs as major determinants of broiler growth performance. Moreover, this comparison is supported by the previously-quoted statement by Rérat (1993) that the 'study of kinetics associated with differences in the relative absorption rate of the various nutrients is necessary for a fuller understanding of variations in the nutritive value of feeds'.

Digestive kinetics is broadly defined as a three-tier process: (i) digestion of protein and starch in the gut lumen, (ii) intestinal uptakes of amino acids and glucose along the small intestine and (iii) their transition across the enterocytes of the gut mucosa into the portal and, ultimately, the systemic circulation. However, tangible proportions of absorbed amino acids and glucose are not transferred into the systemic pool. In the gut mucosa, amino acids may enter anabolic pathways for secretory and structural protein synthesis or catabolic pathways for the provision of energy drive digestive processes. Glucose is also utilised for energy provision to meet the 'high metabolic cost of a functional gut' (van der Schoor et al., 2002). Thus, dietary protein and starch must negotiate complex physiological and biochemical corridors to become available as amino acids and glucose at sites of protein synthesis. Therefore, the focus of protein/amino acids and starch/glucose digestive kinetics is on their digestion and intestinal uptakes resulting in the post-enteral, bilateral bioavailability of amino acid and glucose. The fundamental principle is that a correctly balanced blend of amino acids coupled with appropriate supplies of glucose must be available at sites of protein synthesis to achieve effective growth performance. Amino acids are the 'building blocks' of protein but a 5.35 kJ energy input is required to drive the accretion of 1 g of protein in broiler chickens (Aoyagi et al., 1988); this energy is mainly derived from glucose. Importantly, in addition to being the protein building blocks, amino acids are required to fulfil numerous functional roles, which impacts on the dietary requirements of the so-called essential and non-essential amino acids for broiler chickens beyond their incorporation into body protein (Wu et al., 2013). A core tenet of feedstuff digestive kinetics is that starch is digested too rapidly and protein too slowly; therefore, a closer alignment of these two kinetic parameters should advantage broiler growth performance.

2 Digestion and intestinal uptakes of nutrients

A brief summation of protein and starch digestion, intestinal uptakes of amino acids and glucose and their transition across the gut mucosa before their digestive kinetics can be discussed. Protein digestion is initiated by pepsin and HCl in the proventriculus where dietary proteins are converted into polypeptides. Pancreatic enzymes (trypsin, chymotrypsin, carboxypeptidase, elastase) released into the duodenum convert polypeptides into smaller peptides, which are then converted to di- and tri-peptides (or oligopeptides) by amino peptidases in the apical membrane of enterocytes (Moran, 2016). Intestinal uptakes of oligopeptides are conducted via the Pept-1 transporter in connection with the Na^+/H^+ exchanger, NHE (Spanier, 2014). Pept-1 gene expression has been shown to be upregulated by lysine–glutamine dipeptides in poultry (Tokutake et al., 2021). Alternatively, intestinal uptakes of monomeric or non-bound amino acids are conducted via numerous Na^+-dependent and Na^+-independent transport systems with overlapping affinities (Hyde et al., 2003; Miska and Fetterer, 2019). However, most amino acids are absorbed as oligopeptides (Krehbiel and Matthews, 2003), and their intestinal uptakes via Pept-1 are both rapid and energetically efficient (Daniel, 2004). The difference in intestinal uptake velocities was demonstrated in pigs by Li et al. (1999). The cumulative absorption of lysine derived from a lysine–glycine dipeptide exceeded that of non-bound lysine by a factor of 2.35 (30.30 versus 12.88 μmol) collectively at 30 and 60 min post administration.

As mentioned, digestive kinetics of protein-bound amino acids as opposed to versus non-bound amino acids are inherently different, simply because non-bound amino acids do not require prior digestion and can be absorbed immediately (Wu, 2009). Importantly, rapid intestinal uptakes of non-bound amino acids generate higher amino acid concentrations in systemic plasma. This was demonstrated in humans by Gropper and Acosta (1991) where male subjects were given either intact protein as cottage cheese or an equivalent blend of non-bound amino acids. Collectively, plasma amino acid concentrations derived from the amino acid blend exceeded those of intact protein by 59.9% (676.0 versus 422.7 μmol/L) at 30, 60 and 90 min post prandium.

Amino acid digestion rate constants were determined in broiler chickens offered standard sorghum-based diets by Liu et al. (2013a); the diets contained 4.8 g/kg lysine HCl, 3.4 g/kg DL-methionine (and 1.3 g/kg L-threonine). The observed digestion rate constants of 16 amino acids are shown in Fig. 1 where it is evident that methionine (5.51×10^{-2} min^{-1}) and lysine (4.14×10^{-2} min^{-1}) were the most rapid and comfortably exceeded the mean digestion rate constant of 13 protein-bound amino acids (2.35×10^{-2} min^{-1}). It may be deduced from Liu et al. (2013b) that the digestion rate constants of DL-methionine (8.49×10^{-2} min^{-1}) and lysine HCl (8.78×10^{-2} min^{-1}) were more rapid than protein-bound amino acids by nearly a four-fold factor if protein-bound methionine and lysine fractions had similar digestion rate constants to the average value. The more rapid intestinal uptakes of non-bound amino acids are fundamental to their differences in post-enteral bioavailability compared to protein-bound amino acids (Selle et al., 2022b).

Starch digestion and glucose absorption is much more straightforward as starch digestion in the gut lumen is the province of pancreatic α-amylase (Moran, 1982). Intestinal uptakes of glucose are principally conducted via the Na^+-dependent, sodium–glucose co-transporter, SGLT-1, located in the enterocytes of the gut mucosa (Röder et al., 2014; Shibata et al., 2019). However, apparent digestibility coefficients of starch were negatively correlated with 9 amino acids in distal jejunum, 12 amino acids in proximal ileum and 11 amino acids in distal ileum in broilers offered starch-enriched, maize-based diets in Moss et al. (2018). Given that Vinardell (1990) found there was mutual inhibition between sugars and amino acid for intestinal absorption, this suggests that glucose and amino acids compete for co-absorption with sodium via either SGLT-1 or Na^+-dependent amino acid transporters, respectively, which disrupts intestinal uptakes of glucose and amino acids.

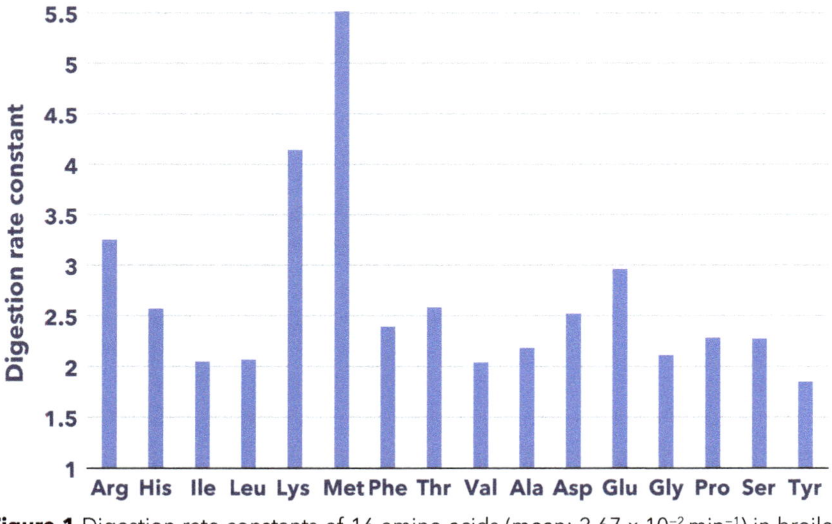

Figure 1 Digestion rate constants of 16 amino acids (mean: 2.67×10^{-2} min^{-1}) in broiler chickens offered sorghum-based diets. Source: taken from Liu et al. (2013a).

Having looked at protein and starch digestion, the following discussion reviews the transition of amino acids and glucose across the gut mucosa. The enterocytes of the small intestinal gut mucosa are the gateway between intestinal uptakes of amino acids and glucose derived from protein and starch digestion and their subsequent post-enteral, bilateral bioavailability. Small intestinal enterocytes have an enormous impact on the transition of amino acids and glucose across the gut mucosa. Amino acids may be 'captured' by anabolic and/or catabolic pathways and denied entry into the portal circulation and, to some extent, glucose is catabolised. In pigs, the net portal flux of 12 amino acids declined by an average of 32.6%, which ranged from 8.0% (arginine) to 54.5% (threonine), due to their entry into catabolic and anabolic pathways (Reeds et al., 1996). The gastrointestinal tract requires around 20% of dietary energy for nutrient digestion and absorption (Cant et al., 1996) and this energy demand is met mainly by the broadly similar oxidation of glucose and glutamine in rats (Fleming et al., 1997). In young pigs (Stoll et al., 1999), enteral (15%) and arterial (29%) glucose contributed to CO_2 production by the portal-drained viscera but this was exceed by glutamate/glutamime with enteral and arterial contributions of 36% and 19%, respectively. These researchers concluded that dietary glutamate is the most important single contributor to mucosal oxidative energy generation in pigs. The proportion of dietary amino acids catabolised in the gut mucosa may approach 20% as net portal ammonia concentrations accounted for 18% of total amino acid nitrogen in young pigs (Stoll et al., 1998). However, glucose and amino acids may not be capable of meeting the energy requirements of the gut entirely; this could result in the oxidation of additional substrates including fatty acids and ketone bodies (Van Der Schoor et al., 2001). The pioneering research in this area was completed by Peter Reeds and his colleagues, mainly in pigs (Reeds et al., 2000), but there is a paucity of corresponding data in poultry. Glucose, glutamine and glutamate are the preferred energy substrates by the gut mucosa in broiler chickens. Watford et al. (1979) and Porteous (1980) found them to be the only individual substrates to stimulate oxidation in avian enterocytes.

Thus, there is effectively a 'catabolic ratio' between amino acids and glucose within enterocytes, which may or may not be open to manipulation by nutritional strategies β a question posed by Reeds et al. (2000). If the catabolic ratio can be modified, it should prove advantageous to tip the ratio towards glucose to spare amino acids from catabolism and energy may be derived more efficiently from glucose (Fleming et al., 1997).

3 Rapidly digestible protein

The distinction between 'fast' and slow' protein was drawn by Beaufrere et al. (2000) in human nutrition. Their proposal was that postprandial protein kinetics and deposition depend on the rate of absorption of amino acids derived from dietary proteins. Decades earlier, Geiger (1950) contended that protein synthesis is dependent not only on the simultaneous presence of essential amino acids but also on the speed with which non-essential amino acids are made available. An opinion that only emphasises the importance of kinetics in the efficient utilisation of amino acids. In other human studies, Boirie et al. (1997) concluded that the speed of protein digestion and amino acid absorption has a major impact on body protein anabolism and Dangin et al. (2001) found that protein digestion rates modulate postprandial protein deposition. Instructively, Daenzer et al. (2001) concluded that whole-body protein homeostasis in rats is better supported by protein-bound amino acids (casein) than non-bound amino acids. The roles of intact proteins, peptides or free amino acids in monogastric farm animals were reviewed by Eugenio et al. (2022); the three categories of 'protein' differ in respect of the velocity of their intestinal uptakes. However, it is evident in this review that a scarcity of research has been completed in poultry in this area relative to pigs.

The digestive dynamics of eight protein-rich feedstuffs were compared in Toghyani et al. (2020). The feedstuffs included blood meal, plasma protein meal, cold-pressed and expeller-pressed canola meals, high- and low-protein soybean meals, lupins and field peas. The average protein digestion rate constant was 0.053/min which ranged from 0.015/min (expeller canola meal) to 0.124/min (blood meal). Protein (N) disappearance rates (g/bird/day) were determined in four small intestinal segments in broiler chickens where significant differences were observed. Ranked in descending order, mean disappearance rates (g/bird/day) declined from 25.51 in high-protein soybean meal, 22.37 in low-protein soybean meal, 18.61 in expeller canola meal, 16.98 in cold-pressed canola meal, 16.36 in blood meal, 15.02 in lupins, 15.01 in peas to 12.41 in plasma protein. Thus, as an example, the mean disappearance rate of high-protein soybean meal was 70.0% (25.51 versus 15.01 g/bird/day) more rapid than lupins.

Whey protein is a more rapidly digested protein source than soybean meal (Dangin et al., 2001), not that whey is incorporated into practical broiler diets. However, the partial replacement of soybean meal with either dry whey powder or whey protein concentrate in broiler diets based on a maize–wheat blend significantly enhanced weight gain and FCR in Pineda-Quiroga et al. (2018). This suggests that 'rapid protein' benefits broiler growth performance. Moreover, Liu et al. (2014a) contended that protein digestion kinetics have a greater bearing on growth performance than starch digestion kinetics in broiler chickens.

The digestion kinetics of nine protein sources generated from an *in vitro* chicken model were quantified in Bryan et al. (2019b). The average predicted rate constant [k_d (h^{-1})] was 0.040, which included, in ascending order, soybean meal, 0.027; canola meal, 0.035; fish meal, 0.040; meat-and-bone meal, 0.046 and blood meal, 0.062. A detailed evaluation of the *in vivo* digestion kinetics of nine protein sources was completed in broiler chickens by Bryan et al. (2019a). Instructively, when one protein source (porcine meal) is not considered, the linear relationship ($r = 0.687$; $P = 0.060$) between CP digestion rates and 14–21-day weight gains approaches significance, based on tabulated mean values.

The rapidly digestible protein concept was investigated in broiler chickens by Truong et al. (2017) in which rapid protein disappearance rates along the small intestine advantaged performance. Dietary rapid protein was achieved by substituting soybean meal with inclusions of casein and non-bound amino acids to formulate foundation, intermediate and summit, maize-based diets with increasing concentrations of rapidly digestible protein. This approach linearly ($r = 0.615$; $P = 0.001$) increased distal ileal protein (N) disappearance rates by up to 6.48% (18.41 versus 17.29 g/bird/day), with corresponding increases of 9.82%, 9.70% and 8.78% were observed in proximal jejunum, distal jejunum and proximal ileum, respectively. In addition, this strategy linearly increased distal ileal disappearance rates of 12 of the 16 amino acids assessed and similar response patterns were observed in three anterior small intestinal segments. Importantly, the transition from foundation to intermediate and summit diets increased weight gains in a linear manner ($r = 0.404$; $P = 0.050$) from 1541 to 1607 and 1610 g/bird, respectively, from 7 to 28 days post hatch and distal ileal protein (N) disappearance rates were linearly related ($r = 0.706$; $P < 0.001$) to 7–28-day weight gains.

Subsequently, Berrocoso et al. (2020) reported that broiler diets containing rapidly digestible proteins supported higher weight gains by 3.80% (2074 versus 1998 g/day) and better FCR by 4.73% (1.510 versus 1.585) to 36 days post hatch than diets containing slowly digestible proteins. However, the protein digestion rates were not determined in this study. Also, the effects of starch, protein and fibre digestive kinetics on broiler growth performance were evaluated by Jaworski et al. (2019) via a Box-Behnken study. Increasing concentrations of rapidly digestible protein enhanced feed intakes and weight gains and rapidly digestible fibre improved FCR.

The protein digestive kinetics of seven meat-and-bone meal samples were compared by Macelline et al. (2020). The seven experimental diets were isonitrogenous (195 g/kg CP) and isoenergetic (12.75 MJ/kg AMEn) and protein/amino acids were derived entirely from meat-and-bone meal and maize as there were not any non-bound amino acid inclusions. Mean protein (N) disappearance rates from four small intestinal segments ranged from 9.18 to 15.13 g/bird/day, a 64.8% difference. Interestingly, proximal jejunal protein (N) disappearance rates were positively correlated ($r = 0.630$; $P < 0.001$) with weight gains, whereas disappearance rates were negatively correlated ($r = -0.520$; $P < 0.001$) with weight gains in the distal ileum. When rapidly digestible protein is defined as the proportion absorbed in the proximal jejunum as opposed to the distal ileum, rapidly digestible protein quadratically increased ($r = 0.759$; $P < 0.001$) weight gains from 24 to 28 days post hatch as shown in Fig. 2. Clearly, these outcomes emphasise the importance of the sites of protein/amino acid uptakes along the small intestine.

3.1 Intact protein versus non-bound amino acids and oligopeptides

The kinetics of amino acid absorption are a key factor in regulating postprandial protein metabolism pursuant to variations in protein intake was the assertion of Fouillet et al. (2009). However, the previous section essentially considered the kinetics of 'intact' protein, which is becoming somewhat redundant by the increasing incorporation of non-bound amino acids to reduce dietary crude CP concentrations. Methionine, usually the first limiting amino acid in broiler diets, was identified in 1922. However, the methodology to synthesise DL-methionine efficiently was developed in Germany during the 1940s and DL-methionine was made available for animals in 1953 (Willke, 2014). Pivotally, this was precisely when amino acids were no longer all created equal (Selle et al., 2022a). This is simply because intestinal uptakes of non-bound methionine are more rapid than methionine derived from intact protein, which was demonstrated in rats

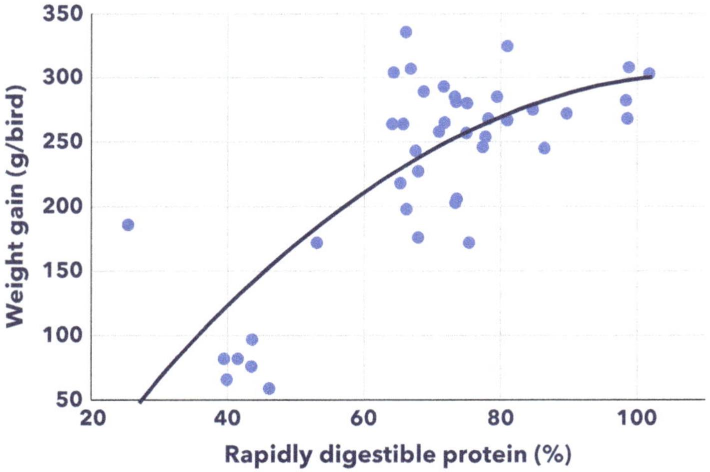

Figure 2 Quadratic relationship ($r = 0.759$; $P < 0.001$) between rapidly digestible protein (proportion absorbed in proximal jejunum) and weight gain from 24 to 28 days post hatch where $y = 7.936 \times \text{proportion} - 0.036 \times \text{proportion}^2 - 137.995$. Source: taken from Macelline et al. (2020).

by Canolty and Nasset (1975). However, di- and tri-peptides, or oligopeptides, provide a third dimension as shown by Rerat (1985), who infused oligopeptides, non-bound amino acids and intact protein (fishmeal) into the duodenum of pigs. Concentrations of α-amino N in portal plasma at 60 min postprandially ranged from 690 mg/L from intact protein, 1103 mg/L from non-bound amino acids and 1678 mg/L from oligopeptides. On this basis, intestinal uptakes of non-bound amino acids were 59.9% more rapid than amino acids derived from intact protein, but intestinal uptakes of oligopeptides were 143% more rapid than intact protein.

Postprandial concentrations of total proteinogenic amino acids in portal plasma in rats were determined by Kodera et al. (2006). Amino acid concentrations derived from soy protein peaked at approximately 3160 μmol/L at 20 min, but an equivalent blend of non-bound amino acids peaked at 5355 μmol/L at 11 min, and a soy protein hydrolysate peaked at 6725 μmol/L at 7 min. The soy protein hydrolysate principally consisted of decapeptides; nevertheless, the data confirm that intestinal uptakes of non-bound amino acids are more rapid than their protein-bound counterparts, but oligopeptides are even more rapid than non-bound amino acids. Also, absorption of the alanyl-glutamine dipeptide was more rapid than glycyl-glutamine in everted jejunal tissue taken from weaner pigs in Wang et al. (2010), but both dipeptides were absorbed more rapidly than non-bound glutamine.

The impact of protein sources with different digestion rates on broiler growth performance was investigated by Macelline et al. (2022) via a triangular response surface design using 10 wheat-based diets with true protein contents of 202.5 g/kg. The three apical diets contained either soybean meal (323 g/kg), non-bound amino acids (66.8 g/kg) or whey powder (161 g/kg) as their principal protein sources. The soybean meal diet, which contained 6.75 g/kg non-bound amino acids, outperformed the non-bound amino acids diet, which contained 162 g/kg soybean meal. Birds offered the soybean meal diet had significantly better weight gains by 4.88% (2021 versus 1927 g/bird) and FCR by 3.84% (1.427 versus 1.484) than their non-bound amino acid counterparts from 14 to 35 days post hatch. Thus, the substitution of soybean meal with high levels of non-bound amino acid inclusions compromised growth performance.

The best growth performance in this study was generated by the diet based on an equal blend of soybean meal and whey powder apical diets. This diet supported better performance than 2019 Aviagen objectives by 13.0% (2089 versus 1849 g/bird) in weight gain and 11.3% (1.401 versus 1.580) in FCR from 14 to 35 days post hatch. This diet contained 663 g/kg wheat, 262 g/kg soybean meal, 181 g/kg whey protein and 13.1 g/kg non-bound amino acids with analysed concentrations of 376 g/kg starch, 211 g/kg CP and a 1.78 dietary starch–protein ratio.

3.2 Post-enteral metabolism and amino acid utilisation

The post-enteral metabolism of amino acids, once they gain entry to the systemic pool, is extremely complex and their efficient utilisation for skeletal protein synthesis drives broiler growth performance. The so-called ideal amino acid ratios relative to lysine, upon which least-cost formulations of broiler diets are largely based, should guarantee efficient amino acid utilisation, but their accuracy is problematic. Just one complicating factor in poultry is that the amino acid composition of feathers is radically different from that of other body proteins (Selle et al., 2023a). An ideal amino acid ratio derived from data generated by Wu (2014) and Maharjan et al. (2021) is shown in Table 1. For comparative purposes, analysed amino acid concentrations, expressed relative to lysine (100), for breasts and thighs (Hamm, 1981) and feathers (Greenhalgh et al., 2020) are tabulated. A perusal of Table 1 reveals that ideal amino acid ratios clearly exceed the composition of breast and thigh proteins (where there are subtle differences), which reflects functional amino acid requirements and the composition of feather proteins are radically different because of very low

Table 1 A comparison of ideal amino acid ratios for broiler chicken diets with analysed amino acid concentrations in selected body parts expressed relative to lysine (100)

Amino acid	Ideal amino acid ratio[a]	Analysed concentrations		
		Breast[b]	Thigh[b]	Feathers[c]
Arginine	103	81	74	488
Histidine	37	53	42	32
Isoleucine	70	51	50	336
Leucine	118	99	95	571
Lysine	100	100	100	100
Methionine	43	40	39	29
Phenylalanine	68	49	49	339
Threonine	69	57	64	335
Valine	80	55	52	549
Alanine	88	69	71	280
Aspartic acid[d]	137	117	118	439
Cysteine	29	8	9	526
Glutamic acid[e]	288	180	187	719
Glycine	119	54	65	507
Proline	139	47	54	727
Serine	72	57	73	855
Tyrosine	45	43	51	170
Average (excluding lysine)	94	66	68	431

[a]Mean of data from Wu (2014) and Maharjan et al. (2021).
[b]Male birds: adapted from Hamm (1981).
[c]Male birds: adapted from Greenhalgh et al. (2020).
[d]Asparagine plus aspartate.
[e]Glutamine plus glutamate.

lysine concentrations. Therefore, it is unlikely that any ideal amino acid ratio can be fully appropriate in all contexts and that the 'first limiting amino acid' will depend on which particular protein is being synthesised.

Free amino acid concentrations in systemic plasma, which are accurate, have been frequently determined in an attempt to identify limiting amino acids, but the interpretation of such data is difficult. As the result of nonstop protein turnover (Swick, 1982; Urdaneta-Rincon and Leeson, 2004; Nieto et al., 1994), the likelihood is that free amino acids in systemic plasma are largely 'endogenous', stemming more from ongoing protein degradation rather than 'exogenous' amino acids derived from the diet. Also, apart from protein turnover, free amino acids in systemic plasma may emanate directly from enterocytes via the venous circulation or from being recycled to the gut mucosa via the arterial circulation. Certainly, fluctuations in free amino acid concentrations in response to different dietary treatments are observed. Free amino acid plasma concentrations in portal and systemic plasma were compared following the transition from a relatively slowly to a rapidly digestible protein diet in Truong et al. (2017). The transition significantly increased concentrations of methionine (44.9%), threonine (42.1%), proline (22.9%), tyrosine (16.6%) and isoleucine (12.8%), as shown in parentheses. The same transition significantly decreased concentrations of asparagine plus aspartate (10.3%), serine (13.0%), histidine (18.8%) and glycine (23.2%). Blood samples were taken once only from birds with access to feed in this study, so sequential blood sampling over time from birds without access to feed may be more instructive. There are, for example, indications that rats can maintain lysine concentrations in the body to a greater extent than threonine when offered diets without either amino acid (Yamashita and Ashida, 1969).

Short-term dynamics in amino acid metabolism was considered by Schreurs et al. (1997) and they proposed that free amino acids in systemic plasma were cleared by either protein synthesis or metabolic degradation. Subsequently, Brosnan (2003) argued that dietary amino acids exceeding requirements for protein synthesis are rapidly catabolised, because the body does not have a store for amino acids. The postprandial oxidation of egg white protein versus a corresponding blend of non-bound amino acids was determined via $^{13}CO_2$ breath tests in rats by Nolles et al. (2009). Postprandial oxidative losses of non-bound leucine (22.5%) were significantly higher than protein-bound leucine (16.3%). This suggests that non-bound amino acids are more likely to undergo postprandial oxidation because of their more rapid entry into the systemic pool. Instructively, dietary non-bound amino acids inclusions in broiler diets were linearly related ($r = 0.546$; $P = 0.019$) to systemic plasma NH_3 concentrations, which indicates that non-bound amino acids are more likely to be deaminated (Macelline et al., 2023). The implication is that imbalances between non-bound and protein-bound amino acids in relation to protein synthesis result in the catabolism of surplus amino acids, which are more likely to be non-bound entities due to rapid intestinal uptakes, which may also mean they are less prone to catabolism in the gut mucosa (Moss et al., 2018).

3.3 Ammonia overload

Systemic amino acid concentrations are sourced from intestinal uptakes of oligopeptides derived from the digestion of intact proteins and/or monomeric amino acids. The digestive kinetics of intact proteins differ which probably applies to monomeric amino acids; therefore, the likelihood is that these disparities generate post-enteral amino acid imbalances (Harperand and Rogers, 1965). Amino acids that are surplus to requirements for the synthesis of body proteins are catabolised, which is thermogenic (Bender, 2012), and the deamination of amino acids results in ammonia (NH_3) production. The intravenous LD_{50} of ammonium acetate in mice is 5.64 mmol/kg, but it is considerably lower in broiler chickens at 2.72 mmol/kg (Wilson et al., 1968); therefore, the systemic presence of NH_3 demands detoxification in poultry. Detoxification is achieved by the condensation of NH_3 and glutamic acid to yield glutamine which enters the Krebs uric acid cycle so that NH_3-N is ultimately excreted as uric acid-N (Stern and Mozdziak, 2019), a process that demands an energy input of 60.7 kJ/g for the synthesis and excretion of uric acid-N (Van Milgen, 2021). Glycine (or glycine equivalents) is a prerequisite for the Krebs uric acid cycle as one mole of glycine is required for every mole of uric acid excreted (Salway, 2018). From a retrospective estimate (Selle et al., 2021a), 14.0–45.4% of dietary glycine equivalents entered the Krebs uric acid cycle for NH_3 detoxification in Chrystal et al. (2021). Glycine and serine concentrations in reduced-CP diets have received considerable attention since the Dean et al. (2006) study and Siegert and Rodehutscord (2019) concluded that the dietary requirement for young broilers is 11–20 g/kg glycine equivalents.

It appears that elevated non-bound amino acid inclusions can become excessive such that NH_3 is not adequately detoxified and the resultant 'NH_3 overload' compromises growth performance. Accordingly, elevated plasma NH_3 concentrations were associated with depressed broiler growth performance in several feeding studies (Namroud et al., 2008; Ospina-Rojas et al., 2013; Ospina-Rojas et al., 2014).

4 Slowly digestible starch

Slowly digestible starch was probably first identified in poultry by Yutste et al. (1991) when they determined proximal and distal ileal digestibilities of purified starch from several sources in broiler chickens. Maize and wheat starch had mean digestibility coefficients of 0.968 in the proximal and 0.977 in the distal ileum; an increase of only 0.93% from starch digested in the ileum. Similarly, cassava and pea starch had corresponding digestibility coefficients of 0.904 and 0.946, or an increase of 4.65%. In contrast, the increase from bean starch digested in the ileum was 20.7% (0.723 versus 0.599), which identified it as a slowly digestible starch source. The kinetics of starch digestion in relevant feedstuffs is notably inconsistent as determined under both *in vitro* (Giuberti et al., 2012) and *in vivo* (Selle et al., 2021b; Weurding et al., 2001b) conditions and in *vitro* determinations are indicative of starch digestion kinetics in poultry (Weurding et al., 2001a). Broiler diets based on twelve different starch sources were evaluated by Weurding et al. (2001b) and the mean fractional starch digestion rate was 1.98 k_d h^{-1}. However, the digestion rates ranged from 0.55 k_d h^{-1} (potato starch) to 4.29 k_d h^{-1} (tapioca starch). A summary of the four cited studies is displayed in Table 2.

Tapioca, or starch from cassava, has very rapid digestion rates, as is evident in Table 2, but these can be modified by processing methods and conditioning temperatures as reported by Wang et al. (2023). These researchers found a mean proportion of 62.5% rapidly digested starch (starch digested in proximal jejunum) in broiler chickens offered diets containing 560 g/kg cassava. However, this proportion varied from 58.5% to 65.7% depending on the nine different processing methods applied. Instructively, increasing proportions of rapidly digested starch linearly compromised FCR from 1 to 21 ($r = 0.661$; $P = 0.052$) days and from 22 to 42 days post hatch ($r = 0.794$; $P = 0.011$), based on tabulated mean values in the Wang et al. (2023) report.

Alternatively, sorghum has relatively slow starch digestion rates as shown in Table 2. Seven graded inclusions (0.0–5.25 g/kg) of the sulphite reducing agent, sodium metabisulphite (SMBS), were evaluated in sorghum-based diets

Table 2 Summary of starch digestion rates *in vitro* and in broiler chickens (*in vivo*)

	Starch digestion rate (per minute)		Fractional starch digestion rate (k_d)	
Feedstuff	*In vitro* Giuberti et al. (2012)	*In vivo* Selle et al. (2021)	*In vitro*Weurding et al. (2001b)	*In vivo* Weurding et al. (2001a)
Maize	0.017 (14)[a]	0.086 (2)[a]	1.29	2.54
Barley	0.024 (14)	0.104 (3)	1.25	2.51
Wheat	0.035 (12)	0.117 (4)	1.59	2.51
Triticale	0.036 (11)	0.093 (2)	1.30	2.00
Rice	0.057 (9)	0.075 (7)	1.11	1.81
Sorghum	0.018 (11)		0.65	1.03
Oats	0.042 (9)		5.31	4.29
Peas			0.34	0.55
Tapioca				
Potato				

[a]Number of samples in parentheses.

(623 g/kg) by Liu et al. (2014b) and Selle et al. (2014). Dietary sodium levels were maintained constant as SMBS essentially replaced sodium bicarbonate. Sulphite reducing agents have the potential to cleave the disulphide cross-linkages in the β- and γ-kafirin fractions of grain sorghum, which was confirmed analytically and SMBS linearly increased (r = 0.943; P < 0.001) the *in vitro* protein solubility of the experimental diets. SMBS linearly improved FCR (r = −0.346; P = 0.014) from 10 to 24 days post hatch with a maximum response of 3.02% (1.414 versus 1.458) observed with a 3.75 g/kg SMBS inclusion. However, sulphite reducing agents also have the capacity to depolymerise starch via oxidative–reductive reactions (Paterson et al., 1996). These outcomes were not anticipated, but possibly as a consequence, SMBS linearly (r = −0.413; P = 0.003) decreased starch disappearance rates in the proximal jejunum (PJ) by up to 19.3% (410 versus 508 g/bird) at 1.50 g/kg, but linearly increased disappearance rates (r = 0.483; P < 0.001) by up to 65.2% (256 versus 155 g/kg g/bird) at 5.25 g/kg in distal jejunum (DJ), proximal (PI) and distal (DI) ileum. Collectively, positive SMBS inclusions condensed rapidly (PJ) to gradually (DJ, PI, DI) digested starch ratios from 3.28 to 2.22. Moreover, SMBS-induced disappearance of gradually digestible starch linearly (r = −0.436; P = 0.003) improved FCR as shown in Fig. 3.

The above studies demonstrate the benefits of slowly digestible starch in broiler chickens that were, importantly, offered essentially identical diets apart from processing methods or SMBS inclusions. Moreover, these outcomes

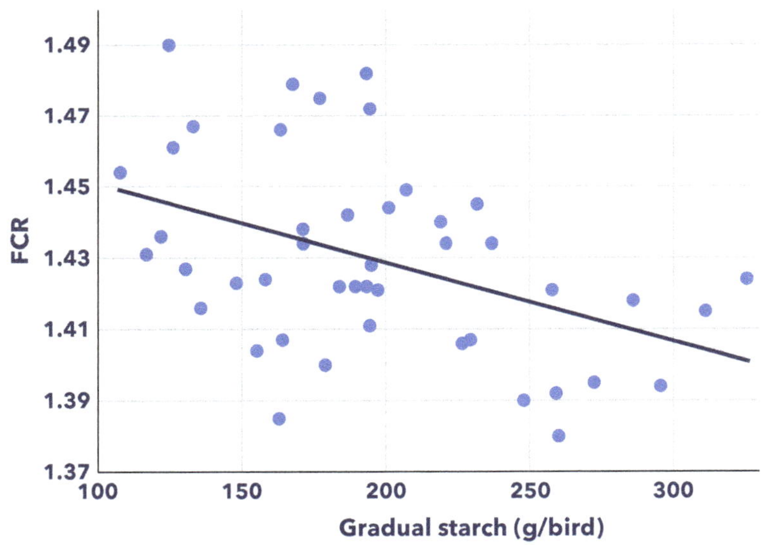

Figure 3 Linear relationship (r = −0.436; P = 0.003) between disappearance of gradually digestible starch (g/bird) and FCR from 10 to 24 days post hatch. Source: taken from Liu et al. (2014) and Selle et al. (2014).

emphasise the relevance of the sites of starch digestion and glucose absorption along the small intestine without the confounding effects of experimental diets with differing compositions.

The growth performance of broilers offered diets formulated to fractional starch digestion rates of either 1.18 or 3.23 k_d h^{-1} was compared by Weurding et al. (2003a). Slowly digestible starch was derived from peas and rapidly digestible starch from tapioca; thus, the two experimental diets differed. Nevertheless, slowly digestible starch diets generated increased weight gains by 5.44% (1823 versus 1729 g/bird; $P < 0.01$) and improved FCR by 2.42% (1.734 versus 1.777; $P = 0.010$) to 38 days post hatch. In a further study, semi-purified starch derived from wheat or peas were compared as rapidly and slowly digestible sources, respectively, in otherwise similar diets based on purified starch and soybean meal by Herwig et al. (2019). The experimental diets contained an average of 535 g/kg wheat or pea starch with six rapidly to slowly starch digestible ratios ranging from 0 to 20, 40, 60, 80 and 100% pea starch. The 25% pea starch inclusion supported the minimum FCR and increasing slowly digestible pea starch inclusions linearly increased breast meat yields.

Weurding et al. (2003a) advanced some possible mechanisms whereby slowly digestible starch *per se* promotes broiler growth performance. Slowly digestible starch may (i) reduce the conversion of glucose to lactate, thereby conserving energy, (ii) generate a more sustained insulin release, thereby enhancing protein deposition and (iii) spare amino acid from catabolism in the caudal part of the small intestine by the provision of glucose as an alternative energy substrate. Subsequent publications (Weurding et al., 2003b; Enting et al., 2005) lend some support to the proposal that slowly digestible starch may spare amino acids from catabolism in the gut mucosa. It is possible that slowly digestible starch would generate a beneficial, more sustained, insulin release; however, the starch–glucose–insulin–glucagon axis in poultry is different from mammalian species and is not equally understood. Blood glucose levels in chickens are considerably higher than in mammals and chickens exhibit a relative resistance to insulin and higher levels are required to obtain hypoglycemic effects than in mammals (Tesseraud et al., 2007). Different sources of starch have been shown to influence glucose absorption and insulin response in pigs (Regmi et al., 2010), but relationships between starch digestion kinetics and postprandial glucose in broiler chickens were not observed in Adeleye et al. (2016). Clearly, further research in this area in poultry is needed as Yaman et al. (2000) linked increases in plasma and glucose concentrations after re-feeding to the stimulation of breast muscle protein synthesis in broiler chickens.

In the context of reduced-CP diets, maize-based diets generated heavier relative fat-pad weights than wheat-based diets in Chrystal et al. (2021). One possible explanation is that glucose derived from rapidly digestible wheat starch is more readily catabolised to generate energy directly, whereas glucose from slowly digestible maize starch is more likely to be converted to glycogen and then fat via *de novo* lipogenesis (Selle et al., 2022c). Another possibility is that starch digestion kinetics may influence the activation of the ileal brake in broiler chickens and promote feed efficiency (Herwig et al., 2020). Finally, exogenous amylases have the potential to accelerate starch digestion rates, which may be inappropriate. The inclusion of 80 KNU/kg α-amylase activity in maize-based diets increased distal jejunal starch digestibility by 6.38% (0.850 versus 0.799) at 21 days post hatch in Aderibigbe et al. (2020) and axiomatically accelerated starch digestion rates.

5 Starch-protein digestive dynamics

The growth performance of broiler chickens may be sub-optimal because starch is digested too rapidly and protein too slowly; therefore, a closer alignment of their relative digestive kinetics should be advantageous. Certainly, starch is digested more extensively than protein based on apparent ileal digestibility coefficients, which is largely, but not entirely, a consequence of endogenous amino acid flows arising from cell desquamation, secretion of mucin and digestive enzymes into the gut lumen (Ravindran, 2021). The asynchronous provision of amino acids and glucose has been shown to increase amino acid oxidation and substantially reduce protein utilisation in pigs (Van den Borne et al., 2007). The negative impacts of asynchrony may be even more profound if a similar study were to be completed in broiler chickens, hence the importance of bilateral protein and starch digestive kinetics in poultry.

In a pivotal study, albeit in pigs, interactions between starch and protein kinetics were demonstrated by Van der Meulen et al. (1997) where the impact of the dietary transition from rapid (maize) to slow (peas) starch on the net portal flux of amino acids was determined. The net portal flux of lysine was significantly increased ($P < 0.05$) by 28.6% (31.0 versus 24.1 mmol) and tyrosine by 23.7% (16.7 versus 13.5 mmol) following the transition from rapidly to slowly digestible starch diets. Additionally, the net portal flux of eight amino acids tended to be ($P < 0.10$) collectively increased by 20.2% (17.43 versus 14.50 mmol) and the combined net portal flux of glutamine plus glutamate was negative in diets based on both maize (−73.6 mm) and peas (−71.7 mm). This outcome indicates that glutamine/glutamate were extensively catabolised in the gut mucosa for energy provision. However, the increases in net portal fluxes of ten amino acids may indicate that these amino acids were being spared from catabolism in the gut mucosa by the more sustained availability of glucose along the small intestine as an alternative energy substrate. There are suggestions in the broiler chicken studies by Weurding et al. (2003b) and Enting et al. (2005) that amino acids may have been similarly spared from catabolism in the gut mucosa by slowly digestible starch. In Enting et al. (2005), diets were formulated to contain either rapidly (tapioca–maize) or slowly (peas–maize) digestible starch with five tiers of digestible lysine (8.5–11.0 g/kg) with equalised amino acid levels. An interaction ($P < 0.05$) between starch source and digestible lysine for FCR was observed from 9 to 18 days post hatch, as illustrated in Fig. 4. At 11.0 g/kg lysine, slow starch diets were favoured by 2.91% (1.334 versus 1.374) but advantaged by 4.33% (1414 versus 1.478) at 8.5 g/kg lysine. This indicates that slowly digestible starch does have a positive impact on amino acid utilisation sparing and the genesis could be that amino acids are spared from catabolism in the gut mucosa. The determination of the net portal flux of amino acids in broiler chickens would be extremely difficult. In the Van der Meulen et al. (1997) study, pigs had catheters inserted in the portal vein, mesenteric artery and mesenteric vein and it may just not be possible to replicate this in broiler chickens.

The relevance of starch–protein digestive dynamics to broiler growth performance was demonstrated in Sydenham et al. (2017) where the objective was to formulate experimental diets with different digestive kinetics. To this end, fish meal partially replaced soybean meal in the basal sorghum–soybean meal diet, then maize starch partially replaced sorghum and finally both substitutions were made in the fourth diet. Instructively, starch–protein disappearance rate ratios in the proximal jejunum were quadratically related to weight gain (r = 0.849; $P < 0.001$) and FCR (r = 0.838; $P < 0.001$). It may be deduced from the relevant quadratic equations that a starch–protein disappearance rate ratio of 3.59:1 would support the maximum weight gain of 1265 g/bird and a ratio of 3.88:1 would generate the minimum FCR of 1.287 from 15 to 28 days post hatch.

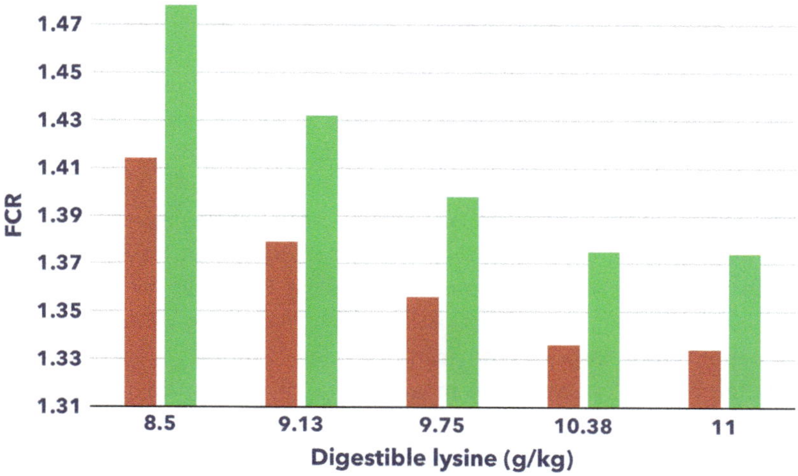

Figure 4 Interaction between diets with five tiers of digestible lysine (and amino acid concentrations) containing either *slowly* (red) or *rapidly* (green) digestible starch and FCR from 9 to 18 days post hatch. Source: Adapted from: Enting et al. (2005).

These predicted performance parameters exceeded the average parameters observed in the study by 6.12% (1265 versus 1192 g/bird) in weight gain and by 5.09% (1.287 versus 1.356) in FCR. Effectively, there was an ideal balance between intestinal uptakes of amino acids and glucose in the proximal jejunum to support optimal weight gain and FCR. Overall, 51.0% of distal ileal digestible starch was digested in the proximal jejunum and the corresponding figure for protein (N) was 51.9%. Interestingly, significant treatment interactions were observed for starch–protein disappearance rate ratios in the distal jejunum, proximal and distal ileum. The confounding effect in this study is the inherent differences between fish meal and soybean meal and with those between sorghum and maize starch; therefore, the findings cannot be solely attributed to the digestive kinetics of feedstuffs.

Jejunal and ileal protein and starch disappearance rates were thoroughly investigated by Pedersen et al. (2021), which involved the incorporation of ten standard feedstuffs into complete broiler diets. Unsurprisingly, starch digestion kinetics were more rapid than protein (N) in this study. Protein (N) disappearance rates were influenced by single feedstuffs and by interactions amongst feedstuffs, whereas starch disappearance rates were not influenced by feedstuff interactions to any extent. The outcomes indicated that one feedstuff can influence protein and starch disappearance rates of other feedstuffs in complete broiler diets. In this study, mean apparent jejunal digestibility coefficients of protein and starch in broilers offered eight complete diets were 0.527 and 0.845, respectively, and the corresponding ileal means were 0.751 and 0.925. However, starch and protein apparent digestibility coefficient ratios ranged from 1.41 to 1.91 in the jejunum and from 1.16 to 1.34 in the ileum in broilers offered eight complete diets. Pedersen et al. (2021) stressed the importance of understanding the protein and starch digestion kinetics of different feedstuffs and their additive and non-additive influences in complex broiler diets.

Many dietary factors may influence digestive dynamics including the variety of cereal grains, feed manufacturing process and particle size. These three factors were investigated in a series of broiler studies in sorghum-based diets and were extensively reviewed in Liu and Selle (2015). Inclusions of exogenous, feed enzymes in broiler diets, especially phytate- and NSP-degrading enzymes, are now routinely practiced in global chicken-meat production. However, these feed enzymes impact on starch–protein digestive dynamics by enhancing digestibilities of protein, amino acids and starch to fluctuating extents. Positive growth performance responses to feed enzymes frequently observed may be attributed in part to favourable modifications of starch–protein digestive kinetics.

The acceptance of phytase feed enzymes by the global chicken-meat industry is, at present, nearly unanimous and their contributions to sustainable production have been immense (Selle et al., 2023b). The potent anti-nutritive properties of the substrate, phytate, are multifaceted but are counteracted by phytate-degrading enzymes to varying extents. For example, the inclusion of 1000 FTU/kg phytase in maize-based diets increased weight gain by 12.3% (1387 versus 1235 g/bird) and improved FCR by 4.02% (1.457 versus 1.518) to 27 days post hatch in Liu et al. (2014c). However, in association with this enhanced growth performance, phytase significantly decreased retention times of digesta in four small intestinal segments by 17.5% (221 versus 268 min). Therefore, phytase was impacting on the digestive kinetics of feedstuffs in broiler chickens and the likelihood is that dietary phytate is a restraint on feed intake.

The inclusion of 500 FTU/kg in maize-based, broiler diets increased weight gain by 11.8% (2937 versus 2627 g/bird) from 1 to 40 days post hatch in Liu et al. (2015). Apparent amino acid and starch digestibility coefficients were subsequently determined in four small intestinal segments in Truong et al. (2015). Phytase increased average digestibility coefficients of 16 amino acids by 49.7% (0.720 versus 0.481) in proximal jejunum, 20.2% (0.802 versus 0.667) in distal jejunum, 9.07% (0.878 versus 0.805) in proximal ileum and 7.24% (0.904 versus 0.843) in distal ileum. Corresponding increases in starch digestibility were 26.6% (0.681 versus 0.538), 3.84% (0.865 versus 0.833), 7.24% (0.948 versus 0.884) and 2.46% (0.959 versus 0.936). Consequently, phytase significantly condensed starch–protein disappearance rate ratios in distal jejunum (2.78 versus 3.22), proximal ileum (2.79 versus 2.90) and distal ileum (2.73 versus 2.85). Indeed, a negative, linear relationship ($r = -0.766$; $P = 0.003$) may be detected between proximal ileal, starch–protein disappearance rate ratios and weight gains to 40 days post hatch. The multiple linear regression equation ($r = 0.936$; $P < 0.001$) between weight gain and individual protein and starch disappearance rates was as follows:

$$y_{\text{weight gain}} = 857.53 + 109.12 \times \text{protein}_{(\text{g/bird/day})} - 13.34 \times \text{starch}_{(\text{g/bird/day})}.$$

It is evident from the equation that increases in protein disappearance rates benefit weight gains, in contrast, starch disappearance rate increases would have a detrimental impact. In essence, exogenous phytase generated a 'proximal shift' in sites of intestinal uptakes of amino acids to a more pronounced extent than intestinal uptakes of glucose and this modification to digestive kinetics was to the advantage of weight gains.

An NSP-degrading enzyme with β-glucanase and xylanase activities was evaluated in wheat–rye broiler diets by Pettersson and Åman (1989). The feed enzyme improved weight gain by 10.1% (577 versus 524 g/bird) and FCR by 3.17% (1.82 versus 1.89) from 15 to 27 days post hatch at the highest inclusion rate. Interestingly, the feed enzyme increased starch digestibility by 3.39% (0.885 versus 0.856) and protein (N) digestibility by 11.4% (0.663 versus 0.595) in the mid-section of the small intestine. It may be deduced from the study that the starch–protein disappearance rate ratio in this segment was condensed from 2.35 to 1.63 in association with the improvements in growth performance.

More recently an exogenous xylanase was shown to increase starch and protein digestibility in the proximal small intestine of broilers offered wheat-based diets, thereby improving growth performance in Matthiesen et al. (2021). Collectively, xylanase increased jejunal starch digestibility by 24.0% (0.750 versus 0.605) and jejunal protein digestibility by 36.8% (0.650 versus 0.475). This feed enzyme-induced adjustment to starch–protein digestive dynamics was associated with a significant improvement of 2.14% (1.37 versus 1.40) in FCR to 35 days post hatch. The researchers suggested that improved growth performance was caused by multiple factors including viscosity reduction, cell wall degradation and changes in intestinal sites of digestion.

6 Conclusion

Interest in the digestive kinetics of the protein and starch components of feedstuffs used in chicken meat production has increased in recent years (Liu and Selle, 2015; Selle and Liu, 2019). To some extent, this has been made necessary by the trend towards increasing inclusions of non-bound amino acids in reduced-CP diets (Liu and Selle, 2017), in addition to non-bound methionine, lysine and threonine, which have been routine additions in poultry diets for decades (Pesti, 2009). However, as is plainly evident in this chapter, our comprehension of protein and starch digestive kinetics in broiler chickens is still in its infancy.

The potential for taking digestive kinetics into consideration in the least-cost formulation of broiler diets has been demonstrated by Liu et al. (2020). In this study, several feedstuffs (wheat, sorghum, soybean meal, canola meal, whole canola seed), with predetermined starch and protein digestion rates, were used to formulate six diets which were offered to broilers from 7 to 35 days post hatch. Starch to protein digestion rate ratios ranged from 1.567 to 2.453, but dietary amino acid and energy densities were similar. The average weight gain exceeded the 2022 Aviagen performance objective by 10.5% (2461 versus 2228 g/bird), without significant treatment differences between birds. There was a quadratic relationship ($r = 0.648$; $P < 0.001$) between starch–protein digestion rate ratios and FCR where:

$$y_{(FCR)} = 1.712 - 0.316 \times \text{ratio} + 0.095 \times \text{ratio}^2.$$

It may be deduced that the minimum FCR of 1.450 was generated by a 1.663 starch–protein digestion rate ratio. Thus, the present study confirmed the importance of protein and starch digestive kinetics in practical diets and showed the potential of incorporating digestive kinetics into least-cost feed formulations for broiler chickens. Certainly, further research in this exciting area is clearly justified to harness the potential of the digestive kinetics of feedstuffs in practical chicken meat production on a global scale.

7 References

Adeleye, O. O., Ologhobo, A. D. and Oje-Adetule, B. T. (2016). Influence of carbohydrate source on digesta kinetics and postprandial glucose responses of broiler chicks, *Livest. Sci.* 188, 37–42.

Aderibigbe, A., Cowieson, A., Sorbara, J. O. and Adeola, O. (2020). Intestinal starch and energy digestibility in broiler chickens fed diets supplemented with α-amylase, *Poult. Sci.* 99(11), 5907–5914.

Aoyagi, Y., Tasaki, I., Okumura, J.-i. and Muramatsu, T. (1988). Energy cost of whole-body protein synthesis measured in vivo in chicks, *Comp. Biochem. Physiol. A, Comp. Physiol.* 91(4), 765–768.

Beaufrere, B., Dangin, M. and Boirie, Y. The 'fast'and 'slow'protein concept, *Nestle Nutr. Workshop Ser. Clin. Perform. Programme* 2000(3), 121–31; discussion 131.

Bender, D. A. (2012). The metabolism of "surplus" amino acids, *Br. J. Nutr.* 108 (Suppl. 2), S113–S121.

Berrocoso, J. D., García-Ruiz, A., Page, G. and Jaworski, N. W. (2020). The effect of added oat hulls or sugar beet pulp to diets containing rapidly or slowly digestible protein sources on broiler growth performance from 0 to 36 days of age, *Poult. Sci.* 99(12), 6859–6866.

Boirie, Y., Dangin, M., Gachon, P., Vasson, M. P., Maubois, J. L. and Beaufrère, B. (1997). Slow and fast dietary proteins differently modulate postprandial protein accretion, *Proc. Natl. Acad. Sci. U. S. A.* 94(26), 14930–14935.

Brosnan, J. T. (2003). Interorgan amino acid transport and its regulation, *J. Nutr.* 133(6) (Suppl. 1), 2068S–2072S.

Bryan, D. D. S. L., Abbott, D. A., Van Kessel, A. G. and Classen, H. L. (2019a). In vivo digestion characteristics of protein sources fed to broilers, *Poult. Sci.* 98(8), 3313–3325.

Bryan, D. D. S. L., Abbott, D. A. and Classen, H. L. (2019b). Digestion kinetics of protein sources determined using an in vitro chicken model, *Anim. Feed Sci. Technol.* 248, 106–113.

Canolty, N. L. and Nasset, E. S. (1975). Intestinal absorption of free and protein-bound dietary methionine in the rat, *J. Nutr.* 105(7), 867–877.

Cant, J. P., Mcbride, B. W. and Croom Jr., W. J. (1996). The regulation of intestinal metabolism and its impact on whole animal energetics, *J. Anim. Sci.* 74(10), 2541–2553.

Chrystal, P. V., Greenhalgh, S., Mcinerney, B. V., Mcquade, L. R., Akter, Y., De Paula Dorigam, J. C., Selle, P. H. and Liu, S. Y. (2021). Maize-based diets are more conducive to crude protein reductions than wheat-based diets for broiler chickens, *Anim. Feed Sci. Technol.* 275, 114867.

Daenzer, M. and Petzke, K. J. Daenzer, M., Petzke, K. J., Bequette, B. J. and Metges, C. C. (2001). Whole-body nitrogen and splanchnic amino acid metabolism differ in rats fed mixed diets containing casein or its corresponding amino acid mixture, *J. Nutr.* 131(7), 1965–1972.

Dangin, M., Boirie, Y., Garcia-Rodenas, C., Gachon, P., Fauquant, J., Callier, P., Ballèvre, O. and Beaufrère, B. (2001). The digestion rate of protein is an independent regulating factor of postprandial protein retention, *Am. J. Physiol. Endocrinol. Metab.* 280(2), E340–E348.

Daniel, H. (2004). Molecular and integrative physiology of intestinal peptide transport, *Annu. Rev. Physiol.* 66, 361–384.

Dean, D. W., Bidner, T. D. and Southern, L. L. (2006). Glycine supplementation to low protein, amino acid-supplemented diets supports optimal performance of broiler chicks, *Poult. Sci.* 85(2), 288–296.

Del Alamo, A. G., Verstegen, M. W., Den Hartog, L. A., De Ayala, P. P. and Villamide, M. J. (2009). Wheat starch digestion rate affects broiler performance, *Poult. Sci.* 88(8), 1666–1675.

Enting, H., Pos, J., Weurding, R. and Veldman, A. Starch digestion rate affects broiler performance. In: Proceedings of the Australian Poultry Science Symposium, Sydney, New South Wales, Australia (7-9 February 2005), 17–20.

Eugenio, F., Van Milgen, J., Duperray, J., Sergheraert, R. and Le Floc'h, N. (2022), Feeding intact proteins, peptides, or free amino acids to monogastric farm animals, *Amino Acids* 54(2), 157–168.

Fleming, S. E., Zambell, K. L. and Fitch, M. D. (1997). Glucose and glutamine provide similar proportions of energy to mucosal cells of rat small intestine, *Am. J. Physiol.* 273(4), G968–G978.

Fouillet, H., Juillet, B., Gaudichon, C., Mariotti, F., Tomé, D. and Bos, C. (2009). Absorption kinetics are a key factor regulating postprandial protein metabolism in response to qualitative and quantitative variations in protein intake, *Am. J. Physiol. Regul. Integr. Comp. Physiol.* 297(6), R1691–R1705.

Geiger, E. (1950). The role of the time factor in protein synthesis, *Science* 111(2892), 594–599.

Giuberti, G., Gallo, A., Cerioli, C. and Masoero, F. (2012). In vitro starch digestion and predicted glycemic index of cereal grains commonly utilized in pig nutrition, *Anim. Feed Sci. Technol.* 174(3–4), 163–173.

Greenhalgh, S., Mcinerney, B. V., Mcquade, L. R., Chrystal, P. V., Khoddami, A., Zhuang, M. A. M., Liu, S. Y. and Selle, P. H. (2020). Capping dietary starch: protein ratios in moderately reduced crude protein, wheat-based diets showed promise but further reductions generated inferior growth performance in broiler chickens, *Anim. Nutr.* 6(2), 168–178.

Gropper, S. S. and Acosta, P. B. (1991). Effect of simultaneous ingestion of l-amino acids and whole protein on plasma amino acid and urea nitrogen concentrations in humans, *JPEN J. Parenter. Enteral Nutr.* 15(1), 48–53.

Gutierrez Del Alamo, A., Pérez De Ayala, P., Den Hartog, L. A., Verstegen, M. W. and Villamide, M. J. (2009). Wheat starch digestion rate in broiler chickens is affected by cultivar but not by wheat crop nitrogen fertilisation, *Br. Poult. Sci.* 50(3), 341–349.

Hamm, D. (1981). Amino acid composition of breast and thigh meat from broilers produced in four locations of the United States, *J. Food Sci.* 46(4), 1122–1124.

Harperand, A. E. and Rogers, Q. R. (1965). Amino acid imbalance, *Proc. Nut.r Soc.* 24(2), 173–190.

Herwig, E., Abbott, D., Schwean-Lardner, K. V. and Classen, H. L. (2019). Effect of rate and extent of starch digestion on broiler chicken performance, *Poult. Sci.* 98(9), 3676–3684.

Herwig, E., Schwean-Lardner, K., Van Kessel, A., Savary, R. K. and Classen, H. L. (2020). Assessing the effect of starch digestion characteristics on ileal brake activation in broiler chickens, *PLOS ONE* 15(2), e0228647.

Hyde, R., Taylor, P. M. and Hundal, H. S. (2003). Amino acid transporters: roles in amino acid sensing and signalling in animal cells, *Biochem. J.* 373(1), 1–18.

Jaworski, N. W., Boerboom, G., Jacobs, M., Alfonso, C., Geerse, K., Dijksalg, A., Ramaekers, P. and Smits, P. C. M. (2019). Effects of different levels of fast digestible starch, protein, and fibre on broiler growth performance. Proceedings, Australian Poultry Science *Symposium* 30, 11–14.

Kodera, T., Hara, H., Nishimori, Y. and Nio, N. (2006). Amino acid absorption in portal blood after duodenal infusions of a soy protein hydrolysate prepared by a novel soybean protease D3, *J. Food Sci.* 71(7), S517–S525.

Krehbiel, C. and Matthews, J. (2003). Absorption of amino acids and peptides, *Amino Acids in Animal Nutrition*. CABI Publishing, Wallingford UK.

Li, D., Zhao, X. H., Yang, T. B., Johnson, E. W. and Thacker, P. A. (1999). A comparison of the intestinal absorption of amino acids in piglets when provided in free form or as a dipeptide, *Asian-Australas. J. Anim. Sci.* 12(6), 939–943.

Liu, S. Y. and Selle, P. H. (2015). A consideration of starch and protein digestive dynamics in chicken-meat production, *Worlds Poult. Sci. J.* 71(2), 297–310.

Liu, S. Y. and Selle, P. H. (2017). Starch and protein digestive dynamics in low-protein diets supplemented with crystalline amino acids, *Anim. Prod. Sci.* 57(11), 2250–2256.

Liu, S. Y., Selle, P. H., Court, S. G. and Cowieson, A. J. (2013a). Protease supplementation of sorghum-based broiler diets enhances amino acid digestibility coefficients in four small intestinal sites and accelerates their rates of digestion, *Anim. Feed Sci. Technol.* 183(3–4), 175–183.

Liu, S. Y., Selle, P. H. and Cowieson, A. J. (2013b). The kinetics of starch and nitrogen digestion regulate growth performance and nutrient utilisation of broilers fed coarsely ground, sorghum-based diets, *Anim. Prod. Sci.* 53(10), 1033–1040.

Liu, S. Y., Khoddami, A., Roberts, T. H. and Selle, P. H. (2014a). Phenolic compounds and phytate influence starch and protein digestion rates in sorghum-based broiler diets. Proceedings, Australian Poultry Science Symposium 25, 45–48.

Liu, S. Y., Selle, P. H., Khoddami, A., Roberts, T. H. and Cowieson, A. J. (2014b). Graded inclusions of sodium metabisulphite in sorghum-based diets: II. Modification of starch pasting properties in vitro and beneficial impacts on starch digestion dynamics in broiler chickens, *Anim. Feed Sci. Technol.* 190, 68–78.

Liu, S. Y., Cadogan, D. J., Péron, A., Truong, H. H. and Selle, P. H. (2014c). Effects of phytase supplementation on growth performance, nutrient utilization and digestive dynamics of starch and protein in broiler chickens offered maize-, sorghum-and wheat-based diets, *Anim. Feed Sci. Technol.* 197, 164–175.

Liu, S. Y., Bold, R. M., Plumstead, P. W. and Selle, P. H. (2015). Effects of 500 and 1000 FTU/kg phytase supplementation of maize-based diets with two tiers of nutrient specifications on performance of broiler chickens, *Anim. Feed Sci. Technol.* 207, 159–167.

Liu, S. Y., Chrystal, P. and Selle, P. (2020). Pre-determined starch and protein digestion rates attain optimal feed conversion ratios in broiler chickens. In: Proceedings of the Australian Poultry Science Symposium 90.

Macelline, S. P., Mcquade, L. R., Mclnerney, B. V., Moss, A. F., Selle, P. H. and Liu, S. Y. (2020). Protein digestive dynamics of meat and bone meals in broiler chickens, *Anim. Nutr.* 6(4), 521–528.

Macelline, S. P., Chrystal, P. V., Selle, P. H. and Liu, S. Y. (2022). Protein sources and starch-protein digestive dynamics manipulate growth performance in broiler chickens defined by an equilateral-triangle response surface design, *Anim. Nutr.* 9, 204–213.

Macelline, S. P., Chrystal, P. V., Toghyani, M., Selle, P. H. and Liu, S. Y. (2023). Dietary crude protein reductions in wheat-based diets with two energy densities compromised performance of broiler chickens from 15 to 36 days post-hatch, *Poult. Sci.* 102(10), 102932.

Maharjan, P., Martinez, D. A., Weil, J., Suesuttajit, N., Umberson, C., Mullenix, G., Hilton, K. M., Beitia, A. and Coon, C. N. (2021). Physiological growth trend of current meat broilers and dietary protein and energy management approaches for sustainable broiler production, *Animal* 15(Suppl. 1), 100284.

Matthiesen, C. F., Pettersson, D., Smith, A., Pedersen, N. R. and Storm, A. C. (2021). Exogenous xylanase improves broiler production efficiency by increasing proximal small intestine digestion of crude protein and starch in wheat-based diets of various viscosities, *Anim. Prod. Sci.* 272, 114739.

Miska, K. B. and Fetterer, R. H. (2019). Expression of amino acid and sugar transporters, aminopeptidase, and the di-and tri-peptide transporter PepT1; differences between modern fast growing broilers and broilers not selected for rapid growth, *Poult. Sci.* 98(5), 2272–2280.

Moran, E. T. (1982). Starch digestion in fowl, *Poult. Sci.* 61(7), 1257–1267.

Moran, E. T. (2016). Gastric digestion of protein through pancreozyme action optimizes intestinal forms for absorption, mucin formation and villus integrity, *Anim. Feed Sci. Technol.* 221, 284–303.

Moss, A. F., Sydenham, C. J., Khoddami, A., Naranjo, V. D., Liu, S. Y. and Selle, P. H. (2018). Dietary starch influences growth performance, nutrient utilisation and digestive dynamics of protein and amino acids in broiler chickens offered low-protein diets, *Anim. Feed Sci. Technol.* 237, 55–67.

Namroud, N. F., Shivazad, M. and Zaghari, M. (2008). Effects of fortifying low crude protein diet with crystalline amino acids on performance, blood ammonia level, and excreta characteristics of broiler chicks, *Poult. Sci.* 87(11), 2250–2258.

Nieto, R., Palmer, R. M., Fernandez-Figares, I., Perez, L. and Prieto, C. (1994). Effect of dietary protein quality, feed restriction and short-term fasting on protein synthesis and turnover in tissues of the growing chicken, *Br. J. Nutr.* 72(4), 499–507.

Nolles, J. A., Verreijen, A. M., Koopmanschap, R. E., Verstegen, M. W. and Schreurs, V. V. (2009). Postprandial oxidative losses of free and protein-bound amino acids in the diet: interactions and adaptation, *J. Anim. Physiol. Anim. Nutr.* 93(4), 431–438.

Ospina-Rojas, I. C., Murakami, A. E., Moreira, I., Picoli, K. P., Rodrigueiro, R. J. and Furlan, A. C. (2013). Dietary glycine+ serine responses of male broilers given low-protein diets with different concentrations of threonine, *Br. Poult. Sci.* 54(4), 486–493.

Ospina-Rojas, I. C., Murakami, A. E., Duarte, C. R., Eyng, C., Oliveira, C. A. and Janeiro, V. (2014). Valine, isoleucine, arginine and glycine supplementation of low-protein diets for broiler chickens during the starter and grower phases, *Br. Poult. Sci.* 55(6), 766–773.

Paterson, L., Mitchell, J. R., Hill, S. E. and Blanshard, J. M. (1996). Evidence for sulfite induced oxidative reductive depolymerisation of starch polysaccharides, *Carbohydr. Res.* 292, 143–151.

Pedersen, N. B., Hanigan, M., Zaefarian, F., Cowieson, A. J., Nielsen, M. O. and Storm, A. C. (2021). The influence of feed ingredients on CP and starch disappearance rate in complex diets for broiler chickens, *Poult. Sci.* 100(5), 101068.

Pesti, G. M. (2009). Impact of dietary amino acid and crude protein levels in broiler feeds on biological performance, *J. Appl. Poult. Res.* 18(3), 477–486.

Pettersson, D. and Åman, P. (1989). Enzyme supplementation of a poultry diet containing rye and wheat, *Br. J. Nutr.* 62(1), 139–149.

Pineda-Quiroga, C., Camarinha-Silva, A., Borda-Molina, D., Atxaerandio, R., Ruiz, R. and García-Rodríguez, A. (2018). Feeding broilers with dry whey powder and whey protein concentrate affected productive performance, ileal digestibility of nutrients and cecal microbiota community, *Animal* 12(4), 692–700.

Porteous, J. W. (1980). Glutamate, glutamine, aspartate, asparagine, glucose and ketone-body metabolism in chick intestinal brush-border cells, *Biochem. J.* 188(3), 619–632.

Ravindran, V. (2021). Progress in ileal endogenous amino acid flow research in poultry, *J. Anim. Sci. Biotechnol.* 12(1), 5.

Reeds, P. J., Burrin, D. G., Jahoor, F., Wykes, L., Henry, J. and Frazer, E. M. (1996). Enteral glutamate is almost completely metabolized in first pass by the gastrointestinal tract of infant pigs, *Am. J. Physiol.* 270(3 Pt 1), E413–E418.

Reeds, P. J., Burrin, D. G., Stoll, B. and Van Goudoever, J. B. (2000). Role of the gut in the amino acid economy of the host, In: Nestle Nutr. Workshop Ser. Clin. Perform. Programme, Karger, Basel, Switzerland, 25–40; discussion 40.

Regmi, P. R., Matte, J. J., Van Kempen, T. A. T. G. and Zijlstra, R. T. (2010). Starch chemistry affects kinetics of glucose absorption and insulin response in swine, *Livest. Sci.* 134(1–3), 44–46.

Rerat, A. A. (1985). Intestinal absorption of end products from digestion of carbohydrates and proteins in the pig, *Arch. Tierernähr.* 35(7), 461–480.

Rérat, A. A. (1993). Nutritional supply of proteins and absorption of their hydrolysis products: consequences on metabolism, *Proc. Nutr. Soc.* 52(2), 335–344.

Röder, P. V., Geillinger, K. E., Zietek, T. S., Thorens, B., Koepsell, H. and Daniel, H. (2014). The role of SGLT1 and GLUT2 in intestinal glucose transport and sensing, *PLOS ONE* 9(2), e89977.

Salway, J. G. (2018). The Krebs uric acid cycle: a forgotten Krebs cycle, *Trends Biochem. Sci.* 43(11), 847–849.

Schreurs, V. V., Koopmanschap, R. E. and Boekholt, H. A. (1997). Short-term dynamics in protein and amino acid metabolism, *Z. Ernahrungswiss.* 36(4), 336–339.

Selle, P. H. and Liu, S. Y. (2019). The relevance of starch and protein digestive dynamics in poultry, *J. Appl. Poult. Res.* 28(3), 531–545.

Selle, P. H., Liu, S. Y., Cai, J. and Cowieson, A. J. (2012). Steam-pelleting and feed form of broiler diets based on three coarsely ground sorghums influences growth performance, nutrient utilisation, starch and nitrogen digestibility, *Anim. Prod. Sci.* 52(9), 842–852.

Selle, P. H., Liu, S. Y., Cai, J., Caldwell, R. A. and Cowieson, A. J. (2014). Graded inclusions of sodium metabisulphite in sorghum-based diets: I. Reduction of disulphide cross-linkages in vitro and enhancement of energy utilisation and feed conversion efficiency in broiler chickens. *A Anim. Feed, Sci. Technol.* 190, 59–67.

Selle, P. H., Cantor, D. I., Mcquade, L. R., Mcinerney, B. V., De Paula Dorigam, J. C., Macelline, S. P., Chrystal, P. V. and Liu, S. Y. (2021a). Implications of excreta uric acid concentrations in broilers offered reduced-crude protein diets and dietary glycine requirements for uric acid synthesis, *Anim. Nutr.* 7(4), 939–946.

Selle, P. H., Moss, A. F., Khoddami, A., Chrystal, P. V. and Liu, S. Y. (2021b). Starch digestion rates in multiple samples of commonly used feed grains in diets for broiler chickens, *Anim. Nutr.* 7(2), 450–459.

Selle, P. H., Macelline, S. P., Chrystal, P. V. and Liu, S. Y. (2022a). Amino acids were not all created equal. In: Proceedings of the Arkansas Nutrition Conference 9.
Selle, P. H., Macelline, S. P., Chrystal, P. V. and Liu, S. Y. (2022b). The impact of digestive dynamics on the bioequivalence of amino acids in broiler chickens, *Front. Biosci. (Landmark Ed)* 27(4), 126.
Selle, P. H., Macelline, S. P., Chrystal, P. V. and Liu, S. Y. (2022c). Fat deposition in broiler chickens offered reduced-crude-protein diets. In: Proceedings of the Australian Poultry Science Symposium 33, 22–25.
Selle, P. H., Macelline, S. P., Chrystal, P. V. and Liu, S. Y. (2023a). A reappraisal of amino acids in broiler chicken nutrition, *World Poult. Sci. J.* 79, 429-447
Selle, P. H., Macelline, S. P., Chrystal, P. V. and Liu, S. Y. (2023b). The contribution of phytate-degrading enzymes to chicken-meat production, *Animals* 13(4), 603.
Shibata, M., Takahashi, T., Kozakai, T., Kakudo, M., Kasuga, S., Azuma, Y. and Kurose, Y. (2019). Active transport of glucose across the jejunal epithelium decreases with age in broiler chickens, *Poult. Sci.* 98(6), 2570–2576.
Siegert, W. and Rodehutscord, M. (2019). The relevance of glycine and serine in poultry nutrition: a review, *Br. Poult. Sci.* 60(5), 579–588.
Spanier, B. (2014). Transcriptional and functional regulation of the intestinal peptide transporter PepT1, *J. Physiol.* 592(5), 871–879.
Stern, R. A. and Mozdziak, P. E. (2019). Differential ammonia metabolism and toxicity between avian and mammalian species, and effect of ammonia on skeletal muscle: A comparative review, *J. Anim. Physiol. Anim. Nutr.* 103(3), 774–785.
Stoll, B., Henry, J., Reeds, P. J., Yu, H., Jahoor, F. and Burrin, D. G. (1998). Catabolism dominates the first-pass intestinal metabolism of dietary essential amino acids in milk protein-fed piglets, *J. Nutr.* 128(3), 606–614.
Stoll, B., Burrin, D. G., Henry, J., Yu, H., Jahoor, F. and Reeds, P. J. (1999). Substrate oxidation by the portal drained viscera of fed piglets, *Am. J. Physiol.* 277(1), E168–E175.
Svihus, B. and Itani, K. (2019). Intestinal passage and its relation to digestive processes, *J. Appl. Poult. Res.* 28(3), 546–555.
Swick, R. W. (1982). Growth and protein turnover in animals, *Crit. Rev. Food Sci. Nutr.* 16(2), 117–126.
Sydenham, C. J., Truong, H. H., Moss, A. F., Selle, P. H. and Liu, S. Y. (2017). Fishmeal and maize starch inclusions in sorghum-soybean meal diets generate different responses in growth performance, nutrient utilisation, starch and protein digestive dynamics of broiler chickens, *Anim. Feed Sci. Technol.* 227, 32–41.
Tesseraud, S., Métayer, S., Duchêne, S., Bigot, K., Grizard, J. and Dupont, J. (2007). Regulation of protein metabolism by insulin: value of different approaches and animal models, *Domest. Anim. Endocrinol.* 33(2), 123–142.
Toghyani, M., Mcquade, L. R., McInerney, B. V., Moss, A. F., Selle, P. H. and Liu, S. Y. (2020). Initial assessment of protein and amino acid digestive dynamics in protein-rich feedstuffs for broiler chickens, *PLOS ONE* 15(9), e0239156.
Tokutake, Y., Taciak, M., Sato, K., Toyomizu, M. and Kikusato, M. (2021). Effect of dipeptide on intestinal peptide transporter 1 gene expression: an evaluation using primary cultured chicken intestinal epithelial cells, *Anim. Sci. J.* 92(1), e13604.
Truong, H. H., Bold, R. M., Liu, S. Y. and Selle, P. H. (2015). Standard phytase inclusion in maize-based broiler diets enhances digestibility coefficients of starch, amino acids and sodium in four small intestinal segments and digestive dynamics of starch and protein, *Anim. Feed Sci. Technol.* 209, 240–248.

Truong, H. H., Chrystal, P. V., Moss, A. F., Selle, P. H. and Liu, S. Y. (2017). Rapid protein disappearance rates along the small intestine advantage poultry performance and influence the post-enteral availability of amino acids, *Br. J. Nutr.* 118(12), 1031–1042.

Urdaneta-Rincon, M. and Leeson, S. (2004). Muscle (pectoralis major) protein turnover in young broiler chickens fed graded levels of lysine and crude protein, *Poult. Sci.* 83(11), 1897–1903.

Van Den Borne, J. J., Schrama, J. W., Heetkamp, M. J., Verstegen, M. W. and Gerrits, W. J. (2007). Synchronising the availability of amino acids and glucose increases protein retention in pigs, *Animal* 1(5), 666–674.

Van Der Meulen, J., Bakker, G. C., Bakker, J. G., De Visser, H., Jongbloed, A. W. and Everts, H. (1997). Effect of resistant starch on net portal-drained viscera flux of glucose, volatile fatty acids, urea, and ammonia in growing pigs, *J. Anim. Sci.* 75(10), 2697–2704.

Van Der Schoor, S. R., Van Goudoever, J. B., Stoll, B., Henry, J. F., Rosenberger, J. R., Burrin, D. G. and Reeds, P. J. (2001). The pattern of intestinal substrate oxidation is altered by protein restriction in pigs, *Gastroenterology* 121(5), 1167–1175.

Van Der Schoor, S. R., Reeds, P. J., Stoll, B., Henry, J. F., Rosenberger, J. R., Burrin, D. G. and Van Goudoever, J. B. (2002). The high metabolic cost of a functional gut, *Gastroenterology* 123(6), 1931–1940.

Van Milgen, J. (2021). The role of energy, serine, glycine, and 1-carbon units in the cost of nitrogen excretion in mammals and birds, *Animal* 15(5), 100213.

Vinardell, M. P. (1990). Mutual inhibition of sugars and amino acid intestinal absorption, *Comp. Biochem. Physiol. A* 95(1), 17–21.

Wang, H., Jia, G., Huang, L., Wu, C. and Wang, K. (2010). Study on the absorption and transport of different glutamine dipeptides in small intestine of weaned piglets, *J. Anim. Plant Sci.* 7(1), 751–759.

Wang, X., Du, B., Nian, F., Ru, Y., Sun, L., Qin, S. and Tang, D. (2023). Effects of processing methods and conditioning temperatures on the cassava starch digestibility and growth performance of broilers, *Animals* 13(8), 1373.

Watford, M., Lund, P. and Krebs, H. A. (1979). Isolation and metabolic characteristics of rat and chicken enterocytes, *Biochem. J.* 178(3), 589–596.

Weurding, R. E., Veldman, A., Veen, W. A., Van Der Aar, P. J. and Verstegen, M. W. (2001a). In vitro starch digestion correlates well with rate and extent of starch digestion in broiler chickens, *J. Nutr.* 131(9), 2336–2342.

Weurding, R. E., Veldman, A., Veen, W. A., Van Der Aar, P. J. and Verstegen, M. W. (2001b). Starch digestion rate in the small intestine of broiler chickens differs among feedstuffs, *J. Nutr.* 131(9), 2329–2335.

Weurding, R. E., Enting, H. and Verstegen, M. W. A. (2003a). The effect of site of starch digestion on performance of broiler chickens, *Anim. Feed Sci. Technol.* 110(1–4), 175–184.

Weurding, R. E., Enting, H. and Verstegen, M. W. (2003b). The relation between starch digestion rate and amino acid level for broiler chickens, *Poult. Sci.* 82(2), 279–284.

Willke, T. (2014). Methionine production: a critical review, *Appl. Microbiol. Biotechnol.* 98(24), 9893–9914.

Wilson, R. P., Muhrer, M. E. and Bloomfield, R. A. (1968). Comparative ammonia toxicity, *Comp. Biochem. Physiol.* 25(1), 295–301.

Wu, G. (2009). Amino acids: metabolism, functions, and nutrition, *Amino Acids* 37(1), 1–17.

Wu, G. (2014). Dietary requirements of synthesizable amino acids by animals: a paradigm shift in protein nutrition, *J. Anim. Sci. Biotechnol.* 5(1), 34.

Wu, G., Wu, Z., Dai, Z., Yang, Y., Wang, W., Liu, C., Wang, B., Wang, J. and Yin, Y. (2013). Dietary requirements of "nutritionally non-essential amino acids" by animals and humans, *Amino Acids* 44(4), 1107–1113.

Yaman, M. A., Kita, K. and Okumura, J. (2000). Different responses of protein synthesis to refeeding in various muscles of fasted chicks, *Br. Poult. Sci.* 41(2), 224–228.

Yamashita, K. and Ashida, K. (1969). Lysine metabolism in rats fed lysine-free diet, *J. Nutr.* 99(3), 267–273.

Yutste, P., Longstaff, M. A., Mcnab, J. M. and Mccorquodale, C. (1991). The digestibility of semipurified starches from wheat, cassava, pea, bean and potato by adult cockerels and young chicks, *Anim. Feed Sci. Technol.* 35(3–4), 289–300.

Chapter 4

Advances in understanding poultry feed intake regulation

Jalila S. Dridi, Université d'Orléans, France; and Sami Dridi, University of Arkansas, USA

1 Introduction

Driven by innovative genetic selection programs,[1–3] contributing to global production in 2022 of 135.93 million metric tons of meat and over 87 million metric tons of eggs,[4] the poultry sector has succeeded in providing a highly efficient, affordable and nutritious source of animal protein. It also supports the livelihoods and food security of billions of people worldwide.[5] However, based on the prediction of Food and Agriculture Organization (FAO) of the United Nations, this production needs to increase annually by over 200 million tons of meat and at least 50 million hens to feed the future[3] under the most challenging environmental conditions[6,7] and already scarce agricultural and natural resources (land, energy, water).[8–11]

Rapid climate change and global warming are some of the most challenging conditions for livestock, in general, and for poultry production sustainability, in particular.[12] In fact, global warming and drought are affecting

http://dx.doi.org/10.19103/AS.2024.0143.04

every aspect of animal lives and their very existence.[13] Besides its direct detrimental physiological, neuroendocrinological, behavioral, health, and well-being effects on the animal,[14–17] drought affects crops (quantity and quality) that are essential for animal nutrition.[12,18,19] Furthermore, increased use of food crops for biofuel production, which currently represents about 10% of global production, could add more complexity and have serious implications for future food security. Approximately, 800 million tons or one-third of total grains produced are used for animal feed; however, with a projected growing global demand for high-quality animal proteins due to human population and economic growth, this quantity will need to increase by 38% by 2050,[20,21] which will be very challenging within the abovementioned planetary boundaries. Siegel[22] reported that raising a 2.44 kg broiler requires 3.66 kg of feed, and Chan and coworkers outlined that producing 9.25 billion broilers annually in the USA necessitates 75 577 km^2 of land.[23] Feed accounts for up to 70% of poultry production cost,[20] hence improving feed efficiency (FE = gain : feed) is critical. Consequently, it is primordial to understand the mechanisms involved in the regulation of feed intake and growth: the two key components of the FE trait.

Feed intake is a complex interplay of physiologic, social, psychological, behavioral, and genetic factors that influence meal frequency, meal timing, food preference, food selection, and quantity of food ingested.[13,24–26] Although commercial broilers do not have the luxury to select or choose food, the questions that occur often and remain to be answered are whether chickens have a body weight (BW) set point and do they regulate their feed intake?

As we described earlier,[27,28] several studies from the 1990s showed that chickens were able to return to their initial BW after feed restriction or force-feeding episodes. This suggests that chickens probably have set BW points which have likely been lost with intensive genetic selection for high growth rate (associated with hyperphagia). However, despite the hyperphagic genotype, modern broilers are still able to control their feed intake via a series of highly integrated peripheral and central regulatory mechanisms.[27–31]

In a continuum to our previous chapters, where we described classical and new peripheral and central pathways involved in the regulation of feed intake in chickens,[27–29] we will focus in this chapter on some new advances mainly related to the factors influencing feeding behavior, including managerial and environmental factors. Understanding the motivational factors driving feed intake, food choice, and their underlying molecular mechanisms at both peripheral and central levels is important to addressing the current challenges to the poultry production sustainability. Collectively, by offering a review of the current understanding of feed intake regulation, the present chapter provides a new fundamental framework for future investigations into the control of energy homeostasis in commercially important poultry as well as other avian species.

2 Factors affecting feed intake: overview

Feed intake is affected (stimulated or inhibited) by various internal and external factors including the properties of the food itself, the animal conditions, and environmental and managerial factors. Before describing these factors, we need to clarify the meaning of feed intake, appetite, and hunger which are often interchangeably and confusingly used. Hunger is defined as the internal physiological urge or drive to eat;[32] however, appetite is the desire to eat and comes to awareness much quicker than hunger.[33] Although hunger often stimulates appetite, someone may be hungry without feeling a desire to eat, or someone may desire to eat something even in the absence of a need for it. Seeing, smelling, and/or thinking about food could be external cues for appetite. Feed intake, however, is the ingestion of food. In contrast to humans, it is difficult to distinguish between appetite and hunger in animals and we will use mainly feed intake terms in this chapter.

Although the spectrum of the factors affecting feed intake is very broad, we will focus here briefly on management, environmental, and physical factors that are mostly associated with stress and then describe in more detail the heat stress-, hypoxia-, and nutrient-sensing factors.

3 Management factors affecting feed intake: effects of flock size, density and lighting

3.1 Flock size and stocking density

In studies on different strains of both layers and broilers, overall and daily feed intake were less in large than in small flocks.[34,35] Furthermore, feed intake and efficiency in broilers were adversely affected by high- compared to moderate-stocking density even under conditions of sufficient feeder spaces.[36,37] Although it is well accepted that social facilitation is a feeding motivation enhancer, in which an animal's motivation to eat is stimulated by the sight and sound of other animals eating,[38] Buijs and colleagues reported that elevated stocking density and group size increase competition among animals and reduce feed intake.[39] This can be caused by psychological and physiological stress that negatively affects the welfare and consequently the feed intake of chickens.[39]

3.2 Lighting

In humans, the evident effects of light on appetite and feed intake are supported by the seasonal affective disorder.[40] In chickens, recent studies have examined an alternate lighting system with intermittent lighting (IL) in comparison with continuous lighting (CL)[41]. Although feed intake was reduced under the alternate lighting system, authors showed that birds had better FE and greater BW than the control group with CL[41]. They attributed this improvement in performance to the birds resting during the dark period and hence to lower maintenance energy requirement, which allowed more energy intake to be used for growth.[41] It is also possible that when the digestive tract is emptying during the dark period, feed is better digested, absorbed, and partitioned.

Downs and colleagues have shown that reducing light intensity stimulated feed intake and BW as compared to high intensity.[42] However, other studies reported that the reduction in light intensity did not influence BW, feed conversion, or mortality rate.[43,44] This disparity was probably associated with experimental designs and/or bird genetics, and further studies are needed.

Light color and wavelength have different penetration power and are important components of the physical light environment that affect broiler feed intake, growth, performance, and welfare.[45] It has been suggested that blue, green, and white light enhance broiler feed intake and growth.[46–48] The blue and green lights were supposed to keep birds calm by reducing hypothalamic gonadotropin production and stress and hence decreasing energy losses via hyperactivity and expenditure. They have been also suggested to stimulate myofiber growth through testosterone production.[46] The white light, however, was supposed to help birds identify feed texture differences.

At the molecular level, the light(active)/dark(rest) cycle parallels endogenous hormone rhythms. For instance, basal corticosterone concentrations in Gambel's white-crowned sparrow have been shown to reach peak levels at the end of the inactive period and return to trough levels just after the active period has begun.[49] In quail, the peak in plasma and adrenal corticosterone occurs during the second half of the dark span of the photoperiod, which is also the resting time of the animals.[50] In humans, cortisol secretion spikes in the early morning directly before awakening and then drops throughout the day, reaching its nadir around the time of sleep onset.[51] These circadian rhythms or clocks, which are driven by an autoregulatory feedback loop of transcriptional activators and repressors,[52] influence many physiological functions, including feed intake. In mammals, CLOCK (circadian locomotor output cycles kaput) and BMAL1 (brain and muscle arnt-like protein 1) form heterodimers that induce expression of the period (PER1-3) and cryptochrome (CRY1 and 2) through E-box enhancers[53] (Fig. 1). PER and CRY proteins accumulate in the cytoplasm throughout the circadian day. Upon reaching critical levels, they form a complex that translocates back to the nucleus to associate with CLOCK and BMAL1 and repress their own transcription.[54] This process takes approximately 24 h to complete a full cycle. Although the brain (the suprachiasmatic nucleus, SCN) is considered as the master, circadian oscillators are also present in most, if not all, tissues of multicellular organisms.[55] The SCN entrains the peripheral clocks via neural and hormonal signals.[56]

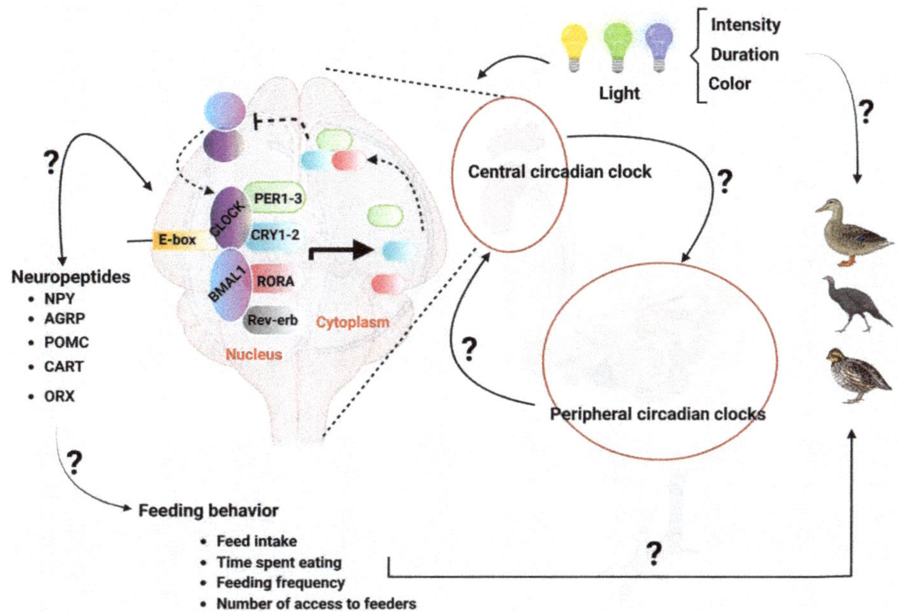

Figure 1 Simplified illustration of the molecular mechanisms of the central and peripheral circadian clocks and their potential role in regulating feeding behavior in avian species**.** The circadian clock is driven by an autoregulatory feedback loop. CLOCK and BMAL1 form a nuclear heterodimer that induces expression of PER and CRY through E-box enhancers. PER and CRY proteins accumulate in the cytoplasm throughout the circadian day. Upon reaching critical levels, PER and CRY form a complex that translocates back to the nucleus to associate with CLOCK and BMAL1 and repress their own transcription. This process takes approximately 24 h to complete a full cycle. Additionally, the CLOCK-BMAL1 heterodimer regulates transcription of Rev-erb and Rora, which have feedback effects on Bmal1. The effects of light (duration, color, and intensity) on the central and peripheral circadian clock, hypothalamic feeding-related neuropeptides, and feeding behavior need further in-depth investigation in chickens and other commercial avian species. AGRP, agouti-related peptide; BMAL1, brain and muscle arnt-like protein 1; CART, cocaine- and amphetamine-regulated transcript; CLOCK, circadian locomotor output cycles kaput; CRY, cryptochrome; E-box, enhancer box; NPY, neuropeptide Y; ORX, orexin; PER, period; POMC, proopiomelanocortin; Rev-erb, nuclear receptor subfamily 1 group D member 1; RORA, RAR-related orphan receptor A. The figure was created using BioRender.com and was not drawn to scale.

Light, the most potent synchronizing factor for the SCN, travels directly from intrinsically photosensitive melanopsin-containing retinal ganglion cells through the retinohypothalamic tract to the SCN. Light exposure (timing, duration, intensities, etc.) has been shown to affect PER expression and advance or delay the circadian clock.[57–59] It has been shown that the internal circadian clock modulates hunger and appetite independently of feed intake and other behaviors.[60] Several feeding-related hormones such as glucagon, insulin, ghrelin, and leptin oscillate in a circadian manner.[61,62] Moreover, there is a rhythmic expression of feeding-related hypothalamic (an)orexigenic neuropeptides such as neuropeptide Y (NPY), galanin, and proopiomelanocortin (POMC).[63,64] Rodents harboring mutations in various circadian clock genes, such as CLOCK, displayed profound changes in circadian rhythmicity and circulating leptin as well as disruption in feed intake and body mass.[65] Similarly, disruption of the PER gene has been reported to alter feeding behavior and glucose homeostasis, resulting in significant BW and fat deposition.[66,67] Disruption of other mammalian core circadian clock genes also affect energy homeostasis and metabolism, and the readers can refer to several elegant research papers.[68–71]

In avian species, although the clock genes have been characterized, their interactions with feeding behavior are not well understood. Recently in layers, Wang and colleagues showed that dusk feeding was synchronously shifted with an altered lighting program, along with a downregulation of the hypothalamic expression of PER2, CRY1, POMC, and cocaine- and amphetamine-regulated transcript (CART) genes.[72] Honda and coworkers reported that continuous white light alters the expression of clock genes in both the brain and peripheral tissues in broilers.[73] Yi-Lin et al.[74] have shown that dynamic feeding a different methionine-level diets modulated the expression of clock genes in the jejunum of laying hens. Further investigations are needed to determine how central and peripheral clock genes are regulated by different light color and intensities and how they regulate feeding behaviors and energy homeostasis in chickens and in other avian species (Fig. 1).

4 Management factors affecting feed intake: noise-related stress

Although it remains an ambiguous concept,[75] stress is generally defined as the magnitude of external forces to displace or disturb the body system from its ground state,[76] which has been called by Claude Bernard (1865–1961) as 'internal milieu'[77] and by Walter Cannon (1925)[78] as 'homeostasis'. Stress factors in humans, for instance, can broadly fall into four categories: physical, psychological, psychosocial, and psychospiritual. In the case of animals, stress can fall into three types: physical, physiological, and behavioral. Sources of stress can be social (isolation, stocking density, etc.), environment (heat stress, humidity, etc.), management (fear, noise, etc.), feeding (hunger, thirst, diet composition, etc.), or pathology (disease, pain, etc.). However, whatever the source, there are two basic steps in the stress response system. The first is a fast response, in which the sympathetic nervous system (SNS) is activated to trigger a myriad of peripheral actions, both through direct action of sympathetic nerves and via the adrenaline release from the adrenal gland. Second, a slower response via the hypothalamus–pituitary–adrenal (HPA) axis, in which the release of corticotropin-releasing hormone (CRH) from the hypothalamic paraventricular nucleus (PVN) induces the release of adrenocorticotropic hormone (ACTH) from the anterior pituitary gland, which in turn triggers the release of cortisol or corticosterone from the adrenal cortex.[79]

4.1 Noise and sounds

It is well accepted that noise is a negative stressor for farm animals.[80,81] Although it has been little investigated in poultry, noise has been reported to adversely affect behavior and reproductive performance in chickens.[82] The critical levels at which chickens are considered stressed are thought to be 85 decibels (dB) or above, and the depressive effects on feed intake and growth would be more drastic with prolonged exposure to noise. Chickens can be exposed to noise stress generated either from external sources such as aircraft or from internal source, including ventilation equipment and power generators. Although the maximum noise level is not strictly established in poultry, the EU Council Directive 2007/43/EC required that noise and sound levels should be minimized, and noise sources (fans, machineries, etc.) should be constructed, placed, operated, and maintained in a such a way that causes the least possible amount of noise.[83]

The positive physiological and psychological effects of music on humans are well known;[84] however, such studies are scarce in animals. Christensen and Knight reported that music did not have any significant effects on feed intake in broilers.[85] Gvaryahu et al.[86] showed, however, that intermittent use of classical music at low levels reduced fear and increased feed intake and BW in broilers. Ladd et al.[87] reported that regular exposure to country, classical, or jazz music reduced heterophil-to-lymphocyte ratio and grooming and stimulated feed intake and head shaking in layers. Using a tonic immobility test, Campo and colleagues have shown that specific classical music reduced stress in four Spanish breeds of chickens.[88] Together, these studies suggested that the effect of music on chicken feed intake could be genre-, volume-, and/or speed-dependent, which merit further investigations.

Although the exact molecular mechanisms by which noise/music affects feed intake in chickens is not known yet, it has been reported that distraction and modulation of mood states are some of the potential mechanisms in humans. It has been also suggested that music can affect taste and flavor perception (see Section 7.1) and thereby affects feeding and drinking behavior.[89,90]

5 Management factors affecting feed intake: heat and cold stress

One of the main feed intake regulation's postulate is the thermostatic theory, originally postulated by Brobeck in 1948,[91] which indicated that the body experiences a temperature-dependent variation in energy needs that should be reflected in feed intake. In other words, endothermic animals eat less when the environmental temperature is high than when it is low.

Chickens are endothermic homeotherms that regulate their body core temperature by reducing feed intake during heat load, which has been extensively reported by many reports, including our own.[92–94] The reduced energy intake is an adaptive mechanism to cope with a hot environment by limiting the metabolic heat production and reducing the thermic effect of food, which is also known as diet-induced thermogenesis or specific dynamic action.

Physiologically, the avian body temperature is higher during the active phase compared to the rest phase, which indicates that it is marked by a circadian rhythm[95] and suggests that it may interact with body clocks. In fact, Takahashi's group has shown that the mammalian body's thermostat also controls body clocks.[96] However, this association between body thermostat and the circadian clock has to be assessed and studied in avian species.

Neurons relating to temperature detection (cold/warm) and regulation have been shown to be developed during the perinatal period.[97,98] Thermal (cold/warm) sensations are detected by thermoreceptors that are widely distributed in the brain, deep tissues, and peripheral regions such as the skin, beak, and tongue. Changes in body temperatures can also be detected by osmo- and baro-receptors.[99] These thermo-sensory signals are then transmitted in a feedforward manner to a second-order thermal sensory neurons in the spinal and trigeminal dorsal horn (Fig. 2), which project to third-order sensory neurons in the dorsal subnucleus of the lateral parabrachial nucleus (LPB) of the brainstem. The thermo-sensory information is, subsequently, dispatched from the LPB to the preoptic anterior area (POA) of the hypothalamus[100] (Fig. 2). Once activated, these neurons are depolarized through permeability to calcium and sodium ions, and ion influx, and thereby provide command signals descending to central and peripheral effectors and evoke behavioral, autonomic, somatic, and hormonal responses[101] (Fig. 2). It is worth mentioning that cold and warm sensory pathways are mediated by separate populations of neurons in the LPB.[102] Furthermore, there is another subset of neurons, termed temperature guardian neurons, that are active when temperature approaches near-extreme

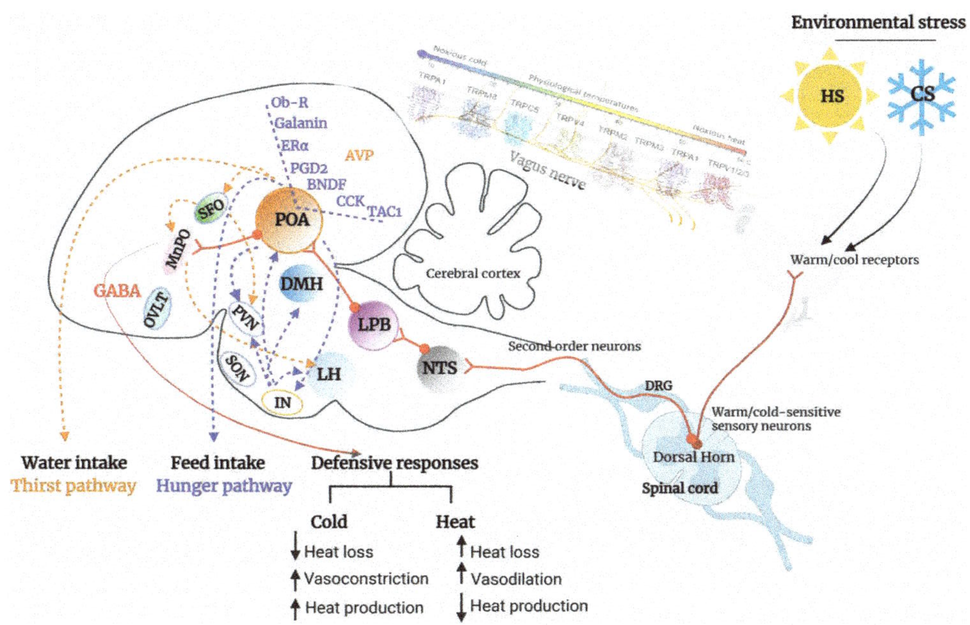

Figure 2 Schematic drawing to depict potential neural interactions between sensory (thermoregulation), hunger, and thirst signals. Based on mammalian studies, the thermal (cold or heat) sensations are detected by thermoreceptors that are widely distributed (brain, skin, etc.). These thermo-sensory signals are transmitted in a feedforward manner to a 2nd-order thermal sensory neurons in the spinal and trigeminal dorsal horn, which project to third-order sensory neurons in the brainstem (NTS-LPB-POA). The activation of these neurons provides command signals and evoke various responses, such as defensive responses and feeding- and drinking-behaviors. AVP, arginine vasopressin; BNDF, brain-derived neurotrophic factor; CCK, cholecystokinin; CS, cold stress; DMH, dorsomedial hypothalamus; ERα, estrogen receptor alpha; GABA, gamma-aminobutyric acid; HS, heat stress; IN, infundibular nucleus (ARC in mammals); LH, lateral hypothalamus; LPB, lateral parabrachial nucleus; MnPO, median preoptic nucleus; NTS, nucleus of the stria terminalis; OVLT, vascular organ of lamina terminalis; PGD2, prostaglandin; POA, preoptic area; PVN, paraventricular nucleus; SFO, subfornical organ; SON, supraoptic nucleus; TAC1, tachykinin 1; TRPA, transient receptor potential cation channel subfamily A member; TRPC, transient receptor potential cation channel subfamily C member; TRPV, transient receptor potential cation channel subfamily V member. The figure was created using BioRender.com and was not drawn to scale.

ends of the spectrum and are thought to be responsible for initiating more robust thermoregulatory mechanisms.[103] Although it is still unclear why or how, it is important to note also that duck temperature-sensitive neurons differentiate and shift from cold (E28–d5 post hatch) to warm sensitivity (d5–d10 post hatch)[14].

At the cellular level, David Julius, the winner of the 2021 Nobel Prize in Physiology or Medicine, made a breakthrough by discovering and cloning the capsaicin receptor[104] and stimulated subsequent research in hunting for other temperature- and pain-sensitive ion channels. Several mammalian thermally sensitive transient receptor potential vanilloids (TRPVs), transient receptor potential ankyrin (TRPAs), and transient receptor potential cation channel subfamily M (TRPMs) were characterized (for review see[105]). The relative activation of these receptors depends on the temperature threshold (Fig. 2). Although the molecular mechanisms of thermo-sensation in avian species are not well defined, several TRPVs, TRPA1, and TRPMs were identified or predicted. Saito et al.[106] have shown that TRPA1 is expressed in chicken skin, and it is responsive to temperature marginally below the average core body temperature (41–42°C). Mahmoud et al.,[107] on the other hand, have found that TRPV1 is expressed in the majority of heat-sensitive neurons in the dorsal root ganglia and activated by high temperature (>45°C). Unfortunately, there is very little information regarding thermo-sensation in birds and how it interacts with the feeding behavior.

In mammals, the thermo-TRPs were found in the feeding-related brain nuclei. For instance, TRPV1 was found in the rodent hypothalamic paraventricular nucleus (PVN), dorsomedial nucleus (DMN), arcuate nucleus (ARC), and anterior (AH) and lateral hypothalamus (LH), though this localization was specie-dependent.[108–110] Regardless of the species-dependence, these findings restated the research question of whether feed intake regulates body temperature as postulated by the thermostatic theory, or the hypothalamus responds to thermal stimuli by modulating (increasing or decreasing) feed intake or via a more complex feedback loop between thermoregulation and energy intake.

Interestingly, Jeong and coworkers have excellently shown, by using various independent techniques, that ARC^{POMC} neurons expressed the TRPV1 subunit-containing receptors.[111] Pharmacological activation of these neurons reduced feed intake.[111] Furthermore, the transmission of increased temperature to the ARC reduced feed intake, which was blocked by a melanocortin receptor antagonist.[111] Together, these results indicated that TRPV directly controls feed intake. In human studies, dietary supplementation of a TRPV1 agonist (capsaicin, capsiate) decreased energy intake,[112–114] and these effects were suggested to be associated with the interaction of TRPV1 and hormonal and/or gastrointestinal vagal afferents. Conversely, other studies from humans and animals reported that capsaicin supplementation had no effects on energy intake (for review see[115]). These controversies could be associated with experimental designs.

5.1 Potential heat stress - central signaling in regulating energy intake

As described earlier, the POA is known to be critical for the regulation of body temperature homeostasis. Due to its interconnections to many brain areas, it is also involved in many physiological and behavioral processes, including modulation of feed intake in response to temperature changes, although this area is highly unexplored. In the following section, we will integrate some recent data from diverse studies, attempting to provide new opportunities and directions for future investigations in avian species.

As the POA encloses the vascular organ of the lamina terminalis (VOLT) and it is connected with the subfornical organ (SFO), it is well positioned for bidirectional communication with the circulation[116]. In addition, several POA subareas (median preoptic nucleus, MnPO; medial POA, MPA; and ventrolateral preoptic nucleus, VLPO) are demarcated by neurons that express neuropeptides such as galanin[117] and hormone receptors such as leptin receptor,[118] both of which are involved in feeding behavior regulation.[119,120] Moreover, POA neurons innervate several brain areas that are involved in the modulation of feeding behavior, such as the ARC, PVN, LHA, and DMN. For instance, the VLPO is involved in the regulation of sleep-wake cycles and expresses the inhibitory neurotransmitter GABA and galanin.[121] These GABAergic VLPO projections innervate orexin/hypocretin neurons in the LHA to initiate and maintain sleep. Moreover, it was suggested that the stimulation of $GABA_A$ receptors in the LHA reduced feed intake through inhibition of orexin neurons.[122]

Yang and colleagues[123] have shown that cold-responsive neurons in the MPA make excitatory synapses onto ARC^{AgRP} neurons. Inhibition of either ARC^{AgRP} neurons or ARC-projecting MPA neurons attenuated cold-evoked feeding, while activation of the MPA-to-ARC projection increased feed intake. These findings revealed a critical role of MPA-ARC neuronal pathway in modulating thermoregulation-evoked feeding behavior. Several studies have shown that local thermal manipulations (heating, cooling, or lesions) on the MPA in various species (goats, rats, or pigs) caused rapid feeding behavior or cessation, regardless of their original dietary states.[124–126]

Recently, Qian et al.[127] have elegantly shown that two neighboring but distinct neuronal populations in MPA mediated feeding behavior by receiving anatomical inputs from ARC and PVN nuclei. The ARC-projecting neurons in MPA can sense low temperature and promote feed intake; however, the PVN-projecting neurons in MPA are primarily sensitive to high temperature and suppressed feed intake.[127] Interestingly, the reproductive hormone estrogen suppresses feed intake via estrogen receptor alpha (ERα1)-expressing neurons, including ERα1 neurons in the POA.[128] Additionally, several POA neurons co-expressed leptin receptor and ERα1[129], suggesting a potential integration of estrogen and leptin signaling in the POA for energy intake control (Fig. 2).

Wang and coworkers mapped the whole brain and found that AgRP and POMC neurons both send axon fibers to the POA, where they can act via their melanocortin 4 receptor.[130] Scammell et al.[131] reported that prostaglandin D2 (PGD2) is bound to its receptor within POA area, and regulated sleep. PGD2 is also associated with an increase in rodent NPY secretion, which is known as a potent orexigenic neuropeptide.[132] Tan et al.[133] have shown that the POA warm-sensitive neurons coexpressed both of the neuropeptide brain-derived neurotrophic factor (BDNF) and adenylate-cyclase-activating polypeptide 1 (ADCYAP1), and both neuropeptides are well known to regulate feeding behavior.[134,135] Yao et al.[136] reported that CRH was expressed in POA, and the role of CRH in the regulation of feeding behavior was well described.[137]

Using retrograde labeling, bidirectional optogenetic manipulation, and single-cell RNA sequencing, Chung and colleagues[138] found that cholecystokinin (CCK)- and tachykinin 1(TAC1)-expressing POA neurons promoted sleep, and both CCK and TAC1 are well-known regulators of feeding behavior.[139,140] Several other feeding-related neuropeptides, such as neurotensin,[141] CART,[142] pre-proenkephalin (PENK),[143] thyrotropin-releasing hormone (TRH),[144] arginine vasopressin (AVP),[145] oxytocin,[146] vasoactive intestinal peptide (VIP),[147] somatostatin,[148] and kisspeptin,[149] to mention a few, were found in the POA region.[150–152] Together these studies suggested a key but complex role of POA in the regulation of energy homeostasis in mammals; however, such studies are very scarce in aves and therefore further in-depth investigations are warranted.

In a recent study, Dridi's group at the University of Arkansas assessed the effect of chronic cyclic HS on feeding behavior, growth, and on the expression of hypothalamic neuropeptides. They compared unselected jungle fowl (JF) and three broiler lines from different stages of genetic selection: a slow-growing Athens Canadian Random Bred ACRB, a moderate growth type from 1995 and a fast-growing type from 2015.[93] Data showed that HS increased core body temperature and reduced feed intake in modern broilers but not in the JF chickens, indicating that unselected chickens are thermotolerant/thermoresistant.[93] Although only gene expression was analyzed, the molecular data revealed no changes in the expression of major classical neuropeptides (NPY, AgRP, POMC, and CART); however, the hypothalamic melanocortin 1 receptor (MC1R) exhibited a line-dependent decreasing trend from JF to modern broilers under both environmental conditions.[93] It is worth noting that studies involving the effects of acute HS on central NPY expression in different chicken genetic stocks revealed controversial results, including upregulation,[153] downregulation,[154] or no change.[155] Bohler et al.[154] reported a diminished orexigenic response to heat-stressed and ICV-injected broilers; however, Bahry et al.[156] reported a luck of such effect in layers, although NPY reduced body temperature in layers.[157] It would be interesting in the future to determine the mRNA and protein expression profile in various hypothalamic nuclei, including the POA. Similarly, with the improvement of genetic targeting tools, electrophysiology and fast retrograde labeling studies are needed to define and map the projection neuron populations in avian brains and determine the neuron circuits underlying both thermoregulation and feeding behavior.

Cline's group has shown that acute exposure to high ambient temperature upregulated CRH mRNA expression in the chicken PVN, but it downregulated it in the nucleus of the hippocampal commissure (NHpC),[154] which is considered to be the area responsible for initiating the stress response in birds.[158] While the inter-trajectories of these NHpC-CRF neurons are unclear, it is possible that projections from the POA extend to the NHpC. In fact, it has been reported in the *Cyanistes coeruleus* (blue tit) that the POA-AVT-and POA-VIP-releasing neurons terminated in the NHpC.[159] Avian AVT acts as an anorexic agent,[160] and plasma AVT has been shown to be induced by heat exposure in chickens,[161] and high AVT has been reported to elicit a hypothermic response in pigeon.[162]

Heat stress has been reported to alter brain serotonin, dopamine, norepinephrine, and catecholamine in broilers,[163] and these monoamines have been shown to regulate feeding behavior in both mammals and aves.[164–167] High throughput analysis showed that HS modulates the hypothalamic expression of neuromedin U (NMU), mesotocin, and thyroid-stimulating hormone beta (TSHβ), which all have been reported also to be directly or indirectly involved in the regulation of energy homeosatsis.[168–170] Similarly, HS has been reported by our group and others to alter the hypothalamic (chemo)cytokine profile and heat shock protein (HSP) expression.[171] Among the inflammatory cytokines, IL-6, for example, has been shown to reduce feed intake and induce thermogenesis.[172] HSPs, however, were originally described as proteins involved in the heat-shock response.[173] Although their direct function in feed intake regulation is still unknown, their critical role in inflammation, immune response, autophagy, and modulation of the nervous system is on the rise.[174] Further studies are needed to fine-tune the intracellular function of central described neuropeptides and HSPs in the regulation of energy homeostasis under heat stress conditions. This would certainly contribute to the panel of remedies to alleviate heat stress and improve feed efficiency.

5.2 Potential heat stress - peripheral signaling in regulating energy intake

As we described previously, environmental (heat or cold) stress triggers a complex peripheral program of gene expression and biochemical adaptive responses.[173,175] In this section, we will briefly summarize the profile of known genes involved in HS/CS responses and then describe in detail those associated with feeding behavior. Conventional techniques, chip array, and omics identified many genes involved in environmental stress responses. Among these genes, HSPs and their transcription factors HSFs are well known to be affected by both HS and CS in peripheral tissues.[176–179] Interestingly, heat therapy has been shown to transiently induce HSP72 in peripheral tissues, reduce inflammation and mitogen-activated protein kinase (MAPK) expression, and thereby protect against diet- or obesity-induced hyperglycemia and insulin resistance in rodents,[180] although the underlying mechanisms remain unknown. HS alters also the expression of genes involved in DNA synthesis, transcription, RNA processing and translation, cell cycle, denaturation and mis-aggregation of proteins, proteasomal-, autophagy-, and lysosomal-pathways, cytoskeletal components, energy metabolism, and membrane permeability[181]. To cite a few examples, environmental stress affected the expression of superoxide dismutase (SOD-1, MnSOD), cytokines, ribosomal RNA, annexin, dual-specificity phosphatases (DUSPs), MAPKs, vascular endothelial growth factor (VEGF), transforming growth factor (TGF) (for review see[181]).

It has been shown that environmental stress affects genes and proteins associated with energy intake and feeding behavior. For instance, it has been reported that HS increased peripheral and circulating BDNF,[182] which is essential for vagal afferent innervation of the stomach wall and has also been shown to play a key role in meal-taking behavior.[183] Although its effect on feed intake is different between mammals[184] and chickens,[185] ghrelin, the other important signal originating in oxyntic gland cells of the stomach, has been reported to be altered by HS.[186] Similarly, HS has been shown to affect the intestinal CCK expression in chickens,[155] and peripheral CCK has been reported to regulate feed intake in aves.[187] El-Sa'adawy et al.[188] have demonstrated that HS exposure downregulated the intestinal expression of the glucagon-like peptide 1 (GLP-1) gene, and GLP-1 has been reported to control feed intake in chickens, although the data were controversial between layers and broilers and between avian species (quail vs. chickens).[189–191] While further studies are needed, CRFergic and noradrenergic systems are potential mediators of GLP-1 (an) orexigenic effects.[192,193] Similarly, Honda et al.[194] have shown that intravascular administration of chicken GLP-2 suppressed feed intake in chickens, however, mammalian GLP-2 did not elicit any effect on energy intake in both chickens and quails,[190,195] and environmental stress has been reported to alter intestinal GLP-2 expression.[196] He et al.[197] showed that HS exposure upregulated the intestinal expression of two feeding-related peptides, namely peptide YY (PYY) and somatostatin. PYY has affinities for NPY2 receptor, which is also expressed in some vagal afferents[198], and PYY has been shown to stimulate firing of gastric vagal afferents.[199] HS alters the expression of several gene-coding hormones such as insulin,[200] amylin,[201] glucagon,[202] leptin,[203] cytokines (TNFα, IL-6, etc.),[204,205] adiponectin,[206] visfatin,[207] orexin,[208] steroid hormones,[209] and thyroid hormones.[210] All these hormones regulate feeding behavior (for review see[211–213]).

Of particular interest, adenosine monophosphate (AMP)-activated protein kinase (AMPK), the master energy sensor,[214] has been found to integrate nutrient and hormonal signals to regulate feed intake and energy balance through effects in both the brain and peripheral tissues (for review see[25]). In the periphery, AMPK mediates the action of several abovementioned feeding-related hormones such as leptin, ghrelin, adiponectin, resistin, orexin, cannabinoids, thyroid hormones, etc.[25] Besides its pivotal role in switching off anabolic to catabolic pathways when cellular energy is depleted as during HS conditions, AMPK has been found recently to be involved in nutrient sensing (see Section 7.2) and circadian rhythms,[215] both of which regulate feeding behavior. Newly, Zapata et al.[216] identified, by using a series of genetic tools from invertebrates (*C. elegans*) to vertebrates (mise and humans), that adipocyte-vacuolar-type H^+ ATPase (ATP6v0a1) is a regulator of peripheral metabolic memory that controls feed intake. These studies highlighted the potential role of mitochondrial (dys)function in the regulation of energy intake and homeostasis. Importantly, recent studies showed that skeletal muscle mitochondrial stress via respiratory uncoupling promotes a growth differentiation factor 15 (GDF15)-dependent daytime-restricted anorectic response to control whole-body energy metabolism.[217] The effects of GDF15 seemed to be associated with alteration in several feeding-related hormones (leptin, ghrelin, and PYY) and neuropeptides (POMC, AgRP).[217] Furthermore, it has been reported that CRF and CCK mediated the anorectic effect of GDF15.[218,219] It must be noted that environmental stress induced GDF15 levels.[220] Such studies are still lacking in avian species. Moreover, it has been recently reported that the mitochondrial anion carrier protein, uncoupling protein 2 (UCP2, which uncouples the oxidative phosphorylation from ATP production by dissipating the proton gradient generated across the mitochondrial inner membrane), plays a key role in the regulation of energy homeostasis and feed intake. In fact, it has been shown that UCP2 mediated ghrelin's action on NPY/AgRP neurons by lowering free radicals.[221] Inhibition of UCP2 by genipin has been shown to induce the excitation of POMC (glucose-excited, GE) neurons, and this effect was abolished in K_{ATP} channel-deficient POMC neurons,[222] which indicated that UCP2 decreases glucose sensing through the regulation of intracellular ATP level and subsequent modulation of K_{ATP} channel. Interestingly, Andrews et al.[223] showed that UCP2 plays also a role in food motivation and reward as well as in hedonic hunger via modulation of ghrelin's effects on dopamine neurons. Contrary to mammals, avians have only one UCP (av-UCP), which has a high homology with mammalian UCP2 and UCP3, and its role in feeding behavior, motivation, and reward system is still unknown.

MAPKs and MAPK phosphatases (MKPs) have been demonstrated to play a variety of roles in stress and metabolic homeostasis.[224,225] The modulation of these kinases and phosphatases by a broad variety of extracellular stimuli, including environmental stress, has been well documented.[226–228] There is increasing evidence that hypothalamic MAPKs play vital roles in the regulation of feed intake in mammals.[229–231] Naznin et al. have shown that peripheral administration of ghrelin activated MAPK-ERK1/2 and AMPKα in nodose ganglion and increased feed intake in control but not in high-fat diet-induced obese (DIO) mice.[232] That study elegantly highlighted a key role of ERK in mediating the ghrelin-vagal afferent effect on energy intake in control mice and showed that diet-induced obesity causes ghrelin resistance. Similarly, Lawan and colleagues[233] have demonstrated that hepatic MKP-1 serves as a selective regulator of MAPK-dependent signals that contributes to the maintenance of glucose homeostasis and peripheral tissue energy balance. Several other studies showed that DUSP6/MKP-3, DUSP7/MPK-X, and DUSP9/MPK-4 play a critical role in the regulation of energy homeostasis.[234,235] Some of these effects are mediated via the microbiota[236] which will be discussed later (Section 7.3).

It has been shown that HS upregulated hepatic acute-phase C-reactive protein (CRP) levels in rodent,[237] and individuals with high CRP levels had a significantly lower intake of grain but higher intake of fluids,[238] which suggests that CRP regulates both feed and fluid intake, though the underlying mechanisms need to be defined. Similarly, the acute-phase orosomucoid1 (Orm1) has been reported to be affected by HS[237] and to regulate feed intake and energy homeostasis via the leptin receptor signaling pathway in rodents.[239]

6 Management factors affecting feed intake: hypoxia

Hypoxia is a state in which oxygen is not available in sufficient amounts at the tissue level to maintain adequate homeostasis. This can result from inadequate oxygen delivery to the tissues either due to low blood supply or low oxygen content in the blood (hypoxemia).[240] In poultry, hypoxia can be induced by HS,[241] low temperature,[242] high altitude,[243] and/or fast growth,[244] and it can vary in intensity from mild to severe and can present in acute or chronic.

Sensing oxygen (dys)homeostasis occurs via specialized chemoreceptor cells that regulate cardiovascular and ventilatory rates. In addition, all nucleated cells sense O_2 levels and respond to reduced O_2 availability acutely and/or chronically through the activation of preexisting proteins and regulation of gene transcription, respectively.[245] This transcriptional response is largely mediated by the action of hypoxia-inducible factor-1 (HIF-1), a transcription factor that interacts with various pathways including PI3K/Akt, VEGF, CBP/p300, etc. and regulates more than 300 genes (depending on cell- and tissue-type)[246] and a variety of physiological responses from cardiac and vascular development, survival, to feed intake and energy homeostasis regulation. It should be pointed out that O_2 limitation is generally considered as an impairment of mitochondrial respiration under hypoxia. In fact, mitochondria are the main source of ATP in normal cells, and it is commonsensical that hypoxia affects feeding behavior and energy balance in vertebrate species.

Studies performed at high altitudes, in conditions of hypobaric hypoxia, have shown a progressive loss of BW induced by depressed feed intake.[247] This suggested that hypoxia could (in)directly control energy intake at peripheral and/or central levels.

At the central level, HIF-1α and HIF-2α were found to be highly expressed in the hypothalamic ARC within the POMC neurons.[248] General neuronal-activation of HIF-1α through deletion of asparaginyl hydroxylase factor inhibiting HIF-1α (FIH) reduced body weight in mice.[249] However, HIF loss-of-function specifically in ARC^{POMC} neurons caused an increase in feed intake and promoted obesity development in rodents.[248] Additionally, Varela et al.[250] have shown that knockdown of endothelium HIF-1α resulted in impairment of glucose uptake in the ARC and decreased POMC neuronal activity, indicating that central HIF-1α not only regulates feed intake via POMC but also controls ARC-glucose sensing and uptake. Using bioinformatic analysis, Gaspar et al.[251] found that HIF-1α transcripts are directly correlated with hypothalamic genes involved in inflammation and autophagy, both of which are known to regulate feed intake. Ferguson et al.[252] have shown that FIH is a negative regulator of the suppressor of cytokine signaling (SOCS), upstream inhibitors of leptin cascade.[253] Chu and coworkers have reported that HIF-1α regulates feed intake via NFκB/NPY/POMC pathways.[254,255] Lim et al.[256] have shown that HIF-1α regulates central dopamine, which is a known regulator of feeding behavior. Furthermore, HIF-1α has been reported to interact with central cytokines such as IL-6 and TNFα,[251] known regulators of feeding behavior. The feeding-related hormones, leptin, insulin, and orexin have been reported to interact with central HIF-1α.[257] According to the existing studies, there is a relationship between PI3K/Akt/mTOR pathway and HIF. This pathway is considered as an important part of energy homeostasis regulated by the hypothalamus because it is stimulated by insulin, leptin, and other hormones and has the function of regulating glucose/lipid homeostasis, body weight, and energy consumption.[258–261] mTOR can directly improve the translation rate of HIF-1α mRNA through 5′oligopyrimidine nucleotide sequence, thus promoting the expression of HIF-1α. mTOR/S6K can promote HIF-2α protein synthesis, as the increased S6K activity will lead to the hypothalamic up-regulation of HIF-2α when glucose is provided. In addition, mTOR can phosphorylate the binding protein 1 (4E-BP1) of eukaryotic initiation factor 4E (eIF4E) to terminate the inhibition of the latter by the former, which in turn, promotes HIF expression.

Bigard's group has elegantly shown that hypoxia leads to an early and transient decrease in hypothalamic AMPK and acetyl-CoA carboxylase (ACC) phosphorylation, concomitant with feed intake depression and alteration of leptin, insulin and neuropeptide expression as well as metabolite levels.[262] The study indicated that hypoxia regulates feed intake via hypothalamic AMPK and suggested a potential interaction between AMPK and HIF[263] (Fig. 3), although proof and further mechanistic studies are needed. The involvement of AMPK, ACC, fatty acid synthase (FAS), and other lipid metabolism-associated factors in the regulation of feeding behavior has been previously described by several researchers, including ourself.[27,28,264–266] Of particular significance, the involvement of AMPK (the master energy gauge) supports our abovementioned statement regarding the role of mitochondrial impairment induced by hypoxia. The mitochondrial respiratory chain complex is well known as a source of reactive oxygen species (ROS), which has recently attracted attention as a regulator of metabolism and feed intake.[267] ROS has been reported to affect feeding behavior through the activation of hypothalamic feeding-related (an) orexigenic neuropeptides (NPY/AgRP and POMC)[221] (Fig. 3). Indeed, increasing H_2O_2 can cause depolarization of POMC neurons, and ICV injection of H_2O_2 can cause significant anorexia.[267] As one of the oxygen free radicals, ROS can also regulate HIF. Under normoxic conditions, prolyl hydroxylase domain (PHD) enzymes are active and can hydroxylate conserved proline residues of the HIF-α subunit using alpha-ketoglutarate and molecular oxygen as co-substrates, which then lead to HIF-α's proteasomal degradation.[268] Under hypoxic conditions, however, ROS elevation helps stabilizing HIF-α protein, mainly by oxidizing central Fe(II) to Fe(III) to promote PHD enzyme inactivation.[269] Several studies, by knocking down various subunits of different mitochondrial complexes such as GRIM-19 (NDUFA13, complex I), SDHB (complex II), or RISP (complex III), have shown that mitochondrial ROS formation can stabilize HIF.[270–272]

Many studies have demonstrated that HIF can mediate the activity of the glucose transporters (GLUTs). Several GLUTs are expressed in the brain. For instance, GLUT1 was found in microvascular endothelial cells and astrocytes and is responsible for promoting glucose transport through blood–brain barrier (BBB).[273] GLUT3 and GLUT4 were found in neurons.[274] Once glucose is absorbed by neurons via GLUTs, it is metabolized into ATP which then regulates AMPK and provides information regarding intracellular energy levels and glucose concentration, a phenomenon known as 'neuron sensor'[275] (see Section 7). Hypoxia induced the expression of both central HIF-1α and GLUT1[276], and HIF-1α has been shown to be a transcription factor of GLUT1, GLUT3, and an activator of GLUT4[277,278] (Fig. 3). Moreover, inhibition of GLUT2 in the brain alters satiety and feeding behavior.[279] Mice that did not express GLUT-2 in the brain had glucagon imbalance, and the re-expression of GLUT-2 in their astrocytes restored the correct control of glucagon levels.[280] Similarly, Ren et al.[281] have

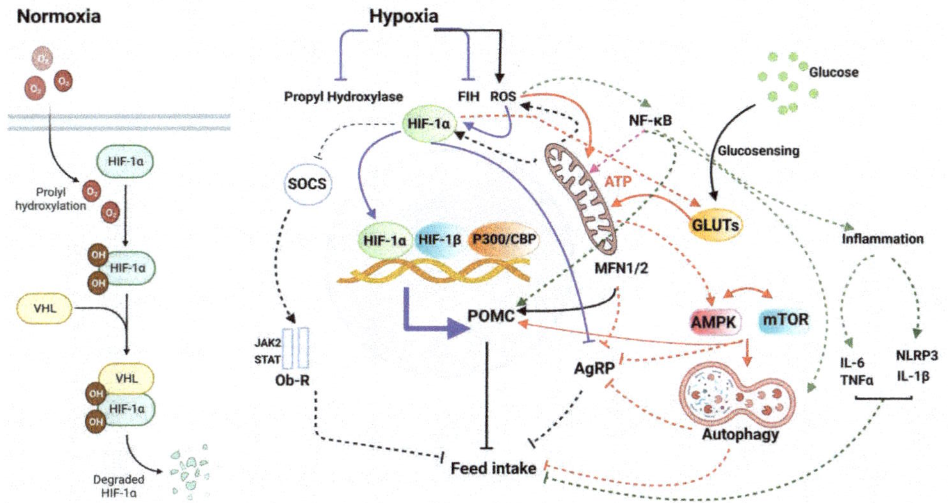

Figure 3 Schematic illustration of potential central mechanisms by which hypoxia regulated feeding behavior. Under normoxia (left panel), HIF-1α is hydroxylated on proline residues 402 and 564 by prolyl hydroxylases (PHDs). This hydroxylation is a target for von Hippel-Lindau protein (VHL) binding, a E3 ligase that targets HIF-1α for ubiquitination and proteasomal degradation. Under hypoxia, PDH and FIH are inhibited and HIF-1α is stabilized to dimerize with HIF-1β and binds to target genes at the DNA, recruits p300/CBP, and activates transcription of several genes involved in various cellular metabolism, including the anorexigenic neuropeptide POMC. Hypoxia can alter mitochondrial function and induce ROS, which in turn affect feed intake via modulation of hypothalamic neuropeptides. Hypoxia can also affect feed intake through AMPK, mTOR, cytokine production, and autophagy. AgRP, agouti-related protein; AMPK, adenosine monophosphate-activated protein kinase; FIH, factor inhibiting HIF-1α; GLUT, glucose transporters; HIF, hypoxia-inducible factor; IL, interleukin; JAK, janus kinase; MFN1/2, mitofusin 1/2; mTOR, mechanistic target of rapamycin; NLRP3, NLR Family Pyrin Domain Containing 3; Ob-R, leptin receptor; POMC, proopiomelanocortin; ROS, reactive oxygen species; STAT, signal transducer and activator of transcription; VHL, von Hippel-Lindau. The figure was created using BioRender .com and was not drawn to scale.

shown that hypothalamic ablation of GLUT4 induced anorexia and impaired glucose metabolism. More recently, Ardanaz and colleagues[282] have shown that GLUT1 ablation in astrocytes affected feed intake and improved central and peripheral glucose metabolism via enhanced insulin-stimulated ATP release. Collectively, these studies showed a key role of HIF and GLUT in central- and peripheral-energetics and feeding behavior.

Although such studies are very rare in poultry, Druyan's group[283] has shown that daily exposure (12 h/day) to hypoxia during chorioallantoic membrane development (E5 to E12) affected feed intake at week 3 post-hatch and growth performance of male broilers. Huang et al.,[243] on the other hand, have demonstrated that high-altitude hypoxia depressed feed intake in Tibetan chickens. Solis de los Santos et al.[284] showed that exposure to hypoxia reduced body weight in commercial broilers, though feed intake has not been reported. These effects on feeding behavior and growth performance were accompanied by alterations in gut weight and morphometry.[284] It is well known that during HS, birds divert blood to the periphery (skin) to maximize radiant heat dissipation, while vasoconstriction occurs in the internal organs to reprioritize blood flow. This, in turn, would create a hypoxia-like status in internal organs, including the brain and the gut.[241] Furthermore, HS has been shown to induce ROS production,[285], which as described previously regulates feeding behavior. Recently, Dridi's group[171] has shown that hypothalamic expression of HIF-1α was higher in heat-stressed JF chickens compared to modern broilers. It is noteworthy to mention that JF chickens eat less feed and are more resistant to HS compared to the modern broilers. Furthermore, Dridi's group[171] showed that HS altered the hypothalamic expression of AMPK, NF-κB, HSP, and oxygen homeostasis-related genes (HBA1, HBBR, HBE, HBZ, HJV) in a genetic line-dependent manner. Although it is still not known at this time whether the effect of HS is direct or is mediated by hypoxia, these data suggested a potential interaction between hypoxia pathways and feeding behavior in birds that need future in-depth investigations.

7 Management factors affecting feed intake: nutritional stress and nutrient sensing

As described earlier, the ingestive behavior is tightly regulated both by peripheral sensory inputs and hypothalamic appetite signals. Significant progress has been made in our fundamental understanding of basic taste qualities, the cellular and molecular basis of taste sensing, and the neuron-sensing systems and circuits that control feeding behavior. In this section, after a brief description of taste processing, we will discuss the interaction between peripheral- and neuron-sensing nutrients in the regulation of energy intake and homeostasis.

7.1 Taste receptors and detection

One of the key components that controls feed intake is appetite, which is related not only to physiological need (hunger) but also to the pleasure of eating-flavor, taste, aroma, texture, sight of food, and a variety of social and psychological factors. In fact, the sensory properties of food is largely, if not primarily, determinant of food choice and intake.[286]

In mammals, the peripheral gustatory system is composed of specialized epithelial cells located on the tongue, soft palate, pharynx, epiglottis, larynx, and upper third of the esophagus.[287] The tongue taste buds (5000 average number), which contain sensory cells with microvilli, can distinguish five basic tastes: salt, sweet, bitter, sour, and umami (the Japanese for savory) (Fig. 4). An additional taste mechanism for dietary fat with particular sensitivity for essential fatty acids has been postulated.[288] Sweet is associated with sugar and it is sensed by specific sets of G protein-coupled receptors (GPCR) called T1Rs (for taste receptor type 1, T1R2 and T1R3)[289] (Fig. 4). Umami is associated with L-amino acids and it is sensed by T1R1 and T1R3.[290,291] It is important to note that T1R3 functions as a co-receptor for both sweet and umami tastes because T1R3 KO animals showed drastically reduced sensitivity to both sweet and umami.[291] Bitter taste is generally evoked by toxic chemicals that are hazardous to animals and is recognized by T2Rs,[292] another family of GPCR. Depending on the concentration, salt at low levels is sensed by the attractive ENaC receptor (for epithelial sodium channels); however, at high levels it involves other aversive (bitter) receptors.[293] Sour taste is sensed by PKD2L1 (for polycystic kidney disease 2-like 1 protein)[294,295] and otopetrin 1.[296] Intriguingly, it has been reported that water can stimulate taste nerves in vertebrates, and a recent study using optogenetic stimulation of PKD2L1-expressing taste receptor cells (TRCs) triggered attractive licking behavior in thirsty mice. This raises the question of how and why water and sour tastes are encoded by the same TRC population.

In avian species (chicken, turkey, pigeon, etc.), taste buds are located mainly in the posterior tongue and pharynx as well as in the upper palate and base of the tongue (Fig. 4) but not in the highly keratinized anterior and central tongue as is the case of mammals.[297] Several studies showed that the taste sense played a pivotal role in feed choice and intake in chickens.[298,299] For instance, supplementation of jojoba oilseed to the diet, which presumably caused bitter taste reduced feed intake in chickens.[300] Similarly, increased sucrose, coniferyl- or cinnamyl-derivatives (bitter compounds for avian species), has been reported to induce avoidance and repellency in birds.[301,302] In addition, sense for amino acids (umami taste) and for fatty acids have been reported in birds.[303,304] Recently, using a homology-based approach, Lagerstrom et al.[305] identified several hundred GPCR genes in the chicken genome, with 40 that might be directly associated with taste and nutrient sensing. Three bitter taste

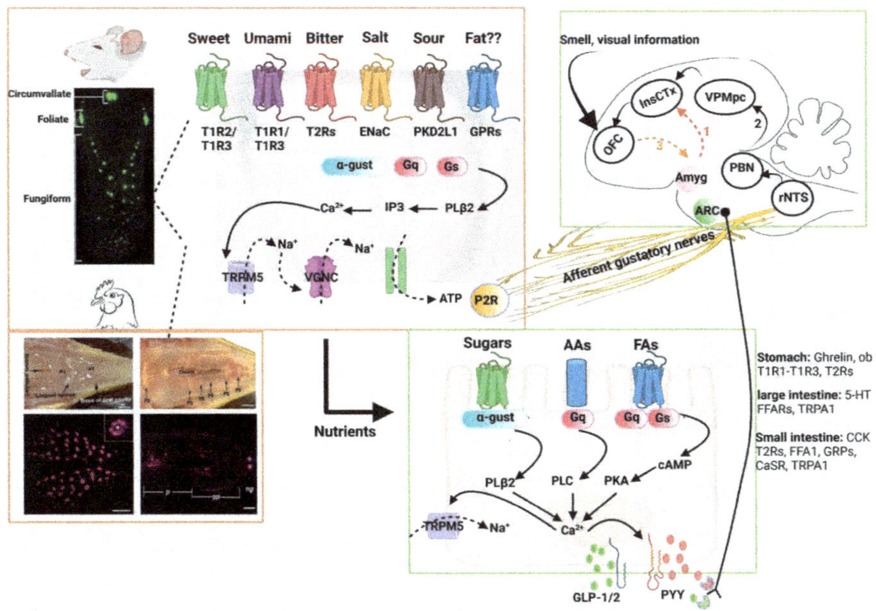

Figure 4 Simplified peripheral and central model of taste receptor signaling in the regulation of feed intake. Taste buds are broadly distributed on the tongue and soft palate in mice (right upper panel)[287] and in chickens (right lower panel).[297] Sweet, umami, bitter, salt, sour, and fat are detected by taste receptors. Taste receptor binding leads to activation of gustatory G-proteins, release of intracellular Ca,$^{2+}$ activation of TRPM5, depolarization, activation of VGCN, and release of ATP, which activates purinergic receptors on afferent gustatory nerves leading to taste perception by the brain. The activation of these taste receptors in the gut leads to the release of gut hormones and peptides that reach the brain and regulate feed intake. AAs, amino acids; ARC, arcuate nucleus; Amyg, amygdala; ATP, adenosine triphosphate; cAMP, cyclic AMP; ENaC, epithelial sodium channel; FAs, fatty acids; GLP1/2, glucagon-like peptide 1/2; GPR, G-protein receptors; InsCTx, insular cortex; NTS, nucleus of the stria terminalis ; OFC, orbitofrontal cortex; PBN, parabrachial nucleus; PKA, protein kinase A; PKD2L1, polycystic kidney disease 2-like 1; PLβ2, phospholipase beta 2; PYY, peptide YY; T1R, taste receptor; VGCN, voltage-gated Na^+ channel; VPMpc, ventroposterior medial nucleus. The figure was created using BioRender.com and was not drawn to scale.

receptors, T2R1, T2R2, and T2R7, were characterized in birds.[306–309] It is important to note that there is a wide variation in the number of bitter taste receptors (T2Rs) between avian species, varying from one reported for domestic pigeons to 18 for the white-throated sparrow.[310] This indicates the diversity of nutritional needs under diverse ecological niches.[311] Although responses of birds to sugar have been reported,[312,313] chickens seemed to lack T1R2 receptor; however, they do have the glucose transporter SGLT1,[314] which has been reported to be a sweet taste receptor.[315] In addition, it has been shown that some birds such as the hummingbird use the umami T1R1 receptor to sense sweetness.[316] Several other taste receptors for fatty acids (G-protein receptors GPR41, GPR43, GPR120, and cluster of differentiation 36, CD36), amino acids (GPR92, GPR139, GPRC6A, T1R1, T1R3, mGluR), minerals (CaSR for ca, ENac for Na^+), and organic acids (PKD2L1, HCN) (for review see[317]) have been characterized in birds.

Although the discipline is at its beginning and it is an emerging area, lagging behind the mammalian knowledges, the vast repertoire of avian taste receptors opens new vista and opportunities for future research to help understand the mechanisms by which olfaction and nutrient sensing regulate energy homeostasis in poultry. This would help to develop science- and mechanism-based approaches for precision nutrition in poultry.

7.2 Integrated peripheral and central nutrient sensing systems in feed intake regulation

Evidence obtained in the past few years supports the complex interaction and integration of peripheral and central nutrient-sensing systems in the regulation of feed intake and energy homeostasis. Indeed, the detection of increased levels of nutrients in both the periphery and the hypothalamus by nutrient (glucose-, fatty acid-, and amino acid-) sensing capacities modulates the hypothalamic signaling and regulates feed intake via (an)orexigenic neuropeptide expression. Moreover, central nutrient sensing modulates functions in the periphery and elicits metabolic changes (hormonal secretion, etc.) to counter-regulate changes in nutrient levels detected by the CNS.

The first step in nutrient sensing occurs in the oral cavity that involves lingual taste receptors as described previously. Recent studies unequivocally showed that the gastrointestinal system is also capable of sensing nutrients. Raybould[318] reported that nutrients in the intestinal lumen are detected by specific taste sensors that respond to sweet, umami, and bitter compounds, as well as both long- and short-chain fatty acids (Fig. 4). Many studies have identified members of T1Rs, CaSRs, and GPRs receptors as well as alpha gustducin in enteroendocrine cells.[319–321]. This intestinal nutrient sensation is essential for eliciting the appropriate neuronal and hormonal responses following meal ingestion. The subsequent post-ingestion release of gut hormones has a variety of roles that affect metabolism, gut motility, satiety, hunger, and glucose and lipid homeostasis (Fig. 4). For instance, activation of the abovesaid taste receptors leads, through cAMP/PKA or PLC, to an increase in intracellular Ca^{2+} concentration, that in turn induces GLP-1, GLP-2, CCK, and PYY release from the enteroendocrine cells in the lamina propria[322] (Fig. 4). Moreover, ghrelin-P/D1 cells sense and respond to nutrients via T1R3, GPR120, and α-gustducin.[323–325] Similarly, the chemical content of ingested foods stimulates the release of gastric acid pepsinogen and mucus, which assist in digestion.[326] Sensation via GPR92 helps breaking down protein and release somatostatin.[327] These released hormones can have paracrine effects on neighboring enterocytes, activate the enteric nervous system (ENS), activate the gut–brain axis by vagal afferent signaling, and/or enter the blood circulation to target other peripheral organs and the brain.

Delineation of the mechanisms underlying CNS sensing and responses to nutrient and gut hormones has been a subject of intense interest and active research. Besides the activation of the gut–brain axis,[328,329] it has been shown that lingual TRCs, when activated by specific tastants, transmit information via sensory afferent fibers and taste ganglions to specific taste perception areas (rNTS, PBN, VPMpc, InsCTx, OFC) in the brain[330–332] (Fig. 4). For example, although the molecular identity of individual gustatory neurons in the brain has rarely been discovered, rNTS-Pdyn-expressing neurons respond solely to sour taste and evoke aversion,[333]; however, PBN-Satb2-expressing neurons selectively respond to sweet taste and transmit the information to the VPMpc to evoke appetitive licking behavior.[334] Interestingly, and in support of the feeding (lipostatic, glucostatic/glucodynamic, and aminostatic) theories, the macronutrients (glucose, fatty acids, and amino acids) were found to provide direct information on nutrient availability to the brain (Fig. 5). Furthermore, central administration of glucose, fatty acids, or amino acids affects feeding behavior,[259,335,336] and these effects seemed to be mediated by hypothalamic (an)orexigenic neuropeptides[275,337–339] (Fig. 5). Interestingly, nutrient transporters were found to be expressed in the brain (Fig. 5). For instance, depletion of CD36 in the VMH of DIO rats fed 45% high-fat diet increased their food intake and body weight gain compared to DIO controls.[340] Similarly, key genes involved in *de novo* fatty acid synthesis (lipogenesis) are expressed in the brain. Acetyl-CoA carboxylase (ACC) carboxylates acetyl-CoA (which comes from citrate produced in the mitochondria by the citric acid cycle) to form malonyl-CoA. The malonyl-CoA is further converted, by fatty acid synthase (FAS), to form long-chain fatty acids (LCFAs). Conversely, malonyl-CoA can be decarboxylated into acetyl-CoA and CO_2 by malonyl-CoA decarboxylase (MCD) and results in a decrease in the *de novo* fatty acid synthesis. Several studies, including our own, showed that central administration of fatty acid inhibitors (C75 or cerulenin) increased hypothalamic malonyl-CoA concentration, reduced food intake and

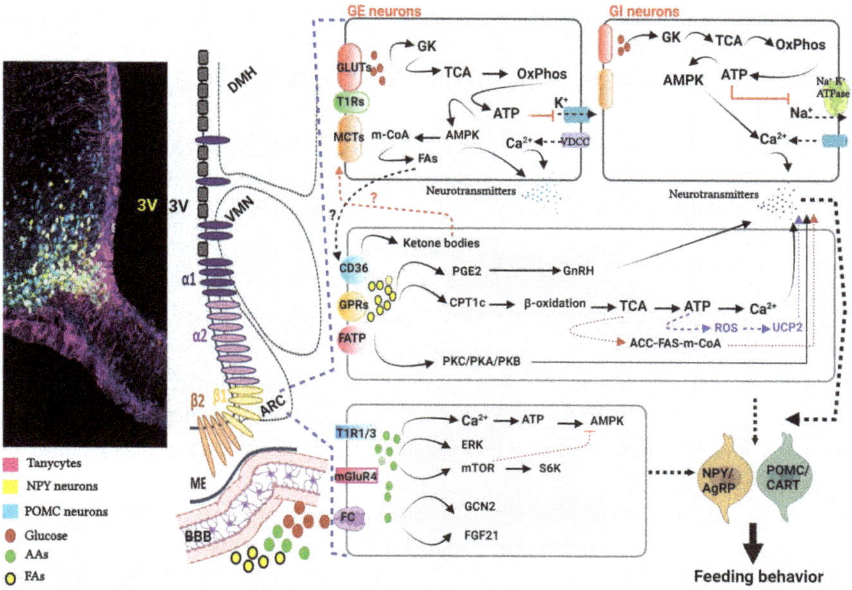

Figure 5 Simplified potential neural model of nutrient-sensing signaling in the regulation of feed intake. Immunofluorescence staining shows tanycytes (purple) interacting with hypothalamic feeding-related neuropeptides (NPY in yellow and POMC in blue) (https://www.openaccessgovernment.org/article/the-watch-project-tanycytes-in-health-and-disease/150441/). Glucose, fatty acids, and amino acids can be transported into the neuron through their appropriate transporters. Inside the neuron, glucose phosphorylation by GK acts as a rate-limiting step proportional to extracellular glucose concentrations. Lactate can be taken up by the neuron via MCTs to serve as a glycolytic substrate and metabolized into pyruvate. Mitochondrial uptake and metabolism of pyruvate lead to an increased ATP-to-ADP ratio and inhibition of ATP-dependent K+ (KATP) channels, resulting in plasma membrane depolarization and neurotransmitter release. Glucose can also regulate feed intake via modulation of AMPK and m-CoA. Neural fatty acids can regulate feed intake via ketone production, PGE2-GnRHA, CPT1c, ROS-UCP2, and/or PKA/PKB pathways. In excess of AA levels, feed intake is regulated by AMPK, ERK, and/or mTOR pathways. However, in low AAs, feed intake is regulated by GCN2 and/or FGF21 pathways. AAs, amino acids; ACC, acetyl-CoA carboxylase; AMPK, AMP-activated protein kinase; ATP, adenosine triphosphate; BBB, blood–brain barrier; CD36, cluster of differentiation 36; CPT1c, carnitine palmitoyl-transferase 1c; ERK, extracellular signal-regulated kinases; FAS, fatty acid synthase; FATP, fatty acid transport protein; FGF21, fibroblast growth factor 21; GCN2, general control nonderepressible 2; GE, glucose excited; GI, glucose inhibited; GK, glucokinase; GLUT, glucose transporter; GnRH, gonadotropin-releasing hormone; GPRs, G-protein receptors; m-CoA, malonyl-CoA; MCTs, monocarboxylate transporters; mGluR4, metabotropic glutamate receptor 4; mTOR, mechanistic target of rapamycin; OxPhos, oxidative phosphorylation; PGE2, prostaglandin; PKA/B/C, protein kinase A/B/C; ROS, reactive oxygen species; S6K, ribosomal S6 protein kinase; T1Rs, taste receptors; TCA, tricarboxylic acid cycle; UCP2, uncoupling protein 2. The figure was created using BioRender.com and was not drawn to scale.

body weight, and increased energy expenditure.[266,341,342] These effects were mediated through the feeding-related hypothalamic neuropeptides.[341,342] ICV administration of 5-(tetradecyloxy)-2-furoic acid (TOFA), an inhibitor of ACC, prior to injection of C75, blocked the anorectic effect of C75. Although both TOFA and C75 inhibit fatty acid synthesis, TOFA antagonizes the anorectic effect of C75. This led to the hypothesis that hypothalamic malonyl-CoA plays a key role in the control of feeding behavior, which has been confirmed by many subsequent studies.[343–345] A recent study has characterized a novel carnitine palmitoyl-transferase (CPT1c) isoform predominantly expressed in the brain[346] and plays a key role in the regulation of feed intake.[347] In addition, lipolysis-associated factors, such as lipoprotein lipase (LPL), have been shown to be localized in the brain[348] and regulate feeding behavior.[349,350] Collectively, these data indicated that genes involved in fatty acid metabolism (lipogenesis, lipolysis, and beta-oxidation) are expressed in the brain and regulate energy homeostasis (Fig. 5).

Central glucose metabolism has been shown also to regulate feeding behavior. First, glucose-sensing neurons have been identified in brain nuclei[351] (Fig. 5). Second, glucose-excited (GE) and glucose-inhibited (GI) neurons were characterized and found to belong to ARC-POMC and ARC-AgRP, respectively (for review see[352]). Additionally, it has been shown that glucose levels in the cerebrospinal fluid (CSF) change proportionally to variations in blood glucose concentrations, implying the presence of a mechanism that transfers glucose from blood to CSF.[353] Furthermore, several genes involved in glucose transport and metabolism, such as GLUT3, GLUT2, GLUT4, glucokinase, and regulatory glucokinase protein (GKRP), were found in the brain and hypothalamic tanycytes.[354–357] Uranga et al.[358] have shown that adenovirus-mediated suppression of hypothalamic glucokinase affects feeding behavior. Barahona et al.,[279] on the other hand, have reported that glial hypothalamic inhibition of GLUT2 expression impacted satiety, and eating behavior. More importantly, it has been shown that glucose activates the sweet taste receptors and increases Ca^{2+} concentrations in tanycytes[359] (Fig. 5). Together, these studies indicated a complex interaction between central glucose metabolism, taste receptors, and (an)orexigenic neuropeptides in the regulation of energy homeostasis in mammals. It is worth noting that, compared to mammals, chickens are constitutively hyperglycemic and insulin resistant.[360,361] This peculiarity suggests potential peripheral/central glucose metabolism in the regulation of feeding behavior in aves that merit future investigations.

Comparable to glucose and fatty acid, a relationship between serum amino acid levels and appetite has been shown for a long time ago.[362] Using deficient- or excess-protein diets, it has been shown that the satiety effects of proteins are source-dependent in both animals and humans.[363,364] Several mechanisms underlying the anorectic effects of protein have been postulated, including, low palatability and chemo-sensation,[365] thermogenic effects,[366] and gluconeogenesis.[367] Several amino-acid taste receptors (Tas1r1/Tas1r3) and umami taste receptors (mGLuR4) were detected in tanycytes, indicating a key role for tanycytes in amino acid sensing and regulation of feed intake (Fig. 5). At the molecular levels, the central effect of amino acids seemed to be mediated via the general control of amino-acid synthesis 2 (GCN2) and mTOR pathways. In fact, the GCN2 pathway is an evolutionarily conserved pathway identified in yeast to mediate the detection of amino acid deficiency. When cellular amino acid levels fall, uncharged tRNAs accumulate in the cell, bind to GCN2 that displays kinase activity toward eiF2α (eukaryotic initiation factor 2α), causing a global suppression of translation, but increased transcription of starvation relevant transcripts.[368–371] When amino acids are abundant, following central administration of leucine for example, the anorectic effect was mediated through mTOR pathway.[259,372,373] The analysis of meal patterns showed that the reduction in food intake measured in food-deprived rodents who received central injections of leucine resulted from the alteration of various components of the feeding sequence: increased first meal latency (and therefore decreased orexigenic tone), decreased meal size, and decreased meal frequency. One circuit implicated in the acute reduction in meal size following hypothalamic leucine sensing involves hypothalamic POMC neurons, their melanocortinergic projections to PVH oxytocin neurons and oxytocinergic projections to the nucleus of the solitary tract in the caudomedial brainstem.[372] It is also possible that leucine regulates feed intake through its metabolic products. Leucine undergoes intracellular metabolism, producing α-ketoisocaproic acid (KIC) and isovaleryl-CoA, respectively by branched-chain amino acid transferase (BCAT) and branched-chain ketoacid dehydrogenase (BCKDH), leading to the production of TCA cycle intermediates and ATP production. Central injection of KIC or BCKDK activator suppressed feeding through a decreased meal number, without affecting meal size. Furthermore, deletion of astrocyte BCATm that increased brain leucine concentration affects food choice with a preference for a low BCAA diet over a normal diet.[374]

Although studies in chickens are scarce, Byerly and colleagues[375] found that the umami taste receptor (T1R1) is present in the hypothalamus. Cheled-Shoval et al. demonstrated the expression of T1R and T2R in the chicken GIT[376]. Fatty acid taste receptors (GPR43, GPR120, and CD36) as well as α-gustducin and α-transducin were found also in chicken gut.[377] Bungo's group has reported that central administration of leucine, but not isoleucine or valine, stimulated feeding behavior in chickens.[378] Denbow and colleagues, on the other hand, reported that ICV administration of glucose did not affect feed intake in layers.[379] Avian species, including chickens, seem to have well-developed molecular systems (taste receptors, neuropeptides, nutrient transporters, etc.) that have many similarities and some peculiarities compared to mammals.

Further functional studies delineating the mechanisms and causal interaction between these factors in the regulation of feed choice and intake in chickens would help developing a precision nutrition and improving poultry production sustainability.

7.3 Microbiota

As nutritional stress has been reported to affect microbiota and gut integrity,[380] it is logical to speculate that microbiota may regulate feeding behavior. Evidence from mammalian studies suggests that gut microbial abundance and richness are related to the leptin signaling, which is a key regulator of energy balance. In fact, it has been reported that high circulating leptin in obese individuals was associated with low bacterial richness.[381] Moreover, *in vivo* and *in vitro* studies showed that the translocation of living gut microbiota to adipose tissues induced by increased intestinal permeability influences energy metabolism by inhibiting the leptin signaling in obese humans and mice.[382,383] Schéle et al.[384] have shown that gut microbiota reduces leptin sensitivity and the expression of the obesity-suppressing neuropeptides proglucagon and BDNF in the central nervous system. Interestingly, gut microbiota depletion inhibits feed intake in mice fed a normal or HFD diet, and this effect is mediated by leptin only under a normal diet.[385] Several studies showed that gut microbial metabolites, such as acetate and SCFA, regulated feed intake through ghrelin signaling pathway[386,387] (Fig. 6). As for leptin, deletion of gut microbiota induced insulin sensitivity and affected feed intake.[388] Together, these studies indicated that gut microbiota regulates feeding behavior via peripheral-derived hormones (Fig. 6).

The activation of taste receptors was recently proposed as an additional mechanism by which gut metabolites regulate feeding behavior. Indeed, it has been shown that SCFAs exhibited their appetite-related functions by binding to various taste receptors, including free fatty acid receptors (GRP43 also known as FFAR2 and GRP41 known as FFAR3).[389] Frost et al.[390] have recently shown that circulating gut-derived SCFAs cross the BBB and directly affect feed intake via a central homeostatic mechanism involving hypothalamic POMC and AgRP.

Of particular interest, gut microbial can produce GABA, BCAAs, and peptides that, in turn, can regulate energy homeostasis (Fig. 6). For instance, it has been shown that most lactobacilli and bifidobacteria strains have the Gad genes that can encode glutamate decarboxylase to synthesize GABA,[391] which is strongly associated with appetite- and feed intake-control. The gut microbiota exhibits also enriched genes related to BCAA biosynthesis (*Prevotella copri* and *Bacteroides vulgatus*), degradation (*Bacteroides thetaiotaomicron* and *Dorea longicatena*), and uptake (*Butyrivibrio crossotus* and *Eubacterium siraeum*)[392], and the role of BCAAs in the regulation of feed intake has been already reported

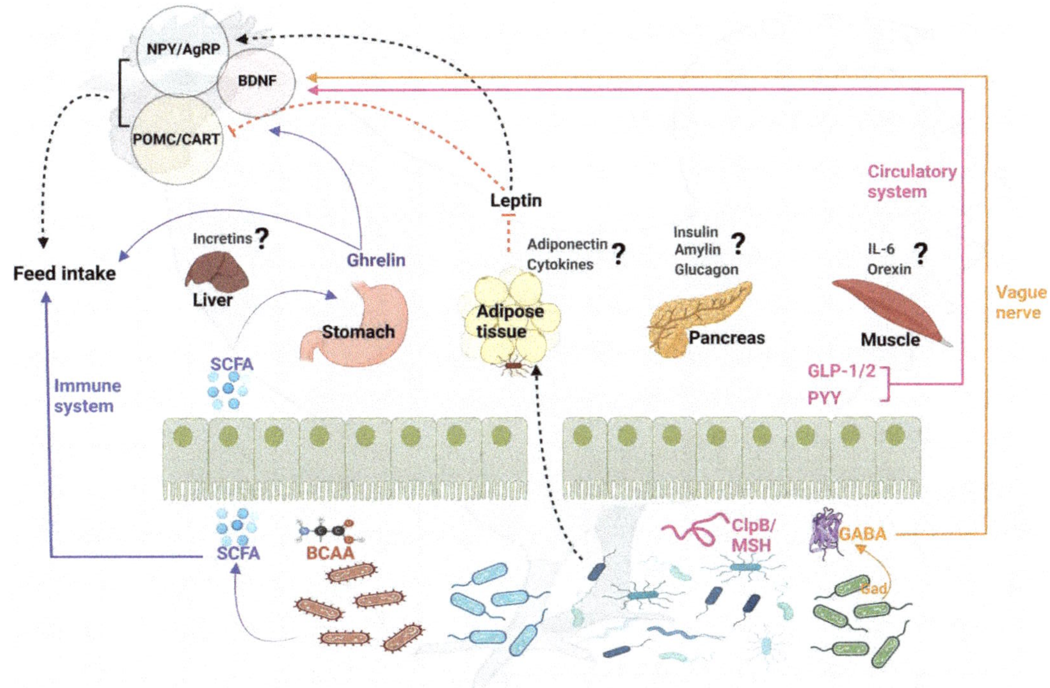

Figure 6 Potential interactions of microbiome and the CNS in the regulation of feed intake. Gut microbial metabolites can stimulate enteroendocrine cells to release (an)orexigenic hormones (PYY, GLP-1, and CCK) and neurotransmitter (5-HT, GABA) and promote the secretion of peripheral hormones (leptin, ghrelin, and insulin). Gut microbiota can also produce identical protein sequences with appetite-regulating peptides, such as ClpB (MSH), that might directly act on anorexigenic neurons or bind to Igs to modulate the secretion of anorexigenic hormones from enteroendocrine L cells. Gut microbiota produces also BCAA and SCFA that regulate feed intake. BCAA, branched-chain amino acids; BDNF, brain-derived neurotrophic factor; ClpB, caseinolytic peptidase B protein homolog; GLP-1/2, glucagon-like peptide 1/2; MSH, melanocyte-stimulating hormone; SCFA, short chain fatty acid. The figure was created using BioRender.com and was not drawn to scale.

as described earlier. Arnoriaga-Rodriguez et al.[393] have shown that ClpB, the best-studied bacterial protein, can act as a mimetic of alpha-melanocyte-stimulating hormone (α-MSH) to result in similar anorexigenic effects. Breton and colleagues[394] have recently shown that ClpB derived from gut commensal *E. Coli* was capable of displaying the α-MSH-like functions, such as enhancing PYY and GLP-1 secretion, and directly activating POMC anorexigenic neurons, and subsequently inducing satiety.

In chickens, although the data are very limited, bacteria belonging to *Bacteroides*, *Clostridium*, *Ruminococcus*, *Faecalibacterium*, and *Lactobacillus* have been positively associated with feed efficiency in chickens. However, bacteria belonging to *Enterobacteriaceae* were described to affect feed efficiency negatively. In addition, Siegerstetter et al.[395] reported an association between the *Lactobacillus* genus and two *Lactobacillus crispatus* taxa and a high feed intake in female chickens. More recently, Dridi's group analyzed the ileal microbial composition and profile in four genetically distinct chicken lines reared under thermoneutral or heat stress conditions and found that the relative abundance of *lactobacillus* was significantly higher in the modern broilers compared to their unselected JF counterparts. This result supported the abovementioned notion that the high abundance of *Lactobacillus* is probably associated with high feed efficiency in the modern broilers.

8 Conclusion

Feeding behaviors can be classified into homeostatic feeding to maintain body weight and metabolic function or hedonic feeding driven by sensory perception or pleasure. Homeostatic and hedonic feeding systems are activated separately or simultaneously depending on various situations related to feeding. Interestingly, the activation pattern of these two systems may shift according to the taste of the food (palatable or aversive) and also according to the physiological state of the animal (hunger or satiety). Further investigations are needed to dissect the homeostatic, hedonic, and rewarding systems as well as their interactions in avian species.

9 Abbreviations

4E-BP1, eukaryotic translation initiation factor 4E binding protein 1
ACC, acetyl-CoA carboxylase
ACTH, adrenocorticotropic hormone
ADCYAP1, adenylate cyclase-activating polypeptide 1
AMPK, adenosine monophosphate-activated protein kinase
ARC, arcuate nucleus
AVP, arginine vasopressin

BBB, blood–brain barrier
BCAT, branched-chain-amino-acid aminotransferase
BDNF, brain-derived neurotrophic factor
BMAL1, brain and muscle arnt-like protein 1
CART, cocaine- and amphetamine-regulated transcript
CCK, cholecystokinin
CD36, cluster of differentiation 36
CLOCK, circadian locomotor output cycle kaput
CNS, central nervous system
CRH, corticotropin-releasing hormone
CRP, C-reactive protein
CRY, cryptochrome
CS, cold stress
DMN, hypothalamic dorsomedial nucleus
DUSP, dual-specificity phosphatase
EIF4E, eukaryotic translation initiation factor 4E
ER, estrogen receptor
FAS, fatty acid synthase
FE, feed efficiency
FIH, factor inhibiting HIF-1
GABA, gamma aminobutyric acid
GCN2, general control nonderepressible 2
GDF15, growth differentiation factor 15
GLUT, glucose transporter
HBA1, hemoglobin subunit alpha 1
HBBR, hemoglobin beta subunit rho
HBE, hemoglobin E
HBZ, hemoglobin subunit zeta
HIF, hypoxia-inducible factor
HJV, hemojuvelin
HPA, hypothalamus–pituitary–adrenal
HS, heat stress
HSF, heat shock factor
HSP, heat shock protein
IL, interleukin
InsCtx, insular cortex
KIC, α-ketoisocaproate
LCFA, long-chain fatty acid
LHA, lateral hypothalamus
LPB, lateral parabrachial nucleus
LPL, lipoprotein lipase
MAPK, mitogen-activated protein kinase

MC1R, melanocortin 1 receptor
MCD, malonyl-CoA decarboxylase
MnPO, median preoptic nucleus
MPA, medial POA
mTOR, mechanistic target of rapamycin
NF-κB, nuclear factor kappa B
NHpC, nucleus of hippocampal commissure
NMU, neuromedin U
NPY, neuropeptide Y
OFC, orbitofrontal cortex
Orm1, orosomucoid 1
PENK, pre-proenkephalin
PER, period
PHD, propyl hydroxylase domain
PGD2, prostaglandin D2
PKD2L1, polycystic kidney disease 2-like 1 protein
POA, preoptic area
POMC, proopiomelanocortin
PVN, hypothalamic paraventricular nucleus
PYY, peptide YY
rNTS, rostral part of the nucleus tractus solitarius
ROS, reactive oxygen species
SFO, subfornical organ
SNS, sympathetic nervous system
SOCS, suppressor of cytokine signalling
SOD, superoxide dismutase
T1R, taste receptor
TAC1, tachykinin 1
TNFα, tumor necrosis factor alpha
TOFA, 5-tetradecyloxy-2-furoic acid
TRC, taste receptor cell
TSHβ, thyroid-stimulating hormone beta
VIP, vasoactive intestinal peptide
VOLT, vascular organ of the lamina terminalis
VPMpc, ventral posteromedial thalamic nucleus

10 Where to look for further information

- Cline, M., Denbow, M., Gilbert, E. and Dridi S. Food intake regulation in book "Sturkie's Avian Physiology". 7th Edition, p. 1405. Edited by Scanes, C. and Dridi, S. Elsevier, 2021. ISBN: 9780128197707.

- Dridi, S. Book "Nutritional Biochemistry: From the Classroom to the Research Bench". 1st Edition. p. 272. Edited by Dridi, S. Bentham Science Publisher, U.A.E., 2022. DOI: 10.2174/97898150515751220101.
- Dridi, S. Understanding feed and water intake in poultry in book "Achieving sustainable production of poultry meat". Vol. 2, 1st Edition, p. 342. Edited by Applegate, T. Burleigh Dodds Science Publishing, London. 2017. https://doi.org/10.4324/9781351114295.
- Dridi, S. The use of nutrigenomics in poultry breeding for sustainable production in book "Advances in poultry genetics and genomics". 1st Edition, pp. 479–496. Edited by Aggrey, Zhou, Tixier-Boichard, and Rhoads. Burleigh Dodds Science Publishing, London. 2020. DOI: 10.19103/AS.2020.0065.27. https://www.ncbi.nlm.nih.gov/pubmed.

11 References

1. Tallentire, C. W., Leinonen, I. and Kyriazakis, I. Breeding for efficiency in the broilerchicken: a review. *Agron Sustain Dev* 36(4), 66 (2016).
2. Rubin, C. J., Zody, M. C., Eriksson, J., Meadows, J. R., Sherwood, E., Webster, M. T., Jiang, L., Ingman, M., Sharpe, T., Ka, S., Hallböök, F., Besnier, F., Carlborg, O., Bed'hom, B., Tixier-Boichard, M., Jensen, P., Siegel, P., Lindblad-Toh, K. and Andersson, L. Whole-genome resequencing reveals loci under selection during chicken domestication. *Nature* 464(7288), 587–591, doi:10.1038/nature08832 (2010).
3. Preisinger, R. Innovative layer genetics to handle global challenges in egg production. *Br Poult Sci* 59(1), 1–6, doi:10.1080/00071668.2018.1401828 (2018).
4. Shahbandeh, M. *Production of Meat Worldwide from 2016 to 2022 (in Million Metric Tons)* (vol. 194) (OECD, & FAO 2022).
5. Grossen, G. Poultry and Eggs. *USDA Economic Research Service* United States Department of Agriculture (2023). https://www.ers.usda.gov/topics/animal-products/poultry-eggs.
6. Steel, D., DesRoches, C. T. and Mintz-Woo, K. Climate change and the threat to civilization. *Proc Natl Acad Sci USA* 119(42), e2210525119, doi:10.1073/pnas.2210525119 (2022).
7. Tai, A. P. K., Martin, M. V. and Heald, C. L. Threat to future global food security from climate change and ozone air pollution. *Nat Clim Change* 4(9), 817–821 (2014).
8. Crew, B. Solving the energy crisis. *Nature* 609(7926) (Suppl.1), doi:10.1038/d41586-022-02830-5 (2022).
9. DeFries, R., Fanzo, J., Remans, R., Palm, C., Wood, S. and Anderman, T. L. Global nutrition. Metrics for land-scarce agriculture. *Science* 349(6245), 238–240, doi:10.1126/science.aaa5766 (2015).
10. He, C., Liu, Z., Wu, J., Pan, X., Fang, Z., Li, J. and Bryan, B. A. Future global urban water scarcity and potential solutions. *Nat Commun* 12(1), 4667, doi:10.1038/s41467-021-25026-3 (2021).
11. Jury, W. A. and Vaux, H., Jr. The role of science in solving the world's emerging water problems. *Proc Natl Acad Sci USA* 102(44), 15715–15720, doi:10.1073/pnas.0506467102 (2005).

12. Godde, C. M., Mason-D'Croz, D., Mayberry, D. E., Thornton, P. K. and Herrero, M. Impacts of climate change on the livestock food supply chain; a review of the evidence. *Glob Food Sec* 28, 100488, doi:10.1016/j.gfs.2020.100488 (2021).
13. Berthoud, H. R. Multiple neural systems controlling food intake and body weight. *Neurosci Biobehav Rev* 26(4), 393–428, doi:10.1016/s0149-7634(02)00014-3 (2002).
14. Bohler, M. W., Chowdhury, V. S., Cline, M. A. and Gilbert, E. R. Heat stress responses in birds: a review of the neural components. *Biology* 10(11), doi:10.3390/biology10111095 (2021), 1-17.
15. Gonzalez-Rivas, P. A., Chauhan, S. S., Ha, M., Fegan, N., Dunshea, F. R. and Warner, R. D. Effects of heat stress on animal physiology, metabolism, and meat quality: a review. *Meat Sci* 162, 108025, doi:10.1016/j.meatsci.2019.108025 (2020).
16. Marder, J. and Arad, Z. Panting and acid-base regulation in heat stressed birds. *Comp Biochem Physiol A Comp Physiol* 94(3), 395–400, doi:10.1016/0300-9629(89)90112-6 (1989).
17. Lara, L. J. and Rostagno, M. H. Impact of heat stress on poultry production. *Animals* 3(2), 356–369, doi:10.3390/ani3020356 (2013).
18. Webber, H., Ewert, F., Olesen, J. E., Müller, C., Fronzek, S., Ruane, A. C., Bourgault, M., Martre, P., Ababaei, B., Bindi, M., Ferrise, R., Finger, R., Fodor, N., Gabaldón-Leal, C., Gaiser, T., Jabloun, M., Kersebaum, K. C., Lizaso, J. I., Lorite, I. J., Manceau, L., Moriondo, M., Nendel, C., Rodríguez, A., Ruiz-Ramos, M., Semenov, M. A., Siebert, S., Stella, T., Stratonovitch, P., Trombi, G. and Wallach, D. Diverging importance of drought stress for maize and winter wheat in Europe. *Nat Commun* 9(1), 4249, doi:10.1038/s41467-018-06525-2 (2018).
19. Wang, X. and Liu, F. Effects of elevated CO(2) and heat on wheat grain quality. *Plants* 10(5), doi:10.3390/plants10051027 (2021).
20. Makkar, H. P. S. Review: feed demand landscape and implications of food-not feed strategy for food security and climate change. *Animal* 12(8), 1744–1754, doi:10.1017/S175173111700324X (2018).
21. Ledvinka, H. D., Toghyani, M., Tan, D. K. Y., Khoddami, A., Godwin, I. D. and Liu, S. Y. The impact of drought, heat and elevated carbon dioxide levels on feed grain quality for poultry production. *Agriculture* 12(11), 1913–1929 (2022).
22. Siegel, P. B. Evolution of the modern broiler and feed efficiency. *Annu Rev Anim Biosci* 2, 375–385, doi:10.1146/annurev-animal-022513-114132 (2014).
23. Chan, I., Franks, B. and Hayek, M. N. The 'sustainability gap' of US broiler chicken production: trade-offs between welfare, land use and consumption. *R Soc Open Sci* 9(6), 210478, doi:10.1098/rsos.210478 (2022).
24. Minokoshi, Y., Alquier, T., Furukawa, N., Kim, Y. B., Lee, A., Xue, B., Mu, J., Foufelle, F., Ferré, P., Birnbaum, M. J., Stuck, B. J. and Kahn, B. B. AMP-kinase regulates food intake by responding to hormonal and nutrient signals in the hypothalamus. *Nature* 428(6982), 569–574, doi:10.1038/nature02440 (2004).
25. Xue, B. and Kahn, B. B. AMPK integrates nutrient and hormonal signals to regulate food intake and energy balance through effects in the hypothalamus and peripheral tissues. *J Physiol* 574(1), 73–83, doi:10.1113/jphysiol.2006.113217 (2006).
26. Verhagen, J. V. The neurocognitive bases of human multimodal food perception: consciousness. *Brain Res Rev* 53(2), 271–286, doi:10.1016/j.brainresrev.2006.09.002 (2007).
27. Dridi, S. *Achieving Sustainable Production of Poultry Meat* (vol. 2) (ed T. Applegate), Ch. 5, 319 (Burleigh Dodds Science Publishing 2017).

28. Dridi, S. *Nutritional Biochemistry: from the Classroom to the Research Bench* (ed S. Dridi) Ch. 3, 263 (Bentham Science Publishers 2022).
29. Cline, M. A., Denbow, M., Gilbert, E. and Dridi, S. *Sturkie's Avian Physiology* (eds C. G. Whittow and S. Dridi) Ch. 28 (Elsevier 2022).
30. Denbow, D. M. Peripheral regulation of food intake in poultry. *J Nutr* 124(8) (Suppl.), 1349S–1354S, doi:10.1093/jn/124.suppl_8.1349S (1994).
31. Richards, M. P. and Proszkowiec-Weglarz, M. Mechanisms regulating feed intake, energy expenditure, and body weight in poultry. *Poult Sci* 86(7), 1478–1490, doi:10.1093/ps/86.7.1478 (2007).
32. Mayer, J. and Thomas, D. W. Regulation of food intake and obesity. *Science* 156(3773), 328–337, doi:10.1126/science.156.3773.328 (1967).
33. Kristensen, S. T. Social and cultural perspectives on hunger, appetite and satiety. *Eur J Clin Nutr* 54(6), 473–478, doi:10.1038/sj.ejcn.1601015 (2000).
34. El-Tahawy, A. A. S., Taha, A. E. and Adel, S. A. Effect of flock size on the productive and economic efficiency of Ross 308 and Cobb 500 broilers. *Europ Poult Sci* 81, 1–10 (2017).
35. Simsek, U. G. and Ozhan, N. Effects of flock size in broilers reared in a floor system on performance, some blood parameters, bone quality and musculus pectoralis pH level. *Ann Anim Sci* 15(2), 505–516 (2015).
36. Ghosh, S., Majumder, D. and Goswami, R. Broiler performance at different stocking density. *Indian J Anim Res* 46, 381–384 (2012).
37. Sarıca, M., Karakoç, K. and Erensoy, K. Effects of varying group sizes on performance, body defects, and productivity in broiler chickens. *Arch Anim Breed* 65(2), 171–181 (2022).
38. Ginane, C., Baumont, R. and Favreau-Peigne, A. Perception and hedonic value of basic tastes in domestic ruminants. *Physiol Behav* 104(5), 666–674, doi:10.1016/j.physbeh.2011.07.011 (2011).
39. Buijs, S., Keeling, L., Rettenbacher, S., Van Poucke, E. and Tuyttens, F. A. Stocking density effects on broiler welfare: identifying sensitive ranges for different indicators. *Poult Sci* 88(8), 1536–1543, doi:10.3382/ps.2009-00007 (2009).
40. Rosenthal, N. E., Genhart, M., Jacobsen, F. M., Skwerer, R. G. and Wehr, T. A. Disturbances of appetite and weight regulation in seasonal affective disorder. *Ann N Y Acad Sci* 499, 216–230, doi:10.1111/j.1749-6632.1987.tb36213.x (1987).
41. Rodrigues, I. and Choct, M. Feed intake pattern of broiler chickens under intermittent lighting: do birds eat in the dark? *Anim Nutr* 5(2), 174–178, doi:10.1016/j.aninu.2018.12.002 (2019).
42. Downs, K. M., Lien, R. J., Hess, J. B., Bilgili, S. F. and Dozier, W. A. The effects of photoperiod length, LightIntensity, and feed energy on GrowthResponses and meat yield of broilers. *J Appl Poult Res* 15(3), 406–416 (2006).
43. Scheideler, S. E. Research Note: effect of various lightsources on broiler performance and efficiency of production undercommercial conditions. *Poult Sci* 69(6), 1030–1033 (1990).
44. Newberry, R. C., Hunt, J. R. and Gardiner, E. E. Light intensity effects on performance, activity, leg disorders, and sudden death syndrome of roaster chickens. *Poult Sci* 65(12), 2232–2238, doi:10.3382/ps.0652232 (1986).
45. Olanrewaju, H. A., Purswell, J. L., Maslin, W. R., Collier, S. D. and Branton, S. L. Effects of color temperatures (kelvin) of LED bulbs on growth performance, carcass

characteristics, and ocular development indices of broilers grown to heavy weights. *Poult Sci* 94(3), 338–344, doi:10.3382/ps/peu082 (2015).
46. Cao, J., Liu, W., Wang, Z., Xie, D., Jia, L. and Chen, Y. Green and blue monochromatic lights promote growth and development of broilers via stimulating testosterone secretion and myofiber growth. *J Appl Poult Res* 17(2), 211–218 (2008).
47. Mohamed, R. A. and Eltholth, M. M., Saidy El, N. Rearing broiler chickens under monochromatic blue light improve performance and reduce fear and stress during pre-slaughter handling and transportation. *Biotechnol Anim Husbandry* 30, 457–471 (2014).
48. Yang, Y., Yu, Y., Pan, J., Ying, Y. and Zhou, H. A new method to manipulate broiler chicken growth and metabolism: response to mixed LED light system. *Sci Rep* 6, 25972, doi:10.1038/srep25972 (2016).
49. Breuner, C. W., Wingfield, J. C. and Romero, L. M. Diel rhythms of basal and stress-induced corticosterone in a wild, seasonal vertebrate, Gambel's white-crowned sparrow. *J Exp Zool* 284(3), 334–342 (1999).
50. Boissin, J. and Assenmacher, I. Circadian rhythms in adrenal cortical activity in the quail. *J Interdiscip Cycle Res* 1(3), 251–265 (1970).
51. Bailey, S. L. and Heitkemper, M. M. Circadian rhythmicity of cortisol and body temperature: morningness-eveningness effects. *Chronobiol Int* 18(2), 249–261, doi:10.1081/cbi-100103189 (2001).
52. Reppert, S. M. and Weaver, D. R. Coordination of circadian timing in mammals. *Nature* 418(6901), 935–941, doi:10.1038/nature00965 (2002).
53. Gekakis, N., Staknis, D., Nguyen, H. B., Davis, F. C., Wilsbacher, L. D., King, D. P., Takahashi, J. S. and Weitz, C. J. Role of the CLOCK protein in the mammalian circadian mechanism. *Science* 280(5369), 1564–1569, doi:10.1126/science.280.5369.1564 (1998).
54. Kume, K., Zylka, M. J., Sriram, S., Shearman, L. P., Weaver, D. R., Jin, X., Maywood, E. S., Hastings, M. H. and Reppert, S. M. mCRY1 and mCRY2 are essential components of the negative limb of the circadian clock feedback loop. *Cell* 98(2), 193–205, doi:10.1016/s0092-8674(00)81014-4 (1999).
55. Mohawk, J. A., Green, C. B. and Takahashi, J. S. Central and peripheral circadian clocks in mammals. *Annu Rev Neurosci* 35, 445–462, doi:10.1146/annurev-neuro-060909-153128 (2012).
56. McNamara, P., Seo, S. B., Rudic, R. D., Sehgal, A., Chakravarti, D. and FitzGerald, G. A. Regulation of CLOCK and MOP4 by nuclear hormone receptors in the vasculature: a humoral mechanism to reset a peripheral clock. *Cell* 105(7), 877–889, doi:10.1016/s0092-8674(01)00401-9 (2001).
57. Albrecht, U., Sun, Z. S., Eichele, G. and Lee, C. C. A differential response of two putative mammalian circadian regulators, mper1 and mper2, to light. *Cell* 91(7), 1055–1064, doi:10.1016/S0092-8674(00)80495-X (1997).
58. Shigeyoshi, Y., Taguchi, K., Yamamoto, S., Takekida, S., Yan, L., Tei, H., Moriya, T., Shibata, S., Loros, J. J., Dunlap, J. C. and Okamura, H. Light-induced resetting of a mammalian circadian clock is associated with rapid induction of the mPer1 transcript. *Cell* 91(7), 1043–1053, doi:10.1016/s0092-8674(00)80494-8 (1997).
59. Schwartz, W. J., Tavakoli-Nezhad, M., Lambert, C. M., Weaver, D. R. and de la Iglesia, H. O. Distinct patterns of Period gene expression in the suprachiasmatic nucleus underlie circadian clock photoentrainment by advances or delays. *Proc Natl Acad Sci USA* 108(41), 17219–17224, doi:10.1073/pnas.1107848108 (2011).

60. Scheer, F. A., Morris, C. J. and Shea, S. A. The internal circadian clock increases hunger and appetite in the evening independent of food intake and other behaviors. *Obesity* 21(3), 421–423, doi:10.1002/oby.20351 (2013).
61. Kalsbeek, A., Fliers, E., Romijn, J. A., La Fleur, S. E., Wortel, J., Bakker, O., Endert, E. and Buijs, R. M. The suprachiasmatic nucleus generates the diurnal changes in plasma leptin levels. *Endocrinology* 142(6), 2677–2685, doi:10.1210/endo.142.6.8197 (2001).
62. Ruiter, M., La Fleur, S. E., van Heijningen, C., van der Vliet, J., Kalsbeek, A. and Buijs, R. M. The daily rhythm in plasma glucagon concentrations in the rat is modulated by the biological clock and by feeding behavior. *Diabetes* 52(7), 1709–1715, doi:10.2337/diabetes.52.7.1709 (2003).
63. Jhanwar-Uniyal, M., Beck, B., Burlet, C. and Leibowitz, S. F. Diurnal rhythm of neuropeptide Y-like immunoreactivity in the suprachiasmatic, arcuate and paraventricular nuclei and other hypothalamic sites. *Brain Res* 536(1–2), 331–334, doi:10.1016/0006-8993(90)90045-d (1990).
64. Xu, B., Kalra, P. S., Farmerie, W. G. and Kalra, S. P. Daily changes in hypothalamic gene expression of neuropeptide Y, galanin, proopiomelanocortin, and adipocyte leptin gene expression and secretion: effects of food restriction. *Endocrinology* 140(6), 2868–2875, doi:10.1210/endo.140.6.6789 (1999).
65. Turek, F. W., Joshu, C., Kohsaka, A., Lin, E., Ivanova, G., McDearmon, E., Laposky, A., Losee-Olson, S., Easton, A., Jensen, D. R., Eckel, R. H., Takahashi, J. S. and Bass, J. Obesity and metabolic syndrome in circadian Clock mutant mice. *Science* 308(5724), 1043–1045, doi:10.1126/science.1108750 (2005).
66. Grimaldi, B., Bellet, M. M., Katada, S., Astarita, G., Hirayama, J., Amin, R. H., Granneman, J. G., Piomelli, D., Leff, T. and Sassone-Corsi, P. PER2 controls lipid metabolism by direct regulation of PPARgamma. *Cell Metab* 12(5), 509–520, doi:10.1016/j.cmet.2010.10.005 (2010).
67. Yang, S., Liu, A., Weidenhammer, A., Cooksey, R. C., McClain, D., Kim, M. K., Aguilera, G., Abel, E. D. and Chung, J. H. The role of mPer2 clock gene in glucocorticoid and feeding rhythms. *Endocrinology* 150(5), 2153–2160, doi:10.1210/en.2008-0705 (2009).
68. Marcheva, B., Ramsey, K. M., Buhr, E. D., Kobayashi, Y., Su, H., Ko, C. H., Ivanova, G., Omura, C., Mo, S., Vitaterna, M. H., Lopez, J. P., Philipson, L. H., Bradfield, C. A., Crosby, S. D., JeBailey, L., Wang, X., Takahashi, J. S. and Bass, J. Disruption of the clock components CLOCK and BMAL1 leads to hypoinsulinaemia and diabetes. *Nature* 466(7306), 627–631, doi:10.1038/nature09253 (2010).
69. McDearmon, E. L., Patel, K. N., Ko, C. H., Walisser, J. A., Schook, A. C., Chong, J. L., Wilsbacher, L. D., Song, E. J., Hong, H. K., Bradfield, C. A. and Takahashi, J. S. Dissecting the functions of the mammalian clock protein BMAL1 by tissue-specific rescue in mice. *Science* 314(5803), 1304–1308, doi:10.1126/science.1132430 (2006).
70. Paschos, G. K., Ibrahim, S., Song, W. L., Kunieda, T., Grant, G., Reyes, T. M., Bradfield, C. A., Vaughan, C. H., Eiden, M., Masoodi, M., Griffin, J. L., Wang, F., Lawson, J. A. and Fitzgerald, G. A. Obesity in mice with adipocyte-specific deletion of clock component Arntl. *Nat Med* 18(12), 1768–1777, doi:10.1038/nm.2979 (2012).
71. Mieda, M. and Sakurai, T. Bmal1 in the nervous system is essential for normal adaptation of circadian locomotor activity and food intake to periodic feeding. *J Neurosci* 31(43), 15391–15396, doi:10.1523/JNEUROSCI.2801-11.2011 (2011).

72. Wang, X., Liu, Z., Zhao, J., Jiao, H. and Lin, H. Dusk feeding in laying hens is shifted by light program via involvement of clock genes. *J Anim Physiol Anim Nutr* 105(6), 1103–1112 (2021).
73. Honda, K., Kondo, M., Hiramoto, D., Saneyasu, T. and Kamisoyama, H. Effects of continuous white light and 12 h white-12 h blue light-cycles on the expression of clock genes in diencephalon, liver, and skeletal muscle in chicks. *Comp Biochem Physiol A* 207, 73–78 (2017).
74. Yi-lin, L., Zhang,K., Dan, W. and Xin, Z. Dynamic feeding low and high methionine diets affect the diurnal rhythm of amino acid transporters and clock related genes in jejunum of laying hens. *Biol Rhythm Res* 49, 671–679 (2017).
75. Kagan, J. (2016). Why stress remains an ambiguous concept: reply to McEwen & McEwen and Kagan, J. *Perspect Psychol Sci* 11(4), 464–465, doi:10.1177/1745691616649952 (2016).
76. Selye, H. *The Stress of Life*, 1–313 (McGraw-Hill Book Company 1950).
77. Bernard, C. *An Introduction to the Study of Experimental Medicine* (Dover Publications, INC. 1957), 1–196.
78. Cannon, W. B. *Bodily Changes in Pain, Hunger, Fear, and Rage: an Account of Recent Researches into the Function of Emotional Excitement*, 1–305 (D. Appleton and Company 1925).
79. de Kloet, E. R., Karst, H. and Joels, M. Corticosteroid hormones in the central stress response: quick-and-slow. *Front Neuroendocrinol* 29(2), 268–272, doi:10.1016/j.yfrne.2007.10.002 (2008).
80. Talling, J. C., Waran, N. K., Wathes, C. M. and Lines, J. A. Sound avoidance by domestic pigs depends upon characteristics of the signal. *Appl Anim Behav Sci* 58(3–4), 255–266 (1998).
81. Waynert, D. F., Stookey, J. M., Schwartkopf-Genswein, K. S., Watts, J. M. and Waltz, C. S. The response of beef cattle to noise during handling. *Appl Anim Behav Sci* 62(1), 27–42 (1999).
82. McAdie, T. M., Foster, T. M., Temple, W. and Matthews, L. R. A method for measuring the aversiveness of sounds to domestic hens. *Appl Anim Behav Sci* 37(3), 223–238 (1993).
83. Chloupek, P., Voslářová, E., Chloupek, J., Bedáňová, I., Pištěková, V. and Veče, V. Stress in broiler chickens due to acute noise exposure. *Acta Vet* 78(1), 93–98 (2009).
84. Stroebele, N. and de Castro, J. M. Listening to music while eating is related to increases in people's food intake and meal duration. *Appetite* 47(3), 285–289, doi:10.1016/j.appet.2006.04.001 (2006).
85. Christensen, A. C. and Knight, A. D. Observations on the effects of music exposure to growing performance of meat-type chicks. *Poult Sci* 54(2), 619–621 (1975).
86. Gvaryahu, G., Cunningham, D. L. and Van Tienhoven, A. Filial imprinting, environmental enrichment, and music application effects on behavior and performance of meat strain chicks. *Poult Sci* 68(2), 211–217 (1989).
87. Ladd, J. K., Albright, J. L., Beck, A. M. and Ladd, B. T. Behavioural and physiological studies on the effect of music on animals. *J Anim Sci* 70, 170 (1992).
88. Campo, J. L., Gil, M. G. and Dávila, S. G. Effects of specific noise and music stimuli on stress and fear levels of laying hens of several breeds. *Appl Anim Behav Sci* 91(1–2), 75–84 (2005).
89. North, A. C. The effect of background music on the taste of wine. *Br J Psychol* 103(3), 293–301, doi:10.1111/j.2044-8295.2011.02072.x (2012).

90. Spence, C. Auditory contributions to flavour perception and feeding behaviour. *Physiol Behav* 107(4), 505–515, doi:10.1016/j.physbeh.2012.04.022 (2012).
91. Brobeck, J. R. Food intake as a mechanism of temperature regulation. *Yale J Biol Med* 20(6), 545–552 (1948).
92. Emami, N. K., Greene, E. S., Kogut, M. H. and Dridi, S. Heat stress and feed restriction distinctly affect performance, carcass and meat yield, intestinal integrity, and inflammatory (chemo)cytokines in broiler chickens. *Front Physiol* 12, 707757, doi:10.3389/fphys.2021.707757 (2021).
93. Brugaletta, G., Greene, E., Tabler, T., Orlowski, S., Sirri, F. and Dridi, S. Effect of cyclic heat stress on feeding-related hypothalamic neuropeptides of three broiler populations and their ancestor jungle fowl. *Front Physiol* 12, 809341, doi:10.3389/fphys.2021.809341 (2021).
94. Baxter, M. F. A., Greene, E. S., Kidd, M. T., Tellez-Isaias, G., Orlowski, S. and Dridi, S. Water amino acid-chelated trace mineral supplementation decreases circulating and intestinal HSP70 and proinflammatory cytokine gene expression in heat-stressed broiler chickens. *J Anim Sci* 98(3), doi:10.1093/jas/skaa049 (2020).
95. Prinzinger, R., Preßmar, A. and Schleucher, E. Body temperature in birds. *Comp Biochem Physiol A Comp Physiol* 99(4), 499–506 (1991).
96. Buhr, E. D., Yoo, S. H. and Takahashi, J. S. Temperature as a universal resetting cue for mammalian circadian oscillators. *Science* 330(6002), 379–385, doi:10.1126/science.1195262 (2010).
97. Tzschentke, B. and Basta, D. Early development of neuronal hypothalamic thermosensitivity in birds: influence of epigenetic temperature adaptation. *Comp Biochem Physiol A Mol Integr Physiol* 131(4), 825–832, doi:10.1016/s1095-6433(02)00020-x (2002).
98. Nichelmann, M. and Tzschentke, B. Ontogeny of thermoregulation during the prenatal period in birds. *Ann NY Acad Sci* 813, 78–86, doi:10.1111/j.1749-6632.1997.tb51676.x (1997).
99. Ruuskanen, S., Hsu, B. Y. and Nord, A. Endocrinology of thermoregulation in birds in a changing climate. *Mol Cell Endocrinol* 519, 111088, doi:10.1016/j.mce.2020.111088 (2021).
100. Simon, E., Martin, R. and Simon-Oppermann, C. Central nervous versus total body thermosensitivity of the duck. *Int J Biometeorol* 25(3), 249–256, doi:10.1007/BF02184526 (1981).
101. Tan, C. L. and Knight, Z. A. Regulation of body temperature by the nervous system. *Neuron* 98(1), 31–48, doi:10.1016/j.neuron.2018.02.022 (2018).
102. Nakamura, K. and Morrison, S. F. A thermosensory pathway that controls body temperature. *Nat Neurosci* 11(1), 62–71, doi:10.1038/nn2027 (2008).
103. Basta, D., Tzschentke, B. and Nichelmann, M. Temperature guardian neurons in the preoptic area of the hypothalamus. *Brain Res* 767(2), 361–362, doi:10.1016/s0006-8993(97)00499-x (1997).
104. Caterina, M. J., Schumacher, M. A., Tominaga, M., Rosen, T. A., Levine, J. D. and Julius, D. The capsaicin receptor: a heat-activated ion channel in the pain pathway. *Nature* 389(6653), 816–824, doi:10.1038/39807 (1997).
105. Cao, E. Structural mechanisms of transient receptor potential ion channels. *J Gen Physiol* 152(3), doi:10.1085/jgp.201811998 (2020).
106. Saito, S., Banzawa, N., Fukuta, N., Saito, C. T., Takahashi, K., Imagawa, T., Ohta, T. and Tominaga, M. Heat and noxious chemical sensor, chicken TRPA1, as a target of bird

repellents and identification of its structural determinants by multispecies functional comparison. *Mol Biol Evol* 31(3), 708–722, doi:10.1093/molbev/msu001 (2014).

107. Mahmoud, M. E., Shimizu, Y., Shiina, T., Nikami, H., Dosoky, R. M., Ahmed, M. M. and Takewaki, T. Involvement of a capsaicin-sensitive TRPV1-independent mechanism in lipopolysaccharide-induced fever in chickens. *Comp Biochem Physiol A Mol Integr Physiol* 148(3), 578–583, doi:10.1016/j.cbpa.2007.07.012 (2007).
108. Cavanaugh, D. J., Chesler, A. T., Jackson, A. C., Sigal, Y. M., Yamanaka, H., Grant, R., O'Donnell, D., Nicoll, R. A., Shah, N. M., Julius, D. and Basbaum, A. I. Trpv1 reporter mice reveal highly restricted brain distribution and functional expression in arteriolar smooth muscle cells. *J Neurosci* 31(13), 5067–5077, doi:10.1523/JNEUROSCI.6451-10.2011 (2011).
109. Molinas, A. J. R., Desmoulins, L. D., Hamling, B. V., Butcher, S. M., Anwar, I. J., Miyata, K., Enix, C. L., Dugas, C. M., Satou, R., Derbenev, A. V. and Zsombok, A. Interaction between TRPV1-expressing neurons in the hypothalamus. *J Neurophysiol* 121(1), 140–151, doi:10.1152/jn.00004.2018 (2019).
110. Mezey, E., Tóth, Z. E., Cortright, D. N., Arzubi, M. K., Krause, J. E., Elde, R., Guo, A., Blumberg, P. M. and Szallasi, A. Distribution of mRNA for vanilloid receptor subtype 1 (VR1), and VR1-like immunoreactivity, in the central nervous system of the rat and human. *Proc Natl Acad Sci USA* 97(7), 3655–3660, doi:10.1073/pnas.97.7.3655 (2000).
111. Jeong, J. H., Lee, D. K., Liu, S. M., Chua, S. C., Schwartz, G. J. and Jo, Y. H. Activation of temperature-sensitive TRPV1-like receptors in ARC POMC neurons reduces food intake. *PLoS Biol* 16(4), e2004399, doi:10.1371/journal.pbio.2004399 (2018).
112. Janssens, P. L., Hursel, R. and Westerterp-Plantenga, M. S. Capsaicin increases sensation of fullness in energy balance, and decreases desire to eat after dinner in negative energy balance. *Appetite* 77, 44–49, doi:10.1016/j.appet.2014.02.018 (2014).
113. Ludy, M. J. and Mattes, R. D. The effects of hedonically acceptable red pepper doses on thermogenesis and appetite. *Physiol Behav* 102(3–4), 251–258, doi:10.1016/j.physbeh.2010.11.018 (2011).
114. Westerterp-Plantenga, M. S., Smeets, A. and Lejeune, M. P. Sensory and gastrointestinal satiety effects of capsaicin on food intake. *Int J Obes* 29(6), 682–688, doi:10.1038/sj.ijo.0802862 (2005).
115. Christie, S., Wittert, G. A., Li, H. and Page, A. J. Involvement of TRPV1 channels in energy homeostasis. *Front Endocrinol* 9, 420, doi:10.3389/fendo.2018.00420 (2018).
116. Langlet, F., Mullier, A., Bouret, S. G., Prevot, V. and Dehouck, B. Tanycyte-like cells form a blood-cerebrospinal fluid barrier in the circumventricular organs of the mouse brain. *J Comp Neurol* 521(15), 3389–3405, doi:10.1002/cne.23355 (2013).
117. Kroeger, D., Absi, G., Gagliardi, C., Bandaru, S. S., Madara, J. C., Ferrari, L. L., Arrigoni, E., Münzberg, H., Scammell, T. E., Saper, C. B. and Vetrivelan, R. Galanin neurons in the ventrolateral preoptic area promote sleep and heat loss in mice. *Nat Commun* 9(1), 4129, doi:10.1038/s41467-018-06590-7 (2018).
118. Yu, S., Qualls-Creekmore, E., Rezai-Zadeh, K., Jiang, Y., Berthoud, H. R., Morrison, C. D., Derbenev, A. V., Zsombok, A. and Münzberg, H. Glutamatergic preoptic area neurons that express leptin receptors drive temperature-dependent body weight homeostasis. *J Neurosci* 36(18), 5034–5046, doi:10.1523/JNEUROSCI.0213-16.2016 (2016).

119. Meister, B. Control of food intake via leptin receptors in the hypothalamus. *Vitam Horm* 59, 265–304, doi:10.1016/s0083-6729(00)59010-4 (2000).
120. Bergonzelli, G. E., Pralong, F. P., Glauser, M., Cavadas, C., Grouzmann, E. and Gaillard, R. C. Interplay between galanin and leptin in the hypothalamic control of feeding via corticotropin-releasing hormone and neuropeptide Y. *Diabetes* 50(12), 2666–2672, doi:10.2337/diabetes.50.12.2666 (2001).
121. Sherin, J. E., Shiromani, P. J., McCarley, R. W. and Saper, C. B. Activation of ventrolateral preoptic neurons during sleep. *Science* 271(5246), 216–219, doi:10.1126/science.271.5246.216 (1996).
122. Yonemochi, N., Ardianto, C., Ueda, D., Kamei, J. and Ikeda, H. GABAergic function in the lateral hypothalamus regulates feeding behavior: possible mediation via orexin. *Neuropsychopharmacol Rep* 39(4), 289–296, doi:10.1002/npr2.12080 (2019).
123. Yang, S., Tan, Y. L., Wu, X., Wang, J., Sun, J., Liu, A., Gan, L., Shen, B., Zhang, X., Fu, Y. and Huang, J. An mPOA-ARC(AgRP) pathway modulates cold-evoked eating behavior. *Cell Rep* 36(6), 109502, doi:10.1016/j.celrep.2021.109502 (2021).
124. Spector, N. H., Brobeck, J. R. and Hamilton, C. L. Feeding and core temperature in albino rats: changes induced by preoptic heating and cooling. *Science* 161(3838), 286–288, doi:10.1126/science.161.3838.286 (1968).
125. Hamilton, C. L. and Brobeck, J. R. Food intake and temperature regulation in rats with rostral hypothalamic lesions. *Am J Physiol* 207, 291–297, doi:10.1152/ajplegacy.1964.207.2.291 (1964).
126. Hrvatin, S., Sun, S., Wilcox, O. F., Yao, H., Lavin-Peter, A. J., Cicconet, M., Assad, E. G., Palmer, M. E., Aronson, S., Banks, A. S., Griffith, E. C. and Greenberg, M. E. Neurons that regulate mouse torpor. *Nature* 583(7814), 115–121, doi:10.1038/s41586-020-2387-5 (2020).
127. Qian, S., Yan, S., Pang, R., Zhang, J., Liu, K., Shi, Z., Wang, Z., Chen, P., Zhang, Y., Luo, T., Hu, X., Xiong, Y. and Zhou, Y. A temperature-regulated circuit for feeding behavior. *Nat Commun* 13(1), 4229, doi:10.1038/s41467-022-31917-w (2022).
128. Santollo, J., Torregrossa, A. M. and Eckel, L. A. Estradiol acts in the medial preoptic area, arcuate nucleus, and dorsal raphe nucleus to reduce food intake in ovariectomized rats. *Horm Behav* 60(1), 86–93, doi:10.1016/j.yhbeh.2011.03.009 (2011).
129. Kim, J. S., Rizwan, M. Z., Clegg, D. J. and Anderson, G. M. Leptin signaling is not required for anorexigenic estradiol effects in female mice. *Endocrinology* 157(5), 1991–2001, doi:10.1210/en.2015-1594 (2016).
130. Wang, D., He, X., Zhao, Z., Feng, Q., Lin, R., Sun, Y., Ding, T., Xu, F., Luo, M. and Zhan, C. Whole-brain mapping of the direct inputs and axonal projections of POMC and AgRP neurons. *Front Neuroanat* 9, 40, doi:10.3389/fnana.2015.00040 (2015).
131. Scammell, T., Gerashchenko, D., Urade, Y., Onoe, H., Saper, C. and Hayaishi, O. Activation of ventrolateral preoptic neurons by the somnogen prostaglandin D2. *Proc Natl Acad Sci USA* 95(13), 7754–7759, doi:10.1073/pnas.95.13.7754 (1998).
132. Gerald, C., Walker, M. W., Criscione, L., Gustafson, E. L., Batzl-Hartmann, C., Smith, K. E., Vaysse, P., Durkin, M. M., Laz, T. M., Linemeyer, D. L., Schaffhauser, A. O., Whitebread, S., Hofbauer, K. G., Taber, R. I., Branchek, T. A. and Weinshank, R. L. A receptor subtype involved in neuropeptide-Y-induced food intake. *Nature* 382(6587), 168–171, doi:10.1038/382168a0 (1996).

133. Tan, C. L., Cooke, E. K., Leib, D. E., Lin, Y. C., Daly, G. E., Zimmerman, C. A. and Knight, Z. A. Warm-sensitive neurons that control body temperature. *Cell* 167(1), 47–59 e15, doi:10.1016/j.cell.2016.08.028 (2016).
134. Xu, B., Goulding, E. H., Zang, K., Cepoi, D., Cone, R. D., Jones, K. R., Tecott, L. H. and Reichardt, L. F. Brain-derived neurotrophic factor regulates energy balance downstream of melanocortin-4 receptor. *Nat Neurosci* 6(7), 736–742, doi:10.1038/nn1073 (2003).
135. Maunze, B. and Choi, S. Hypothalamic PACAP/PAC1R involvement in feeding and body weight regulation. *Endocrinology* 164(5), doi:10.1210/endocr/bqad044 (2023).
136. Yao, M., Schulkin, J. and Denver, R. J. Evolutionarily conserved glucocorticoid regulation of corticotropin-releasing factor expression. *Endocrinology* 149(5), 2352–2360, doi:10.1210/en.2007-1551 (2008).
137. Spina, M., Merlo-Pich, E., Chan, R. K., Basso, A. M., Rivier, J., Vale, W. and Koob, G. F. Appetite-suppressing effects of urocortin, a CRF-related neuropeptide. *Science* 273(5281), 1561–1564, doi:10.1126/science.273.5281.1561 (1996).
138. Chung, S., Weber, F., Zhong, P., Tan, C. L., Nguyen, T. N., Beier, K. T., Hörmann, N., Chang, W. C., Zhang, Z., Do, J. P., Yao, S., Krashes, M. J., Tasic, B., Cetin, A., Zeng, H., Knight, Z. A., Luo, L. and Dan, Y. Identification of preoptic sleep neurons using retrograde labelling and gene profiling. *Nature* 545(7655), 477–481, doi:10.1038/nature22350 (2017).
139. Sturdevant, R. A. and Goetz, H. Cholecystokinin both stimulates and inhibits human food intake. *Nature* 261(5562), 713–715, doi:10.1038/261713a0 (1976).
140. Maguire, C. A., Leon, S., Carroll, R. S., Kaiser, U. B. and Navarro, V. M. Altered circadian feeding behavior and improvement of metabolic syndrome in obese Tac1-deficient mice. *Int J Obes* 41(12), 1798–1804, doi:10.1038/ijo.2017.185 (2017).
141. Schroeder, L. E. and Leinninger, G. M. Role of central neurotensin in regulating feeding: implications for the development and treatment of body weight disorders. *Biochim Biophys Acta Mol Basis Dis* 1864(3), 900–916, doi:10.1016/j.bbadis.2017.12.036 (2018).
142. Rogge, G., Jones, D., Hubert, G. W., Lin, Y. and Kuhar, M. J. CART peptides: regulators of body weight, reward and other functions. *Nat Rev Neurosci* 9(10), 747–758, doi:10.1038/nrn2493 (2008).
143. Mendez, I. A., Ostlund, S. B., Maidment, N. T. and Murphy, N. P. Involvement of endogenous enkephalins and beta-endorphin in feeding and diet-induced obesity. *Neuropsychopharmacology* 40(9), 2103–2112, doi:10.1038/npp.2015.67 (2015).
144. Lin, M. T., Chu, P. C. and Leu, S. Y. Effects of TSH, TRH, LH and LHRH on thermoregulation and food and water intake in the rat. *Neuroendocrinology* 37(3), 206–211, doi:10.1159/000123544 (1983).
145. Pei, H., Sutton, A. K., Burnett, K. H., Fuller, P. M. and Olson, D. P. AVP neurons in the paraventricular nucleus of the hypothalamus regulate feeding. *Mol Metab* 3(2), 209–215, doi:10.1016/j.molmet.2013.12.006 (2014).
146. Onaka, T. and Takayanagi, Y. Role of oxytocin in the control of stress and food intake. *J Neuroendocrinol* 31(3), e12700, doi:10.1111/jne.12700 (2019).
147. Talbot, J., Hahn, P., Kroehling, L., Nguyen, H., Li, D. and Littman, D. R. Feeding-dependent VIP neuron-ILC3 circuit regulates the intestinal barrier. *Nature* 579(7800), 575–580, doi:10.1038/s41586-020-2039-9 (2020).

148. Stengel, A. and Tache, Y. Central somatostatin signaling and regulation of food intake. *Ann N Y Acad Sci* 1455(1), 98–104, doi:10.1111/nyas.14178 (2019).
149. Stengel, A., Wang, L., Goebel-Stengel, M. and Tache, Y. Centrally injected kisspeptin reduces food intake by increasing meal intervals in mice. *NeuroReport* 22(5), 253–257, doi:10.1097/WNR.0b013e32834558df (2011).
150. Kalil, B., Ribeiro, A. B., Leite, C. M., Uchôa, E. T., Carolino, R. O., Cardoso, T. S., Elias, L. L., Rodrigues, J. A., Plant, T. M., Poletini, M. O. and Anselmo-Franci, J. A. The increase in signaling by kisspeptin neurons in the preoptic area and associated changes in clock gene expression that trigger the LH surge in female rats are dependent on the facilitatory action of a noradrenaline input. *Endocrinology* 157(1), 323–335, doi:10.1210/en.2015-1323 (2016).
151. Tsuneoka, Y., Yoshida, S., Takase, K., Oda, S., Kuroda, M. and Funato, H. Neurotransmitters and neuropeptides in gonadal steroid receptor-expressing cells in medial preoptic area subregions of the male mouse. *Sci Rep* 7(1), 9809, doi:10.1038/s41598-017-10213-4 (2017).
152. Herget, U. and Ryu, S. Coexpression analysis of nine neuropeptides in the neurosecretory preoptic area of larval zebrafish. *Front Neuroanat* 9, 2, doi:10.3389/fnana.2015.00002 (2015).
153. Tu, W. L., Cheng, C. Y., Wang, S. H., Tang, P. C., Chen, C. F., Chen, H. H., Lee, Y. P., Chen, S. E. and Huang, S. Y. Profiling of differential gene expression in the hypothalamus of broiler-type Taiwan country chickens in response to acute heat stress. *Theriogenology* 85(3), 483–494 e488, doi:10.1016/j.theriogenology.2015.09.028 (2016).
154. Bohler, M., Gilbert, E. R. and Cline, M. A. Reduced food intake during exposure to high ambient temperatures is associated with molecular changes in the nucleus of the hippocampal commissure and the paraventricular and arcuate hypothalamic nuclei. *Gen Comp Endocrinol* 298, 113576, doi:10.1016/j.ygcen.2020.113576 (2020).
155. Lei, L., Hepeng, L., Xianlei, L., Hongchao, J., Hai, L., Sheikhahmadi, A., Yufeng, W. and Zhigang, S. Effects of acute heat stress on gene expression of brain-gut neuropeptides in broiler chickens. *J Anim Sci* 91(11), 5194–5201, doi:10.2527/jas.2013-6538 (2013).
156. Bahry, M. A., Chowdhury, V. S., Yang, H., Tran, P. V., Do, P. H., Han, G., Ikeda, H., Cockrem, J. F. and Furuse, M. Central administration of neuropeptide Y differentially regulates monoamines and corticosterone in heat-exposed fed and fasted chicks. *Neuropeptides* 62, 93–100, doi:10.1016/j.npep.2016.11.008 (2017).
157. Eltahan, H. M., Bahry, M. A., Yang, H., Han, G., Nguyen, L. T. N., Ikeda, H., Ali, M. N., Amber, K. A., Furuse, M. and Chowdhury, V. S. Central NPY-Y5 sub-receptor partially functions as a mediator of NPY-induced hypothermia and affords thermotolerance in heat-exposed fasted chicks. *Physiol Rep* 5(23), doi:10.14814/phy2.13511 (2017).
158. Nagarajan, G., Kang, S. W. and Kuenzel, W. J. Functional evidence that the nucleus of the hippocampal commissure shows an earlier activation from a stressor than the paraventricular nucleus: implication of an additional structural component of the avian hypothalamo-pituitary-adrenal axis. *Neurosci Lett* 642, 14–19, doi:10.1016/j.neulet.2017.01.064 (2017).
159. Montagnese, C. M., Szekely, T., Csillag, A. and Zachar, G. Distribution of vasotocin- and vasoactive intestinal peptide-like immunoreactivity in the brain of blue tit (Cyanistes coeruleus). *Front Neuroanat* 9, 90, doi:10.3389/fnana.2015.00090 (2015).

160. Tachibana, T., Saito, E. S., Saito, S., Tomonaga, S., Denbow, D. M. and Furuse, M. Comparison of brain arginine-vasotocin and corticotrophin-releasing factor for physiological responses in chicks. *Neurosci Lett* 360(3), 165–169, doi:10.1016/j.neulet.2004.02.055 (2004).
161. Wang, S., Bottje, W. G., Kinzler, S., Neldon, H. L. and Koike, T. I. Effect of heat stress on plasma levels of arginine vasotocin and mesotocin in domestic fowl (Gallus domesticus). *Comp Biochem Physiol A Comp Physiol* 93(4), 721–724, doi:10.1016/0300-9629(89)90492-1 (1989).
162. John, T. M. and George, J. C. Effects of arginine vasotocin on cardiorespiratory and thermoregulatory responses in the pigeon. *Comp Biochem Physiol C Comp Pharmacol Toxicol* 102(3), 353–359, doi:10.1016/0742-8413(92)90126-r (1992).
163. Calefi, A. S., Fonseca, J. G. D. S., Nunes, C. A. Q., Lima, A. P. N., Quinteiro-Filho, W. M., Flório, J. C., Zager, A., Ferreira, A. J. P. and Palermo-Neto, J. Heat stress modulates brain monoamines and their metabolites production in broiler chickens co-infected with clostridium perfringens type A and Eimeria spp. *Vet Sci* 6(1), doi:10.3390/vetsci6010004 (2019).
164. Tecott, L. H., Sun, L. M., Akana, S. F., Strack, A. M., Lowenstein, D. H., Dallman, M. F. and Julius, D. Eating disorder and epilepsy in mice lacking 5-HT2c serotonin receptors. *Nature* 374(6522), 542–546, doi:10.1038/374542a0 (1995).
165. Zendehdel, M., Sardari, F., Hassanpour, S., Rahnema, M., Adeli, A. and Ghashghayi, E. Serotonin-induced hypophagia is mediated via alpha(2) and beta(2) adrenergic receptors in neonatal layer-type chickens. *Br Poult Sci* 58(3), 298–304, doi:10.1080/00071668.2017.1278626 (2017).
166. Zendehdel, M., Hasani, K., Babapour, V., Mortezaei, S. S., Khoshbakht, Y. and Hassanpour, S. Dopamine-induced hypophagia is mediated by D1 and 5HT-2c receptors in chicken. *Vet Res Commun* 38(1), 11–19, doi:10.1007/s11259-013-9581-y (2014).
167. Denbow, D. M., Cherry, J. A., Siegel, P. B. and Van Krey, H. P. Eating, drinking and temperature response of chicks to brain catecholamine injections. *Physiol Behav* 27(2), 265–269, doi:10.1016/0031-9384(81)90268-7 (1981).
168. Kamisoyama, H., Honda, K., Saneyasu, T., Sugahara, K. and Hasegawa, S. Central administration of neuromedin U suppresses food intake in chicks. *Neurosci Lett* 420(1), 1–5, doi:10.1016/j.neulet.2007.03.062 (2007).
169. Masunari, K., Khan, M. S., Cline, M. A. and Tachibana, T. Central administration of mesotocin inhibits feeding behavior in chicks. *Regul Pept* 187, 1–5, doi:10.1016/j.regpep.2013.10.004 (2013).
170. Ao, Y., Go, V. L., Toy, N., Li, T., Wang, Y., Song, M. K., Reeve, J. R., Liu, Y. and Yang, H. Brainstem thyrotropin-releasing hormone regulates food intake through vagal-dependent cholinergic stimulation of ghrelin secretion. *Endocrinology* 147(12), 6004–6010, doi:10.1210/en.2006-0820 (2006).
171. Brugaletta, G., Greene, E., Ramser, A., Maynard, C. W., Tabler, T. W., Sirri, F., Anthony, N. B., Orlowski, S. and Dridi, S. Effect of cyclic heat stress on hypothalamic oxygen homeostasis and inflammatory state in the jungle fowl and three broiler-based research lines. *Front Vet Sci* 9, 905225, doi:10.3389/fvets.2022.905225 (2022).
172. Mishra, D., Richard, J. E., Maric, I., Porteiro, B., Häring, M., Kooijman, S., Musovic, S., Eerola, K., López-Ferreras, L., Peris, E., Grycel, K., Shevchouk, O. T., Micallef, P., Olofsson, C. S., Wernstedt Asterholm, I., Grill, H. J., Nogueiras, R. and Skibicka, K. P. Parabrachial interleukin-6 reduces body weight and food intake and increases

thermogenesis to regulate energy metabolism. *Cell Rep* 26(11), 3011–3026.e5, doi:10.1016/j.celrep.2019.02.044 (2019).
173. Lindquist, S. The heat-shock response. *Annu Rev Biochem* 55, 1151–1191, doi:10.1146/annurev.bi.55.070186.005443 (1986).
174. Dukay, B., Csoboz, B. and Toth, M. E. Heat-shock proteins in neuroinflammation. *Front Pharmacol* 10, 920, doi:10.3389/fphar.2019.00920 (2019).
175. Fujita, J. Cold shock response in mammalian cells. *J Mol Microbiol Biotechnol* 1(2), 243–255 (1999).
176. Pirkkala, L., Nykanen, P. and Sistonen, L. Roles of the heat shock transcription factors in regulation of the heat shock response and beyond. *FASEB J* 15(7), 1118–1131, doi:10.1096/fj00-0294rev (2001).
177. Laios, E., Rebeyka, I. M. and Prody, C. A. Characterization of cold-induced heat shock protein expression in neonatal rat cardiomyocytes. *Mol Cell Biochem* 173(1–2), 153–159, doi:10.1023/a:1006844114348 (1997).
178. Nguyen, P., Greene, E., Ishola, P., Huff, G., Donoghue, A., Bottje, W. and Dridi, S. Chronic mild cold conditioning modulates the expression of hypothalamic neuropeptide and intermediary metabolic-related genes and improves growth performances in young chicks. *PLoS ONE* 10(11), e0142319, doi:10.1371/journal.pone.0142319 (2015).
179. Greene, E. S., Adeogun, E., Orlowski, S. K., Nayani, K. and Dridi, S. Effects of heat stress on cyto(chemo)kine and inflammasome gene expression and mechanical properties in isolated red and white blood cells from 4 commercial broiler lines and their ancestor jungle fowl. *Poult Sci* 101(6), 101827, doi:10.1016/j.psj.2022.101827 (2022).
180. Chung, J., Nguyen, A. K., Henstridge, D. C., Holmes, A. G., Chan, M. H., Mesa, J. L., Lancaster, G. I., Southgate, R. J., Bruce, C. R., Duffy, S. J., Horvath, I., Mestril, R., Watt, M. J., Hooper, P. L., Kingwell, B. A., Vigh, L., Hevener, A. and Febbraio, M. A. HSP72 protects against obesity-induced insulin resistance. *Proc Natl Acad Sci USA* 105(5), 1739–1744, doi:10.1073/pnas.0705799105 (2008).
181. Sonna, L. A., Fujita, J., Gaffin, S. L. and Lilly, C. M. Invited review: effects of heat and cold stress on mammalian gene expression. *J Appl Physiol (1985)* 92(4), 1725–1742, doi:10.1152/japplphysiol.01143.2001 (2002).
182. Goekint, M., Roelands, B., Heyman, E., Njemini, R. and Meeusen, R. Influence of citalopram and environmental temperature on exercise-induced changes in BDNF. *Neurosci Lett* 494(2), 150–154, doi:10.1016/j.neulet.2011.03.001 (2011).
183. Fox, E. A. A genetic approach for investigating vagal sensory roles in regulation of gastrointestinal function and food intake. *Auton Neurosci* 126–127, 9–29, doi:10.1016/j.autneu.2006.03.005 (2006).
184. Egecioglu, E., Jerlhag, E., Salomé, N., Skibicka, K. P., Haage, D., Bohlooly-Y, M., Andersson, D., Bjursell, M., Perrissoud, D., Engel, J. A. and Dickson, S. L. Ghrelin increases intake of rewarding food in rodents. *Addict Biol* 15(3), 304–311, doi:10.1111/j.1369-1600.2010.00216.x (2010).
185. Geelissen, S. M., Swennen, Q., Geyten, S. V., Kühn, E. R., Kaiya, H., Kangawa, K., Decuypere, E., Buyse, J. and Darras, V. M. Peripheral ghrelin reduces food intake and respiratory quotient in chicken. *Domest Anim Endocrinol* 30(2), 108–116, doi:10.1016/j.domaniend.2005.06.005 (2006).
186. Song, Z., Liu, L., Sheikhahmadi, A., Jiao, H. and Lin, H. Effect of heat exposure on gene expression of feed intake regulatory peptides in laying hens. *J Biomed Biotechnol* 2012, 484869, doi:10.1155/2012/484869 (2012).

187. Rodriguez-Sinovas, A., Fernandez, E., Manteca, X., Fernandez, A. G. and Gonalons, E. CCK is involved in both peripheral and central mechanisms controlling food intake in chickens. *Am J Physiol* 272(1 Pt 2), R334–R340, doi:10.1152/ajpregu.1997.272.1.R334 (1997).
188. El-Sa'adawy, H. A., Kamal, T. K., Khalaf, A. A. Evaluation of the protective activities of dietary turmeric powder on growth performance, biochemical parameters, antioxidant status, and gene expression in heat-stressed broiler. *J Adv Vet Res* 13, 214–221 (2023).
189. Tachibana, T., Matsumoto, M., Furuse, M., Hasegawa, S., Yoshizawa, F. and Sugahara, K. Central, but not peripheral, glucagon-like peptide-1 inhibits crop emptying in chicks. *Comp Biochem Physiol A* 134(4), 777–781, doi:10.1016/S1095-6433(03)00009-6 (2003).
190. Shousha, S., Nakahara, K., Nasu, T., Sakamoto, T. and Murakami, N. Effect of glucagon-like peptide-1 and -2 on regulation of food intake, body temperature and locomotor activity in the Japanese quail. *Neurosci Lett* 415(2), 102–107, doi:10.1016/j.neulet.2007.01.007 (2007).
191. Furuse, M., Bungo, T., Shimojo, M., Masuda, Y., Saito, N., Hasegawa, S. and Sugahara, K. Effects of various N-terminal fragments of glucagon-like peptide-1(7-36) on food intake in the neonatal chick. *Brain Res* 807(1–2), 214–217, doi:10.1016/s0006-8993(98)00798-7 (1998).
192. Tachibana, T., Sato, M., Oikawa, D. and Furuse, M. Involvement of CRF on the anorexic effect of GLP-1 in layer chicks. *Comp Biochem Physiol A Mol Integr Physiol* 143(1), 112–117, doi:10.1016/j.cbpa.2005.11.007 (2006).
193. Bungo, T., Kawakami, S. I., Ohgushi, A., Sashihara, K., Saito, N., Sugahara, K., Hasegawa, S., Denbow, D. M. and Furuse, M. Intracerebroventricular injection of fusaric acid attenuates the anorexia by glucagon-like peptide-1 in the neonatal chick. *Pharmacol Biochem Behav* 70(2–3), 251–255, doi:10.1016/s0091-3057(01)00613-x (2001).
194. Honda, K., Shimatani, T., Aoki, K., Yamaguchi, T., Kondo, M., Saneyasu, T. and Kamisoyama, H. Glucagon-like Peptide-2 Functions as an Anorexigenic Peptide not only in the central nervous system but also in the peripheral Circulation in Broiler Chicks. *J Poult Sci* 52(3), 183–187 (2015).
195. Hu, X. F., Guo, Y. M., Huang, B. Y., Bun, S., Zhang, L. B., Li, J. H., Liu, D., Long, F. Y., Yang, X. and Jiao, P. The effect of glucagon-like peptide 2 injection on performance, small intestinal morphology, and nutrient transporter expression of stressed broiler chickens. *Poult Sci* 89(9), 1967–1974, doi:10.3382/ps.2009-00547 (2010).
196. Goel, A., Ncho, C. M., Kim, B. J., Jeong, C. M., Gupta, V., Jung, J. Y., Ha, S. Y., Yang, J. K. and Choi, Y. H. Dietary shredded steam-exploded pine particle supplementation as a strategy to mitigate chronic cyclic heat stress by modulating gut microbiota in broilers. *Sci Rep* 12(1), 19704, doi:10.1038/s41598-022-24031-w (2022).
197. He, X., Lu, Z., Ma, B., Zhang, L., Li, J., Jiang, Y., Zhou, G. and Gao, F. Effects of chronic heat exposure on growth performance, intestinal epithelial histology, appetite-related hormones and genes expression in broilers. *J Sci Food Agric* 98(12), 4471–4478, doi:10.1002/jsfa.8971 (2018).
198. Zhang, X., Shi, T., Holmberg, K., Landry, M., Huang, W., Xiao, H., Ju, G. and Hökfelt, T. Expression and regulation of the neuropeptide Y Y2 receptor in sensory and autonomic ganglia. *Proc Natl Acad Sci USA* 94(2), 729–734, doi:10.1073/pnas.94.2.729 (1997).

199. Koda, S., Date, Y., Murakami, N., Shimbara, T., Hanada, T., Toshinai, K., Niijima, A., Furuya, M., Inomata, N., Osuye, K. and Nakazato, M. The role of the vagal nerve in peripheral PYY3-36-induced feeding reduction in rats. *Endocrinology* 146(5), 2369–2375, doi:10.1210/en.2004-1266 (2005).
200. Wang, G., Li, X., Zhou, Y., Feng, J. and Zhang, M. Effects of heat stress on gut-microbial metabolites, gastrointestinal peptides, glycolipid metabolism, and performance of broilers. *Animals* 11(5), doi:10.3390/ani11051286 (2021).
201. Hekman, C. M., DeMond, W. S., Kelley, P. J., Mauch, S. F. and Williams, J. D. Isolation and identification of cyclic imide and deamidation products in heat stressed Pramlintide injection drug product. *J Pharm Biomed Anal* 20(5), 763–772, doi:10.1016/s0731-7085(99)00075-8 (1999).
202. Sands, J. S. S. and Smith, M. O. S. Effects of dietary manganese proteinate or chromium picolinate supplementation on plasma insulin, glucagon, glucose and serum lipids in broiler chickens reared under thermoneutral or heat stress conditions. *Int J Poult Sci* 1(5), 145–149 (2002).
203. Dridi, S., Temim, S., Derouet, M., Tesseraud, S. and Taouis, M. Acute cold- and chronic heat-exposure upregulate hepatic leptin and muscle uncoupling protein (UCP) gene expression in broiler chickens. *J Exp Zool A Ecol Genet Physiol* 309(7), 381–388, doi:10.1002/jez.461 (2008).
204. Prakasam, R., Fujimoto, M., Takii, R., Hayashida, N., Takaki, E., Tan, K., Wu, F., Inouye, S. and Nakai, A. Chicken IL-6 is a heat-shock gene. *FEBS Lett* 587(21), 3541–3547, doi:10.1016/j.febslet.2013.09.012 (2013).
205. Li, G. M., Liu, L. P., Yin, B., Liu, Y. Y., Dong, W. W., Gong, S., Zhang, J. and Tan, J. H. Heat stress decreases egg production of laying hens by inducing apoptosis of follicular cells via activating the FASL/Fas and TNF-alpha systems. *Poult Sci* 99(11), 6084–6093, doi:10.1016/j.psj.2020.07.024 (2020).
206. Morera, P., Basirico, L., Hosoda, K. and Bernabucci, U. Chronic heat stress up-regulates leptin and adiponectin secretion and expression and improves leptin, adiponectin and insulin sensitivity in mice. *J Mol Endocrinol* 48(2), 129–138, doi:10.1530/JME-11-0054 (2012).
207. Bülent Bayraktar, B., Emre Tekce, E. , Sevil Bayraktar, S., Gülbahar Böyük, G., Takma, C., Aksakal, V., Genç, M., Kaya, H., Ülker, U. and Gürbüz, A. B. Investigation of endocrine response of thyroid and intestinal and adipose tissues due to the addition of Moringa oleifera essential oil in diet for quails exposed to heat stress. *R Bras Zootec* 52, e20210040 (2023).
208. Greene, E., Khaldi, S., Ishola, P., Bottje, W., Ohkubo, T., Anthony, N. and Dridi, S. Heat and oxidative stress alter the expression of orexin and its related receptors in avian liver cells. *Comp Biochem Physiol A Mol Integr Physiol* 191, 18–24, doi:10.1016/j.cbpa.2015.08.016 (2016).
209. Yoshida, N., Fujita, M., Nakahara, M., Kuwahara, T., Kawakami, S. I. and Bungo, T. Effect of high environmental temperature on egg production, serum lipoproteins and follicle steroid hormones in laying hens. *J Poult Sci* 48(3), 207–211 (2011).
210. Jiang, S., Mohammed, A. A., Jacobs, J. A., Cramer, T. A. and Cheng, H. W. Effect of Synbiotics on thyroid hormones, intestinal histomorphology, and heat shock protein 70 expression in broiler chickens reared under cyclic heat stress. *Poult Sci* 99(1), 142–150, doi:10.3382/ps/pez571 (2020).

211. Lenard, N. R. and Berthoud, H. R. Central and peripheral regulation of food intake and physical activity: pathways and genes. *Obesity* 16 (Suppl. 3), S11–22, doi:10.1038/oby.2008.511 (2008).
212. Hussain, S. S. and Bloom, S. R. The regulation of food intake by the gut-brain axis: implications for obesity. *Int J Obes* 37(5), 625–633, doi:10.1038/ijo.2012.93 (2013).
213. Kim, K. S., Seeley, R. J. and Sandoval, D. A. Signalling from the periphery to the brain that regulates energy homeostasis. *Nat Rev Neurosci* 19(4), 185–196, doi:10.1038/nrn.2018.8 (2018).
214. Hardie, D. G., Ross, F. A. and Hawley, S. A. AMPK: a nutrient and energy sensor that maintains energy homeostasis. *Nat Rev Mol Cell Biol* 13(4), 251–262, doi:10.1038/nrm3311 (2012).
215. Lamia, K. A., Sachdeva, U. M., DiTacchio, L., Williams, E. C., Alvarez, J. G., Egan, D. F., Vasquez, D. S., Juguilon, H., Panda, S., Shaw, R. J., Thompson, C. B. and Evans, R. M. AMPK regulates the circadian clock by cryptochrome phosphorylation and degradation. *Science* 326(5951), 437–440, doi:10.1126/science.1172156 (2009).
216. Zapata, R. C., Carretero, M., Reis, F. C. G., Chaudry, B. S., Ofrecio, J., Zhang, D., Sasik, R., Ciaraldi, T., Petrascheck, M. and Osborn, O. Adipocytes control food intake and weight regain via vacuolar-type H(+) ATPase. *Nat Commun* 13(1), 5092, doi:10.1038/s41467-022-32764-5 (2022).
217. Ost, M., Igual Gil, C., Coleman, V., Keipert, S., Efstathiou, S., Vidic, V., Weyers, M. and Klaus, S. Muscle-derived GDF15 drives diurnal anorexia and systemic metabolic remodeling during mitochondrial stress. *EMBO Rep* 21(3), e48804, doi:10.15252/embr.201948804 (2020).
218. Worth, A. A., Shoop, R., Tye, K., Feetham, C. H., D'Agostino, G., Dodd, G. T., Reimann, F., Gribble, F. M., Beebe, E. C., Dunbar, J. D., Alexander-Chacko, J. T., Sindelar, D. K., Coskun, T., Emmerson, P. J. and Luckman, S. M. The cytokine GDF15 signals through a population of brainstem cholecystokinin neurons to mediate anorectic signalling. *eLife* 9, doi:10.7554/eLife.55164 (2020).
219. Cimino, I., Kim, H., Tung, Y. C. L., Pedersen, K., Rimmington, D., Tadross, J. A., Kohnke, S. N., Neves-Costa, A., Barros, A., Joaquim, S., Bennett, D., Melvin, A., Lockhart, S. M., Rostron, A. J., Scott, J., Liu, H., Burling, K., Barker, P., Clatworthy, M. R., Lee, E. C., Simpson, A. J., Yeo, G. S. H., Moita, L. F., Bence, K. K., Jørgensen, S. B., Coll, A. P., Breen, D. M. and O'Rahilly, S. Activation of the hypothalamic-pituitary-adrenal axis by exogenous and endogenous GDF15. *Proc Natl Acad Sci USA* 118(27), e2106868118. doi:10.1073/pnas.2106868118 (2021).
220. Keipert, S. and Ost, M. Stress-induced FGF21 and GDF15 in obesity and obesity resistance. *Trends Endocrinol Metab* 32(11), 904–915, doi:10.1016/j.tem.2021.08.008 (2021).
221. Andrews, Z. B., Liu, Z. W., Walllingford, N., Erion, D. M., Borok, E., Friedman, J. M., Tschöp, M. H., Shanabrough, M., Cline, G., Shulman, G. I., Coppola, A., Gao, X. B., Horvath, T. L. and Diano, S. UCP2 mediates ghrelin's action on NPY/AgRP neurons by lowering free radicals. *Nature* 454(7206), 846–851, doi:10.1038/nature07181 (2008).
222. Parton, L. E., Ye, C. P., Coppari, R., Enriori, P. J., Choi, B., Zhang, C. Y., Xu, C., Vianna, C. R., Balthasar, N., Lee, C. E., Elmquist, J. K., Cowley, M. A. and Lowell, B. B. Glucose sensing by POMC neurons regulates glucose homeostasis and is impaired in obesity. *Nature* 449(7159), 228–232, doi:10.1038/nature06098 (2007).

223. Andrews, Z. B., Erion, D., Beiler, R., Liu, Z. W., Abizaid, A., Zigman, J., Elsworth, J. D., Savitt, J. M., DiMarchi, R., Tschoep, M., Roth, R. H., Gao, X. B. and Horvath, T. L. Ghrelin promotes and protects nigrostriatal dopamine function via a UCP2-dependent mitochondrial mechanism. *J Neurosci* 29(45), 14057–14065, doi:10.1523/JNEUROSCI.3890-09.2009 (2009).
224. Gehart, H., Kumpf, S., Ittner, A. and Ricci, R. MAPK signalling in cellular metabolism: stress or wellness? *EMBO Rep* 11(11), 834–840, doi:10.1038/embor.2010.160 (2010).
225. Dickinson, R. J. and Keyse, S. M. Diverse physiological functions for dual-specificity MAP kinase phosphatases. *J Cell Sci* 119(22), 4607–4615, doi:10.1242/jcs.03266 (2006).
226. Gorostizaga, A., Brion, L., Maloberti, P., Cornejo Maciel, F., Podestá, E. J. and Paz, C. Heat shock triggers MAPK activation and MKP-1 induction in Leydig testicular cells. *Biochem Biophys Res Commun* 327(1), 23–28, doi:10.1016/j.bbrc.2004.11.129 (2005).
227. Li, J., Gorospe, M., Hutter, D., Barnes, J., Keyse, S. M. and Liu, Y. Transcriptional induction of MKP-1 in response to stress is associated with histone H3 phosphorylation-acetylation. *Mol Cell Biol* 21(23), 8213–8224, doi:10.1128/MCB.21.23.8213-8224.2001 (2001).
228. Yu, J., Yin, P., Yin, J., Liu, F., Zhu, X., Cheng, G., Guo, K., Yin, Y. and Xu, J. Involvement of ERK1/2 signalling and growth-related molecules' expression in response to heat stress-induced damage in rat jejunum and IEC-6 cells. *Int J Hyperthermia* 26(6), 538–555, doi:10.3109/02656736.2010.481276 (2010).
229. Rahmouni, K., Sigmund, C. D., Haynes, W. G. and Mark, A. L. Hypothalamic ERK mediates the anorectic and thermogenic sympathetic effects of leptin. *Diabetes* 58(3), 536–542, doi:10.2337/db08-0822 (2009).
230. Balland, E., Dam, J., Langlet, F., Caron, E., Steculorum, S., Messina, A., Rasika, S., Falluel-Morel, A., Anouar, Y., Dehouck, B., Trinquet, E., Jockers, R., Bouret, S. G. and Prévot, V. Hypothalamic tanycytes are an ERK-gated conduit for leptin into the brain. *Cell Metab* 19(2), 293–301, doi:10.1016/j.cmet.2013.12.015 (2014).
231. Unger, E. K., Piper, M. L., Olofsson, L. E. and Xu, A. W. Functional role of c-Jun-N-terminal kinase in feeding regulation. *Endocrinology* 151(2), 671–682, doi:10.1210/en.2009-0711 (2010).
232. Naznin, F., Toshinai, K., Waise, T. M., NamKoong, C., Md Moin, A. S., Sakoda, H. and Nakazato, M. Diet-induced obesity causes peripheral and central ghrelin resistance by promoting inflammation. *J Endocrinol* 226(1), 81–92, doi:10.1530/JOE-15-0139 (2015).
233. Lawan, A., Zhang, L., Gatzke, F., Min, K., Jurczak, M. J., Al-Mutairi, M., Richter, P., Camporez, J. P., Couvillon, A., Pesta, D., Roth Flach, R. J., Shulman, G. I. and Bennett, A. M. Hepatic mitogen-activated protein kinase phosphatase 1 selectively regulates glucose metabolism and energy homeostasis. *Mol Cell Biol* 35(1), 26–40, doi:10.1128/MCB.00503-14 (2015).
234. Feng, B., Jiao, P., Helou, Y., Li, Y., He, Q., Walters, M. S., Salomon, A. and Xu, H. Mitogen-activated protein kinase phosphatase 3 (MKP-3)-deficient mice are resistant to diet-induced obesity. *Diabetes* 63(9), 2924–2934, doi:10.2337/db14-0066 (2014).
235. Emanuelli, B., Eberle, D., Suzuki, R. and Kahn, C. R. Overexpression of the dual-specificity phosphatase MKP-4/DUSP-9 protects against stress-induced

insulin resistance. *Proc Natl Acad Sci USA* 105(9), 3545–3550, doi:10.1073/pnas.0712275105 (2008).
236. Ruan, J. W., Statt, S., Huang, C. T., Tsai, Y. T., Kuo, C. C., Chan, H. L., Liao, Y. C., Tan, T. H. and Kao, C. Y. Dual-specificity phosphatase 6 deficiency regulates gut microbiome and transcriptome response against diet-induced obesity in mice. *Nat Microbiol* 2, 16220, doi:10.1038/nmicrobiol.2016.220 (2016).
237. Yiangou, M., Paraskeva, E., Hsieh, C. C., Markou, E., Victoratos, P., Scouras, Z. and Papaconstantinou, J. Induction of a subgroup of acute phase protein genes in mouse liver by hyperthermia. *Biochim Biophys Acta* 1396(2), 191–206, doi:10.1016/s0167-4781(97)00188-7 (1998).
238. Qureshi, M. M., Singer, M. R. and Moore, L. L. A cross-sectional study of food group intake and C-reactive protein among children. *Nutr Metab* 6, 40, doi:10.1186/1743-7075-6-40 (2009).
239. Sun, Y., Yang, Y., Qin, Z., Cai, J., Guo, X., Tang, Y., Wan, J., Su, D. F. and Liu, X. The acute-phase protein orosomucoid regulates food intake and energy homeostasis via leptin receptor signaling pathway. *Diabetes* 65(6), 1630–1641, doi:10.2337/db15-1193 (2016).
240. Luo, Z., Tian, M., Yang, G., Tan, Q., Chen, Y., Li, G., Zhang, Q., Li, Y., Wan, P. and Wu, J. Hypoxia signaling in human health and diseases: implications and prospects for therapeutics. *Signal Transduct Target Ther* 7(1), 218, doi:10.1038/s41392-022-01080-1 (2022).
241. Dadgar, S., Greene, E., Dhamad, A., Mallmann, B., Dridi, S. and Rajaram, N. Diffuse reflectance spectroscopy reveals heat stress-induced changes in hemoglobin concentration in chicken breast. *Sci Rep* 11(1), 3649, doi:10.1038/s41598-021-83293-y (2021).
242. Akşit, M., Altan, O., Karul, A. B., Balkaya, M. and Özdemir, D. Effects of cold temperature and vitamin E supplementation on oxidative stress, troponin-T level, and other ascites-related traits in broilers. *Arch Geflügelk* 72, 221–230 (2008).
243. Huang, S., Zhang, L., Rehman, M. U., Iqbal, M. K., Lan, Y., Mehmood, K., Zhang, H., Qiu, G., Nabi, F., Yao, W., Wang, M. and Li, J. High altitude hypoxia as a factor that promotes tibial growth plate development in broiler chickens. *PLoS ONE* 12(3), e0173698, doi:10.1371/journal.pone.0173698 (2017).
244. Olkowski, A. A., Duke, T. and Wojnarowicz, C. The aetiology of hypoxaemia in chickens selected for rapid growth. *Comp Biochem Physiol A Mol Integr Physiol* 141(1), 122–131, doi:10.1016/j.cbpb.2005.04.011 (2005).
245. Semenza, G. L. Oxygen-regulated transcription factors and their role in pulmonary disease. *Respir Res* 1(3), 159–162, doi:10.1186/rr27 (2000).
246. Benita, Y., Kikuchi, H., Smith, A. D., Zhang, M. Q., Chung, D. C. and Xavier, R. J. An integrative genomics approach identifies hypoxia inducible factor-1 (HIF-1)-target genes that form the core response to hypoxia. *Nucleic Acids Res* 37(14), 4587–4602, doi:10.1093/nar/gkp425 (2009).
247. Westerterp-Plantenga, M. S., Westerterp, K. R., Rubbens, M., Verwegen, C. R., Richelet, J. P. and Gardette, B. Appetite at "high altitude" [Operation Everest III (Comex-'97)]: a simulated ascent of Mount Everest. *J Appl Physiol (1985)* 87(1), 391–399, doi:10.1152/jappl.1999.87.1.391 (1999).
248. Zhang, H., Zhang, G., Gonzalez, F. J., Park, S. M. and Cai, D. Hypoxia-inducible factor directs POMC gene to mediate hypothalamic glucose sensing and energy balance regulation. *PLoS Biol* 9(7), e1001112, doi:10.1371/journal.pbio.1001112 (2011).

249. Zhang, N., Fu, Z., Linke, S., Chicher, J., Gorman, J. J., Visk, D., Haddad, G. G., Poellinger, L., Peet, D. J., Powell, F. and Johnson, R. S. The asparaginyl hydroxylase factor inhibiting HIF-1alpha is an essential regulator of metabolism. *Cell Metab* 11(5), 364–378, doi:10.1016/j.cmet.2010.03.001 (2010).
250. Varela, L., Suyama, S., Huang, Y., Shanabrough, M., Tschöp, M. H., Gao, X. B., Giordano, F. J. and Horvath, T. L. Endothelial HIF-1alpha enables hypothalamic glucose uptake to drive POMC neurons. *Diabetes* 66(6), 1511–1520, doi:10.2337/db16-1106 (2017).
251. Gaspar, J. M., Mendes, N. F., Corrêa-da-Silva, F., Lima-Junior, J. C., Gaspar, R. C., Ropelle, E. R., Araujo, E. P., Carvalho, H. M. and Velloso, L. A. Downregulation of HIF complex in the hypothalamus exacerbates diet-induced obesity. *Brain Behav Immun* 73, 550–561, doi:10.1016/j.bbi.2018.06.020 (2018).
252. Ferguson, J. E., 3rd, Wu, Y., Smith, K., Charles, P., Powers, K., Wang, H. and Patterson, C. ASB4 is a hydroxylation substrate of FIH and promotes vascular differentiation via an oxygen-dependent mechanism. *Mol Cell Biol* 27(18), 6407–6419, doi:10.1128/MCB.00511-07 (2007).
253. Howard, J. K. and Flier, J. S. Attenuation of leptin and insulin signaling by SOCS proteins. *Trends Endocrinol Metab* 17(9), 365–371, doi:10.1016/j.tem.2006.09.007 (2006).
254. Chu, S. C., Chen, P. N., Yu, C. H., Hsieh, Y. S. and Kuo, D. Y. Role of hypoxia-inducible factor-1alpha in regulating oxidative stress and hypothalamic neuropeptides-mediated appetite control. *Brain Res* 1721, 146329, doi:10.1016/j.brainres.2019.146329 (2019).
255. Shi, X., Wang, X., Li, Q., Su, M., Chew, E., Wong, E. T., Lacza, Z., Radda, G. K., Tergaonkar, V. and Han, W. Nuclear factor kappaB (NF-kappaB) suppresses food intake and energy expenditure in mice by directly activating the POMC promoter. *Diabetologia* 56(4), 925–936, doi:10.1007/s00125-013-2831-2 (2013).
256. Lim, J., Kim, H. I., Bang, Y., Seol, W., Choi, H. S. and Choi, H. J. Hypoxia-inducible factor-1alpha upregulates tyrosine hydroxylase and dopamine transporter by nuclear receptor ERRgamma in SH-SY5Y cells. *NeuroReport* 26(6), 380–386, doi:10.1097/WNR.0000000000000356 (2015).
257. Du, D., Zhang, Y., Zhu, C., Chen, H. and Sun, J. Metabolic regulation of hypoxia-inducible factors in hypothalamus. *Front Endocrinol* 12, 650284, doi:10.3389/fendo.2021.650284 (2021).
258. Hu, F., Xu, Y. and Liu, F. Hypothalamic roles of mTOR complex I: integration of nutrient and hormone signals to regulate energy homeostasis. *Am J Physiol Endocrinol Metab* 310(11), E994–E1002, doi:10.1152/ajpendo.00121.2016 (2016).
259. Cota, D., Proulx, K., Smith, K. A., Kozma, S. C., Thomas, G., Woods, S. C. and Seeley, R. J. Hypothalamic mTOR signaling regulates food intake. *Science* 312(5775), 927–930, doi:10.1126/science.1124147 (2006).
260. Xu, Y., Hill, J. W., Fukuda, M., Gautron, L., Sohn, J. W., Kim, K. W., Lee, C. E., Choi, M. J., Lauzon, D. A., Dhillon, H., Lowell, B. B., Zigman, J. M., Zhao, J. J. and Elmquist, J. K. PI3K signaling in the ventromedial hypothalamic nucleus is required for normal energy homeostasis. *Cell Metab* 12(1), 88–95, doi:10.1016/j.cmet.2010.05.002 (2010).
261. Tavares, M. R., Lemes, S. F., de Fante, T., Saenz de Miera, C., Pavan, I. C. B., Bezerra, R. M. N., Prada, P. O., Torsoni, M. A., Elias, C. F. and Simabuco, F. M.

Modulation of hypothalamic S6K1 and S6K2 alters feeding behavior and systemic glucose metabolism. *J Endocrinol* 244(1), 71–82, doi:10.1530/JOE-19-0364 (2020).

262. Simler, N., Malgoyre, A., Koulmann, N., Alonso, A., Peinnequin, A. and Bigard, A. X. Hypoxic stimulus alters hypothalamic AMP-activated protein kinase phosphorylation concomitant to hypophagia. *J Appl Physiol (1985)* 102(6), 2135–2141, doi:10.1152/japplphysiol.01150.2006 (2007).
263. Chen, S., Yin, C., Lao, T., Liang, D., He, D., Wang, C. and Sang, N. AMPK-HDAC5 pathway facilitates nuclear accumulation of HIF-1alpha and functional activation of HIF-1 by deacetylating Hsp70 in the cytosol. *Cell Cycle* 14(15), 2520–2536, doi:10.1080/15384101.2015.1055426 (2015).
264. Bouchard, C. Inhibition of food intake by inhibitors of fatty acid synthase. *N Engl J Med* 343(25), 1888–1889, doi:10.1056/NEJM200012213432511 (2000).
265. Obici, S., Feng, Z., Arduini, A., Conti, R. and Rossetti, L. Inhibition of hypothalamic carnitine palmitoyltransferase-1 decreases food intake and glucose production. *Nat Med* 9(6), 756–761, doi:10.1038/nm873 (2003).
266. Loftus, T. M., Jaworsky, D. E., Frehywot, G. L., Townsend, C. A., Ronnett, G. V., Lane, M. D. and Kuhajda, F. P. Reduced food intake and body weight in mice treated with fatty acid synthase inhibitors. *Science* 288(5475), 2379–2381, doi:10.1126/science.288.5475.2379 (2000).
267. Drougard, A., Fournel, A., Valet, P. and Knauf, C. Impact of hypothalamic reactive oxygen species in the regulation of energy metabolism and food intake. *Front Neurosci* 9, 56, doi:10.3389/fnins.2015.00056 (2015).
268. Zhao, L., Liu, Z., Yang, F., Zhang, Y., Xue, Y., Miao, H., Liao, X., Huang, H. and Li, G. Intrabody against prolyl hydroxylase 2 promotes angiogenesis by stabilizing hypoxia-inducible factor-1alpha. *Sci Rep* 9(1), 11861, doi:10.1038/s41598-019-47891-1 (2019).
269. Fuhrmann, D. C. and Brune, B. Mitochondrial composition and function under the control of hypoxia. *Redox Biol* 12, 208–215, doi:10.1016/j.redox.2017.02.012 (2017).
270. Guzy, R. D., Sharma, B., Bell, E., Chandel, N. S. and Schumacker, P. T. Loss of the SdhB, but not the SdhA, subunit of complex II triggers reactive oxygen species-dependent hypoxia-inducible factor activation and tumorigenesis. *Mol Cell Biol* 28(2), 718–731, doi:10.1128/MCB.01338-07 (2008).
271. Klimova, T. and Chandel, N. S. Mitochondrial complex III regulates hypoxic activation of HIF. *Cell Death Differ* 15(4), 660–666, doi:10.1038/sj.cdd.4402307 (2008).
272. Brunelle, J. K., Bell, E. L., Quesada, N. M., Vercauteren, K., Tiranti, V., Zeviani, M., Scarpulla, R. C. and Chandel, N. S. Oxygen sensing requires mitochondrial ROS but not oxidative phosphorylation. *Cell Metab* 1(6), 409–414, doi:10.1016/j.cmet.2005.05.002 (2005).
273. Garcia-Caceres, C., Quarta, C., Varela, L., Gao, Y., Gruber, T., Legutko, B., Jastroch, M., Johansson, P., Ninkovic, J., Yi, C. X., Le Thuc, O., Szigeti-Buck, K., Cai, W., Meyer, C. W., Pfluger, P. T., Fernandez, A. M., Luquet, S., Woods, S. C., Torres-Alemán, I., Kahn, C. R., Götz, M., Horvath, T. L. and Tschöp, M. H. Astrocytic insulin signaling couples brain glucose uptake with nutrient availability. *Cell* 166(4), 867–880, doi:10.1016/j.cell.2016.07.028 (2016).
274. Maher, F., Vannucci, S. J. and Simpson, I. A. Glucose transporter proteins in brain. *FASEB J* 8(13), 1003–1011, doi:10.1096/fasebj.8.13.7926364 (1994).

275. Burdakov, D., Luckman, S. M. and Verkhratsky, A. Glucose-sensing neurons of the hypothalamus. *Philos Trans R Soc Lond B Biol Sci* 360(1464), 2227–2235, doi:10.1098/rstb.2005.1763 (2005).
276. Harik, N., Harik, S. I., Kuo, N. T., Sakai, K., Przybylski, R. J. and LaManna, J. C. Time-course and reversibility of the hypoxia-induced alterations in cerebral vascularity and cerebral capillary glucose transporter density. *Brain Res* 737(1–2), 335–338, doi:10.1016/0006-8993(96)00965-1 (1996).
277. Denko, N. C. Hypoxia, HIF1 and glucose metabolism in the solid tumour. *Nat Rev Cancer* 8(9), 705–713, doi:10.1038/nrc2468 (2008).
278. Sakagami, H., Makino, Y., Mizumoto, K., Isoe, T., Takeda, Y., Watanabe, J., Fujita, Y., Takiyama, Y., Abiko, A. and Haneda, M. Loss of HIF-1alpha impairs GLUT4 translocation and glucose uptake by the skeletal muscle cells. *Am J Physiol Endocrinol Metab* 306(9), E1065–E1076, doi:10.1152/ajpendo.00597.2012 (2014).
279. Barahona, M. J., Llanos, P., Recabal, A., Escobar-Acuña, K., Elizondo-Vega, R., Salgado, M., Ordenes, P., Uribe, E., Sepúlveda, F. J., Araneda, R. C. and García-Robles, M. A. Glial hypothalamic inhibition of GLUT2 expression alters satiety, impacting eating behavior. *Glia* 66(3), 592–605, doi:10.1002/glia.23267 (2018).
280. Marty, N., Dallaporta, M., Foretz, M., Emery, M., Tarussio, D., Bady, I., Binnert, C., Beermann, F. and Thorens, B. Regulation of glucagon secretion by glucose transporter type 2 (glut2) and astrocyte-dependent glucose sensors. *J Clin Invest* 115(12), 3545–3553, doi:10.1172/JCI26309 (2005).
281. Ren, H., Lu, T. Y., McGraw, T. E. and Accili, D. Anorexia and impaired glucose metabolism in mice with hypothalamic ablation of Glut4 neurons. *Diabetes* 64(2), 405–417, doi:10.2337/db14-0752 (2015).
282. Ardanaz, C. G., de la Cruz, A., Elizalde-Horcada, M., Puerta, E., Ramírez, M. J., Ortega, J. E., Urbiola, A., Ederra, C., Ariz, M., Ortiz-de-Solórzano, C., Fernández- Irigoyen, J., Santamaría, E., Karsenty, G., Brüning, J. C. and Solas, M. GLUT1 ablation in astrocytes paradoxically improves central and peripheral glucose metabolism via enhanced insulin-stimulated ATP release. *bioRxiv*, 1–70, doi:10.1101/2022.10.06.511112 (2022).
283. Druyan, S., Ruzal, M., Shinder, D. and Haron, A. Effects of low oxygen during chorioallantoic membrane development on post-hatch growing performance of broiler chickens. *Poult Sci* 97(6), 1961–1967, doi:10.3382/ps/pey052 (2018).
284. Solis de los Santos, F., Farnell, M. B., Téllez, G., Balog, J. M., Anthony, N. B., Torres-Rodriguez, A., Higgins, S., Hargis, B. M. and Donoghue, A. M. Effect of prebiotic on gut development and ascites incidence of broilers reared in a hypoxic environment. *Poult Sci* 84(7), 1092–1100, doi:10.1093/ps/84.7.1092 (2005).
285. Yang, L., Tan, G. Y., Fu, Y. Q., Feng, J. H. and Zhang, M. H. Effects of acute heat stress and subsequent stress removal on function of hepatic mitochondrial respiration, ROS production and lipid peroxidation in broiler chickens. *Comp Biochem Physiol C Toxicol Pharmacol* 151(2), 204–208, doi:10.1016/j.cbpc.2009.10.010 (2010).
286. Forde, C. G. and de Graaf, K. Influence of sensory properties in moderating eating behaviors and food intake. *Front Nutr* 9, 841444, doi:10.3389/fnut.2022.841444 (2022).
287. Yarmolinsky, D. A., Zuker, C. S. and Ryba, N. J. Common sense about taste: from mammals to insects. *Cell* 139(2), 234–244, doi:10.1016/j.cell.2009.10.001 (2009).

288. Gilbertson, T. A., Fontenot, D. T., Liu, L., Zhang, H. and Monroe, W. T. Fatty acid modulation of K+ channels in taste receptor cells: gustatory cues for dietary fat. *Am J Physiol* 272(4 Pt 1), C1203–C1210, doi:10.1152/ajpcell.1997.272.4.C1203 (1997).
289. Nelson, G., Hoon, M. A., Chandrashekar, J., Zhang, Y., Ryba, N. J. and Zuker, C. S. Mammalian sweet taste receptors. *Cell* 106(3), 381–390, doi:10.1016/s0092-8674(01)00451-2 (2001).
290. Nelson, G., Chandrashekar, J., Hoon, M. A., Feng, L., Zhao, G., Ryba, N. J. and Zuker, C. S. An amino-acid taste receptor. *Nature* 416(6877), 199–202, doi:10.1038/nature726 (2002).
291. Zhao, G. Q., Zhang, Y., Hoon, M. A., Chandrashekar, J., Erlenbach, I., Ryba, N. J. and Zuker, C. S. The receptors for mammalian sweet and umami taste. *Cell* 115(3), 255–266, doi:10.1016/s0092-8674(03)00844-4 (2003).
292. Mueller, K. L., Hoon, M. A., Erlenbach, I., Chandrashekar, J., Zuker, C. S. and Ryba, N. J. The receptors and coding logic for bitter taste. *Nature* 434(7030), 225–229, doi:10.1038/nature03352 (2005).
293. Oka, Y., Butnaru, M., von Buchholtz, L., Ryba, N. J. and Zuker, C. S. High salt recruits aversive taste pathways. *Nature* 494(7438), 472–475, doi:10.1038/nature11905 (2013).
294. Huang, A. L., Chen, X., Hoon, M. A., Chandrashekar, J., Guo, W., Tränkner, D., Ryba, N. J. and Zuker, C. S. The cells and logic for mammalian sour taste detection. *Nature* 442(7105), 934–938, doi:10.1038/nature05084 (2006).
295. Ishimaru, Y., Inada, H., Kubota, M., Zhuang, H., Tominaga, M. and Matsunami, H. Transient receptor potential family members PKD1L3 and PKD2L1 form a candidate sour taste receptor. *Proc Natl Acad Sci USA* 103(33), 12569–12574, doi:10.1073/pnas.0602702103 (2006).
296. Ye, W., Chang, R. B., Bushman, J. D., Tu, Y. H., Mulhall, E. M., Wilson, C. E., Cooper, A. J., Chick, W. S., Hill-Eubanks, D. C., Nelson, M. T., Kinnamon, S. C. and Liman, E. R. The K+ channel KIR2.1 functions in tandem with proton influx to mediate sour taste transduction. *Proc Natl Acad Sci USA* 113(2), E229–E238, doi:10.1073/pnas.1514282112 (2016).
297. Rajapaksha, P., Wang, Z., Venkatesan, N., Tehrani, K. F., Payne, J., Swetenburg, R. L., Kawabata, F., Tabata, S., Mortensen, L. J., Stice, S. L., Beckstead, R. and Liu, H. X. Labeling and analysis of chicken taste buds using molecular markers in oral epithelial sheets. *Sci Rep* 6, 37247, doi:10.1038/srep37247 (2016).
298. Kare, M. R. and Pick, H. L. The influence of the sense of taste on feed and fluid consumption. *Poult Sci* 39(3), 697–706 (1960).
299. Gentle, M. J. Taste preference in the chicken (Gallus domesticus L.). *Br Poult Sci* 13(2), 141–155, doi:10.1080/00071667208415928 (1972).
300. Vermaut, S., De Coninck, K., Flo, G., Cokelaere, M., Onagbesan, M. and Decuypere, E. Effect of deoiled jojoba meal on feed intake in chickens: satiating or taste effect? *J Agric Food Chem* 45(8), 3158–3163 (1997).
301. Jakubas, W. J., Shah, P. S., Mason, J. R. and Norman, D. M. Avian repellency of coniferyl and cinnamyl derivatives. *Ecol Appl* 2(2), 147–156, doi:10.2307/1941771 (1992).
302. Brugger, K. E., Nol, P. and Phillips, C. I. Sucrose repellency to European starlings: will high-sucrose cultivars deter bird damage to fruit? *Ecol Appl* 3(2), 256–261, doi:10.2307/1941828 (1993).

303. Edmonds, M. S. and Baker, D. H. Comparative effects of individual amino acid excesses when added to a corn-soybean meal diet: effects on growth and dietary choice in the chick. *J Anim Sci* 65(3), 699–705, doi:10.2527/jas1987.653699x (1987).
304. Picard, M. L., Uzu, G., Dunnington, E. A. and Siegel, P. B. Food intake adjustments of chicks: short term reactions to deficiencies in lysine, methionine and tryptophan. *Br Poult Sci* 34(4), 737–746, doi:10.1080/00071669308417632 (1993).
305. Lagerstrom, M. C., Hellström, A. R., Gloriam, D. E., Larsson, T. P., Schiöth, H. B. and Fredriksson, R.. The G protein-coupled receptor subset of the chicken genome. *PLoS Comput Biol* 2(6), e54, doi:10.1371/journal.pcbi.0020054 (2006).
306. Cheled-Shoval, S., Behrens, M., Korb, A., Di Pizio, A., Meyerhof, W., Uni, Z. and Niv, M. Y. From cell to beak: in-vitro and in-vivo characterization of chicken bitter taste thresholds. *Molecules* 22(5), doi:10.3390/molecules22050821 (2017). doi: 10.3390/molecules22050821.
307. Hirose, N., Kawabata, Y., Kawabata, F., Nishimura, S. and Tabata, S. Bitter taste receptor T2R1 activities were compatible with behavioral sensitivity to bitterness in chickens. *Biochem Biophys Res Commun* 460(2), 464–468, doi:10.1016/j.bbrc.2015.03.056 (2015).
308. Dey, B., Kawabata, F., Kawabata, Y., Yoshida, Y., Nishimura, S. and Tabata, S. Identification of functional bitter taste receptors and their antagonist in chickens. *Biochem Biophys Res Commun* 482(4), 693–699, doi:10.1016/j.bbrc.2016.11.096 (2017).
309. Behrens, M., Korsching, S. I. and Meyerhof, W. Tuning properties of avian and frog bitter taste receptors dynamically fit gene repertoire sizes. *Mol Biol Evol* 31(12), 3216–3227, doi:10.1093/molbev/msu254 (2014).
310. Davis, J. K., Lowman, J. J., Thomas, P. J., ten Hallers, B. F., Koriabine, M., Huynh, L. Y., Maney, D. L., de Jong, P. J., Martin, C. L., NISC Comparative Sequencing Program and Thomas, J. W. Evolution of a bitter taste receptor gene cluster in a New World sparrow. *Genome Biol Evol* 2, 358–370, doi:10.1093/gbe/evq027 (2010).
311. Li, D. and Zhang, J. Diet shapes the evolution of the vertebrate bitter taste receptor gene repertoire. *Mol Biol Evol* 31(2), 303–309, doi:10.1093/molbev/mst219 (2014).
312. Harriman, A. E. and Milner, J. S. Preference for sucrose solutions by Japanese quail (Coturnix coturnix japonica) in two-bottle drinking tests. *Am Midl Nat* 81(2), 575–578 (1969).
313. Kare, M. R. and Medway, W. Discrimination between carbohydrates by the fowl. *Poult Sci* 38(5), 1119–1127 (1959).
314. Gal-Garber, O., Mabjeesh, S. J., Sklan, D. and Uni, Z. Partial sequence and expression of the gene for and activity of the sodium glucose transporter in the small intestine of fed, starved and refed chickens. *J Nutr* 130(9), 2174–2179, doi:10.1093/jn/130.9.2174 (2000).
315. Stearns, A. T., Balakrishnan, A., Rhoads, D. B. and Tavakkolizadeh, A. Rapid upregulation of sodium-glucose transporter SGLT1 in response to intestinal sweet taste stimulation. *Ann Surg* 251(5), 865–871, doi:10.1097/SLA.0b013e3181d96e1f (2010).
316. Baldwin, M. W., Toda, Y., Nakagita, T., O'Connell, M. J., Klasing, K. C., Misaka, T., Edwards, S. V. and Liberles, S. D. Sensory biology. Evolution of sweet taste perception in hummingbirds by transformation of the ancestral umami receptor. *Science* 345(6199), 929–933, doi:10.1126/science.1255097 (2014).

317. Niknafs, S. and Roura, E. Nutrient sensing, taste and feed intake in avian species. *Nutr Res Rev* 31(2), 256–266, doi:10.1017/S0954422418000100 (2018).
318. Raybould, H. E. Gut chemosensing: interactions between gut endocrine cells and visceral afferents. *Auton Neurosci* 153(1–2), 41–46, doi:10.1016/j.autneu.2009.07.007 (2010).
319. Jang, H. J., Kokrashvili, Z., Theodorakis, M. J., Carlson, O. D., Kim, B. J., Zhou, J., Kim, H. H., Xu, X., Chan, S. L., Juhaszova, M., Bernier, M., Mosinger, B., Margolskee, R. F. and Egan, J. M. Gut-expressed gustducin and taste receptors regulate secretion of glucagon-like peptide-1. *Proc Natl Acad Sci USA* 104(38), 15069–15074, doi:10.1073/pnas.0706890104 (2007).
320. Margolskee, R. F., Dyer, J., Kokrashvili, Z., Salmon, K. S., Ilegems, E., Daly, K., Maillet, E. L., Ninomiya, Y., Mosinger, B. and Shirazi-Beechey, S. P. T1R3 and gustducin in gut sense sugars to regulate expression of Na+-glucose cotransporter 1. *Proc Natl Acad Sci USA* 104(38), 15075–15080, doi:10.1073/pnas.0706678104 (2007).
321. Wu, S. V., Rozengurt, N., Yang, M., Young, S. H., Sinnett-Smith, J. and Rozengurt, E. Expression of bitter taste receptors of the T2R family in the gastrointestinal tract and enteroendocrine STC-1 cells. *Proc Natl Acad Sci USA* 99(4), 2392–2397, doi:10.1073/pnas.042617699 (2002).
322. Gerspach, A. C., Steinert, R. E., Schonenberger, L., Graber-Maier, A. and Beglinger, C. The role of the gut sweet taste receptor in regulating GLP-1, PYY, and CCK release in humans. *Am J Physiol Endocrinol Metab* 301(2), E317–E325, doi:10.1152/ajpendo.00077.2011 (2011).
323. Hass, N., Schwarzenbacher, K. and Breer, H. T1R3 is expressed in brush cells and ghrelin-producing cells of murine stomach. *Cell Tissue Res* 339(3), 493–504, doi:10.1007/s00441-009-0907-6 (2010).
324. Janssen, S., Laermans, J., Iwakura, H., Tack, J. and Depoortere, I. Sensing of fatty acids for octanoylation of ghrelin involves a gustatory G-protein. *PLoS ONE* 7(6), e40168, doi:10.1371/journal.pone.0040168 (2012).
325. Janssen, S., Laermans, J., Verhulst, P. J., Thijs, T., Tack, J. and Depoortere, I. Bitter taste receptors and alpha-gustducin regulate the secretion of ghrelin with functional effects on food intake and gastric emptying. *Proc Natl Acad Sci USA* 108(5), 2094–2099, doi:10.1073/pnas.1011508108 (2011).
326. Feng, J., Petersen, C. D., Coy, D. H., Jiang, J. K., Thomas, C. J., Pollak, M. R. and Wank, S. A. Calcium-sensing receptor is a physiologic multimodal chemosensor regulating gastric G-cell growth and gastrin secretion. *Proc Natl Acad Sci USA* 107(41), 17791–17796, doi:10.1073/pnas.1009078107 (2010).
327. Gribble, F. M., Reimann, F. and Cells, E. Chemosensors in the intestinal epithelium. *Annu Rev Physiol* 78, 277–299, doi:10.1146/annurev-physiol-021115-105439 (2016).
328. Lassman, D. J., McKie, S., Gregory, L. J., Lal, S., D'Amato, M., Steele, I., Varro, A., Dockray, G. J., Williams, S. C. and Thompson, D. G. Defining the role of cholecystokinin in the lipid-induced human brain activation matrix. *Gastroenterology* 138(4), 1514–1524, doi:10.1053/j.gastro.2009.12.060 (2010).
329. Krieger, J. P., Arnold, M., Pettersen, K. G., Lossel, P., Langhans, W. and Lee, S. J. Knockdown of GLP-1 receptors in vagal afferents affects normal food intake and glycemia. *Diabetes* 65(1), 34–43, doi:10.2337/db15-0973 (2016).
330. Pritchard, T. C., Hamilton, R. B. and Norgren, R. Neural coding of gustatory information in the thalamus of Macaca mulatta. *J Neurophysiol* 61(1), 1–14, doi:10.1152/jn.1989.61.1.1 (1989).

331. Rolls, E. T., Yaxley, S. and Sienkiewicz, Z. J. Gustatory responses of single neurons in the caudolateral orbitofrontal cortex of the macaque monkey. *J Neurophysiol* 64(4), 1055–1066, doi:10.1152/jn.1990.64.4.1055 (1990).
332. Yaxley, S., Rolls, E. T. and Sienkiewicz, Z. J. Gustatory responses of single neurons in the insula of the macaque monkey. *J Neurophysiol* 63(4), 689–700, doi:10.1152/jn.1990.63.4.689 (1990).
333. Zhang, J., Jin, H., Zhang, W., Ding, C., O'Keeffe, S., Ye, M. and Zuker, C. S. Sour sensing from the tongue to the brain. *Cell* 179(2), 392–402 e315, doi:10.1016/j.cell.2019.08.031 (2019).
334. Fu, O., Iwai, Y., Kondoh, K., Misaka, T., Minokoshi, Y. and Nakajima, K. I. SatB2-expressing neurons in the parabrachial nucleus encode sweet taste. *Cell Rep* 27(6), 1650–1656.e4, doi:10.1016/j.celrep.2019.04.040 (2019).
335. Obici, S., Feng, Z., Morgan, K., Stein, D., Karkanias, G. and Rossetti, L. Central administration of oleic acid inhibits glucose production and food intake. *Diabetes* 51(2), 271–275, doi:10.2337/diabetes.51.2.271 (2002).
336. Blouet, C. and Schwartz, G. J. Brainstem nutrient sensing in the nucleus of the solitary tract inhibits feeding. *Cell Metab* 16(5), 579–587, doi:10.1016/j.cmet.2012.10.003 (2012).
337. Jo, Y. H., Su, Y., Gutierrez-Juarez, R. and Chua, S., Jr. Oleic acid directly regulates POMC neuron excitability in the hypothalamus. *J Neurophysiol* 101(5), 2305–2316, doi:10.1152/jn.91294.2008 (2009).
338. Schwinkendorf, D. R., Tsatsos, N. G., Gosnell, B. A. and Mashek, D. G. Effects of central administration of distinct fatty acids on hypothalamic neuropeptide expression and energy metabolism. *Int J Obes* 35(3), 336–344, doi:10.1038/ijo.2010.159 (2011).
339. Karnani, M. M., Apergis-Schoute, J., Adamantidis, A., Jensen, L. T., de Lecea, L., Fugger, L. and Burdakov, D. Activation of central orexin/hypocretin neurons by dietary amino acids. *Neuron* 72(4), 616–629, doi:10.1016/j.neuron.2011.08.027 (2011).
340. Le Foll, C., Dunn-Meynell, A. A. and Levin, B. E. Role of FAT/CD36 in fatty acid sensing, energy, and glucose homeostasis regulation in DIO and DR rats. *Am J Physiol Regul Integr Comp Physiol* 308(3), R188–R198, doi:10.1152/ajpregu.00367.2014 (2015).
341. Shimokawa, T., Kumar, M. V. and Lane, M. D. Effect of a fatty acid synthase inhibitor on food intake and expression of hypothalamic neuropeptides. *Proc Natl Acad Sci USA* 99(1), 66–71, doi:10.1073/pnas.012606199 (2002).
342. Dridi, S., Ververken, C., Hillgartner, F. B., Arckens, L., Van der Gucht, E., Cnops, L., Decuypere, E. and Buyse, J. FAS inhibitor cerulenin reduces food intake and melanocortin receptor gene expression without modulating the other (an)orexigenic neuropeptides in chickens. *Am J Physiol Regul Integr Comp Physiol* 291(1), R138–R147, doi:10.1152/ajpregu.00899.2005 (2006).
343. Gao, S., Moran, T. H., Lopaschuk, G. D. and Butler, A. A. Hypothalamic malonyl-CoA and the control of food intake. *Physiol Behav* 122, 17–24, doi:10.1016/j.physbeh.2013.07.014 (2013).
344. Wolf, G. The regulation of food intake by hypothalamic malonyl-coenzyme A: the MaloA hypothesis. *Nutr Rev* 64(8), 379–383, doi:10.1111/j.1753-4887.2006.tb00223.x (2006).
345. Hu, Z., Dai, Y., Prentki, M., Chohnan, S. and Lane, M. D. A role for hypothalamic malonyl-CoA in the control of food intake. *J Biol Chem* 280(48), 39681–39683, doi:10.1074/jbc.C500398200 (2005).

346. Price, N., van der Leij, F., Jackson, V., Corstorphine, C., Thomson, R., Sorensen, A. and Zammit, V. A novel brain-expressed protein related to carnitine palmitoyltransferase I. *Genomics* 80(4), 433–442, doi:10.1006/geno.2002.6845 (2002).
347. Wolfgang, M. J., Kurama, T., Dai, Y., Suwa, A., Asaumi, M., Matsumoto, S., Cha, S. H., Shimokawa, T. and Lane, M. D. The brain-specific carnitine palmitoyltransferase-1c regulates energy homeostasis. *Proc Natl Acad Sci USA* 103(19), 7282–7287, doi:10.1073/pnas.0602205103 (2006).
348. Wang, H. and Eckel, R. H. Lipoprotein lipase in the brain and nervous system. *Annu Rev Nutr* 32, 147–160, doi:10.1146/annurev-nutr-071811-150703 (2012).
349. Laperrousaz, E., Moullé, V. S., Denis, R. G., Kassis, N., Berland, C., Colsch, B., Fioramonti, X., Philippe, E., Lacombe, A., Vanacker, C., Butin, N., Bruce, K. D., Wang, H., Wang, Y., Gao, Y., Garcia-Caceres, C., Prévot, V., Tschöp, M. H., Eckel, R. H., Le Stunff, H., Luquet, S., Magnan, C. and Cruciani-Guglielmacci, C. Lipoprotein lipase in hypothalamus is a key regulator of body weight gain and glucose homeostasis in mice. *Diabetologia* 60(7), 1314–1324, doi:10.1007/s00125-017-4282-7 (2017).
350. Picard, A., Rouch, C., Kassis, N., Moullé, V. S., Croizier, S., Denis, R. G., Castel, J., Coant, N., Davis, K., Clegg, D. J., Benoit, S. C., Prévot, V., Bouret, S., Luquet, S., Le Stunff, H., Cruciani-Guglielmacci, C. and Magnan, C. Hippocampal lipoprotein lipase regulates energy balance in rodents. *Mol Metab* 3(2), 167–176, doi:10.1016/j.molmet.2013.11.002 (2014).
351. Oomura, Y., Ono, T., Ooyama, H. and Wayner, M. J. Glucose and osmosensitive neurones of the rat hypothalamus. *Nature* 222(5190), 282–284, doi:10.1038/222282a0 (1969).
352. Routh, V. H. Glucose sensing neurons in the ventromedial hypothalamus. *Sensors* 10(10), 9002–9025, doi:10.3390/s101009002 (2010).
353. Shram, N. F., Netchiporouk, L. I., Martelet, C., Jaffrezic-Renault, N. and Cespuglio, R. Brain glucose: voltammetric determination in normal and hyperglycaemic rats using a glucose microsensor. *NeuroReport* 8(5), 1109–1112, doi:10.1097/00001756-199703240-00009 (1997).
354. Dunn-Meynell, A. A., Routh, V. H., Kang, L., Gaspers, L. and Levin, B. E. Glucokinase is the likely mediator of glucosensing in both glucose-excited and glucose-inhibited central neurons. *Diabetes* 51(7), 2056–2065, doi:10.2337/diabetes.51.7.2056 (2002).
355. Kang, L., Routh, V. H., Kuzhikandathil, E. V., Gaspers, L. D. and Levin, B. E. Physiological and molecular characteristics of rat hypothalamic ventromedial nucleus glucosensing neurons. *Diabetes* 53(3), 549–559, doi:10.2337/diabetes.53.3.549 (2004).
356. Garcia, M., Millán, C., Balmaceda-Aguilera, C., Castro, T., Pastor, P., Montecinos, H., Reinicke, K., Zúñiga, F., Vera, J. C., Oñate, S. A. and Nualart, F. Hypothalamic ependymal-glial cells express the glucose transporter GLUT2, a protein involved in glucose sensing. *J Neurochem* 86(3), 709–724, doi:10.1046/j.1471-4159.2003.01892.x (2003).
357. Salgado, M., Tarifeño-Saldivia, E., Ordenes, P., Millán, C., Yañez, M. J., Llanos, P., Villagra, M., Elizondo-Vega, R., Martínez, F., Nualart, F., Uribe, E. and de Los Angeles García-Robles, M. Dynamic localization of glucokinase and its regulatory protein in hypothalamic tanycytes. *PLoS ONE* 9(4), e94035, doi:10.1371/journal.pone.0094035 (2014).
358. Uranga, R. M., Millán, C., Barahona, M. J., Recabal, A., Salgado, M., Martinez, F., Ordenes, P., Elizondo-Vega, R., Sepúlveda, F., Uribe, E. and García-Robles, M. L. Á.

Adenovirus-mediated suppression of hypothalamic glucokinase affects feeding behavior. *Sci Rep* 7(1), 3697, doi:10.1038/s41598-017-03928-x (2017).

359. Benford, H., Bolborea, M., Pollatzek, E., Lossow, K., Hermans-Borgmeyer, I., Liu, B., Meyerhof, W., Kasparov, S. and Dale, N. A sweet taste receptor-dependent mechanism of glucosensing in hypothalamic tanycytes. *Glia* 65(5), 773–789, doi:10.1002/glia.23125 (2017).
360. Dupont, J., Dagou, C., Derouet, M., Simon, J. and Taouis, M. Early steps of insulin receptor signaling in chicken and rat: apparent refractoriness in chicken muscle. *Domest Anim Endocrinol* 26(2), 127–142, doi:10.1016/j.domaniend.2003.09.004 (2004).
361. Seki, Y., Sato, K., Kono, T., Abe, H. and Akiba, Y. Broiler chickens (Ross strain) lack insulin-responsive glucose transporter GLUT4 and have GLUT8 cDNA. *Gen Comp Endocrinol* 133(1), 80–87, doi:10.1016/s0016-6480(03)00145-x (2003).
362. Mellinkoff, S. M., Frankland, M., Boyle, D. and Greipel, M. Relationship between serum amino acid concentration and fluctuations in appetite. *J Appl Physiol* 8(5), 535–538, doi:10.1152/jappl.1956.8.5.535 (1956).
363. Anderson, G. H., Tecimer, S. N., Shah, D. and Zafar, T. A. Protein source, quantity, and time of consumption determine the effect of proteins on short-term food intake in young men. *J Nutr* 134(11), 3011–3015, doi:10.1093/jn/134.11.3011 (2004).
364. Li, E. T. and Anderson, G. H. Meal composition influences subsequent food selection in the young rat. *Physiol Behav* 29(5), 779–783, doi:10.1016/0031-9384(82)90324-9 (1982).
365. Wang, Y., Chandra, R., Samsa, L. A., Gooch, B., Fee, B. E., Cook, J. M., Vigna, S. R., Grant, A. O. and Liddle, R. A. Amino acids stimulate cholecystokinin release through the Ca2+-sensing receptor. *Am J Physiol Gastrointest Liver Physiol* 300(4), G528–G537, doi:10.1152/ajpgi.00387.2010 (2011).
366. Crovetti, R., Porrini, M., Santangelo, A. and Testolin, G. The influence of thermic effect of food on satiety. *Eur J Clin Nutr* 52(7), 482–488, doi:10.1038/sj.ejcn.1600578 (1998).
367. Azzout-Marniche, D., Gaudichon, C., Blouet, C., Bos, C., Mathé, V., Huneau, J. F. and Tomé, D. Liver glyconeogenesis: a pathway to cope with postprandial amino acid excess in high-protein fed rats? *Am J Physiol Regul Integr Comp Physiol* 292(4), R1400–R1407, doi:10.1152/ajpregu.00566.2006 (2007).
368. Dong, J., Qiu, H., Garcia-Barrio, M., Anderson, J. and Hinnebusch, A. G. Uncharged tRNA activates GCN2 by displacing the protein kinase moiety from a bipartite tRNA-binding domain. *Mol Cell* 6(2), 269–279, doi:10.1016/s1097-2765(00)00028-9 (2000).
369. Zhang, P., McGrath, B. C., Reinert, J., Olsen, D. S., Lei, L., Gill, S., Wek, S. A., Vattem, K. M., Wek, R. C., Kimball, S. R., Jefferson, L. S. and Cavener, D. R. The GCN2 eIF2alpha kinase is required for adaptation to amino acid deprivation in mice. *Mol Cell Biol* 22(19), 6681–6688, doi:10.1128/MCB.22.19.6681-6688.2002 (2002).
370. Hao, S., Sharp, J. W., Ross-Inta, C. M., McDaniel, B. J., Anthony, T. G., Wek, R. C., Cavener, D. R., McGrath, B. C., Rudell, J. B., Koehnle, T. J. and Gietzen, D. W. Uncharged tRNA and sensing of amino acid deficiency in mammalian piriform cortex. *Science* 307(5716), 1776–1778, doi:10.1126/science.1104882 (2005).
371. Maurin, A. C., Jousse, C., Averous, J., Parry, L., Bruhat, A., Cherasse, Y., Zeng, H., Zhang, Y., Harding, H. P., Ron, D. and Fafournoux, P. The GCN2 kinase biases feeding

behavior to maintain amino acid homeostasis in omnivores. *Cell Metab* 1(4), 273–277, doi:10.1016/j.cmet.2005.03.004 (2005).
372. Blouet, C., Jo, Y. H., Li, X. and Schwartz, G. J. Mediobasal hypothalamic leucine sensing regulates food intake through activation of a hypothalamus-brainstem circuit. *J Neurosci* 29(26), 8302–8311, doi:10.1523/JNEUROSCI.1668-09.2009 (2009).
373. Morrison, C. D., Xi, X., White, C. L., Ye, J. and Martin, R. J. Amino acids inhibit Agrp gene expression via an mTOR-dependent mechanism. *Am J Physiol Endocrinol Metab* 293(1), E165–E171, doi:10.1152/ajpendo.00675.2006 (2007).
374. Purpera, M. N., Shen, L., Taghavi, M., Münzberg, H., Martin, R. J., Hutson, S. M. and Morrison, C. D. Impaired branched chain amino acid metabolism alters feeding behavior and increases orexigenic neuropeptide expression in the hypothalamus. *J Endocrinol* 212(1), 85–94, doi:10.1530/JOE-11-0270 (2012).
375. Byerly, M. S., Simon, J., Cogburn, L. A., Le Bihan-Duval, E., Duclos, M. J., Aggrey, S. E. and Porter, T. E. Transcriptional profiling of hypothalamus during development of adiposity in genetically selected fat and lean chickens. *Physiol Genomics* 42(2), 157–167, doi:10.1152/physiolgenomics.00029.2010 (2010).
376. Cheled-Shoval, S. L., Druyan, S. and Uni, Z. Bitter, sweet and umami taste receptors and downstream signaling effectors: expression in embryonic and growing chicken gastrointestinal tract. *Poult Sci* 94(8), 1928–1941, doi:10.3382/ps/pev152 (2015).
377. Mazzoni, M., Bombardi, C., Vallorani, C., Sirri, F., De Giorgio, R., Caio, G., Grandis, A., Sternini, C. and Clavenzani, P. Distribution of alpha-transducin and alpha-gustducin immunoreactive cells in the chicken (Gallus domesticus) gastrointestinal tract. *Poult Sci* 95(7), 1624–1630, doi:10.3382/ps/pew057 (2016).
378. Izumi, T., Kawamura, K., Ueda, H. and Bungo, T. Central administration of leucine, but not isoleucine and valine, stimulates feeding behavior in neonatal chicks. *Neurosci Lett* 354(2), 166–168, doi:10.1016/j.neulet.2003.09.071 (2004).
379. Denbow, D. M., Van Krey, H. P., Lacy, M. P. and Skewes, P. A. Effect of intracerebroventricular glucose injection on feeding in Leghorn chicks. *Poult Sci* 64(3), 550–554, doi:10.3382/ps.0640550 (1985).
380. Grosso, G. Nutritional psychiatry: how diet affects brain through gut microbiota. *Nutrients* 13(4), doi:10.3390/nu13041282 (2021). doi: 10.3390/nu13041282.
381. Le Chatelier, E., Nielsen, T., Qin, J., Prifti, E., Hildebrand, F., Falony, G., Almeida, M., Arumugam, M., Batto, J. M., Kennedy, S., Leonard, P., Li, J., Burgdorf, K., Grarup, N., Jørgensen, T., Brandslund, I., Nielsen, H. B., Juncker, A. S., Bertalan, M., Levenez, F., Pons, N., Rasmussen, S., Sunagawa, S., Tap, J., Tims, S., Zoetendal, E. G., Brunak, S., Clément, K., Doré, J., Kleerebezem, M., Kristiansen, K., Renault, P., Sicheritz-Ponten, T., de Vos, W. M., Zucker, J. D., Raes, J., Hansen, T., MetaHIT consortium, Bork, P., Wang, J., Ehrlich, S. D. and Pedersen, O. Richness of human gut microbiome correlates with metabolic markers. *Nature* 500(7464), 541–546, doi:10.1038/nature12506 (2013).
382. Amar, J., Chabo, C., Waget, A., Klopp, P., Vachoux, C., Bermúdez-Humarán, L. G., Smirnova, N., Bergé, M., Sulpice, T., Lahtinen, S., Ouwehand, A., Langella, P., Rautonen, N., Sansonetti, P. J. and Burcelin, R. Intestinal mucosal adherence and translocation of commensal bacteria at the early onset of type 2 diabetes: molecular mechanisms and probiotic treatment. *EMBO Mol Med* 3(9), 559–572, doi:10.1002/emmm.201100159 (2011).
383. Massier, L., Chakaroun, R., Tabei, S., Crane, A., Didt, K. D., Fallmann, J., von Bergen, M., Haange, S. B., Heyne, H., Stumvoll, M., Gericke, M., Dietrich, A., Blüher, M.,

Musat, N. and Kovacs, P. Adipose tissue derived bacteria are associated with inflammation in obesity and type 2 diabetes. *Gut* 69(10), 1796–1806, doi:10.1136/gutjnl-2019-320118 (2020).

384. Schele, E., Grahnemo, L., Anesten, F., Hallén, A., Bäckhed, F. and Jansson, J. O. The gut microbiota reduces leptin sensitivity and the expression of the obesity-suppressing neuropeptides proglucagon (Gcg) and brain-derived neurotrophic factor (Bdnf) in the central nervous system. *Endocrinology* 154(10), 3643–3651, doi:10.1210/en.2012-2151 (2013).
385. Backhed, F., Ding, H., Wang, T., Hooper, L. V., Koh, G. Y., Nagy, A., Semenkovich, C. F. and Gordon, J. I. The gut microbiota as an environmental factor that regulates fat storage. *Proc Natl Acad Sci USA* 101(44), 15718–15723, doi:10.1073/pnas.0407076101 (2004).
386. Perry, R. J., Peng, L., Barry, N. A., Cline, G. W., Zhang, D., Cardone, R. L., Petersen, K. F., Kibbey, R. G., Goodman, A. L. and Shulman, G. I. Acetate mediates a microbiome-brain-beta-cell axis to promote metabolic syndrome. *Nature* 534(7606), 213–217, doi:10.1038/nature18309 (2016).
387. Torres-Fuentes, C., Golubeva, A. V., Zhdanov, A. V., Wallace, S., Arboleya, S., Papkovsky, D. B., Aidy, S. E., Ross, P., Roy, B. L., Stanton, C., Dinan, T. G., Cryan, J. F. and Schellekens, H. Short-chain fatty acids and microbiota metabolites attenuate ghrelin receptor signaling. *FASEB J* 33(12), 13546–13559, doi:10.1096/fj.201901433R (2019).
388. Suarez-Zamorano, N., Fabbiano, S., Chevalier, C., Stojanović, O., Colin, D. J., Stevanović, A., Veyrat-Durebex, C., Tarallo, V., Rigo, D., Germain, S., Ilievska, M., Montet, X., Seimbille, Y., Hapfelmeier, S. and Trajkovski, M. Microbiota depletion promotes browning of white adipose tissue and reduces obesity. *Nat Med* 21(12), 1497–1501, doi:10.1038/nm.3994 (2015).
389. Bolognini, D., Barki, N., Butcher, A. J., Hudson, B. D., Sergeev, E., Molloy, C., Moss, C. E., Bradley, S. J., Le Gouill, C., Bouvier, M., Tobin, A. B. and Milligan, G. Chemogenetics defines receptor-mediated functions of short chain free fatty acids. *Nat Chem Biol* 15(5), 489–498, doi:10.1038/s41589-019-0270-1 (2019).
390. Frost, G., Sleeth, M. L., Sahuri-Arisoylu, M., Lizarbe, B., Cerdan, S., Brody, L., Anastasovska, J., Ghourab, S., Hankir, M., Zhang, S., Carling, D., Swann, J. R., Gibson, G., Viardot, A., Morrison, D., Louise Thomas, E. and Bell, J. D. The short-chain fatty acid acetate reduces appetite via a central homeostatic mechanism. *Nat Commun* 5, 3611, doi:10.1038/ncomms4611 (2014).
391. Yunes, R. A., Poluektova, E. U., Dyachkova, M. S., Klimina, K. M., Kovtun, A. S., Averina, O. V., Orlova, V. S. and Danilenko, V. N. GABA production and structure of gadB/gadC genes in Lactobacillus and Bifidobacterium strains from human microbiota. *Anaerobe* 42, 197–204, doi:10.1016/j.anaerobe.2016.10.011 (2016).
392. Pedersen, H. K., Gudmundsdottir, V., Nielsen, H. B., Hyotylainen, T., Nielsen, T., Jensen, B. A., Forslund, K., Hildebrand, F., Prifti, E., Falony, G., Le Chatelier, E., Levenez, F., Doré, J., Mattila, I., Plichta, D. R., Pöhö, P., Hellgren, L. I., Arumugam, M., Sunagawa, S., Vieira-Silva, S., Jørgensen, T., Holm, J. B., Trošt, K., MetaHIT Consortium, Kristiansen, K., Brix, S., Raes, J., Wang, J., Hansen, T., Bork, P., Brunak, S., Oresic, M., Ehrlich, S. D. and Pedersen, O. Human gut microbes impact host serum metabolome and insulin sensitivity. *Nature* 535(7612), 376–381, doi:10.1038/nature18646 (2016).
393. Arnoriaga-Rodriguez, M., Mayneris-Perxachs, J., Burokas, A., Pérez-Brocal, V., Moya, A., Portero-Otin, M., Ricart, W., Maldonado, R. and Fernández-Real, J. M. Gut

bacterial ClpB-like gene function is associated with decreased body weight and a characteristic microbiota profile. *Microbiome* 8(1), 59, doi:10.1186/s40168-020-00837-6 (2020).

394. Breton, J., Tennoune, N., Lucas, N., Francois, M., Legrand, R., Jacquemot, J., Goichon, A., Guérin, C., Peltier, J., Pestel-Caron, M., Chan, P., Vaudry, D., do Rego, J. C., Liénard, F., Pénicaud, L., Fioramonti, X., Ebenezer, I. S., Hökfelt, T., Déchelotte, P. and Fetissov, S. O. Gut commensal E. coli proteins activate Host Satiety Pathways following Nutrient-Induced Bacterial Growth. *Cell Metab* 23(2), 324–334, doi:10.1016/j.cmet.2015.10.017 (2016).
395. Siegerstetter, S. C., Schmitz-Esser, S., Magowan, E., Wetzels, S. U., Zebeli, Q., Lawlor, P. G., O'Connell, N. E. and Metzler-Zebeli, B. U. Intestinal microbiota profiles associated with low and high residual feed intake in chickens across two geographical locations. *PLoS ONE* 12(11), e0187766, doi:10.1371/journal.pone.0187766 (2017).

Chapter 5

Advances in understanding the cellular basis of feed efficiency in poultry muscle

Walter Bottje and Kentu Lassiter, University of Arkansas, USA

1 Introduction: cell and mitochondrial primer

To be able to understand the potential relationship between feed efficiency and mitochondrial physiology and muscle accretion, it will be good to give a short review of fundamental cell and mitochondrial physiology. To function, grow, and thrive, cells must have a constant and ready supply of energy in the form of adenosine triphosphate (ATP) for a myriad of cell processes. In muscle cells, this energy requirement is magnified by the need to carry out muscular contraction.

Approximately 90% of all the energy produced in the cell occurs in mitochondria, the powerhouse of the cell. Mitochondria are believed to have evolved as the result of a commensal relationship when a single-celled organism engulfed (phagocytized) a bacterial cell, but did not digest it. The cell provided a stable environment and nutrients to the engulfed bacteria that in turn produced energy substrates for the host cell to conduct its business. Mitochondria are the only organelle outside of the nucleus to contain its own discrete DNA, a result of the bacterial DNA carried into the single-cell organism. Mitochondrial DNA is responsible for producing (encoding) a few proteins in the mitochondrial electron transport chain (where ATP is produced), that is assembled with other nuclear-encoded proteins that are synthesized in the cytosol and imported into the mitochondria.

The energy that drives many vital processes in the cell is provided in the form of a high-energy phosphate group on ATP. For example, to build a protein

http://dx.doi.org/10.19103/AS.2024.0143.05

composed of a chain of 100 amino acids, 100 molecules of ATP are needed just for the transport and insertion of amino acids by transfer RNA (tRNA) onto the growing protein chain. Roughly, five ATP molecules are needed for each peptide bond between amino acids (protein building blocks). Thus, a typical protein requires over 2000 ATP molecules. Another big energy 'user' is the sodium-potassium ATPase molecule. The action of Na+/K+ ATPase pumps Na+ out of the cell and K+ into the cell. This helps maintain cellular membrane potential and sets up concentration gradients that enables material (e.g. glucose, amino acids) to be transported into the cell that provide building blocks for protein and energy to drive the mitochondrial energy production.

ATP is produced on the electron transport chain (ETC) of mitochondria by oxidative phosphorylation. The ETC consists of a series of multi-protein complexes (I, II, III, and IV) that reside on the inner mitochondrial membrane (shown in the upper left-hand corner of Fig. 1). Electrons are passed from energy substrates from the Kreb cycle and other sources to the ETC then pass from one protein to the next along the ETC to the terminal electron acceptor, oxygen, resulting in the formation of water. As electrons move along the ETC, hydrogen ions are pumped across the membrane producing a proton motive force in the intramembranous space (between the inner and outer mitochondrial membranes. This proton motive force drives ATP synthesis when hydrogen ions pass through ATP synthase (complex V) back into the mitochondrial matrix. The ATP synthase molecule under high magnification looks like a motor that spins when ATP passes through. Thus, it can be visualized as a water wheel that turns to provide energy for milling grain. The high energy phosphate bond on ATP would be confined to the mitochondria and unavailable to the rest of the cell without two carrier/transport systems; adenine nucleotide translocase (ANT) for transport across the inner mitochondrial membrane (TIMM) and voltage-dependent activated channel (VDAC) for transport across the outer mitochondrial membrane (TOMM) into the cytosol. This allows shuttling of ADP and ATP into and out of the mitochondria, respectively. The second way high-energy phosphate can be provided to the cell to conduct 'work' is by a special protein (enzyme) called creatinine kinase (CK). This enables phospho-creatinine to interact with creatinine kinase in the cytosol which donates the phosphate group to rejuvenate ATP from ADP the cell.

The process of electron movement on the electron transport chain is not 100% efficient. While being passed from each redox protein along the ETC, electrons can be 'fumbled' and leak from the ETC. When this occurs, this electron can interact with oxygen to lead to formation of a broad group of very reactive and damaging oxygen-containing compounds called reactive oxygen species (ROS). The mitochondria are well known to be the major site of oxidative stress in the cell. ROS can damage (oxidize) any cell structure it encounters, e.g. proteins, carbohydrates, lipids, and DNA in a process called

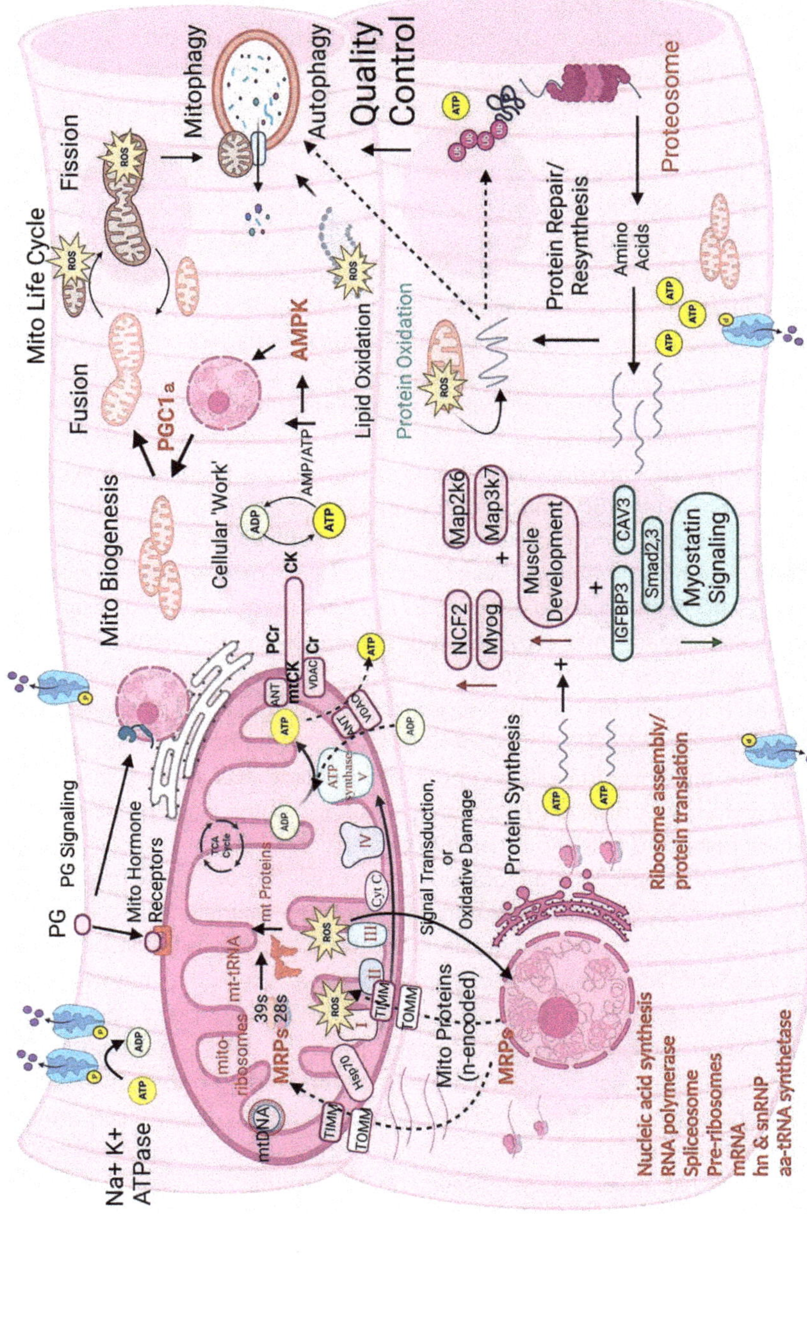

Figure 1 Mechanisms in poultry skeletal muscle that would contribute to feed efficiency at the cellular level.

oxidative stress. This damage to cell structures diminishes their functionality of the compound – so for example, if an enzyme is damaged, the ability of the enzyme to carry out its function can be compromised. Oxidative damage to DNA can cause mutations that may be associated with metabolic diseases such as cancer. Antioxidants are chemicals or enzymes that prevent oxidation in the cell. Whereas high level of ROS production can be detrimental, low levels of ROS play important roles in signal transduction and part of normal cell metabolism.

Although oxidation of structures in the cells is detrimental, the cell has several 'repair shops' to exert an important quality control function in the cell that can either digest damaged parts of the cell (a process called autophagy using an autophagosome) and another that can digest damaged proteins, and recycle amino acids for resynthesis of proteins (proteolysis occurring in a proteosome). When a mitochondrion becomes damaged or is producing excessive amounts of ROS, the damaged portion can be 'pinched off' by mitochondrial fission processes and sent to the autophagosome for digestion in a process called mitophagy.

Originally, mitochondria were thought to be single static structures with anywhere from 10 to 1000 mitochondria per cell. It has now well known that mitochondria move about the cell and can grow or divide in fusion and fission processes. Among other things, these fusion and fission processes are important in sensing and responding to changes in cellular energy status. For example, if there is an increase in energy demand, mitochondrial content will increase in a process termed mitochondrial biogenesis. With low nutrient levels (e.g. starvation), mitochondria will fuse together and elongate and become more efficient at producing ATP. During periods of excess nutrient availability (e.g. obesity), the mitochondria will fragment and become less efficient in energy production to burn off excess nutrients. In this fragmented state, mitochondria will exhibit increased mitochondrial ROS production.

Mitochondria also have a feedback system called proton leak to self-limit mitochondrial ROS production. Proton leak has been estimated to account for 25% of basal metabolic rate in animals. In proton leak, protons (hydrogen ions) pass through the inner mitochondrial membrane at sites other than the ATP synthase. In this case, the ETC keeps running but ATP is not produced. Proton leak reduces the mitochondrial membrane potential thereby reducing ROS production. A great deal of information and understanding of proton leak has been advanced by Dr. Martin Brand (Cambridge University, UK and currently at the Buck Institute on Aging (San Francisco)). There are several proteins within the mitochondria that facilitate proton leak. The first one that was discovered was the classical uncoupling protein (UCP) from in brown fat tissue that essentially generates heat in neonates. Another protein that can uncouple mitochondria is ANT – described above as being instrumental in shuttling adenine nucleotides

(ATP, ADP) in and out of the mitochondria. There is also a process in which mitochondrial ROS mediated lipid oxidation produces an oxidized fatty acid (2 hydroxy-4 nonenal, HNE) that activates uncoupling of the inner mitochondrial membrane and reduces mitochondrial ROS production (Brown and Brand, 1991; Rolfe and Brand, 1997; Skculachev, 1996, 1997). Small differences in proton leak could play a role in the phenotypic expression of feed efficiency (Bottje et al., 2009).

2 Mitochondrial physiology

2.1 Introduction

The discovery and first description of mitochondria is credited to Albert von Kolliker (Lehninger, 1965). Mitochondria are descendants of bacteria that developed a commensal relationship with a host cell several millennia ago and as such, they are the only organelle outside the nucleus with its own discrete pool of mitochondrial DNA (mtDNA) (Lane, 2010). mDNA is a circular molecule containing over 16 000 base pairs for roughly 37 genes that code for 2 ribosomal RNAs, 22 transfer RNAs, and 13 proteins of the respiratory chain that, along with over 70 other nuclear encoded proteins, make up the respiratory (electron transport) chain (Anderson et al., 1981). Nuclear-encoded proteins, such as mitochondrial ribosomal proteins (MRPs) and proteins of the electron transport chain, that are synthesized in the cytosol are transported into the mitochondria by outer (TOMM) and inner (TIMM) membrane translocase proteins. These processes are shown in the upper left-hand corner of Fig. 1.

Mitochondria are part of an active network within the cell that undergo complex fission and fusion processes that enables them to communicate and *'form into local and widespread mitochondrial syncytia within cells'* (Hoppins et al., 2007) (upper right-hand corner of Fig. 1). A mitochondrial life cycle has been described that is instrumental in removing defective mitochondria (see review by Liesa and Shirhai, 2013). With high nutrient availability, inefficient fragmented mitochondria are present whereas during low energy availability, the mitochondria elongate and become very efficient in producing ATP. Fission and fusion are under the control of highly conserved dynamin-related GTPase proteins (DRPs) that regulate membrane-associated cellular processes (Praefke and McMahon, 2004). Mitochondrial fusion is necessary to distribute mtDNA to mitochondria throughout the cell and helps maintain functionally competent mitochondria within a cell. In contrast, mitochondrial fission insures that competent mitochondria are distributed throughout the cell and is important in mitochondrial biogenesis.

2.2 Oxidative phosphorylation

Mitochondria generate 90% of cellular energy by oxidative phosphorylation. The mitochondrial electron transport chain (ETC), first reported by Kennedy and Lehninger (1949), consists of five multi-protein enzyme complexes (CI–CV) and two mobile electrons (e–) carriers, coenzyme Q (Q), and cytochrome *c* (cyt *c*) (Fig. 1). Electrons (e–) enter the ETC at CI (for NADH-linked substrates), or CII (FADH2-linked substrate). Coenzyme Q (CoQ, ubiquinone) carries e– from CI and CII to CIII while cyt *c* shuttles e– from Complex III to Complex
Mitochondria generate 90% of cellular energy by oxidative phosphorylation. The mitochondrial electron transport chain (ETC), first reported by Kennedy and Lehninger (1949), consists of five multi-protein enzyme complexes (CI–CV) and two mobile electrons (e–) carriers, coenzyme Q (Q), and cytochrome *c* (cyt *c*) (Fig. 1). Electrons (e–) enter the ETC at CI (for NADH-linked substrates), or CII (FADH2-linked substrate). Coenzyme Q (CoQ, ubiquinone) carries e– from CI and CII to CIII while cyt *c* shuttles e– from Complex III to Complex IV. Electron movement along the ETC is accompanied by proton pumping into the intermembranous space that creates a proton motive force (Lehninger et al., 1993). Proton motive force, consisting of a membrane potential ($\Delta\psi\mu$) and a pH gradient, provides energy for ATP synthesis as protons flow back into the matrix through (complex V, ATP synthase).

2.3 Oxidative stress

Mitochondria are a major site of endogenous oxidative stress in the cell from ROS production. In this process, O2 is converted to ROS by univalent reduction of O2 to superoxide (O2•-) (Boveris and Chance, 1973; Chance et al., 1979; Turrens and Boveris, 1980). Superoxide dismutase converts O2•- to H2O2. While H2O2 is not very reactive, it can be converted to the highly reactive hydroxyl radical (•OH) in the presence of Fe^{2+} and Cu^{2+}. Due to its lipid solubility, H2O2 can cross membranes and oxidize proteins, DNA, lipids, and carbohydrates throughout the cell. Protection from oxidation in the cell is provided by a host of antioxidants; e.g. superoxide dismutase, glutathione (GSH) and GSH shuttling enzymes (GSH peroxidase and GSH reductase), peroxiredoxin (PRx), and thioredoxin (TRr).

2.4 Proton leak

Protons can cross the mitochondrial membrane at sites other than the ATP synthase that short circuits ATP synthesis in a process termed proton leak (Brand, 1990a, 1990b). Proton leak represents an inefficiency but attenuates mitochondrial ROS production. As proton leak represents up to 50% of oxygen use in a perfused muscle (Brand, 1990a, 1990b; Rolfe et al., 1999), proton leak could contribute as much as 25% of the total basal metabolic rate of an animal (Rolfe and Brand, 1996, 1997). A self-limiting feedback on mitochondrial ROS production by superoxide through mitochondrial uncoupling was shown by Skulachev (1996, 1997).

2.5 Phosphocreatine mitochondrial-cytosol energy transfer system

Cellular work (Fig. 1, upper middle) is carried out using the high-energy phosphate in ATP produced by mitochondrial oxidative phosphorylation or glycolysis (Fig. 1, upper middle). ATP is provided to the cytosol either directly as ATP or indirectly as phosphocreatine (PCr) by a protein transfer system consisting of adenine nucleotide translocase (ANT), voltage-dependent activated channel (VDAC), and cytosolic creatine kinase (CK) and mitochondrial CK (MTCK). Cytosolic CK catalyzes phosphate transfer from PCr to ADP and is crucial for energy metabolism in tissues with high and/or fluctuating energy demand (e.g. skeletal, heart, smooth muscle, and brain). Jacobus and Lehninger (1973) first proposed a mechanism in which the MTCK transfers phosphate groups from ATP to form PCr which is a high energy phosphate reservoir for rapid regeneration of cytosolic ATP. Brain (B) and muscle (M) isoforms of CK exist as dimers in the cytosol (either BB, MM, or MB) whereas mitochondrial CK is found as an octamer present between the inner and outer mitochondrial membranes [e.g. reviews by Brdiczka et al. (2006); Schlattner et al. (2006)]. The transfer of PO4 from mitochondrial ATP to form PCr in the cytosol is carried out by adenine nucleotide translocase (ANT), MTCK, and VDAC. The ANT-CKMT-VDAC system links mitochondrial ATP generation to a PCr reservoir that can be drawn upon when cytosolic ATP levels fall in response to energy demand. A prime example of cellular work in most cells is the requirement to expend energy to drive the Na+/K+ ATPase (Fig. 1, upper left) to maintain low levels of Na+ in the cell necessary for active and facilitated transport of important chemicals (e.g. amino acids, sugars) as well as water by osmosis.

2.6 Mitochondrial biogenesis

Mitochondrial biogenesis is stimulated in response to increased energy demand. Paul and Sperling (1952) observed that there were more mitochondria in breast muscle in active flying pigeons than in the relatively sedentary commercial chicken. Because of its role in mitochondrial biogenesis, peroxisome proliferator-activated receptor gamma coactivator 1-alpha (PGC-1α) has been termed the master regulator of mitochondrial protein synthesis and biogenesis Nisoli et al. (2003, 2004) (Fig. 1, upper middle right). Adenosine monophosphate-activated protein kinase (AMPK) is critical for (a) sensing energy (the AMP to ATP ratio) status and stimulating mitochondrial biogenesis (Zhou et al., 2001; Carling,

2004; Hardie, 2004; Hardie and Sakamoto, 2006; Hardie et al., 2006), and (b) in regulating animal food intake and overall energy balance (Minokoshi et al., 2004). In general, AMPK reduces ATP-utilizing (anabolic) pathways (e.g. fatty acid synthesis) and increases ATP-generating (catabolic) pathways (e.g. fatty acid oxidation, glycolysis). AMPK also up-regulates PGC-1α expression and mitochondrial biogenesis (Ojuka, 2004). In conjunction with thyroid hormone receptor activation, PGC-1α upregulates ANT (Masatoshi et al., 2005). The AMPK/AMPK cascade is sensitive to ROS, and mitochondrial ROS-mediated upregulation of AMPK increased cellular antioxidants (Colombo and Moncada, 2009).

3 Mitochondria and a cellular basis of feed efficiency

3.1 Overview

A mitochondrial-centric view of feed efficiency (FE) developed from studies investigating relationships between mitochondrial function and FE in pedigree broiler males (Bottje et al., 2002). As mitochondria are responsible for producing 90% of energy in the form of ATP for the cell via oxidative phosphorylation, and a large component of the basal metabolic rate of animals is attributed to proton leak in muscle mitochondria (Rolfe and Brand 1990a, 1990b, Rolfe et al., 1999), it is really no surprise that mitochondria would play a major role in feed efficiency. As indicated previously, mitochondria are not static structures but are well-documented to move within the cell and adjust their energetic efficiency by elongation (mitochondrial fusion) or fragmentation in response to low or high energy availability, respectively (Liesa and Shirhai, 2013). These changes in mitochondrial morphology can also be instrumental in the removal of damaged mitochondria by mitophagy to improve energy production efficiency. Mitochondria also have intimate contacts with other organelles in the cell such as the endoplasmic reticulum, golgi apparatus, and nucleus such that communication to and from the mitochondria and channeling of proteins is facilitated for incorporation into the mitochondrial structure. From proteogenomic studies, a cellular architecture appears to exist in high FE (more efficient) compared to low FE (less efficient) muscle that would (a) facilitate the production and export of high energy phosphate groups to support cellular work, (b) enhanced ability to remove damaged components in the cell, and (c) enhanced ribosome construction for protein synthesis. These processes and more will be described in greater detail below in an effort to understand the components of feed efficiency at the cellular/molecular level in poultry muscle. Parts of this have been presented previously (Bottje and Kong, 2017).

Continued improvement in animal agriculture efficiency is critical for maintaining sustainable poultry and livestock production. Because of the cost that feed (60–70% of total production) represents to overall costs of animal production, feed efficiency remains a very important genetic trait in animal agriculture. As such, the efficiency of how that energy is produced and utilized ultimately contributes to cellular utilization efficiency – cellular efficiency. Across the spectrum of animal species, skeletal muscle accounts for approximately 50% of body mass, and when considered as an entire organ, skeletal muscle contributes between 25% and 40% of the overall basal metabolic rate (Brand, 1990a, 1990b; Zurlo et al., 1990; Rolfe et al., 1999). A quarter of total basal metabolic rate in an animal may be assigned to one component of mitochondrial function – proton leak (Rolfe and Brand, 1996, 1997). A clear link between breast muscle mitochondrial function and feed efficiency has been reported (Bottje et al., 2002; Bottje and Carstens, 2009). Thus, mitochondrial function in skeletal muscle can play a substantive role in overall efficiency.

A review of avian muscle physiology, structure, growth, and development in avian species is provided by Velleman and McFarland (2015). Skeletal muscle growth in poultry post-hatch is due to hypertrophy (increased size of existing cells) (Smith, 1963). In skeletal muscle, 'work' can be viewed as mechanics associated with movement and exercise, but in agriculture animals, 'work' can also be considered as processes associated with hypertrophy, protein synthesis, assembly of myofibrils, skeletal fibers. There is a continual need for energy input for protein synthesis, repair, and degradation. There are numerous energy-consuming processes that are needed to maintain cell homeostasis; e.g. cell membrane potential and transport of materials (Na+/K+ ATPase), construction-related processes (muscular and cytoarchitecture), repair processes (e.g. ability to 'fix' damaged structures – digestion of proteins in proteosomes, resynthesis of proteins from component amino acids), etc. How efficiently muscle cells can carry out these processes at the cellular level likely has a large impact on overall production efficiency in an animal.

Figure 1 shows processes/mechanisms that will be subsequently discussed below pertaining to some, but certainly not all, of what may comprise feed efficiency at the cellular level in muscle. An important caveat for these mechanisms is that the studies were conducted on tissue obtained from pedigree male broilers fed the same diet and individually phenotyped for feed efficiency as described (Bottje et al., 2002). A strength of these studies is that the proteogenomic studies upon which the findings were reported (Kong et al., 2011, 2016; Bottje et al., 2012, 2017a, 2017b; Lassiter et al., 2018; Piekarski-Welsher et al., 2018) were all conducted on the same set of tissues that had been collected in Bottje et al. (2009). The weakness of this approach is that inherent natural variability that would come from more 'replications' of the study is avoided. However, it should be pointed out that a number of different studies investigating respiratory chain complex activities in several tissues from different had remarkably similar findings.

In Fig. 1, red type and pink or red structures are used to denote processes that were found to be enhanced in the high feed efficiency pedigree male broiler phenotype whereas green structures or type were enhanced in the low feed efficiency pedigree phenotype that was investigated. (See text for details and explanation.)

Functional studies in isolated mitochondria revealed a link between mitochondrial function and FE and identified site-specific defects in mitochondrial electron transport in muscle (breast and leg), liver, intestines, and heart (Bottje et al., 2002; Bottje and Carstens, 2009). In addition, greater protein oxidation (likely from increased mitochondrial ROS production in low FE animals) was observed in these tissues as well as in lymphocytes and heart muscle. Because mitochondrial ROS can be important in signal transduction and affect gene and protein and translation, proteogenomic studies were conducted in breast muscle (Kong et al., 2011, 2016). These studies used a paradigm in which tissues were obtained from a single pedigree male broiler line – highly selected for growth performance and muscle development – that were (a) fed the same diet, (b) housed under identical conditions, and (c) individually phenotyped for FE. In these studies, a group of 300 males were placed in floor pens and raised to 6 weeks of age. From this group, 100 males were placed in cages and individually phenotyped for FE and those with the highest or lowest FE (6–8 per group) were selected and tissue used for analysis.

The paradigm used in the pedigree broiler male line studies differs from those in which animals were divergently selected for FE (e.g. Aggrey et al., 2010; Lee et al., 2015) or were derived from an outbred commercial broiler line (Zhou et al., 2015). These studies were conducted with breast muscle (with a predominance of type IIa fibers as opposed to muscle containing a mixture of type I and type IIa fibers that is more typical of skeletal muscle. In the proteomics study (Kong et al., 2016), out of 232 mitochondrial proteins that were recognized in the study 150 were upregulated (significantly or numerically) and 71 downregulated in the high FE phenotype compared to the low FE phenotype. Similarly, in a global gene expression study (RNAseq), 475 genes encoding mitochondrial proteins were upregulated and 224 were downregulated (Bottje et al., 2017a). In both cases, the binomial statistical analysis indicated a highly significant skew ($P < 0.00001$) in the high FE phenotype. This is a clear indication that mitochondrial expression in general would be enhanced in the high compared to low FE breast muscle.

3.2 Mitochondrial function, biochemistry, and oxidative stress in feed efficiency

It was determined that mitochondrial ETC chain coupling[1] was lower in low FE muscle mitochondria compared to high FE mitochondria (Bottje et al., 2002). The reduced coupling in low FE was associated with site-specific defects in electron transport, increased mitochondrial ROS production, and increased protein oxidation in muscle mitochondria, heart, liver, and duodenal tissue (Bottje and Carstens, 2009). Decreased respiratory chain complex activities in low FE mitochondria were hypothesized to be due to increased ROS-mediated protein oxidation.

Proton leak attenuates mitochondrial ROS production by a self-limiting feed back mechanism (Skulachev, 1996, 1997; Brand et al., 2004). A comprehensive proton leak kinetic study revealed that proton leak may be enhanced in low compared to high FE muscle mitochondria (Bottje et al., 2009). These results suggest that there are subtle differences in membrane characteristics (e.g. lipids. integral membrane proteins) or integral proteins that affect proton leak and contribute to higher mitochondrial ROS production in this low FE animal model (Bottje et al. 2002; Bottje and Carstens, 2009).

3.3 Enhanced energy production and transfer system in high feed efficiency

In cells with fluctuating energy demand (e.g. skeletal muscle), a transport system across the inner and outer mitochondrial membranes links oxidative phosphorylation to cytosolic phosphorylated creatine (PCr) that is a phosphate reservoir for rapid repletion of cytosolic ATP (Jacobus and Lehninger, 1973). Crucial proteins of this energy transfer system include several creatine kinase (CK) isoforms found in the cytosol and mitochondria. In proteomic studies (Kong et al., 2016), several components of this system were upregulated in high FE compared to low FE breast muscle. These components include adenine nucleotide translocase (ANT), voltage-dependent activated channel 2 (VDAC2), the brain isoform of creatine kinase (CK-B), one form of mitochondrial creatine kinase (CKMT1A), and several proteins of the electron transport chain.

Increased expression of ANT, VDAC, and mitochondrial CK would connect mitochondrial produced ATP to the formation of phosphorylated creatine (PCr) to replenish cytosolic ATP. An example of an essential and extensive ATP consuming reaction is Na^{+}/K^{+} ATPase. In the high FE muscle, increased adenosine kinase (AdK or AK) expression would increase the AMP:ATP ratio (an indicator of increased energy demand in the cell). The increase in AMP:ATP

1 Coupling was determined by measuring oxygen consumption during state III (ADP in excess) and state IV (ADP limiting) respiration rate in isolated mitochondria *in vitro* (Estabrook, 1967).

stimulates AMPK (5′ AMP-activated protein kinase) that in turn would increase mitochondrial ATP output through mitochondrial biogenesis via the activity of peroxisome proliferator-activated receptor coactivator 1-alpha (PGC1α). Due to its role in upregulating key transcription factors, PGC1α has been called the master regulator of mitochondrial biogenesis (Nisoli et al., 2003, 2004). The predicted activation of tri-iodothyronine (T3) (Kong et al., 2016) would also enhance PGC1α activity. The insulin signaling pathway also would be enhanced in the high FE phenotype with the predicted increase in activities of insulin receptor (INSR) and insulin-like growth factor receptor 1 (IGF1R). In this pathway, caveolin 1 (CAV1) and insulin degrading enzyme (IDE) help internalize insulin after binding to the receptor, and in transmitting the signal to the nucleus, respectively. The predicted activation of progesterone (PG) would enhance mitochondrial activity and complement the activity of the insulin signaling pathway.

AMPK not only stimulates mitochondrial biogenesis but also suppresses apoptosis (programmed cell death). Apoptosis inhibition in the high FE muscle phenotype would be complemented by anti-apoptotic protease activating factor 1 (APIP) and protein phosphatase 2A (PP2A) upregulation (Bottje and Kong, 2013). The APIP is highly expressed in skeletal muscle where it inhibits cytochrome c release and APAF-1 mediated activation of caspase-9, critical steps in apoptosis (Cho, et al., 2004). Muscle accounts for up to 50% of total energy consumption in the body (Brand, 1990a, 1990b; Rolfe and Brand, 1996). Thus, increased expression of APIP in muscle tissue would contribute to the high-FE phenotype by decreasing cell turnover.

3.4 Enhanced nucleotide metabolism, ribosome assembly, and protein translation in high feed efficiency

Evidence of enhanced nucleotide, purine, and pyrimidine synthesis was reported to be upregulated in high FE breast muscle tissue (Bottje and Kong, 2017) and is shown in Fig. 1 (bottom left). A connection for enhanced deoxynucleotide synthesis in DNA formation (deoxynucleic acid) in high FE is indicated by the upregulation of glutaredoxin and ribonucleotide synthetase in the high FE muscle. In mitochondria, upregulation of pyrophosphatase 2 (PPA2) that catalyzes the metabolism of pyrophosphate (ppi) to phosphate (pi) would be used as a substrate, along with ADP (transported into the mitochondria through VDAC and ANT channels), by ATP synthase for the synthesis of ATP.

There was evidence of enrichment of transcription and translation in high FE breast muscle (Bottje et al., 2017a) – also shown in Fig. 1 (bottom left). This evidence was obtained by not using a cutoff of significance or fold differences – criteria that are typically used in global gene and protein expression analysis – but rather it was based on analyzing the entire datasets by using a 'yes-no' using binomial distribution analysis that showed a highly significant binomial *P* value (*P* < 0. 00001) of higher or lower expression in the high FE compared to Low FE data. Using a similar similar binomial distribution analyses of transcriptomic and proteomic data, additional information was obtained to paint a more comprehensive of the proteogenomic landscape present in the high FE breast muscle tissue related to the cellular basis of high FE pedigree male broilers (Bottje et al., 2017a; Piekarski et al., 2018). This would not have been possible without the approach taken of analyzing entire global expression datasets to identify regulatory impact factors (Hudson et al., 2009, 2012). Tables 1 to 3 from Bottje et al. (2017a) are reproduced here for convenience. The data were obtained from global expression analysis (microarray, proteomics, and RNAseq) conducted on the same set of muscles from high and low FE pedigree male broilers. Essentially, what these data point to are mechanisms associated with feed efficiency that would be missed when fold difference and significance cut offs are used on global expression datasets. The potential loss of data using cut-offs is exemplified in Fig. 2. Using the binomial distribution analysis approach should be viewed as simply another tool that can be used to understand fundamental mechanisms that complement targeted gene and protein analysis using PCR, western blot analysis, or pathway analysis programs.

Table 1 Numbers of genes and proteins associated with mitochondrial protein synthesis (ribosomes, transcription, and tRNA) as well as transport of molecules into the mitochondria that were higher (H) or lower (L) in breast muscle of high feed efficiency (FE) compared to low FE pedigree broiler male for the terms provided obtained from global expression microarray, proteomics and RNAseq datasets and for combined in all three global expression datasets.

	Microarray[1] ---			**Proteomics[2] ----**			**RNAseq[3] ---**		
Terms	**H**	**L**	**P***	**H**	**L**	**P***	**H**	**L**	**P***
Mitochondrial Ribosomal Proteins-	18	6	0.008	0	1		55	4	8×10^{-13}
TOMM[4]	0	0		2	1		2	2	
TIMM[5]	1	1		0	0		9	0	0.0029
Mitochondrial transcription	2	0		0	0		5	1	0.0938
Mitochondrial tRNA	1	0		1	0		14	2	0.0018
Mitochondrial ribosome, translational, processing, peptidase	4	0	0.0625	0	0		6	2	0.1093

[1]cDNA microarray - Kong et al. (2011).;
[2]Shotgun proteomics - Kong et al. (2016);
[3]RNAseq - Bottje et al. (2017a);
*Binomial (exact) P values for H vs L in each dataset.
[4]transport protein outer mitochondrial membrane
[5]transport protein inner mitochondrial membrane

Table 2 Number of genes and proteins associated with transcription, pre-mRNA processing, and ribosomal assembly that were higher (H) or lower (L) in breast muscle of high feed efficiency (FE) compared to low FE pedigree broiler male for the terms provided obtained from global expression microarray, proteomics, and RNAseq datasets, and for number of all three global expression datasets combined

	Microarray[1]			**Proteomics**[2]			**RNAseq**[3]		
Terms	**H**	**L**	**P***	**H**	**L**	**P***	**H**	**L**	**P***
RNA polymerase	14	13		1	2		28	8	0.0004
Transcription	16	22	0.0812	3	1		94	69	0.0092
TATA Box, T-box	6	4		0	0		19	8	0.0165
Small nuclear (sn) ribonucleoproteins (RNP) and heterogeneous (hn) RNP	10	1	0.0054	4	1		12	1	0.0016
Spliceosome, helicase,	6	1	0.0547	4	0	0.0625	27	11	0.0044
Pre mRNA processing									
RNA binding motif	11	8		0	0		39	13	0.0001
Ribosome-Ribosomal	8	1	0.0017	41	4	4×10^{-9}	57	49	0.0573
Transfer RNA (tRNA)	8	8		6	0	0.0156	64	11	1.0×10^{-10}
Translation (Eukaryotic initiation-elongation)	12	11		12	7	0.0961	43	13	3.0×10^{-5}
Karyopherins (nuclear transport proteins) (nucleoporins, karyopherins, Importins, exportins)	5	6	0.0537	1	0		24	15	0.0457
Small ubiquitin-like modifier (SUMO) (peptidase, isopeptidase, ligase)	1	1		1	1		9	2	0.0269

P values not underlined indicate a significant binomial skew in high FE numbers *P* values that are *underlined* indicate a significant skew in low FE numbers.
P values that are blank are indicative of values 0.10.[1]cDNA microarray (Kong et al.,2011).
[2]Shotgun proteomics (Kong et al., 2016).
[3]RNAseq (Bottje et al., 2017a).
*Binomial (exact) *P* values for H vs L (higher vs. lower) in each dataset.

Table 3 Number of genes and proteins associated with cytoskeletal-muscle fibers, phosphorylation regulation, and cellular quality control mechanisms (proteosome and autophagy) that were higher (H) or lower (L) in breast muscle of high feed efficiency (FE) compared to low FE pedigree broiler male for the terms provided obtained from global expression microarray, proteomics, and RNAseq datasets and for numbers of all three global expression datasets combined

	Microarray			Proteomics			RNAseq			
Term	**H**	**L**	**P***	**H**	**L**	**P***	**H**	**L**	**P***	**H**
Cytoskeletal-muscle fibers[4]	9	20	0.0004	7	21	0.0044	21	42	0.0030	35
Phosphorylation										
Protein tyrosinephosphatase	2	1		2	0		5	27	0.0020	9
Protein phosphatase	4	2		7	1	0.0313	30	24	0.0778	41
Tyrosine kinase	6	4		1	0		11	31	0.001	17
Quality control										
Proteosome	7	3	0.1172	21	4	0.0004	31	4	2E-06	59
Autophagy	6	2	0.1093	4	0	0.0625	36	15	0.0014	49

P Values not underlined indicate a significant binomial skew in high FE P values that are *underlined* indicate a significant skew in low FE
P values that are blank are indicative of values 0.10.[1]cDNA microarray, Kong et al. (2011).
[2]Shotgun proteomics - Kong et al. (2016)
[3]RNAseq - Bottje et al. (2017).[4]*Actin (ACTA), myosin (MYH), troponin (TNNT)*
*Binomial (exact) *P* values for H vs L in each dataset. Blanks are indicative of *P* values 0.10.

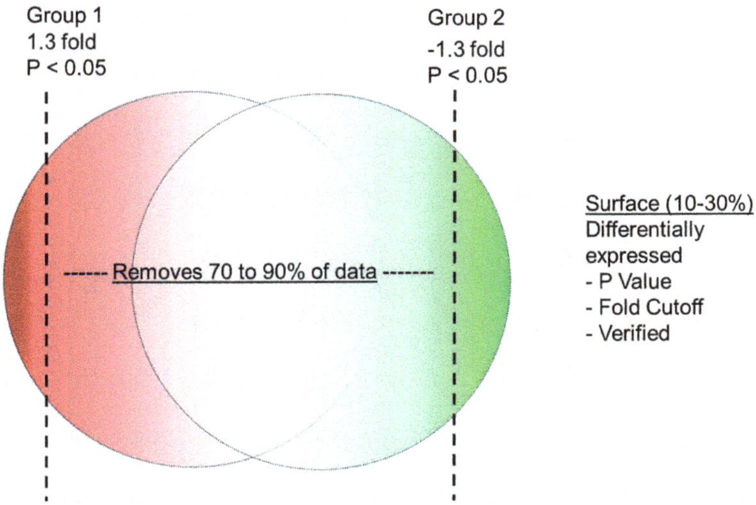

Figure 2 Depiction of the effect of using fold difference in expression and statistical significance cutoff can have on the amount of data discarded in global expression datasets. The darker red and green colors denote groups of the most (highest fold difference) of up- and downregulated gene expression. Hudson et al. (2009) used the entire dataset of global expression analysis to calculate regulatory impact factors. Transcriptomic and proteogenomic studies typically use fold (e.g. 1.3) and *P* value (*P* 0.05) differences resulting in the removal of 70–90% of the global expression data shown in the inset in the figure.

3.5 Muscle development and progesterone signaling

Finally, the last part of the cellular basis of FE in this pedigree male broiler model is shown in the bottom middle part of Fig. 1 that would fall under the umbrella of enhanced muscle development in high FE animals. This is based on several pieces of information that was reported in (Bottje et al. 2017; Lassiter et al., 2019). A major component of this mechanism was the significant downregulation of myostatin (MSTN), a phenomenon that is well known and documented to be associated with extreme muscle hypertrophy (double muscling) (e.g. McPherron and Lee, 1997; McPherron et al., 1997; Kollias and McDermott, 2008). There appeared to be a coordinated reduction in slow twitch isoforms of the contractile components of the muscle observed in breast muscle in broilers. This concurred with an earlier global expression study showing downregulation of many muscle-related proteins (Kong et al., 2011). Using regulatory impact factor analysis, it was determined that progesterone signaling through progesterone receptors was playing an important role in the phenotypic expression of FE (Bottje et al., 2017). This research lead to identification of progesterone receptors as well as several others (estrogen, thyroid hormone, corticosterone, and insulin) on mitochondria both *in vitro* and *in vivo* (Lassiter et al., 2018). Thus, it is possible that there is a hidden world of hormonal signaling going on within the cell in which nuclear regulation of mitochondrial function internally and have a profound effect on energetic efficiency at the cellular level.

4 Conclusion

Returning to Fig. 1, there is evidence to indicate that a cellular landscape that would contribute to enhanced efficiency of operation in muscle could contain the following components:

1. Enhanced energy production and energy transfer mechanisms; energy production occurring in the mitochondria and energy transfer facilitated by increased expression of creatinine kinase, transfer of nucleotides through ANT and VDAC channel proteins. Importantly, efficient mitochondria appear to be ones that have lower ROS production and cause less oxidative damage to proteins.
2. A background of enhanced nucleotide synthesis, ribosome assembly, and protein translation.
3. Enhanced genes associated with muscle development that include downregulation of myostatin signaling.
4. Enhanced quality control mechanisms associated with autophagy and proteosomal activity. Greater quality control and repair of proteins would insure that proteins would be well maintained for peak performance to operate smoothly and efficiently.
5. Potential for progesterone signaling occurring either extra- and/or intracellularly that would enhance muscle development.

It would be nice if a single 'magic' genetic marker could be identified for marker-assisted selection for FE. Unfortunately, this probably won't occur because FE is such a complex trait. Identification of progesterone signaling as a major player in establishing FE phenotypes by regulatory impact factor analysis (Hudson related to FE, based on connectivity and impact on the expression of many genes. However, by the nature of RIF analysis, these factors are ones that are not differentially expressed individually. This suggests that progesterone and possibly other hormones may be operating in the background – under the radar – and having a collective effect on the phenotypic expression of FE. It could be hypothesized that endogenous production of hormones within muscle cells could be playing an important role in the phenotypic expression of FE that has not previously been detected (Bottje et al., 2021). Further investigation is needed to determine if minute amounts of steroid hormones produced within the muscle cell could have direct impact on nuclear or mitochondrial activities.

5 Acknowledgements

Information in this chapter would not have existed without full support and cooperation of a poultry industry partner to provide access to unique genetic broiler lines and monetary support of USDA (USDA-NIFA #2013-01953 and USDA NIFA SAS #2019-69012-29905). I would like to dedicate this writing to the two new loves of my life, Ryan and Issy Bottje, their parents (Tyler and Kelsey) and MeMe (Sari).

6 References

Aggrey, S. E., Karnuah, A. B., Sebastian, B. and Anthony, N. B. 2010. Genetic properties of feed efficiency parameters in meat-type chickens. *Genet. Sel. Evol.* 42:25. doi:10.1186/1297-9686-42-25.

Anderson, S., Bankier, A. T., Barrell, B. G., Debruijn, M. L., Coulson, A. R., Drouin, J., Eperon, I. C., Nierlich, D. P., Roe, B. A., Sanger, F., Schreier, P. H., Smith, A. J. H., Staden, R. and Young, I. G. 1981. Sequence and organization of the human mitochondrial genome. *Nature* 290:457–465.

Bottje, W., Iqbal, M., Tang, Z. X., Cawthon, D., Okimoto, R., Wing, T. and Cooper, M. 2002. Association of mitochondrial function with feed efficiency within a single genetic line of male broilers. *Poult. Sci.* 81:546–555.

Bottje, W. and Carstens, G. E. 2009. Association of mitochondrial function and feed efficiency in poultry and livestock species. *J. Anim. Sci.* 87:E48–E63.

Bottje, W. G., Brand, M. D., Ojano-Dirain, C., Lassiter, K., Toyomizu, M. and Wing, T. 2009. Mitochondrial proton leak kinetics and relationship with feed efficiency within a single genetic line of male broilers. *Poult. Sci.* 88:1683–1693.

Bottje, W. G., Kong, B. W., Song, J. J., Lee, J. Y., Hargis, B. M., Lassiter, K., Wing, T. and Hardiman, J. 2012. Gene expression in breast muscle associated feed efficiency in a single male broiler line using a chicken 44k microarray II. Differentially expressed focus genes. *Poult. Sci.* 91:2576–2587.

Bottje, W. and Kong, B. W. 2013. Feed efficiency: mitochondrial function to global gene expression. *J. Anim. Sci.* 91:1582–1593. doi:10.2527/jas.2012-5787.

Bottje, W. G., Kong, B., Lee, J. Y., Washington, T., Baum, J. I., Dridi, S., Wing, T. and Hardiman, J. 2014. Potential roles of mTOR and protein degradation pathways in the phenotypic expression of feed efficiency in broilers. *J. Biochem. Physiol.* 3:1–8.

Bottje, W. G. and Kong, B. W. 2017. The cellular basis of feed efficiency in poultry muscle: mitochondria and nucleic acid metabolism. In T. Applegate (Ed.), *Achieving sustainable production of poultry meat.* Burleigh Dodds Series in Agricultural Science (No. 14) (pp. 51–69). https://doi.org/10.4324/9781351114325.

Bottje, W. G., Lassiter, K., Piekarski-Welsher, A., Dridi, S., Gomez-Reverter, A., Hudson, N.J. and Kong, B.-W. 2017a. Proteogenomics reveals enriched ribosome assembly and protein translation in *Pectoralis major* of high feed efficiency pedigree broiler males. *Front. Physiol.* 8 (article 306):1–11. https://doi.org/10.3389/fphys.2017.00306.

Bottje, W. G., Lassiter, K., Dridi, S., Hudson, N. and Kong, B.-W. 2017b. Enhanced expression of proteins involved in energy production and transfer in breast muscle of pedigree male broilers exhibiting high feed efficiency. *Poult. Sci.* 96:2454–2458.

Bottje, W., Kong, B. and Lassiter, K. 2021 (Abst). *Endogenous hormone signatures: implications for mitochondria and cell function*. International Congress of Food Animal Endocrinology (October, 2021).

Boveris, A. and Chance, B. 1973. The mitochondrial generation of hydrogen peroxide. General properties and effect of hyperbaric oxygen. *Biochem. J.* 134(3):707–716.

Brand, M. D. 1990a. The contribution of the leak of protons across the mitochondrial inner membrane to standard metabolic rate. *J. Theor. Biol.* 145:267–286.

Brand, M. D. 1990b. The proton leak across the mitochondrial inner membrane. *Biochem. Biophys. Acta.* 1018:128–133.

Brand, M. D., Affourtit, C., Esteves, T. C. K., Kambert, A. J., Miwa, S., Pakay, J. L. and Parker, N. 2004. Mitochondrial superoxide production, biological effects, and activation of uncoupling proteins. *Free Rad. Biol. Med.* 37:755–767.

Brown, G. C. and Brand, M. D. 1991. On the nature of the mitochondrial proton leak. *Biochim. Biophys. Acta.* 1059:55–62.

Brdiczka, D. G., Zorov, D. B. and Sheu, S. 2006. Mitochondrial contact sites: their role in energy metabolism and apoptosis. *Biochem. Biophys. Acta.* 1762:148–163.

Carling, D. 2004. The AMP-activated protein kinase cascade - A unifying system for energy control. *Trends Biochem. Sci.* 29:18–24.

Chance, B., Sies, H. and Boveris, A. 1979. Hydroperoxide metabolism in mammalian organs. *Physiol. Rev.* 59:527–605.

Cho, D. H., Hong, Y. M., Lee, H. J., Woo, H. N., Pyo, J. O., Mak, T. W. and Jung, Y. K. 2004. Induced inhibition of ischemic/hypoxic injury by APIP, a novel Apaf-1-interacting protein. *J. Biol. Chem.* 279:39942–39950. doi:10.1074/jbc.M405747200.

Colombo, S. L. and Moncada, S. 2009. AMPKa1 regulates the antioxidant status of vascular endothelial cells. *Biochem. J.* 421:163–169.

Estabrook, R. W. 1967. Mitochondrial respiratory control and the polarographic measurement of ADP:O ratios. *Method Enzymol.* 10:41–47.

Hardie, D. G. 2004. The AMP-activated protein kinase pathway-new players upstream, and downstream. *J. Cell Sci.* 117:5479–5487.

Hardie, D. G., Hawley, S. A. and Scott, J. W. 2006. AMP-activated protein kinase-development of the energy sensor concept. *J. Physiol.* 574:7–15.

Hardie, D. G. and Sakamoto, K. 2006. AMPK: a key sensor of fuel and energy status in skeletal muscle. *Physiology* 21:48–60.

Hoppins, S., Lackner, L. and Nunnari, J. 2007. The machines that divide and fuse mitochondria. *Ann. Rev. Biochem* 2007. 76:751–780.

Hudson, N. J., Reverter, A. and Dalrymple, B. P. 2009. A differential wiring analysis of expression data correctly identifies the gene containing the causal mutation. *PLoS Computational. Biol.* 5:1–15. doi:10:1371.pcbi.1000382.

Hudson N. J., Dalrymple, B. P. and Reverter, A. 2012. Beyond differential expression: the quest for causal mutations and effector molecules. *BMC Genomics*. 13:356.

Jacobus, W. E. and Lehninger, A. L. 1973. Creatine kinase of rat heart mitochondria: coupling of creatine phosphorylation to electron transport. *J. Biol. Chem.* 248:4803–4810.

Kennedy, E. P. and Lehninger, A. L. 1949. Oxidation of fatty acids and tricarboxylic acid cycle intermediates by isolated rat liver mitochondria. *J. Biol. Chem.* 179:957–963.

Kollias, H. D. and McDermott, J. C. (2008). Transforming growth factor-β and myostatin signaling in skeletal muscle. *J. Appl. Physiol.* 104:579–587.

Kong, B., Song, J., Lee, J. Y., Hargis, B. M., Wing, T., Lassiter, K. and Bottje, W. G. 2011. Gene expression in breast muscle associated feed efficiency in a single male broiler line using a chicken 44k microarray. I. Top differentially expressed genes. *Poult. Sci.* 90:2535–2547.

Kong, B., Lassiter, K., Piekarski-Welsher, A., Dridi, S., Reverter-Gomez, A., Hudson, N. J. and Bottje, W. G. 2016. Proteomics of breast muscle tissue associated with the phenotypic expression of feed efficiency within a pedigree male broiler line: I. Highlight on Mitochondria. *PLoS One.* 11:e0155679. doi:10.1371/journal.pone.0155679.

Lane, N. 2010. *Power, Sex and Suicide. Mitochondria and the Meaning of Life*. Oxford University Press, New York, NY.

Lassiter, K., Dridi, S., Greene, E., Kong, B. and Bottje, W. G. 2018. Identification of mitochondrial hormone receptors in avian muscle cells. *Poult. Sci.* 97:2926–2933.

Lassiter, K., Kong, B., Dridi, S. and Bottje, W. G. 2019. Genes involved in muscle development, muscle differentiation, and insulin signaling may play a role in the phenotypic expression of feed efficiency seen in a single genetic line of male broilers. *Poult. Sci.* 95(1):197.

Lee, J., Karnuah, A. B., Rekaya, R., Anthony, N. A. and Aggrey, S. E. 2015. Transcriptomic analysis to elucidate the molecular mechanisms that underlie feed efficiency in meat-type chickens. *Mol. Genet. Genomics*. doi:10:10.1007/s00438-015-1025-7.

Lehninger, A. L. 1965. *The Mitochondrion*. W. A. Benjamin, Inc., New York, NY.

Lehninger, A. L., Nelson, D. L. and Cox, M. M. eds. 1993. *Principles of Biochemistry*. 2nd ed. Worth Publishers, New York, NY.

Liesa, M. and Shirhai, O. S. 2013. Mitochondria dynamics in the regulation of nutrient utilization and energy expenditure. *Cell Metabolism* 17: 491. doi:10.1016/j.cmet.2013.03.002.

Masatoshi, U., Watanabe, K., Sato, K., Akiba, Y. and Toyomizu, M. 2005. Possible role for avPGC- 1alpha in the control of expression of fiber type, along with avUCP and avANT mRNA's in skeletal muscles of cold-exposed chickens. *FEBS Letters*. 579:11–17.

McPherron, A. C., Lawler, A. M. and Lee, S. J. 1997. Regulation of skeletal muscle mass in mice by a new TGF-beta superfamily member. *Nature*. 387:83–90.

McPherron, A. C. and Lee, S. 1997. Double muscling in cattle due to mutations in the myostatin gene. *Proc. Natl. Acad. Sci. (USA)* 94:12457–12461.

Minokoshi, Y., Alquier, T., Furukawa, N., Kim, Y. B., Lee, A., Xue, B., Mu, J., Foufelle, F., Ferre, P. and Birnbaum, M. J. 2004. AMP-Kinase regulates food intake by responding to hormonal and nutrient signals in the hypothalamus. *Nature* 428:569–574.

Nisoli, E., Clementi, E., Paolucci, C., Cozzi, V., Tonello, C., Sciorati, C., Bracale, R., Valerio, A., Francolini, M., Moncada, S. and Carruba, M. O. 2003. Mitochondrial biogenesis in mammals: the role of endogenous nitric oxide. *Science* 299:896–899.

Nisoli, E., Clementi, E., Moncada, S. and Carruba, M. O. 2004. Mitochondrial biogenesis as a cellular signaling framework. *Biochem. Pharm.* 67:1–15.

Ojuka, E. O. 2004. Role of calcium and AMP kinase in the regulation of mitochondrial biogenesis and GLUT4 levels in muscle. *Proc. Nutr. Soc.* 63:275–278.

Paul, M. H. and Sperling, E. D. 1952. Cyclophorus system-XXIII. Correlation of Cyclophorase activity and mitochondrial density in striated muscle. *Proc. Soc. Exp. Biol. Med.* 79:352–354.

Piekarski-Welsher, A. E. G., Lassiter, K., Kong, B. C., Dridi, S. and Bottje, W. 2018. Enrichment of autophagy and proteosome pathways in breast muscle of feed efficient pedigree male broilers. *Front. Physiol.* 9:Article 1342. doi:10.3389/fphys.2918.01342.

Praefcke, G. J. K. and McMahon, H. T. 2004. The dynamin superfamily: universal membrane tubulation and fission molecules? *Nat. Rev. Mol. Cell Biol.* 5(2): 144–147.

Rolfe, D. F. S., Newman, J. M. B., Buckingham, J. A., Clark, M. G. and Brand, M. D. 1999. Contribution of mitochondrial proton leak to respiration rate in working skeletal muscle and liver and to SMR. *Am. J. Physiol.* 276:C692–C699.

Rolfe, D. F. S. and Brand, M. D. 1997. The physiological significance of mitochondrial proton leak in animal cells and tissues. *Biosci. Rep.* 17:9–16.

Rolfe, D. F. and Brand, M. D. 1996. Contribution of mitochondrial proton leak to skeletal muscle respiration and to standard metabolic rate. *Am. J. Physiol.* 271:1380–1389.

Schlattner, U., Tokarska-Schlattner, M. and Wallimann, T. 2006. Mitochondrial creatinine kinase in health and disease. *Biochem. Biophys. Acta.* 1762:164–180.

Skulachev, V. P. 1996. Role of uncoupled and non-coupled oxidations in maintenance of safely low levels of oxygen and its one-electron reductants. *Q. Rev. Biophys.* 29:169–202.

Skulachev, V. P. 1997. Membrane linked systems preventing superoxide formation. *Biosci. Rep.* 17:347–366.

Smith, J. H. 1963. Relation of body size to muscle size and number in the chicken. *Poult. Sci.* 42:283–290.

Turrens, J. F. and A. Boveris. 1980. Generation of superoxide anion by the NADH dehydrogenase of bovine heart mitochondria. *Biochem. J.* 191:421–427.

Velleman, S. G. and McFarland, S. D. 2015. Skeletal Muscle. In C. G. Scanes (Ed.), *Sturkie's Avian physiology* (6th ed., pp. 379–402). Elesvier Academic Press, San Diego, CA.

Zhou, G., Myers, R., Li, Y., Chen, Y., Shen, X., Fenyk-Melody, J., Wu, M., Ventre, J., Doebber, T., Fujii, M., Hirshman, M. F., Goodyear, L. J. and Moller, D. E. 2001. Role of AMP-activated protein kinase in mechanism of metformin action. *J. Clin. Invest.* 108:1167–1174.

Zhou, N., Lee, W. R. and Abasht, B. 2015. Messenger RNA sequencing and pathway analysis provide novel insights into the biological basis of chickens' feed efficiency. *BMC Genomics.* 16:195–215.

Zurlo, F., Larson, K., Bogardus, C. and Ravussin, E. 1990. Skeletal muscle metabolism is a major determinant of resting energy expenditure. *J. Clin. Invest.* doi:10.1172/jci114857.

Chapter 6

Genetic factors affecting feed efficiency in meat-type chickens

Samuel E. Aggrey and Romdhane Rekaya, University of Georgia, USA

1 Introduction

There will be around 8.6 billion people on the planet by 2031, up from 7.8 billion in 2021 (OECD/FAO, 2022). Increasing urbanization and affluence are leading to an increasing proportion of animal products being incorporated into diets in developing countries. Consequently, the global supply of animal source proteins will have to increase to meet this rising demand. The 2040 Protein Challenge aims to devise the best path to answer the following complex question: 'How do we provide for up to 10 billion people with enough protein in a way that is healthy, affordable and good for the planet?' (https://www.forumforthefuture.org/protein-challenge). Poultry meat accounts for most of the increase in global meat supply (OECD/FAO, 2022). It has been estimated that worldwide poultry production has increased from 115 million metric tons in 2016 to 136 million metric tons in 2022 (18% increase). During the same period, pork production increased from 118 million metric tons to 121 million metric tons (<3% increase). The global consumption of poultry meat is estimated to reach 154 million metric tons in 2031, representing about 41% of the expected

http://dx.doi.org/10.19103/AS.2024.0143.06

total meat consumption (377 million metric tons) (OECD/FAO, 2022). By 2031, poultry meat production will have increased by 17% compared to the 2020 production level of 132 million metric tons.

Natural resources, especially cereals, soybeans and fishmeal, are currently used to formulate an energy and nutrient-rich feed to meet the increased maintenance and production requirements of modern fast-growing and heavier broiler strains. Feed costs constitute about 70% of the cost of poultry meat production (Aggrey et al., 2010). However, growing cereals and soybeans for feed both uses scarce natural resources and competes with growing these crops for human consumption. The use of fish products as ingredient for poultry feed has put pressure on ocean ecosystems. Meeting current and future demand for poultry meat will require an increase in yield per hectare of crops used for feed without adverse environmental impact, innovation in alternative feed ingredients (e.g. insects, algae and seaweeds) and more efficient use of feed by the animal.

2 The concept of feed efficiency: feed conversion ratio, residual feed intake and other measures

Feed efficiency is a measure of how an animal can transform feed into saleable product(s). For the producer, feed efficiency defined as feed conversion ratio (FCR) is an economic value that translates into profit or loss. According to National Chicken Council (http://nationalchickencouncil.org) every 1% improvement in FCR in broiler chickens leads to 0.53 million tons of broiler feed savings in the US. For the producer, therefore, a combination of FCR and the price of feed contributes significantly to profit/loss equations. Over the years, the poultry breeding companies have sought to improve FCR to meet producers' needs. However, FCR as a trait has ambivalent properties.

Feed conversion ratio is defined as: $$FCR = \frac{Feed\,intake(FI)}{Body\,weight\,gain(BWG)} \quad [1]$$

2.1 Properties of feed conversion ratio as a measure of feed efficiency

Although the components of FCR are continuous responses that are normally distributed, their ratio is not normally distributed (Pearson, 1897; Fieller, 1932). The expected genetic gain from selection based on selection differential and heritability is valid only when the trait follows a normal distribution, and the regression of its genetic and phenotypic values is linear (Turner, 1949; Kennedy, 1984). The deviation from the normality of a ratio trait increases with the magnitude of the coefficient of variation of the trait in the denominator

(Atchley et al., 1976). The relative response due to selection on a ratio is not the same for the numerator and denominator traits (Essl, 1989).

This means that actual genetic gain from selection on a ratio trait (e.g. FCR) will not agree with predicted gain due to the disproportionate amount of selection pressure applied to the trait with larger genetic variance (Sutherland, 1965; Kennedy, 1984). For example, selection on a ratio of 1/2 and 2/4 equally will penalize heavier birds as the maintenance requirement for heavier birds is greater than that of smaller birds. A 10% reduction in the numerator of 2/4 (0.50) also changes the ratio to 0.45, whereas the same percentage change in the denominator changes the ratio to 0.56.

There is evidence showing that selection on the linear approximation of the ratio of the traits is a better approach than selecting directly on the ratio, and selection on the ratio should be preferred over a restricted selection index (Lin, 1980; Gunsett, 1984). A change in FCR (ratio) alone is not enough to ascertain the direction of the change in the components of the ratio. In fact:

1 A larger relative increase in BWG compared to FI will show an improvement in FCR;
2 A larger relative decrease in FI compared to BWG will also show an improvement in FCR; and
3 An increase in BWG and a concomitant decrease in FI will lead to an improvement in FCR.

The US broiler industry has shown significant changes in market weight, feed intake (FI) and FCR (Fig. 1). Market weight and FCR continued to improve from 1925 to 2021. However, that was not the case for FI. In fact, FI decreased by 32% from 1925 to 1975, and increased by 37% from 1975 to 2021. The basis for improvement in FCR from 1925 to 1975 is different from that from 1975 to 2021. Even though FCR improved from 1975 onwards with both increasing market

(a)

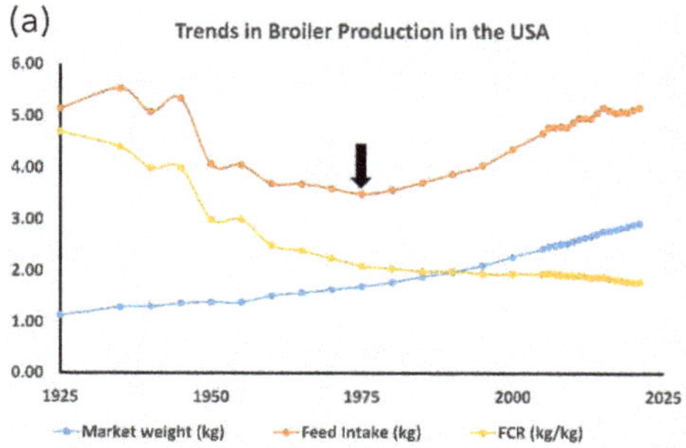

(b) Percentage changes in market weight, feed intake and feed conversion ratio (FCR)

Year	Market weight	Feed intake	FCR
1925-1975	50.40	-31.97	-55.32
1975-2021	48.40	37.45	-8.10
1975-1985	10.26	6.01	-5.00
1985-1995	10.28	8.18	-2.56
1995-2005	13.04	13.35	0.00
2005-2015	12.25	9.65	-3.17
2015-2021	5.26	0.05	-5.59

Figure 1 (a) Changes in market age, feed intake and feed conversion ratio (FCR) from 1925 to 2021; (b) Percent changes in Market weight, feed intake and FCR at different periods from 1925 to 2021. Source: Adapted from: National Chicken Council.

weight and FI, the improvement was the result of relatively higher growth when compared with FI. Different decades since 1975 show fluctuations in the relative changes in market weight and FI. In spite of the drawbacks associated with the use of FCR, producers remain still interested measure economic efficiency using the ratio while researchers continue to investigate alternative ways of assessing feed efficiency.

2.2 Properties of residual feed intake as a measure of feed efficiency

In addition to the drawbacks associated with FCR, variability in maintenance requirements and the composition of growth makes it difficult for FCR to provide an accurate measure of feed efficiency (Aggrey and Rekaya, 2013). Koch et al. (1963) introduced residual feed intake (RFI) as an alternate measure of feed efficiency. Residual feed intake is the difference between the expected and actual FI after accounting for metabolic body weight (MBW = $BW^{0.75}$) and growth (i.e. BWG). The fixed regression model proposed by Koch et al. (1963) was:

$$y_i = a_0 + a_i MBW_i + a_2 BWG_i + e_i \quad [2]$$

where y_i is the FI for the bird i, a_j is a fixed regression coefficient (j = 0, 1, and 2), and e_i is the error term assumed to be normally distributed with a mean of zero and variance σ_e^2. The estimated RFI for a bird i is given by:

$$\widehat{RFI}_i = y_i - \left(\hat{a}_0 + \hat{a}_1 MBW_i + \hat{a}_2 BWG_i\right) \quad [3]$$

From equations [2] and [3], for a single measurement on a bird, the estimated RFI includes the residual term, in addition to the bird effect. This creates a bias proportional to the variance of the error term. To deal with this bias, Aggrey and Rekaya (2013) proposed a mixed regression model to separate the true RFI from the residual terms, thereby eliminating the bias presented by the original method proposed by Koch et al. (1963). The original model presented in equation [2] also combines two components of RFI into a single estimate. Assuming bird-specific regression coefficients instead of population level parameters, estimated RFIs for MBW and BWG could be obtained. Thus, the extension of equation [2] can be written as:

$$y_i = a_0 + \left(a_1 + u_{1i}\right) MBW_i + \left(a_2 + u_{2i}\right) BWG_i + \epsilon_i \quad [4]$$

where a_j is a fixed regression coefficient (j = 0, 1, and 2) as in equation [2] and u_{ji} (j =1, 2) is the random regression specific to bird i, and e_i is the error term assumed to be normally distributed with a mean of zero and variance, σ_ϵ^2. Based on equation [4]:

$$\widehat{RFI}_i = \breve{u}_{1i} MBW_i + \breve{u}_{2i} BWG_i = y_i - \left(a_0 + a_1 MBW_i + a_2 BWG_i\right) \quad [5]$$

$$\widehat{RFI}_i = \widehat{RFI}_{Mi} + \widehat{RFI}_{Gi}$$

where $\widehat{RFI}_M$ and $\widehat{RFI}_G$ are the metabolic and growth RFI's, respectively. However, when there is only one FI observation per bird, and there is no pedigree information, the residual and random regression variance components are confounded, the model in equation [4] is not identifiable and u_{1i} and u_{2i} cannot be separated from each other or from the residual terms.

Regardless of model type (fixed or random), the estimated components of RFI would be phenotypically independent because of the distribution properties of the regression procedure (Netter et al., 2004). Kennedy et al. (1993) demonstrated that RFI is not genetically independent from MBW and BWG. Savietto et al. (2014) presented the random regression type with random intercept. Rekaya and Aggrey (2015) showed that the random regression type model is superior and ranked animals more accurately than using the fixed regression model proposed by Koch et al. (1963).

2.3 Other residual measures of feed efficiency

Romero et al. (2021) and Berry and Crowley (2012) proposed the use of alternative measures of feed efficiency:

- Residual maintenance energy (RMEm);
- Residual gain (RG); and
- Residual intake and gain (RIG).

Residual maintenance energy relies on non-linear regression to estimate the maintenance energy requirement (MER) of each individual after which a linear regression is implemented between metabolic energy for maintenance (MEm) and FI. This approach assumes a linear relationship between FI and production. However, intake and production levels change at different phases of the growth curve.

Residual gain (RG) is the difference between the actual and predicted gain after correcting for BW and FI. According to Berry and Crowley (2012), improved RG is associated with faster growth with no change in FI, and that $RIG\left[(-1*RFI)+RG\right]$ is independent of BW and combines the beneficial characteristics of RFI and RG.

3 Genetic factors affecting feed efficiency: environmental and feed factors

Regardless of how sophisticated a statistical model is in assessing feed efficiency, it will grossly underestimate the biological factors that underlie the trait. Both FCR and residual indices consider FI more broadly, including MBW/maintenance requirement and growth over a specified period. However, the conversion of feed by a bird into growth is a combination of several complex processes. Several management, environmental and animal factors contribute to FI and the animal's ability to convert the ingested feed into output products (e.g. muscle, fat and skeleton). A schematic representation of the process can be found in Fig. 2. This section focusses on environmental and feed factors.

The ambient environment of the poultry house can affect FI. Ambient temperatures above the upper limit of the thermoneutral zone will lead to a reduction in FI and growth. Changes in either or both FI and growth change FCR. Beaumont et al. (1998) estimated the genetic parameters of broiler chickens raised under either 22°C or 32°C. The estimated genetic correlations between body weight (BW) gain and FCR under both temperature settings were 0.73 and 0.74, respectively. They concluded that genes controlling these parameters at both temperatures differed. Variability in ambient environment in the poultry house can therefore significantly affect feed efficiency estimates.

Nutrient deficiency or over supply of a nutrient can lead to changes in consumption, which can affect feed efficiency measures. Lumpkins et al. (2004) demonstrated that growth and feed efficiency of broiler chickens that received a high-density diet were better than their counterparts receiving a low-density diet. Chickens consume more feed in pellet form compared to mash feed (Massuquetto et al. 2020a). The same group showed that the physical qualities

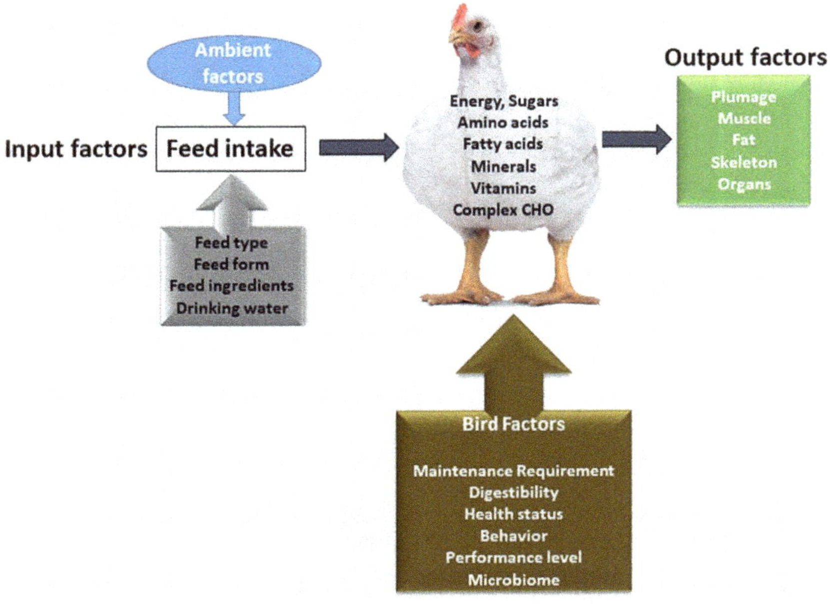

Figure 2 Factors affecting feed utilization efficiency in broiler chickens.

of feed affect FI and growth (Netto et al., 2019; Massuquetto et al. 2020b). In fact, it was reported that a 10% increase in pellet fines (broken pellet fragments) impairs feed efficiency by a factor of 0.01 (De Jong et al., 2017).

Water is an essential nutrient for any living biological organism. Broiler chickens drink about twice as much water than the amount of feed consumed (Kirkpatrick, 2008). There is a strong relationship between feed and water consumption (Aggrey et al., 2023). Restrictions in water supply lead to reduction in FI and significantly affect feed efficiency (Marks, 1981). In addition, ambient temperature and water quality affect water intake in broiler chickens (Larbier and Leclerq, 1992).

Regardless of the method used to assess feed efficiency, the ambient environment, feed properties and, water availability and quality should be the same for every bird in order to reduce cumulative systematic errors in the estimate.

4 Genetic factors affecting feed efficiency: bird factors

When feed efficiency is under consideration, factors that come into play are usually the amount of feed consumed and BW gain. There are several intrinsic factors that control FI by a bird and how that feed is metabolized into BW gain. BW gain comprises of changes in protein accretion, fat content, skeletal and plumage weight.

The maintenance requirement of a bird and the relative efficiencies of the components of BW gain will have a significant effect on the feed efficiency estimates. Chambers et al. (1994) showed that the additive genetic variance that underlies feed efficiency changes with the adjustments of FI data for either initial BW or both initial and final BWs. The adjustment increased the heritability of feed efficiency for three strains of chickens (0.33 to 0.35 vs 0.21 to 0.28). The differences in initial (maintenance requirement) and final body (efficiency of gain) weights formed the basis for subsequent random regression type models (Aggrey and Rekaya, 2013; Savietto et al., 2014; Rekaya and Aggrey, 2015).

Maintenance energy requirement is the amount of energy an animal needs to support energy balance and to account for thermoregulation, involuntary endogenous activities, spontaneous exercise and activity (Kleiber, 1961), and it is often estimated through a regression on metabolic BW ($BW^{0.75}$). This is the minimum amount of feed energy required to maintain BW in the absence of exercise or ambient temperature changes outside the thermoneutral zone, which could increase energy needs. Differences in MER of birds will invariably affect feed efficiency. Using an average initial BW to estimate MBW will definitely bias RFI estimates.

De Verdal et al. (2011a) reported that apparent metabolizable energy corrected for zero nitrogen balance (AMEn) is genetically correlated with feed efficiency and gastrointestinal (GIT) traits. In that study, they reported heritability for AMEn, FCR and RFI was 0.30, 0.21 and 0.46, respectively. The genetic correlation estimates between AMEn and FCR was –0.98, and that between AMEn and RFI was –0.60. The study also showed that AMEn was genetically correlated to both the upper and lower parts of the GIT:

- RFI was correlated with the upper part of the GIT; and
- FCR was correlated with the lower part of the GIT.

Li et al. (2020) reported that low-RFI broilers had significantly lower gross energy excretion, crude protein excretion and AMEn than high-RFI chickens. Thus, genetic difference in digestion, which is reflected in AMEn, may simultaneously improve feed efficiency and control excretion (de Verdal et al., 2011a, 2011b).

Nitsan et al. (1991) found significant differences among different chicken genetic lines for trypsin, chymotrypsin and amylase in the pancreas. Dunnington and Siegel (1995) reported a line-by-age interaction for pancreatic amylase and chymotrypsin. In addition, lipase activity was significantly higher in a high BW line than a low BW line.

There is genetic variation in nutrient requirements for poultry (Nesheim, 1975). These differences are exhibited only when dietary levels are low. When nutrients are in limited supply or alternate nutrients are used, this variability can be exploited to maximize feed efficiency. The differences in nutrient requirements may be related to changes in the efficiency of digestion and absorption.

Tran et al. (2014) detected quantitative trait loci (QTL) controlling digestive efficiency and the anatomy of the digestive tract. On chromosome (GGA) 20, there were significant genome-wide QTL for coefficient of digestive utilization of dry matter (CDUDM) and coefficient for digestive utilization of starch (CDUS). Ankra-Badu et al. (2010) reported that the genetic correlation between FCR and energy balance was –0.25, suggesting that improvement in energy retention will significantly improve feed efficiency.

However, digestive or retention efficiency is a combination of efficiencies of the different nutrients in the diet. Therefore, major genes that underlie the efficiency of a particular nutrient may not significantly affect overall digestive efficiency. From the study of Tran et al. (2014), QTL for CDUS was not significant for AMEn even though the genetic correlation between the two traits is high.

de Verdal (2013a, 2013b) reported that digestive efficiency is genetically independent of post-digestion metabolic traits. They confirmed that selection for improved AMEn reduced the environmental impact on production without a negative impact on growth and body composition. It is possible that AMEn is related to differences in maintenance requirement and not efficiency of use for growth. For simplicity, growth as it is measured is a trait comprising of plumage, muscle, fat, offal and skeleton (Fig. 2). The partitioning of post-digestive nutrients towards the different components of growth is different. Aggrey and Rekaya (2013) demonstrated that two chickens with similar MBW could have significant differences in the efficiency of growth and therefore have different feed efficiency measures. Future studies should explore the genetic relationship of AMEn and efficiencies of bone parameters, protein biosynthesis, fat deposition and plumage changes.

Apparent metabolizable energy corrected for zero nitrogen balance by itself is an aggregation of efficiencies of the different sections of the digestive tract. Li et al. (2020) reported a significant QTL on GGA 5 for metabolizable dry matter and AMEn and the nearest candidate genes were SH3 and multiple ankyrin repeat domain 2 (SHANK2) and ETS homologous factor (EHF).

Transcriptomic analysis also identified key genes associated with specialized sections of the digestive tract. For example, Juanchich et al. (2018) reported a significant enrichment of metabolic processes (fatty acid metabolism, nucleic acid metabolism and beta-oxidation) in the jejunum. There were differentially expressed transcripts in acyl-coA oxidase, enoyl-coA hydratase and acyl-coA transferase. The same authors reported an enrichment of functions related to morphogenesis and development in the gizzard compared to the jejunum.

5 Genetic factors affecting feed efficiency: feed efficiency and output factors

The *MEm*, as a function of BW and energy costs, varies for feather coverage, fat deposition, protein deposition and skeletal changes. These factors among others affect meat yield and feed efficiency. According to Luiting (1990), the main component of RFI variation seems to be the variation of *MEm*: differences in physical activity, feathering density and basal metabolic rate. The degree of non-feathered skin (comb, wattle and legs), body temperature and body composition also affect *MEm*. Therefore, individual variation in *MEm* efficiency can significantly affect feed efficiency.

For the majority of energy partitioning models, the energy retained (*RE*) in body tissues only considers body protein and fat (no plumage, or bones). According to Boekholt et al. (1994), metabolizable energy intake (MEI) can be partitioned into MEm and RE in body fat (RE_f) or body protein (RE_p), where MEI = MEm + ($1/K_f \times RE_f$) + ($1/K_p \times RE_p$); K_f and K_p are the efficiencies of utilization of ME for fat and protein retention, respectively. The efficiencies of energy retention in protein (K_p) and fat (K_f) were estimated to be 0.66 and 0.86, respectively (Boekholt et al., 1994). Based on these efficiency estimates, Lopez and Leeson (2005) showed that the energy required for fat deposition in the body is greater than that for protein (11.1 vs 8 Kcal/g *ME*).

The genetic correlation between carcass/breast meat yield and abdominal fat yield is negative. Le Bihan-Duval et al. (2001) reported a genetic correlation of –0.17 between breast meat yield and abdominal fat yield. Zerehdaran et al. (2004) reported a genetic correlation of –0.55 between carcass yield and abdominal fat yield, and –0.28 between breast meat yield and abdominal fat yield. This implies that an increase in RE_p/MEI relative to RE_f/MEI will improve feed efficiency and vice versa. The relationship between these ratios and feed efficiency may explain the dynamics of FCR in chicken lines selected for lean or fatness (MacLeod et al., 1988; Leclercq and Guy, 1991; Su et al., 2022).

Selecting for improved feed efficiency in broiler chickens produced lean birds with higher metabolizability of the diet (Jørgensen et al., 1990). They showed that an improved FCR line had a slightly higher gross energy metabolizability (0.752 vs 0.725%), retained more protein (0.51 vs 0.46%) and deposited less fat when compared with chickens selected for better growth. Washburn et al. (1975) reported that selection for improved FCR was associated with increased carcass moisture content and lowered lipid. There was a significant decrease in C16:1 and an increase in C18:0 fatty acids in chickens with lower FCR.

Sorensen (1988) concluded that selection for improved feed efficiency reduced fat deposition, and this reduction was primarily in abdominal and subcutaneous fat. Chambers et al. (1984) reported that the genetic correlation between 28- and 49-day feed efficiency measured as BWG per FI and abdominal fat yield was –0.68. In a slow-growing broiler line, N'Dri et al. (2006) reported a genetic relationship of 0.44 between either FCR or RFI and abdominal fat yield. It should be pointed out that low values of FCR are desirable than high values.

There is evidence showing that selection against abdominal fat yield can lead to an improvement in FCR. Using a broiler line divergently selected for low or high abdominal content, Chen et al. (2021) reported that the genetic correlation between abdominal fat yield and RFI and FCR was 0.58 and 0.51, respectively. Indirect selection for reduced muscle or abdominal fatness is as effective as selection for FCR. This relationship has also been reported in farmed rainbow trout (Kause et al., 2016). Whether selection is focussed on directly improving feed efficiency or reducing fatness, genetic selection affects the pattern of RE over time. The RE_p/RE and RE_f/RE in broilers were 51% and 49% in 2008 (Lopez and Leeson, 2005). Recent values estimated by Wu et al. (2019) were 55% and 45% for RE_p/RE and RE_f/RE, respectively.

The depth of plumage coverage affects thermoregulation in chickens. It is therefore expected that MEm requirements will vary with body plumage coverage. Thus, the extent of plumage coverage and non-feathered skin will invariably affect feed efficiency. Heat loss due to poorly feathered areas or to the degree of feather cover increases feed consumption and lowers feed efficiency (Leeson and Morrison, 1978). In cockerels of laying hens, there is a significant correlation between feather thickness and residual FI (Bordas and Merat, 1981). From a study of two commercial male lines differing in feed efficiency, Cahaner et al. (1987) reported that chickens from the more efficient line have less fat, less plumage cover and spent more time sitting than their less efficient counterpart. They concluded that feed efficiency is a function of several factors. It should be pointed out that the independent effects of fatness, plumage cover and activity were not elucidated in the study of Cahaner et al. (1987). Akbar et al. (1985) also concluded from a study of three Leghorn strains that chicks of the more active strain consumed more feed and had poorer feed efficiency than those with the least active strain.

In Single Comb White Leghorn hens, Leeson and Morrison (1978) demonstrated a significant correlation between featherweight and feed efficiency. They showed that FCR deteriorated by 0.04 per gram reduction in featherweight. Glatz (2001) reported that FI from commercial layers with poor feathers was significantly higher than hens with good feather cover. Also, hens with good feather cover produced more eggs and had better feed efficiency than their counterparts with poor feathers. In broiler chickens, Pym and Nicholls (1979) showed that higher MERs, high fasting heat production and ultra-slow feathering are associated with poor feed efficiency. There is a genetic relationship between plumage thickness and feed efficiency; however, the degree of genetic inheritance is yet to be determined.

6 Conclusion

Feed efficiency is clearly an aggregate of several complex and highly interactive components. Each component represents a subsystem with its own complexity. It is unlikely that it will be possible to identify all the underlying drivers. However, major influencing factors, if identified, can be used to improve the trait. Therefore, molecular studies aimed at identifying QTL, genes or polymorphisms associated with feed efficiency should be related to the relevant, individual components of feed efficiency. This will allow for a better understanding of the trait or its utility in genetic improvement programmes.

Improvement of feed efficiency can occur by reducing FI for maintenance, maintaining proper feathering, reducing activities pertaining to feed wastage and/or by reducing body fatness. The use of residual FI as a trait allows for direct selection to reduce maintenance requirement. However, to producers, FCR as a trait, despite its biological limitations, is directly correlated with profit margins. For accurate estimates of genetic parameters of feed efficiency, regardless of the trait used, the ambient factors should be constant for all birds, and parameters of the output factors (fat, plumage, muscles and offal) should be accounted for whenever possible.

7 Where to look for further information

- Marchesi JAP, Ono RK, Cantão ME, Ibelli AMG, Peixoto JO, Moreira GCM, Godoy TF, Coutinho LL, Munari DP, Ledur MC. Exploring the genetic architecture of feed efficiency traits in chickens. Sci Rep. 2021 Feb 25;11(1):4622. doi: 10.1038/s41598-021-84125-9.
- Aggrey, S. E., H. Zhou, M. Tixier-Boichard, D. D. Rhoads, Editors, 2020. Advances in Poultry Genetics and Genomics. Burleigh Dodds Science Publishing, Cambridge, UK, p. 540.
- W. H. Muir and S. E. Aggrey, Editors. 2003. Poultry Genetics, Breeding and Biotechnology. CABI, UK, p. 744.

8 References

Aggrey, S. E., Ghareeb, A. F. A., Milfort, M. C., Ariyo, O. W., Aryal, B., Hartono, E., Kwakye, J., Sovi, S., Hipple, S. A., Stevenson, C., Fuller, A. L., El Sabry, M. I., Stino, F. and Rekaya, R. 2023. Quantitative and molecular aspects of water intake in meat-type chickens. *Poultry Science* 102(11), 102973. https://doi.org/10.1016/j.psj.2023.102973.

Aggrey, S. E., Karnuah, A. B., Sebastian, B. and Anthony, N. B. 2010. Genetic properties of feed efficiency parameters in meat-type chickens. *Genetics, Selection, Evolution* 42(1), 25. https://doi.org/10.1186/1297-9686-42-25.

Aggrey, S. E., Lee, J., Karnuah, A. B. and Rekaya, R. 2014. Transcriptomic analysis of genes in the nitrogen recycling pathway of meat-type chickens divergently selected for feed efficiency. *Animal Genetics* 45(2), 215–222. https://doi.org/10.1111/age.12098.

Aggrey, S. E. and Rekaya, R. 2013. Dissection of Koch's residual feed intake: implications for selection. *Poultry Science* 92(10), 2600–2605. https://doi.org/10.3382/ps.2013-03302.

Akbar, M. K., Gavora, J. S. and Fairfull, R. W. 1985. Relationship between activity at a young age and feed efficiency in chickens. *Poultry Science* 64(7), 1402–1404. https://doi.org/10.3382/ps.0641402.

Ankra-Badu, G. A., Pesti, G. M. and Aggrey, S. E. 2010. Genetic interrelationships among phosphorus, nitrogen, calcium, and energy bioavailability in a growing chicken population. *Poultry Science*, 89(11), 2351–2355. https://doi.org/10.3382/ps.2010-00870.

Atchley, W. R., Gaskins, C. T. and Anderson, D. 1976. Statistical properties of ratios. I. Empirical results. *Systematic Biology* 25, 137–148. https://doi.org/10.2307/2412740.

Beaumont, C., Guillaumin, S., Geraert, P. A., Mignon-Grasteau, S. and Leclercq, B. 1998. Genetic parameters of body weight of broiler chickens measured at 22°C or 32°C. *British Poultry Science* 39,488–491.

Berry, D. P. and Crowley, J. J. 2012. Residual intake and body weight gain: a new measure of efficiency in growing cattle. *Journal of Animal Science*, *90*(1), 109–115. https://doi.org/10.2527/jas.2011-4245.

Boekholt, H. A., van der Grinten, P., Schreurs, V. V., Los, M. J. and Leffering, C. P. 1994. Effect of dietary energy restriction on retention of protein, fat and energy in broiler chickens. *British Poultry Science* 35(4), 603–614. https://doi.org/10.1080/00071669408417725.

Bordas, A. and Merat, P. 1981. Genetic variation and phenotypic correlations of food consumption of laying hens corrected for body weight and production. *British Poultry Science* 22(1), 25–33. https://doi.org/10.1080/00071688108447860

Cahaner, A., Dunnington, E. A., Jones, D. E., Cherry, J. A. and Siegel, P. B. 1987. Evaluation of two commercial broiler male lines differing in efficiency of feed utilization. *Poultry Science* 66(7), 1101–1110. https://doi.org/10.3382/ps.0661101.

Chambers, J. R., Bernon, D. E. and Gavora, J. S. 1984. Synthesis and parameters of new populations of meat-type chickens. *Theoretical and Applied Genetics* 69(1), 23–30. https://doi.org/10.1007/BF00262532.

Chambers, J. R., Wang, L. and McMillan, I. 1994. Genetic variation of broiler feed consumption and efficiency corrected for differences in test body weights. *Poultry Science*, 73(8), 1196–1203. https://doi.org/10.3382/ps.0731196.

Chen, C., Su, Z., Li, Y., Luan, P., Wang, S., Zhang, H., Xiao, F., Guo, H., Cao, Z., Li, H. and Leng, L. 2021. Estimation of the genetic parameters of traits relevant to feed efficiency: result from broiler lines divergent for high or low abdominal fat content. *Poultry Science* 100(2), 461–466. https://doi.org/10.1016/j.psj.2020.10.028.

De Jong, A., DeRouchey, J. M., Tokach, M. D., Goodband, R. D., Woodworth, J. C., Dritz, S. S., Stark, C. R., Jones, C. K., Williams, H. E., Erceg, J., Haberl, B., McKinney, L. J., Smith, G., Van Otterloo, D. and Paulk, C. B. 2017. Formation of pellet fines during the feed manufacturing process, transportation and feed line delivery, and their nutrient composition. *Applied Engineering in Agriculture* 33(6), 921–926. https://doi.org/10.13031/aea.12304.

de Verdal, H., Narcy, A., Bastianelli, D., Chapuis, H., Même, N., Urvoix, S., Le Bihan-Duval, E. and Mignon-Grasteau, S. 2011a. Improving the efficiency of feed utilization in poultry by selection. 1. Genetic parameters of anatomy of the gastro-intestinal tract and digestive efficiency. *BMC Genetics* 12, 59. https://doi.org/10.1186/1471-2156-12-59.

de Verdal, H., Narcy, A., Bastianelli, D., Chapuis, H., Même, N., Urvoix, S., Le Bihan-Duval, E. and Mignon-Grasteau, S. 2011b. Improving the efficiency of feed utilization in poultry by selection. 2. Genetic parameters of excretion traits and correlations with anatomy of the gastro-intestinal tract and digestive efficiency. *BMC Genetics* 12, 71. https://doi.org/10.1186/1471-2156-12-71.

de Verdal, H., Mignon-Grasteau, S., Bastianelli, D., Même, N., Le Bihan-Duval, E. and Narcy, A. 2013a. Reducing the environmental impact of poultry breeding by genetic selection. *Journal of Animal Science* 91(2), 613–622. https://doi.org/10.2527/jas.2012-5572.

de Verdal, H., Narcy, A., Bastianelli, D., Même, N., Urvoix, S., Collin, A., Le Bihan-Duval, E. and Mignon-Grasteau, S. 2013b. Genetic variability of metabolic characteristics in chickens selected for their ability to digest wheat. *Journal of Animal Science* 91(6), 2605–2615. https://doi.org/10.2527/jas.2012-6182.

Dunnington, E. A. and Siegel, P. B. 1995. Enzyme activity and organ development in newly hatched chicks selected for high or low eight-week body weight. *Poultry Science* 74(5), 761–770. https://doi.org/10.3382/ps.0740761.

Essl, A. 1989. Selection for a ratio of two traits: results of a simulation study. *Journal of Animal Breeding and Genetics* 106, 81–88. https://doi.org/10.1111/j.1439-0388.1989.tb00217.x.

Fieller, E. C. 1932. The distribution of the index in a normal bivariate population. *Biometrika* 24, 428–440.

Glatz, P. C. 2001. Effect of poor feather cover on feed intake and production of aged laying hens. *Asian-Australasian Journal of Animal Science* 14(4), 553–558.

Gunsett, F. C. 1984. Linear index selection to improve traits defined as ratios. *Journal of Animal Science* 59, 1185–1193. https://doi.org/10.2527/jas1984.5951185x.

Herd, R. M. and Arthur, P. F. 2009. Physiological basis for residual feed intake. *Journal of Animal Science* 87(14 Suppl), E64–E71. https://doi.org/10.2527/jas.2008-1345.

Herd, R. M., Oddy, V. H. and Richardson, E. C. 2004. Biological basis for variation in residual feed intake in beef cattle. 1. Review of potential mechanisms. *Australian Journal of Experimental Agriculture* 44, 423–430.

Hoque, M. A. and K. Suzuki. 2009. Genetics of residual feed in-take in cattle and pigs: a review. *Asian-Australasian Journal of Animal Science* 22, 747–755.

Jensen, J., Wang, C. S., Sorensen, D. A. and Gianola, D. 1994. Marginal inferences of variance and covariance components for traits influenced by maternal and direct genetic effects using the Gibb sampler. *Acta Agriculturae Scandinavica, Section A-Animal Science* 44, 193–201.

Jørgensen, H., Sørensen, P. and Eggum, B. O. 1990. Protein and energy metabolism in broiler chickens selected for either body weight gain or feed efficiency. *British Poultry Science* 31(3), 517–524. https://doi.org/10.1080/00071669008417283.

Juanchich, A., Hennequet-Antier, C., Cabau, C., Le Bihan-Duval, E., Duclos, M. J., Mignon-Grasteau, S. and Narcy, A. 2018. Functional genomics of the digestive tract in broilers. *BMC Genomics* 19(1), 928. https://doi.org/10.1186/s12864-018-5344-z.

Kause, A., Kiessling, A., Martin, S. A., Houlihan, D. and Ruohonen, K. 2016. Genetic improvement of feed conversion ratio via indirect selection against lipid deposition in farmed rainbow trout (Oncorhynchus mykiss Walbaum). *The British Journal of Nutrition* 116(9), 1656–1665. https://doi.org/10.1017/S0007114516003603.

Kennedy, B. W. 1984. Breeding for feed efficiency: swine and dairy cattle. *Canadian Journal of Animal Science* 64, 505–512.

Kennedy, B. W., van der Werf, J. H. J. and Meuwissen, T. H. E. 1993. Genetic and statistical properties of residual feed intake. *Journal of Animal Science* 71(12), 3239–3250. https://doi.org/10.2527/1993.71123239x.

Kirkpatrick, K. 2008. AviagenBrief_WaterQuality_Feb08-2.pdf

Kleiber, M. 1961. *The fire of life*. New York: John Wiley and Sons.

Koch, R. M., Swiger, L. A., Chambers, D. and Gregory, K. E. 1963. Efficiency of feed use in beef cattle. *Journal of Animal Science* 22, 486–494.

Larbier, M. and Leclerq, B. 1992. Metabolism of water and minerals. InM. Larbier and B. Leclerq (eds.), *Nutrition and feed of Poultry*, pp. 101–118. Nottingham, UK: Nottingham University Press.

Le Bihan-Duval, E., Berri, C., Baeza, E., Millet, N. and Beaumont, C. 2001. Estimation of the genetic parameters of meat characteristics and of their genetic correlations with growth and body composition in an experimental broiler line. *Poultry Science* 80(7), 839–843. https://doi.org/10.1093/ps/80.7.839.

Leclercq, B. and Guy, G. 1991. Further investigations on protein requirement of genetically lean and fat chickens. *British Poultry Science* 32(4), 789–798. https://doi.org/10.1080/00071669108417404.

Leeson, S. and Morrson, W. D. 1978. Effect of feather cover in feed efficiency in laying birds. *Poultry Science* 57(4), 1094–1096. https://doi.org/10.3382/ps.0571094.

Li, W., Liu, R., Zheng, M., Feng, F., Liu, D., Guo, Y., Zhao, G. and Wen, J. 2020. New insights into the associations among feed efficiency, metabolizable efficiency traits and related QTL regions in broiler chickens. *Journal of Animal Science and Biotechnology* 11, 65. https://doi.org/10.1186/s40104-020-00469-8.

Lin, C. Y. 1980. Relative efficiency of selection methods for improvement of feed efficiency. *Journal of Dairy Science* 63, 491–494. https://doi.org/10.3168/jds.S0022-0302(80)82960-2.

Lopez, G. and Leeson, S. 2005. Utilization of metabolizable energy by young broilers and birds of intermediate growth rate. *Poultry science* 84(7), 1069–1076. https://doi.org/10.1093/ps/84.7.1069.

Luiting, P. 1990. Genetic variation of energy partitioning in laying hens: causes of variation in residual feed consumption. *World's Poultry Science Journal* 46(2), 133–152. https://doi.org/10.1079/WPS19900017.

Lumpkins, B. S., Batal, A. B. and Dale, N. M. 2004. Evaluation of distillers dried grains with solubles as a feed ingredient for broilers. *Poultry Science* 83(11), 1891–1896. https://doi.org/10.1093/ps/83.11.1891.

MacLeod, M. G., Whitehead, C. C., Griffin, H. D. and Jewitt, T. R. 1988. Energy and nitrogen retention and loss in broiler chickens genetically selected for leanness and fatness. *British Poultry Science* 29(2), 285–292. https://doi.org/10.1080/00071668808417053.

Marks H. L. 1981. Role of water in regulating feed intake and feed efficiency of broilers. *Poultry Science* 60(4), 698–707. https://doi.org/10.3382/ps.0600698.

Massuquetto, A., Durau, J. F., Ezaki Barrilli, L. N., Fernandes Dos Santos, R. O., Krabbe, E. L. and Maiorka, A. 2020b. Thermal processing of corn and physical form of broiler diets. *Poultry Science* 99(6), 3188–3195. https://doi.org/10.1016/j.psj.2020.01.027.

Massuquetto, A., Panisson, J. C., Schramm, V. G., Surek, D., Krabbe, E. L. and Maiorka, A. 2020a. Effects of feed form and energy levels on growth performance, carcass yield and nutrient digestibility in broilers. *Animal* 14(6), 1139–1146. https://doi.org/10.1017/S1751731119003331.

Mignon-Grasteau, S., Rideau, N., Gabriel, I., Chantry-Darmon, C., Boscher, M. Y., Sellier, N., Chabault, M., Le Bihan-Duval, E. and Narcy, A. 2015. Detection of QTL controlling feed efficiency and excretion in chickens fed a wheat-based diet. *Genetics, Selection, Evolution* 47, 74. https://doi.org/10.1186/s12711-015-0156-y.

N'Dri, A. L., Mignon-Grasteau, S., Sellier, N., Tixier-Boichard, M. and Beaumont, C. 2006. Genetic relationships between feed conversion ratio, growth curve and body composition in slow-growing chickens. *British Poultry Science* 47(3), 273–280. https://doi.org/10.1080/00071660600753664.

Nesheim, M. C. 1975. Genetic variation in nutritional requirements of poultry. In M. L. Sunde (ed.), *Effect of genetic variance on nutritional requirements of animals: Proceedings of a symposium*, pp. 47–87 Washington, DC. The National Academies Press. https://doi.org/10.17226/19.

Netter, J., Wasserman W. and Kutner, M. H. 2004. *Applied linear statistical models*. New York: McGraw-Hill.

Netto, M. V. T., Massuqueto, A., Krabbe, E. L., Surek, D. and Oliveira, S. G. 2019. Effect of conditioning temperature on pellet quality, diet digestibility, and broiler performance. *Journal Applied Poultry Research* 28, 963–973. http://dx.doi.org/10.3382/japr/pfz056.

Nitsan, Z., Dunnington, E. A. and Siegel, P. B. 1991. Organ growth and digestive enzyme levels to fifteen days of age in lines of chickens differing in body weight. *Poultry Science* 70(10), 2040–2048. https://doi.org/10.3382/ps.0702040.

OECD/FAO. 2022. "OECD-FAO Agricultural Outlook", OECD Agriculture statistics (database). http://dx.doi.org/10.1787/agr-outl-data-en (Accessed: April 11, 2023).

Pearson, K. 1897. Mathematical contributions to the theory of evolution-on a form of spurious correlation which may arise when indices are used in the measurement of organs. *Proceedings of the Royal Society (London)* 60, 489–498.

Pym, R. A. E. and Nicholls, P. J. 1979. Selection for food conversion in broilers: direct and correlated responses to selection for body-weight gain, food consumption and food conversion ratio. *British Poultry Science* 20(1), 73–86. https://doi.org/10.1080/00071667908416551.

Rekaya, R. and Aggrey, S. E. 2015. Genetic properties of residual feed intakes for maintenance and growth and the implications of error measurement. *Journal of Animal Science* 93(3), 944–948. https://doi.org/10.2527/jas.2014-8061.

Rekaya, R., Sapp, R. L., Wing, T. and Aggrey, S. E. 2013. Genetic evaluation for growth, body composition, feed efficiency, and leg soundness. *Poultry Science* 92(4), 923–929. https://doi.org/10.3382/ps.2012-02649.

Rodríguez-Romero, J. J., Durán-Castañeda, A. C., Cárdenas-Castro, A. P., Sánchez-Burgos, J. A., Zamora-Gasga, V. M. and Sáyago-Ayerdi, S. G. 2021. What we know about protein gut metabolites: implications and insights for human health and diseases. *Food Chemistry: X* 13, 100195. https://doi.org/10.1016/j.fochx.2021.100195.

Savietto, D., Berry, D. P. and Friggens, N. C. 2014. Towards an improved estimation of the biological components of residual feed intake in growing cattle. *Journal of Animal Science* 92(2), 467–476. https://doi.org/10.2527/jas.2013-6894.

Sorensen, P. 1988. The influence on leanness of selection for feed efficiency. In B. Leclecq and C. C. Whitehead (eds.), *Leanness in domestic birds*, pp. 127–128. London: Butterworth.

Su, Z., Bai, X., Wang, H., Wang, S., Chen, C., Xiao, F., Guo, H., Gao, H., Leng, L. and Li, H. 2022. Identification of biomarkers associated with the feed efficiency by metabolomics profiling: results from the broiler lines divergent for high or low abdominal fat content. *Journal of Animal Science and Biotechnology* 13(1), 122. https://doi.org/10.1186/s40104-022-00775-3.

Sutherland, T. M. 1965. The correlation between feed efficiency and rate of gain: a ratio and its denominator. *Biometrics* 21: 739–749.

Tran, T. S., Narcy, A., Carré, B., Gabriel, I., Rideau, N., Gilbert, H., Demeure, O., Bed'Hom, B., Chantry-Darmon, C., Boscher, M. Y., Bastianelli, D., Sellier, N., Chabault, M., Calenge, F., Le Bihan-Duval, E., Beaumont, C. and Mignon-Grasteau, S. 2014. Detection of QTL controlling digestive efficiency and anatomy of the digestive tract in chicken fed a wheat-based diet. *Genetics, Selection, Evolution* 46(1), 25. https://doi.org/10.1186/1297-9686-46-25.

Turner, H. N. 1949. Ratios as criteria for selection in animal or plant breeding with particular reference to efficiency of feed conversion in sheep. *Australian Journal of Agricultural Research* 10, 565–580.

Washburn, K. W., Guill, R. A. and Edwards, H. M. Jr. 1975. Influence of genetic differences in feed efficiency on carcass composition of young chickens. *The Journal of Nutrition* 105(10), 1311–1317. https://doi.org/10.1093/jn/105.10.1311.

Wu, S. B., Swick, R. A., Noblet, J., Rodgers, N., Cadogan, D. and Choct, M. 2019. Net energy prediction and energy efficiency of feed for broiler chickens. *Poultry science* 98(3), 1222–1234. https://doi.org/10.3382/ps/pey442.

Zerehdaran, S., Vereijken, A. L., van Arendonk, J. A. and van der Waaijt, E. H. 2004. Estimation of genetic parameters for fat deposition and carcass traits in broilers. *Poultry Science* 83(4), 521–525. https://doi.org/10.1093/ps/83.4.521.

Part 2

Feed formulation

Chapter 7

Maintaining the safety of poultry feed

G. Raj Murugesan, dsm-firmenich Nutritional Products, Switzerland; and Chasity M. Pender, dsm-firmenich Nutritional Products NA, USA

1 Introduction

Feed safety is reliant on a multitude of factors and can be affected by both natural and artificial contaminants including plant and fungal toxins, chemicals, and pathogens to name a few. These contaminants may result in deteriorated feed quality, reduced performance, and increased incidence of diseases in poultry, ultimately leading to significant economic loss for producers. Moreover, livestock feed is the beginning of the human food chain. As poultry are raised to produce meat and eggs for human consumption, these contaminants may also represent a human health concern. This chapter focuses on those contaminants considered to pose the most significant risk in poultry and the human food supply: mycotoxins, dioxins, and pathogens. Each section will discuss the negative effects of these factors as well as possible control measures that may be implemented to reduce feed contamination and secure feed and food safety.

2 Mycotoxins

The term 'mycotoxin' is derived from 'mykes' meaning fungi and 'toxicon' meaning poison. Mycotoxins are secondary metabolites of low molecular weight produced by a wide range of fungi, principally moulds. There are over 200 species of moulds that produce mycotoxins. Aflatoxins (AF), zearalenone

http://dx.doi.org/10.19103/AS.2024.0143.09

(ZEN), ochratoxin A (OTA), fumonisins (FUM), and trichothecenes, such as deoxynivalenol (DON) and T-2 toxin, are some of the mycotoxins that can significantly impact the health and productivity of poultry species (Fig. 1).

The conditions under which fungi and mycotoxins are produced in agricultural commodities depend not only on environmental factors such

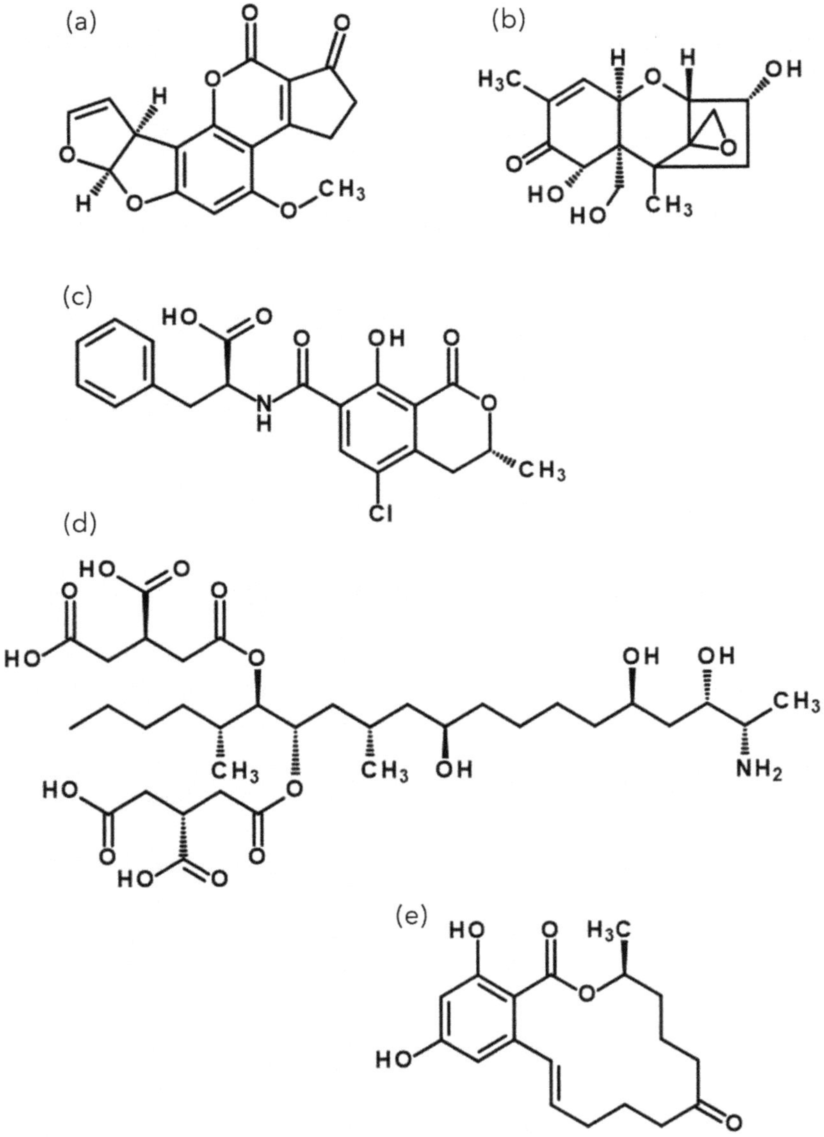

Figure 1 Chemical structures of most prevalent mycotoxins. (a) Aflatoxin B1, (b) deoxynivalenol, (c) ochratoxin A, (d) fumonisin B1, (e) zeralenone.

as availability of moisture and temperature but also slightly elevated CO_2 concentration, which may stimulate the growth of mycotoxigenic fungi (Magan et al., 2011). Extreme weather situations, precipitation, and drought lead to plant stress and hence they become more vulnerable to fungal infection (Wu et al., 2011). A primary challenge is that fungi, for example *Fusarium* species, can produce different mycotoxins with diverse toxigenic potentials.

2.1 Mycotoxicoses

Mycotoxins produce a variety of diseases, collectively called 'mycotoxicoses', directly or in combination with other primary stressors such as pathogens (Raju and Devegowda, 2000). These diseases are exhibited by symptoms and lesions, which can be used to clinically diagnose the presence of mycotoxins. However, these symptoms are not straightforward as co-occurrence is common and one mycotoxin may alter the effects of another. For example, when AF and OTA are co-contaminants of poultry feed, they interact in a synergistic manner (Huff and Doerr, 1981). During dual exposure of these toxins, OTA prevents the major effects of AF (i.e. fatty, yellow, enlarged, and friable liver), thus reducing the ability to diagnose aflatoxicosis in the field.

Acute cases caused by ingestion of high levels of mycotoxins may result in mortality and a marked decline in productivity characterized by obvious clinical signs and post-mortem lesions. However, in most cases, mycotoxicoses are chronic and caused by low-level ingestion of fungal metabolites, resulting in a measurable decline in performance and the occurrence of non-specific changes, including subcutaneous haemorrhage in broilers and immunosuppression (D'mello et al., 1999). Under field conditions, suboptimal performance, in the absence of an obvious infectious, environmental, management, or nutritional facter, suggests the possibility of mycotoxicoses.

2.2 Aflatoxins

AFs are a class of mycotoxins produced by fungal species *Aspergillus flavus*, *Aspergillus nominus*, and *Aspergillus parasiticus*. Most prevalent forms of AF include B_1, B_2, G_1 and G_2, with AF B_1 (AFB_1) being the most common and biologically active component (Busby and Wogan, 1981). Exposure to AF has been shown to have severe toxic and carcinogenic effects in many animal species, including poultry and humans (Monson et al., 2015). AFs cause a multitude of effects in poultry, including decreased weight gain, poor feed efficiency, reduced egg production and egg weight, increased liver fat, changes in organ weights, reduction in serum protein levels, carcass bruising, poor pigmentation, liver damage (pale and enlarged), decreased activities of pancreatic enzymes, and immunosuppression (Edds and Bortell, 1983; Leeson et al., 1995; Devegowda and Murthy, 2005). The immune system may also be negatively impacted as even low concentrations of AF can cause damage to immune tissues, such as the bursa of Fabricius, spleen, and thymus, and suppress innate and adaptive immune responses (Hoerr, 2010). Immunosuppression caused by AF may result in vaccination failures and poor antibody titres as well decreased resistance to viral pathogens (Devegowda and Murthy, 2005). Moreover, birds with aflatoxicosis are more susceptible to secondary bacterial infections and may experience increased severity of viral, protozoal, or fungal infections, thus leading to more dramatic economic impacts (Monson et al., 2015).

2.3 Trichothecenes

Trichothecenes are fungal metabolites with the same basic structure produced by *Fusarium*, *Myrothecium*, *Trichoderma*, *Verticimonosporium*, *Cephalosporium*, *Trichothecium*, and *Stachybotrys*. In terms of their structure, all trichothecenes have a sesquiterpene skeleton and a 12,13-epoxy ring, which is important for their toxic effects and biological activity. Based on different substitutions at five positions of the epoxide ring, trichothecenes are classified into four groups: Type A, B, C, and D, with Types A and B being the most common. Type A trichothecenes include T-2 toxin, HT-2 toxin, neosolaniol, 8-acetylneosolaniol, 4-deacetylneosolaniol, monoacetoxyscirpenol (MAS), and diacetoxyscirpenol (DAS). Type B trichothecenes include deoxynivalenol (DON; also known as vomitoxin), nivalenol, 3-acetyldeoxynivalenol, 15-acetyldeoxynivalenol, 4-acetoxynivalenol (Fusarenon-X), and DON-3-Glucoside (Leeson et al., 1995; Cope, 2018). Trichothecenes are the most potent inhibitors of protein synthesis. The main toxic effect at the cellular level appears to be primary inhibition of protein synthesis followed by a secondary disruption of DNA and RNA syntheses (Leeson et al., 1995).

Toxic effects of trichothecenes include oral lesions, growth retardation, abnormal feathering, decreased egg production and eggshell quality, regression of the bursa of Fabricius, peroxidative changes in liver, abnormal blood coagulation, leucopoenia and proteinaemia, and immunosuppression (Leeson et al., 1995; Dänicke, 2002). Concentrations of T-2 that cause oral lesions are lower (0.4 mg/kg) than concentrations reported to decrease chick performance (3–4 mg/kg; Leeson et al., 1995). In a comprehensive review, Dänicke (2002) concluded that broiler performance is affected at dietary concentrations of 3–4 mg/kg of T-2 toxin, whereas ducks were affected when the dietary concentration was as low as 0.4 mg/kg.

Deoxynivalenol is less toxic than T-2 toxin and the level of DON that affects chick performance is still debated, with some researchers (Huff et al., 1986; Kubena et al., 1988, 1989) reporting toxic effects at 16 mg/kg diet, whereas others (Moran et al., 1982) report no toxic effect until dietary concentrations exceeded

116 mg/kg of DON. Dänicke et al. (2001) summarized results of 49 studies with DON and concluded that a dietary concentration of 5 mg/kg had no negative effects on performance, though this threshold for impacting performance does appear to be lower when a challenge or mycotoxin co-occurrence is present (Liu et al., 2020; Ruhnau et al., 2020; Paraskeuas et al., 2021). Deoxynivalenol has also been reported to have both immunosuppressive and immunomodulating effects in poultry (Dänicke, 2002). Recent studies indicate that DON at concentrations ranging from 1 to 7 mg/kg diet significantly alters several key functions of the intestinal tract including decreasing villus surface area available for absorption, altering the permeability of the intestinal tract, and increased susceptibility to bacterial and coccidial pathogens (Awad et al., 2011; Osselaere et al., 2013; Antonissen et al., 2014; Grenier et al., 2015a; Ruhnau et al., 2020; Ruhnau et al., 2020, Santos et al., 2021). Due to these effects on the intestine, the European Food Safety Authority (EFSA) Panel on Contaminants in the Food Chain (CONTAM) recently published a revised scientific opinion in 2023. In it, they state that the reference point for adverse animal health effects for DON should be lowered from the previously established levels of 5 mg/kg and 7 mg/kg for broilers and turkeys, respectively, to 0.6 mg/kg due to evaluation of research published since the original values were established in 2017 (EFSA CONTAM Panel, 2023).

2.4 Fumonisins

The FUM are a group of mycotoxins produced by moulds of the genus *Fusarium*, while some species of *Alternaria* have also been found to produce FB_1 (Chen et al., 1992). Six different FUM (A_1, A_2, B_1, B_2, B_3, B_4) have been identified and their structures were defined so far (Cawood et al., 1991), and FUM B_1 (FB_1) has been the predominant form (Norred, 1993). While prior work with FUM suggested that chicks and turkeys were relatively resistant to their toxic effects along with low levels of absorption in the bird, recent evidence suggests that FUM can alter the control of inflammatory processes (Grenier and Applegate, 2013), especially in the lower small intestine, thus increasing their susceptibility to some enteric pathogens (Antonissen et al., 2014; Grenier et al., 2015b). Additionally, reductions in body weight gain and liver pathology have also been observed in chicks, ducks, and turkeys (Broomhead et al., 2002; Tran et al., 2005).

The mechanism by which FUM cause toxicity in animals appears to be due to the disruption of sphingolipid metabolism (Wang et al., 1991). The FUM are specific inhibitors of ceramide synthase (sphinganine/sphingosine *N*-acyltransferase), a key enzyme required for the synthesis of ceramide and more complex sphingolipids. Inhibition of this enzyme system leads to an increase in tissue concentrations of the sphingolipids, sphingosine (SO) and sphinganine (SA), and a change in the SA:SO ratio. An increase in the SA:SO ratio has been observed in serum and tissues of broilers, turkeys, and ducklings fed FB_1 (Broomhead et al., 2002; Tran et al., 2005; Grenier et al., 2017). Signs of oxidative stress in the liver, including increases in reactive oxygen species (ROS), thiobarbituric acid reactive substances (TBARS), and glutathione S-transferase (GST) and decreases in catalase and superoxide dismutase, have also been observed (Galli et al., 2020). Due to these effects, the EFSA CONTAM Panel recently revised their reference points for adverse animal health effects for FB1 from 20 mg/kg, established in 2018, to 1 mg/kg for chickens (EFSA CONTAM Panel, 2022).

2.5 Ochratoxins

Ochratoxins are a group of structurally related metabolites that are produced by fungi belonging to the genera *Aspergillus* and *Penicillium*, and OTA is the most prevalent and toxic mycotoxin of this group. Signs of OTA toxicity in poultry include weakness, anaemia, decreased feed consumption, reduced growth rate and egg production, poor feathering, and excessive mortality at high dietary concentrations (Gibson et al., 1989). Increases in relative weights of liver, spleen, pancreas, proventriculus, gizzard, and testes have also been reported in poultry fed OTA (Gibson et al., 1989).

Though the kidney is the main target for OTA-induced toxicity, significant effects on intestinal histomorphology and lesions have been observed. Moreover, exposure to OTA has also been shown to impair intestinal integrity by suppressing expression of tight junction proteins, thus increasing intestinal permeability and subsequent translocation of bacteria from the lumen (Zhai et al., 2021).

OTA consists of an isocoumarin moiety linked through the 7-carboxy group to the amino acid L-β-phenylalanine. At the cellular level, OTA interferes with DNA, RNA, and protein synthesis by inhibiting the enzyme phenylalanine-tRNA synthetase (Marquardt and Frohlich, 1992). OTA also affects renal carbohydrate metabolism through a reduction of renal mRNA coding for phosphoenolpyruvate carboxykinase (PEPCK), a key enzyme in gluconeogenesis (Leeson et al., 1995). The effects of OTA on DNA, RNA, and protein synthesis are thought to be due to the phenylalanine moiety of the toxin competing with phenylalanine in the enzyme catalysed reaction (Marquardt and Frohlich, 1992). OTA also causes hypocarotenidaemia in broilers (Huff and Hamilton, 1975) that is more severe than that caused by AF (Osborne et al., 1982; Schaeffer et al., 1987).

2.6 Interactions among mycotoxins

In general, contaminated feeds usually contain more than one mycotoxin and those co-occurring mycotoxins often cause more adverse effects than a single mycotoxin due to additive or even synergistic interaction (CAST, 2003). Moreover, the combination of mycotoxins, at concentrations that individually should not cause negative effects, may negatively affect animals (Grenier and Oswald, 2011). Another challenge is the global trade of agricultural raw materials used as feed ingredients, which can result in the distribution of mycotoxins across the world (Bryden, 2012). Hence, as part of a proper mycotoxin risk management, surveying the mycotoxin occurrence is very important to allow feed and animal producers to assess the risk of using certain feed ingredients or feeds from different regions.

2.7 Mycotoxin counteracting strategies

It is commonly known that mycotoxins vary in their chemical structures, resulting in vast differences in their chemical, physical, and biochemical properties. While the biochemical properties define the toxicity of mycotoxins, chemical and physical properties determine the methods that can be used to detoxify them. Considering the great variety of mycotoxin structures, it is obvious that there is no single method which can be used to deactivate mycotoxins in feed. Therefore, different strategies must be combined in order to specifically target individual mycotoxins without impacting the quality of feed.

The best-known method for mycotoxin deactivation is 'binding' with the use of binding agents, which are referred to as mycotoxin binders, adsorbents, or enterosorbents. They can be of organic (microbial) or inorganic (mainly clay minerals) in nature. The use of clay-based materials for toxin binding is not new. For centuries, humans and animals have been reported to eat clay minerals for various purposes and, in most cases, it is beneficial to health (Carretero, 2002). Thus, the inclusion of non-nutritive clay minerals in the diet of animals has been widely adopted for reducing toxin bioavailability and exposure from contaminated feeds.

Due to low feed inclusion requirements and easy management of enterosorbents, the widespread acceptance of these products by the farm animal industry has led to the introduction of a variety of diverse materials and/or complex mixtures for binding. These materials (and/or mixtures) are reported to contain smectite clays, zeolites, kaolinite, mica, silica, charcoal, and various biological constituents including chlorophyllins, yeast products, lactic acid bacteria, plant extracts, and algae. Some contain smectite or zeolite minerals that have been amended with natural or synthetic surfactants resulting in hydrophobic organoclays or organozeolites (Miller et al., 2014). There is considerable evidence to indicate that smectite clays are the most effective enterosorbents. The adsorption efficacy of binding agents or enterosorbents is limited to only a few mycotoxins, such as AF, ergot alkaloids, and some other fungal toxins, while they have been shown to be ineffective for trichothecenes, FUMs, and OTA (Murugesan et al., 2015).

The approach to use microorganisms and their enzymes to detoxify specific mycotoxins not only works for non-adsorbable mycotoxins, but for all other toxins for which respective microbes can be isolated from nature. This approach has been known for a long time, even longer than the binder concept. Within a few years after the discovery of AF, the first report of a bacterium capable of detoxifying AF by catabolization was published (Ciegler, 1966). Since then, many microorganisms have been isolated from different habitats such as the gastrointestinal tract of animals, soil, mycotoxin-contaminated materials (e.g. grains), and insects feeding on such materials; however, only a few of these organisms were useful or further investigated for practical applications in animal nutrition.

One of the microorganisms that have been further developed into practical application is *Trichosporon mycotoxinivorans*, a yeast strain capable of detoxifying OTA and ZEN (Molnar et al., 2004). Application of this yeast in poultry diets has been proven to detoxify OTA (Politis et al., 2005). Another organism has been an anaerobic rumen bacterium BBSH 797 (genus novus of family Coriobacteriacae, formerly Eubacterium), which was isolated and developed as a trichothecene detoxifying feed additive (Fuchs et al., 2002; Schatzmayr et al., 2006b). The BBSH 797 detoxifies trichothecenes by cleavage of the 12, 13 epoxide ring resulting in de-epoxy trichothecenes. Several microorganisms, not only aerobic bacteria, but also yeasts, with FUM degradation properties were also explored and isolated to detoxify FUM. However, for various reasons, none of these microorganisms were useful as a mycotoxin-deactivating feed additive (Schatzmayr et al., 2006a). Therefore, the catabolic pathway of FUM degradation was investigated and the gene coding for the key enzyme of FUM detoxification – a carboylesterase (FUM*zyme*®, dsm-firmenich Nutritional Products, Switzerland) – was identified, cloned, and expressed in a yeast strain (Heinl et al., 2010; Hartinger and Moll, 2011). This carboylesterase detoxifies FUM by cleaving its tricarballylic side chains, thus generating the non-toxic metabolite hydrolysed FUM (HFB_1; Grenier et al., 2012, 2013). EFSA published a positive Scientific Opinion on the safety and efficacy of FUM*zyme*® as a technological feed additive for swine in 2014 (EFSA Panel on Additives and Products or Substances used in Animal Feed (FEEDAP, 2014)). This was the first-ever EFSA positive opinion for a purified enzyme capable of biotransforming FUM. This was followed shortly by another positive EFSA Scientific Opinion for all avian species in 2016 (EFSA FEEDAP Panel, 2016). In 2020, a food additive petition was submitted to the United States Food and Drug Administration (FDA) seeking approval for use of FUM*zyme*® in swine, subsequently followed by a submission for use in poultry in 2021. In 2022, FUM*zyme*® received FDA approval for use in poultry and swine for FUM mitigation and became the first mycotoxin degradation ingredient recognized by the FDA, thus carving out a regulatory pathway for future mycotoxin mitigation technologies in the United States (Food and Drug Administration, Department of Health and Human Services, 2022).

3 Dioxins

3.1 Sources and risks of dioxin contamination

The generic term 'dioxins' refers to a large group of structurally and chemically related compounds including polychlorinated dibenzo-*p*-dioxins (PCDDs) and polychlorinated dibenzofurans (PCDFs). Dioxin-like polychlorinated biphenyls (PCBs) are also included under this term due to their similar toxic properties (Fig. 2; Fries, 1995). Approximately 419 types of dioxins have been identified, of which about 30 are highly toxic. Of those, 2, 3, 7, 8-tetrachlorodibenzo-*para*-dioxin (TCDD) is the most toxic congener.

Dioxins are predominantly the by-products of industrial processes (smelting, heat and power generation, waste incineration, etc.) but can also result from natural sources such as volcanic eruptions and forest fires. Due to their long half-life, dioxins are persistent and ubiquitous in the environment with soil being a typical site of accumulation. Since dioxins are lipophilic, they tend to accumulate in the fatty tissues; thus, they are commonly found in fat containing foods such as meat, dairy products, eggs, fish, and shellfish (De Vos et al., 2003; WHO, 2016).

In many instances, dioxin presence in the human food supply is a result of contaminated animal feed. In humans, short-term, high-level exposure of dioxins may result in skin lesions and altered liver function, while chronic exposure may lead to impaired immunity, nervous system development and endocrine and reproductive functions. The dioxin TCDD is also classified by the International Agency for Research on Cancer as a known human carcinogen (WHO, 2016). The major source of dioxin exposure in humans is the food supply with food of animal origin being the predominant source.

The first known incidents of dioxin-related toxicity in chickens were reported in 1957, when millions of broilers died due to ingestion of toxic fatty acids (Firestone, 1973). Dioxins were originally known as 'chicken oedema factor' due to their association with chicken oedema disease. It was not until 1966 that researchers identified dioxins as the causative agent (Cantrell et al., 1969).

Like mammals, the liver and adipose tissue are the major accumulation and storage sites of dioxins in poultry, although avian species appear to be more susceptible to their toxic effects (Bursian et al., 2012; Fulton, 2008). Dioxins cause damage to vascular endothelium leading to vascular leakage and subsequent oedema, hydropericardium, and ascites (Fulton, 2008; Pang

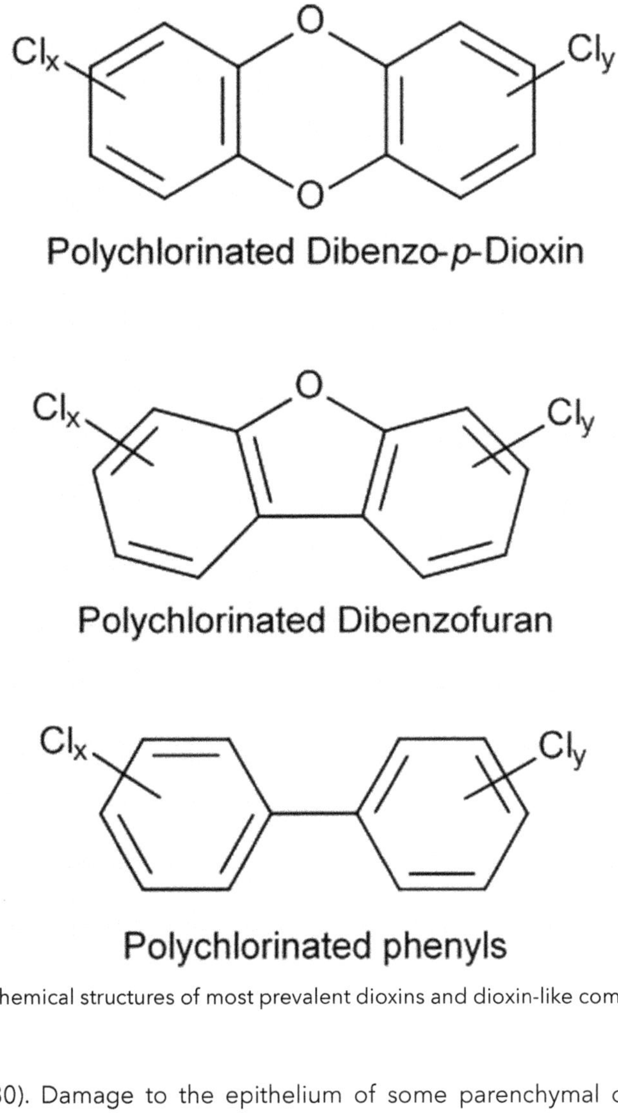

Figure 2 Chemical structures of most prevalent dioxins and dioxin-like compounds.

et al., 1980). Damage to the epithelium of some parenchymal organs and degeneration of heart and skeletal muscle may also be observed. Depending on the level of dioxin contamination, some flocks will show signs of stunting, respiratory distress, weakness, ataxia, and high mortality (Fulton, 2008). Dioxin toxicity has also been shown to result in inhibition of lymphoid development and atrophy of the thymus and bursa of Fabricius leading to immunosuppression (Andersoon et al., 1991; Fox and Grasman, 1999).

Several carry-over studies performed in poultry have concluded that once dioxins have accumulated in the body, concentrations tend to persist for an extended period in tissues and eggs after withdrawal (Hoogenboom et al.,

2004, 2006; Iben et al., 2003). Additionally, it has been shown that PCDD/F and PCB accumulation in poultry is higher than in beef or pork (Huwe et al., 2009).

3.2 Dioxin counteracting strategies

Since food of animal origin is the predominant route of human exposure to dioxins, it is imperative to avoid contamination of animal feed as well as be cognizant of the many other potential sources including litter, soil, plants, worms, and insects (Cardo et al., 2014). In terms of feed manufacturing, good manufacturing practices and Hazard Analysis and Critical Control Point principles should be implemented to control dioxin contamination throughout the process. Feed monitoring systems should also be enforced. Any material used in the production of animal feed should be guaranteed for quality and safety to ensure that it is not a source of contamination. Additionally, efforts should be made to reduce contamination of feed once on the farm (European Commission, 2000; Food and Agriculture Organization, 2006). Since dioxins retain high chemical stability and are not destroyed by microbial, photochemical, chemical, or thermal degradation, measures directed at identifying and eradicating dioxin sources are essential for contamination reduction.

4 Bacterial contaminations

Animal feed has been identified as a notable carrier for numerous bacteria that can cause health issues not only for the animal ingesting the contaminated feed but also for the humans who may use the products from those infected animals. Feed materials can become contaminated by various sources including soil, wind, water, insects, and faeces (Maciorowski et al., 2007).

4.1 Clostridium perfringens

Animal feed is often manufactured with animal by-products such as feathers, bone, and blood, which can significantly reduce feed costs; however, these ingredients provide an ideal environment for bacterial growth, most notably *C. perfringens*. Because *C. perfringens* spores are widespread in the environment, in soil, water, and faeces, it is commonly found in a variety of other feed ingredients including wheat bran, barley, soybeans, corn, and sunflowers, though it has been reported that *C. perfringens* is more prevalent in wheat- and barley-based diets than corn-based diets (Annett et al., 2002).

C. perfringens is a Gram-positive, anaerobic, spore-forming bacterium capable of producing a number of toxins. *C. perfringens* has been identified as the causative agent of necrotic enteritis in poultry. Necrotic enteritis is one of the world's most common and financially crippling poultry diseases, which, when triggered, can cause mortality rates of up to 50%. *C. perfringens* has also been linked to gangrenous dermatitis and cellulitis in broilers and turkeys, which may result in mortality rates of up to 1–2% per week (Thachil et al., 2012). Moreover, according to the Center for Disease Control and Prevention, *C. perfringens* is one of the most common causes of foodborne illness in the United States, with nearly 1 million cases reported every year (CDC, 2023).

Though it is a normal inhabitant of the intestines of humans and animals, some variants of *C. perfringens* could be considered as opportunistic pathogens that may produce toxins associated with disease when the intestinal microbiota become disrupted. Isolates producing alpha and netB toxins are most associated with necrotic enteritis (Kulkarni, et al., 2007; Keyburn et al., 2008).

Several factors have been implicated playing a role in necrotic enteritis outbreaks (Fig. 3). In general, diets rich in indigestible dietary protein, such as that found in animal proteins like meat and bone meal or fishmeal, can predispose birds to necrotic enteritis by altering the gut environment in such a way that it creates a favourable environment *for C. perfringens* growth. Since

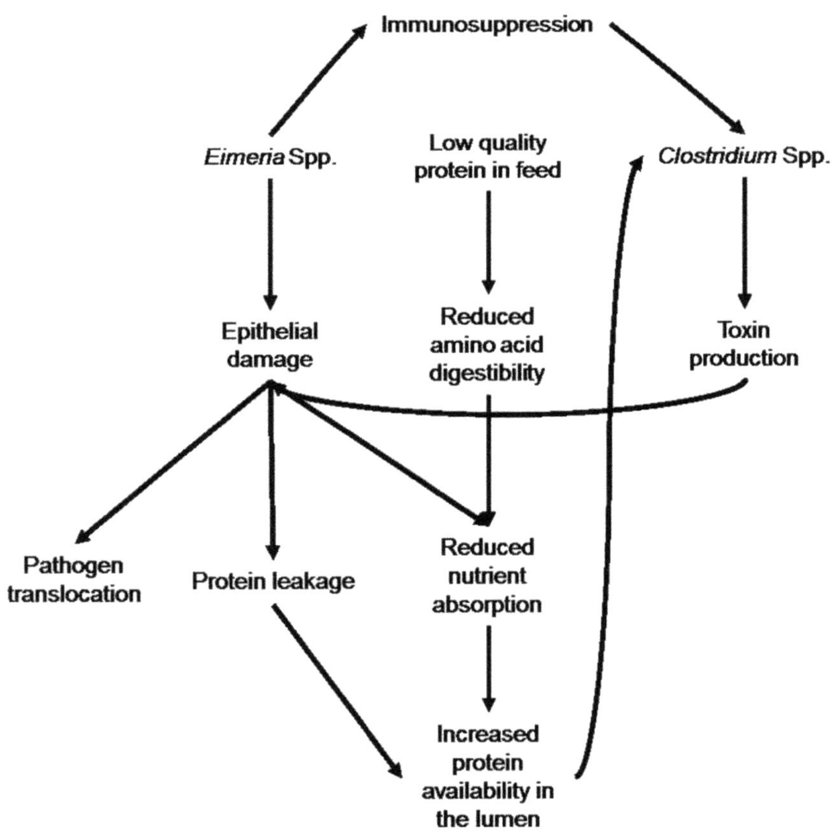

Figure 3 Factors influencing the development of necrotic enteritis.

indigestible dietary protein cannot be digested and absorbed in the upper part of the intestinal tract, it can lead to higher concentrations of protein in the lower portion of the intestinal tract, which can then act as a substrate for the gut microbiota. The fermentation of protein produces unfavourable by-products such as amines and ammonia, and increases intestinal pH, which encourages the proliferation of pathogenic bacteria such as *C. perfringens*. Coccidiosis is another well-known predisposing factor for outbreaks of necrotic enteritis. As *Eimeria* oocysts invade, replicate, and cycle through the intestinal epithelial cells, significant damage occurs. This damage allows the leakage of plasma proteins into the intestinal lumen, which is a rich-nutrient substrate that can be utilized by *C. perfringens* for proliferation and toxin production (Annett et al., 2002; Wu et al., 2015).

4.2 Salmonella

Bacterial contamination of poultry feed is an important consideration when aiming to reduce the spread of *Salmonella* in poultry production and potential food safety hazards. Numerous surveys of feed ingredients have been conducted and show that animal feed is frequently contaminated with *Salmonella* (Crump et al., 2002). As such, several international regulatory agencies, including the FDA and EFSA, have implemented guidelines to reduce feed contamination and abate food safety hazards (Jones and Richardson, 2004).

Salmonellosis is one of the most reported health concerns for humans, causing significant morbidity and mortality. According to the World Health Organization, it is estimated that there are tens of millions of humans cases of salmonellosis worldwide resulting in more than 100,000 deaths annually (WHO, 2018).

The genus *Salmonella* is made up of more than 2,500 different serovars. The most common *Salmonella* serotypes found in poultry are *S. Typhimurium*, *S. Enteritidis*, *S. Kentucky,* and *S. Heidelberg*. Although they are frequently found in poultry populations, these non-host adapted strains (collectively referred to as paratyphoid salmonellae) rarely induce disease in poultry, except in vulnerable young birds, and most birds remain asymptomatic carriers. These serotypes, however, can cause foodborne disease in humans (Foley et al., 2011).

Protein-rich vegetable and animal-derived protein sources, such as meat and fishmeal, have higher risks of becoming contaminated with microbes such as *Salmonella*, as the high levels of protein within these ingredients can act as a substrate (Veldman et al., 1995). Once ingested, these bacteria can multiply within the intestine and subsequently contaminate the environment when excreted in the faeces. *Salmonella* are also capable of translocating across the intestinal epithelia, multiplying in macrophages, and being transported via the bloodstream or lymphatic system to infect vital organs such as lungs, liver, and spleen as well as reproductive organs such as ovaries and oviducts.

4.3 Escherichia coli

E. coli is a Gram-negative, non-spore-forming, facultative anaerobe. It is a normal inhabitant of animal and human intestinal microbiota, and thus are commonly utilized as an indicator of faecal contamination of feed. Since *E. coli* is commonly found in animal feed, it can be surmised that contamination with wildlife faeces or fertilization of crops could possibly be a major transmission source (Maciorowski et al., 2007). In a study conducted by da Costa et al. (2007), *E. coli* was present in 50% of commercial broiler feed and 32% of raw feed ingredients.

Colibacillosis, caused by certain *E. coli* strains designated as avian pathogenic *E. coli* (APEC), is the most common infectious bacterial disease in poultry of all ages. It is known to result in high incidence of morbidity and mortality and is associated with significant economic losses. Colibacillosis can manifest itself in a wide variety of disease syndromes, depending on the *E. coli* strain responsible, such as colisepticaemia, swollen-head syndrome, coliform cellulitis, airsacculitis, and salpingitis. Secondary infections of *E. coli* are also common after bouts of immunosuppression (Barnes et al., 2008; Maciorowski et al., 2007). Currently, there are an estimated 150–200 serotypes of *E. coli* with the most frequently isolated serotypes being O78:K80, O1:K1, and O2:K1 (Kabir, 2010).

4.4 Biogenic amines

It is common practice to include animal by-product meals, including meat and bone meal, blood meal, fishmeal, feather meal, and poultry by-product meal, as a major component of poultry feed to meet protein requirements and reduce feed costs. Unfortunately, due to their high protein content, the use of these animal by-product meals provides an opportunity to produce biogenic amines during putrefaction before rendering and/or when those feed components become spoiled (Tamim and Doerr, 2003).

Biogenic amines (histamine, tyramine, putrescine, cadaverine, gizzerosine, agmatine, spermine, and spermidine) are the result of bacterial decomposition or putrefaction of animal proteins. They are low-molecular-weight compounds, which are formed predominantly by decarboxylation of particular amino acids into their corresponding amines through the action of bacterial enzymes during storage (Friday et al., 1999; Barnes et al., 2001). The presence of biogenic amines has been implicated in causing several detrimental effects in poultry including malabsorption syndrome, decreased body weight gain and feed efficiency, the passage of undigested feed, sloughing of intestinal epithelia, and lesions in the proventriculus, gizzard, and intestines (Bermudez and Firman, 1998; Barnes et al., 2001). Intestinal irritation caused by biogenic amines can also lead to secondary bacterial infections. Moreover, increased carcass contamination may be observed at processing due to intestinal rupture (Barnes et al., 2001).

Toxicity of these compounds varies depending on the particular amine in question. For example, Barnes and colleagues (2001) found that dietary inclusion of 0.1% histidine resulted in reduced body weight and feed conversion in 21-day-old broiler chicks. The toxic effects of spermine have been observed in 14-day-old chicks at levels as low as 0.2% (Sousadias and Smith, 1995). Gizzerosine, on the other hand, is considered to be 10 times as toxic as histamine in terms of its ability to promote hydrochloric acid production in the proventriculus and 300 times more toxic in its ability to cause gizzard erosion (Masumura et al., 1985).

4.5 Solutions to bacterial contaminations

Animal feed is a primary source of bacterial contamination. These pathogens can be very difficult to control and every possible tool to mediate them needs to be utilized to establish an effective control programme. In regard to eliminating the presence of pathogenic bacteria in animal feeds, actions need to be taken to prevent contamination of feed and/or feed ingredients, prevent the growth of bacteria, and kill pathogens that are present.

Measures should be taken to prevent bacterial infections by obtaining pathogen-free, high-quality feed ingredients, particularly animal-derived by-product meals. Regarding specifically reducing biogenic amine production, a quality control programme must be implemented to ensure that raw products are stored properly to maintain ideal temperatures (below 5°C) and prevent bacterial contamination after rendering as these are the most likely sources of biogenic amines (Barnes et al., 2001). It should also be emphasized that animal by-products should be rendered immediately after slaughter to avoid bacterial growth and production of biogenic amines (Feddern et al., 2019).

One of the most common and simplest means of feed decontamination is thermal treatment during pelleting. For example, pelleting has been shown to reduce *Salmonella* isolation rates from 50% to 93% (Jones and Richardson, 2004; Veldman et al., 1995). Despite efforts to reduce bacterial populations using heat treatments, there is always a risk of recontamination. Bactericidal chemical treatments, such as organic acids or formaldehyde, may also be used to kill microbes in feed and reduce recontamination. The antimicrobial effects of short-chain organic acids, such as propionic and formic acids, are derived from the ability of their undissociated forms to penetrate the outer membrane of Gram-negative bacteria and, due to the higher relative pH inside the cell, dissociate within the cell into anion and protons. The increased quantity of protons acidifies the intracellular environment and disrupts many cellular functions while the anion may inhibit DNA synthesis. To maintain a neutral cytoplasmic pH, the cell will actively export excess protons, thus consuming cellular adenosine triphosphate (ATP) and ultimately resulting in energy depletion and cell death (Ricke, 2003).

If pathogenic bacteria are ingested, there are several feed additives that could be utilized to protect intestinal health. A common means of lessening pathogenic threats in the intestinal tract is dietary supplementation of antibiotics. Antibiotic supplementation at sub-therapeutic levels has been a widely used practice for over 50 years to prevent disease and promote growth; however, concerns about antibiotic resistance has led many regulatory agencies and producers to re-evaluate this practice and limit antibiotic use.

Other feed additives are quickly gaining popularity to control pathogenic bacteria. Prebiotics and probiotics, for example, can be supplemented in the feed to reduce or prevent enteric infections and subsequent contamination of poultry products. Prebiotics are generally defined as non-digestible food/feed ingredients that beneficially affect the host by selectively stimulating the growth and/or activity of one or a limited number of bacteria in the intestine (Gibson and Roberfroid, 1995). In other words, prebiotics provide the environment and/or the nutrients for beneficial bacteria to better establish themselves in the intestine. A probiotic is defined as 'a live microbial feed supplement, which beneficially affects the host animal by improving intestinal balance' (Fuller, 1989). Probiotics are used to help maintain an ideal microbial balance within the intestine and promote gut integrity. This is achieved through three fundamental mechanisms: competitive exclusion, bacterial antagonism, and stimulation of the immune system (Ohimain and Ofongo, 2012). It is important to note, however, that probiotic products are very diverse in their makeup; thus, they have varying modes of action depending on the strain(s) being utilized and doses used (Cox and Dalloul, 2015). In recent years, the supplementation of post-biotics, a newly coined category of feed additives, has gained some popularity. Post-biotics, as defined by the International Scientific Association for Probiotics and Prebiotics, are a 'preparation of inanimate microorganisms and/or their components that confers a health benefit on the host'. These preparations contain inactivated microbial cells or cell components and may or may not include metabolic by-products secreted by live bacteria or released after bacterial lysis (Salminen et al., 2021). Post-biotics have been shown to improve immune function, reduce intestinal lesions due to *Eimeria maxima* and *E. tenella* infection, and mitigate colonization of pathogenic bacteria, such as *Salmonella enteritidis* (Chaney et al., 2023).

Phytogenic feed additives, or plant-derived compounds, may also alleviate microbial challenge and biogenic amine formation in the intestine. Herbs and essential oils have been used in humans and animals for centuries as flavour enhancers, food preservatives, and medicines. Furthermore, several plant extracts are known to have antimicrobial, antiviral, anti-coccidial, fungicidal, and/or antioxidant properties, making them ideal as feed additives (Applegate et al., 2010). For example, carvacrol and thymol, the major biologically active constituents found in oregano and thyme, have been shown to have antioxidant and antimicrobial properties in poultry (Hoffman-Pennesi and Wu, 2010). Moreover, they have been shown to control pathogenic bacteria, promote beneficial bacteria, and augment nutrient utilization in the intestine (Mitsch et al., 2004; Murugesan et al., 2015). Overall, phytogenic feed additives, depending on their composition, may help improve performance by sustaining an ideal intestinal environment.

Finally, efforts to keep feed clean should include maintaining good feed practices and storage facilities and ensuring proper biosecurity to prevent access of natural carriers, such as rodents and wild birds, out of feed facilities.

5 Conclusion and future trends

The safety of animal feed plays an important role in maintaining not only the health of production animals but also the health of humans consuming products from those animals. Ultimately, ensuring feed safety requires a multi-factorial approach as it is reliant on a multitude of factors and can be affected by several contaminants including plant and fungal toxins, chemicals, and pathogens.

Mycotoxins, secondary fungal metabolites responsible for mycotoxicoses in livestock and poultry, must be addressed before, during, and after harvest to minimize risks. The most commonly used method of counteracting mycotoxins is the use of binding agents; however, due to the heterogeneity of mycotoxins in terms of their chemical, physical, and biochemical properties, not all mycotoxins can be effectively bound, and other strategies must be utilized. A more novel approach is the use of microorganisms and their enzymes to detoxify specific non-absorbable mycotoxins, such as DON and FUM.

Dioxins and PCBs are harmful environmental contaminants that can enter feedstuffs through air, soil, or sediments. Not only do dioxins cause significant consequences in terms of animal health, but they are also commonly found in the human food supply as a result of contaminated animal feed. To avoid dioxin contamination, it is imperative to be cognizant of their many potential sources. Good manufacturing practices and Hazard Analysis and Critical Control Point principles should be implemented to control dioxin contamination during the feed manufacturing process and feed monitoring systems should also be enforced.

Animal feed is a significant source of numerous bacteria that can cause health issues for animals and humans alike. Bacterial contamination of feedstuffs can be mitigated through a thermal treatment during pelleting or the application of bactericidal chemical treatments such as organic acids or formaldehyde. In instances where pathogenic bacteria are ingested, they can be controlled using antibiotics or natural feed additives such as prebiotics, probiotics, and phytogenic feed additives.

The safety of poultry feed, as a whole, will continue to improve as more sophisticated measures of identification and control are developed. As poultry production continues to evolve, more alternatives for eliminating mycotoxins and pathogens will begin to take position at the forefront. Furthermore, as biotechnology advances, so will our surveillance and control methods. Genomics, transcriptomics, and proteomics are quickly transforming our approaches to the detection, prevention, and treatment of biological feed contaminants. These advances are also allowing the development of control measures and treatments that are more specific in terms of their targets. Feedstuffs play an important role in maintaining the health of production animals as well as humans; therefore, it is essential that producers develop procedures and practices that ensure a supply of safe feed for animal production.

6 Where to look for further information

For an overview of mycotoxins, feed sampling, and mycotoxin analysis, please refer to *Romer Lab's Guide to Mycotoxins* (edited by E.M. Binder and R. Krska; http://www.foodriskmanagement.nl/wp-content/uploads/2013/03/Romer-Labs-Guide-to-Mycotoxin-Book_Original_41686.pdf). An in-depth read regarding masked mycotoxins can be found in *Masked Mycotoxins in Food: Formation, Occurrence and Toxicological Relevance* (edited by Chiara Dall'Asta and Franz Berthiller). The *Mycotoxin Factbook: Food & Feed Topics* (edited by D. Barug) is a great resource if you are interested in information on major mycotoxins and emerging problems in the food chain, the impact of mycotoxins in the feed chain, developments in mycotoxin prevention, trends in mycotoxin analysis, and regulatory issues related to mycotoxins. The journal article 'Prevalence and effects of mycotoxins on poultry health and performance, and recent development in mycotoxin counteracting strategies' (G.R. Murugesan et al., 2015; DOI: 10.3382/ps/pev075) is also a good source of valuable information on the effects of mycotoxins in poultry and various counteracting strategies. The Food and Agriculture Organization of the United Nations provides a great amount of information covering basic facts about dioxins as well as prevention and control measures (http://www.fao.org/ag/againfo/home/en/news_archive/2009_IN_dioxin.html). A good source of information pertaining to *Salmonella* control can be found in this educational document produced by the American Feed Industry Association, (http://ucfoodsafety.ucdavis.edu/files/172958.pdf) as well as in this article, 'A review of practical *Salmonella* control measures in animal feed' by F.T. Jones (*J. Appl. Poult. Res.* (2011) 20 (1): 102–113. doi: 10.3382/japr.2010-00281). For a comprehensive review of animal feed contaminants, their negative effects on both animal and human health, as well as analysis and control measures please refer to *Animal Feed Contamination: Effects on Livestock and Food Safety* (edited by J. Fink-Gremmels).

7 References

Andersson, L., Nikolaidis, E., Brunstrom, B., Bergman, A. and Dencker, L. 1991. Effects of polychlorinated biphenyls with Ah receptor affinity on lymphoid development in the thymus and the bursa of Fabricius of chick embryos in ovo and in mouse thymus anlagen in vitro. *Toxicol. Appl. Pharmacol.* 107: 183–188.

Annett, C. B., Vistea, J. R., Chirino-Trejob, M., Classenc, H. L., Middletona, D. M. and Simkoa, E. 2002. Necrotic enteritis: effect of barley, wheat and corn diets on proliferation of Clostridium perfringens type A. *Avian Pathol.* 31 (6): 598–601.

Antonissen, G., Immerseel, F. V., Pasmans, F., Ducatelle, R., Haesebrouck, F., Timbermont, L., Verlinden, M., Janssens, G. P. J., Eeckhaut, V., Eeckhout, M., De Saeger, S., Hessenberger, S., Martel, A. and Croubels, S. 2014. The Mycotoxin Deoxynivalenol predisposes for the development of Clostridium perfringens-induced Necrotic Enteritis in broiler chickens. *PLoS ONE.* 9 (9): e108775.

Applegate, T. J., Klose, V., Steiner, T., Ganner, A. and Schatzmayr, G. 2010. Probiotics and phytogenics for poultry: myth or reality? *J. Appl. Poult. Res.* 19 (2): 194–210.

Awad, W. A., Hess, M., Twaruzek, M., Grajewski, J., Kosicki, R., Bohm, J. and Zentek, J. 2011. The impact of the *Fusarium* Mycotoxin deoxynivalenol on the health and performance of broiler chicks. *Intl. J. Mol. Sci.* 12: 7996–8012.

Barnes, D. M., Kirby, Y. K. and Oliver, K. G. 2001. Effects of biogenic amines on growth and the incidence of proventricular lesions in broiler chickens. *Poult. Sci.* 80: 906–911.

Barnes, J. H., Nolan, L. K. and Vaillancourt, J. 2008. Colibacillosis. In Y. M. Saif (ed.), *Diseases of Poultry*, pp. 691–732. Ames, IA: Iowa State Press.

Bermudez, A. J. and Firman, J. D. 1998. Effects of biogenic amines in broiler chickens. *Avian Dis.* 42: 199–203.

Broomhead, J. N., Ledoux, D. R., Bermudez, A. J. and Rottinghaus, G. E. 2002. Chronic effects of fumonisin B_1 in broilers and turkeys fed dietary treatments to market age. *Poult. Sci.* 81: 56–61.

Bryden, W. L. 2012. Mycotoxin contamination of the feed supply chain: implications for animal productivity and feed security. *Anim. Feed Sci. Technol.* 173 (1–2): 134–158.

Bursian, S., Newsted, J. and Zwiernik, M. 2012. Polychlorinated biphenyls, polybrominated biphenys, polychlorinated debenzo-p-dioxins, and polychlorinated dibenzofurans. In R. Gupta (ed.), *Veterinary Toxicology: Basic and Clinical Principles*, pp. 776–779. Waltham, MA: Academic Press.

Busby, W. F. Jr. and Wogan, G. N. 1981. Aflatoxins. In R. C. Shank (ed.), *Mycotoxins and N-Nitrosocompounds, Environmental Risks*, Vol. 2, pp. 3–27. Boca Raton, FL: CRC Press Inc.

Cantrell, J. S., Webb, N. C. and Mabis, A. J. 1969. The identification and crystal structure of a hydropericardium-producing factor: 1,2,3,7,8,9-hexachlorodibenzo-p-dioxin. *Acta Cryst.* B25: 150–156.

Cardo, M., Castel-Branco, M., Andreozzi, V. and Bernardo, F. 2014. Dioxins in the food chain: contamination fingerprint analysis in breeding hens, hatching eggs and broilers. *J. Environ. Prot.* 5: 1323--1330.

Carretero, M. 2002. Clay minerals and their beneficial effects upon human health. *Appl. Clay Sci.* 21: 155–163.

CAST. 2003. *Mycotoxins: Risks in Plant, Animal, and Human Systems, Task Force Report No. 139,* pp. 13–85. Ames, IA: Council for Agricultural Science and Technology.

Cawood, M. E., Gelderblom, W. C. A., Vleggaar, R., Behrend, Y., Thiel, P. G. and Marasas, W. F. O. 1991. Isolation of the fumonisin mycotoxins: a quantitative approach. *J. Agric. Food Chem.* 39: 1958–1962.

Centers for Disease Control and Prevention. 2023. *Clostridium Perfringens.* http://www.cdc.gov/foodsafety/diseases/clostridium-perfringens.html.

Chaney, W. E., McBride, H. and Girgis, G. 2023. Effect of a *Saccharomyces cerevisiae* Postbiotic Feed Additive on *Salmonella* Enteritidis Colonization of Cecal and Ovarian Tissues in Directly Challenged and Horizontally Exposed Layer Pullets. Animals. 13(7):1186.

Chen, J., Mirocha, C. J., Xie, W., Hogge, L., and Olson, D. 1992. Production of the mycotoxin fumonisin B_1 by Alternaria alternata f. sp. Lycopersici. *Appl. Environ. Microbiol.* 8: 3928–3931.

Ciegler, A., Lillehoj, B., Peterson, E. and Hall, H. H. 1966. Microbial detoxification of Aflatoxin. *Appl. Microbiol.* 14: 934–939.

Cope, R. B. 2018. Trichothecenes. In R. C. Gupta (ed.), *Veterinary Toxicology*, pp. 1043–1053. Waltham, MA: Academic Press.

Cox, C. M. and Dalloul, R. A. 2015. Immunomodulatory role of probiotics in poultry and potential in ovo application. *Benef. Microbes.* 6 (1): 45–52.

Crump, J. A., Griffen, P. M. and Angulo, F. J. 2002. Bacterial contamination of animal feed and its relationship to human foodborne illness. *Clin. Infect. Dis.* 35 (7): 859–865.

da Costa, P. M., Olivera, M., Bica, A., Vaz-Pires, P. and Bernardo, F. 2007. Antimicrobial resistance in *Enterococcus* spp. and *Escherichia coli* isolated from poultry feed and feed ingredients. *Vet. Microbiol.* 120 (1–2): 122–131.

Dänicke, S., Gareis, M. and Bauer, J. 2001. Orientation values for critical concentrations of deoxynivalenol and zearalenone in diets for pigs, ruminants and gallinaceous poultry. *Proc. Soc. Nutr. Physiol.* 10: 171–174.

Dänicke, S. 2002. Prevention and control of mycotoxins in the poultry production chain: a European view. *World. Poult. Sci. J.* 58: 451–474.

Devegowda, G. and Murthy, T. N. K. 2005. Mycotoxins: their effects in poultry and some practical solutions. In D. E. Diaz (ed.), *The Mycotoxin Blue Book*, pp. 25–56. Nottingham, UK: Nottingham University Press.

De Vos, S., Maervoetb, J., Schepensb, P. and De Schrijver, R. 2003. Polychlorinated biphenyls in broiler diets: their digestibility and incorporation in body tissues. *Chemosphere*. 15 (1): 7–11.

D'mello, J., Placinta, C. and Macdonald, A. 1999. Fusarium mycotoxins: a review of global implications for animal health, welfare and productivity. *Anim. Feed Sci. Technol.* 80: 183–205.

Edds, G. T. and Bortell, R. R. 1983. Biological effects of aflatoxin in poultry. In U. L. Diener, R. L. Asquith, and J. W. Dickens (eds.), *Aflatoxin and Aspergillus flavus* in corn, pp. 55–61. Southern Cooperative Services Bulletin. 279. Auburn, AL: Alabama Agricultural Experiment Station, Auburn University.

European Commission. 2000. Opinion of the Scientific Committee on Animal Nutrition on the dioxin contamination of feedingstuffs and their contribution to the contamination of food of animal origin. http://ec.europa.eu/food/committees/scientific/out55_en.pdf.

EFSA CONTAM Panel (EFSA Panel on Contaminants in the Food Chain), Schrenk, D., Bignami, M., Bodin, L., Chipman, J. K., del Mazo, J., Grasl-Kraupp, B., Hogstrand, C., Leblanc, J. C., Nielsen, E., Ntzani, E. Petersen, Sand, S., Schwerdtle, T., Vleminckx, C., Wallace, H., Daenicke, S., Nebbia, C. S., Oswald, I. P., Rovesti, E., Steinkellner, H. and Hoogenboom, L. R. 2022. Scientific opinion on the assessment of information as regards the toxicity of fumonisins for pigs, poultry and horses. *EFSA J.* 20 (8): 7534.

EFSA CONTAM Panel (EFSA Panel on Contaminants in the Food Chain), Schrenk, D., Bignami, M., Bodin, L., Chipman, J. K., del Mazo, J., Grasl-Kraupp, B., Hogstrand, C., Leblanc, J. C., Nielsen, E., Ntzani, E. Petersen, Sand, S., Schwerdtle, T., Vleminckx, C., Wallace, H., Daenicke, S., Nebbia, C. S., Oswald, I. P., Rovesti, E., Steinkellner, H. and Hoogenboom, L. R. 2023. Scientific opinion on the assessment of information as regards the toxicity of deoxynivalenol for horses and poultry. *EFSA J.* 21 (2): 7806.

EFSA FEEDAP Panel (EFSA Panel on Additives and Products or Substances used in Animal Feed (FEEDAP), Aquilina, G., Bampidis, V., De Lourdes Bastos, M., Costa, L. G., Flachowsky, G., Gralak, M. A., Hogstrand, C., Leng, L., López-Puente, S., Martelli, G., Mayo, B., Ramos, F., Renshaw, D., Rychen, G., Saarela, M., Sejrsen, K., Van Beelen, P., Wallace, R. J. and Westendorf, J. 2014. Scientific opinion on the safety and efficacy of fumonisin esterase (FUMzyme®) as a technological feed additive for pigs. *EFSA J.* 12 (5): 3667.

EFSA FEEDAP Panel (EFSA Panel on Additives and Products or Substances used in Animal Feed), Rychen, G., Aquilina, G., Azimonti, G., Bampidis, V., de Lourdes Bastos, M., Bories, G., Chesson, A., Cocconcelli, P. S., Flachowsky, G., Gropp, J., Kolar, B., Kouba, M., Lopez-Alonso, M., Mantovani, A., Mayo, B., Ramos, F., Saarela, M., Villa, R. E., Wallace, R. J., Wester, P., Martelli, G., Renshaw, D. and Lopez Puente, S. 2016. Scientificopinion on the safety and efficacy of fumonisin esterase (FUMzyme®) as a technological feed additive for all avian species. *EFSA J.* 14 (11): 4617.

Feddern, V., Mazzuco, H., Da Fonseca, F. and Lima, G. 2019. A review on biogenic amines in food and feed: toxicological aspects, impact on health and control measures. *Anim. Prod. Sci.* 59 (4): 608–618.

Firestone, D. 1973. Etiology of chick edema disease. *Environ. Health Perspect.* 5: 59–66.

Foley, S. L., Nayak, R., Hanning, I. B., Johnson, T. J., Han, J. and Ricke, S. C. 2011. Population dynamics of *Salmonella enterica* serotypes in commercial egg and poultry production. *Appl. Environ. Microbiol.* 77 (13): 4273–4279.

Food and Agriculture Organization. 2006. *Code of Practice for the Prevention and Reduction of Dioxin and Dioxin-like PCB Contamination in Foods and Feeds*. Rome: Food and Agriculture Organization.

Food and Drug Administration, Department of Health and Human Services, Code of Federal Regulations. 2022. Fumonisin esterase. Title 21, Sub-Chapter E, Part 573.485. https://www.accessdata.fda.gov/scripts/cdrh/cfdocs/cfcfr/CFRSearch.cfm?fr=573.485

Fox, L. L. and Grasman, K. A. 1999. Effects of PCB 126 on primary immune organ development in chicken embryos. *J. Toxicol. Environ. Health A.* 58: 233–244.

Friday, M. L., Firman, J. D. and Bermudez, A. J. 1999. Effects of biogenic amines on broiler performance. *J. Appl. Poul. Res.* 8: 408–413.
Fries, G. F. 1995. A review of the significance of animal food products as potential pathways of human exposures to dioxins. *J. Anim. Sci.* 73: 1639–1650.
Fuchs, E., Binder, E. M., Heidler, D. and Krska, R. 2002. Structural characterization of metabolites after the microbial degradation of type A trichothecenes by the bacterial strain BBSH 797. *Food Addit. Contam.* 19: 379–386.
Fuller, R. 1989. Probiotics in man and animals. *J. Appl. Bacteriol.* 66: 365–378.
Fulton, M. 2008. Other toxins and poisons. In Y. M. Saif (ed.), *Diseases of Poultry*, pp. 1231–1258. Ames, IA: Iowa State Press.
Galli G., Griss, L., Fortuoso, B., Silva, A., Fracasso, M., Lopes, T., Schetinger, M., Gundel, S., Ourique, A., Carneiro, C., Mendes, R., Boiago, M. and Da Silva, A. 2020. Feed contaminated by fumonisin (Fusarium spp.) in chicks has a negative influence on oxidative stress and performance, and the inclusion of curcumin-loaded nanocapsules minimizes these effects. *Microb. Pathog.* 148: 104496.
Gibson, G. R. and Roberfroid, M. B. 1995. Dietary modulation of the human colonic microbiota: introducing the concept of prebiotics. *J. Nutr.* 125: 1401–412.
Gibson, R. M., Bailey, C. A., Kubena, L. F., Huff, W. E. and Harvey, R. B. 1989. Ochratoxin A and dietary protein. 1. Effects on body weight, feed conversion, relative organ weight, and mortality in three-week-old broilers. *Poult. Sci.* 68: 1658–1663.
Grenier, B. and Oswald, I. P. 2011. Mycotoxin co-contamination of food and feed: meta-analysis of publications describing toxicological interactions. *World Mycotoxin J.* 4 (3): 285–313.
Grenier, B., Bracarense, A. P. F. L., Schwartz, H. E., Trumel, C., Cossalter, A. M., Schatzmayr, G., Kolf-Clauw, M., Moll, W. D. and Oswald, I. P. 2012. The low intestinal and hepatic toxicity of hydrolyzed fumonisin B 1 correlates with its inability to alter the metabolism of sphingolipids. *Biochem. Pharmacol.* 83: 1465–1473.
Grenier, B. and Applegate, T. J. 2013. Invited review -modulation of intestinal functions upon mycotoxin ingestion: meta-analysis of published experiments in animals. *Toxins*. 5: 396–430.
Grenier, B., Bracarense, A. P. F., Schwartz, H. E., Lucioli, J., Cossalter, A.-M., Moll, W.-D., Schatzmayr, G. and Oswald, I. P. 2013. Biotransformation approaches to alleviate the effects induced by Fusarium Mycotoxins in Swine. *J. Agric. Food Chem.* 61: 6711–6719.
Grenier, B., Dohnal, I., Shanmugasundaram, R., Eicher, S. D., Selvaraj, R. K., Schatzmayr, G. and Applegate, T. J. 2015a. Susceptibility of broiler chickens to coccidiosis when fed subclinical doses of deoxynivalenol and fumonisins – special emphasis on the immunological response and the mycotoxin interaction. *Poult. Sci.* 94 (E-Suppl. 1): 83.
Grenier, B., Schwartz-Zimmerman, H. E., Caha, S., Moll, W. D., Schatzmayr, G. and Applegate, T. J. 2015b. Dose-dependent effects on sphingoid bases and cytokines in chickens fed diets prepared with Fusarium verticilloides culture material containing fumonisins. *Toxins* 7: 1253–1272.
Grenier, B., Schwartz-Zimmermann, H. E., Gruber-Dorninger, C., Dohnal, I., Aleschko, M., Schatzmayr, G., Moll, W. D. and Applegate, T. J. 2017. Enzymatic hydrolysis of fumonisins in the gastrointestinal tract of broiler chickens. *Poult. Sci.* 96: 4342–4351.
Hartinger, D. and Moll, W. D. 2011. Fumonisin elimination and prospects for detoxification by enzymatic transformation. *World Mycotoxin J.* 4: 271–283.

Heinl, S., Hartinger, D., Thamhesl, M., Vekiru, E., Krska, R., Schatzmayr, G., Moll, W. D. and Grabherr, R. 2010. Degradation of fumonisin B1 by the consecutive action of two bacterial enzymes. *J. Biotechnol.* 145: 120–129.

Hoerr, F. J. 2010. Clinical aspects of immunosuppression in poultry. *Avian Dis.* 54: 2–15.

Hoffman-Pennesi, D. and Wu, C. 2010. The effect of thymol and thyme oil feed supplementation on growth performance, serum antioxidant levels, and cecal *Salmonella* population in broilers. *J. Appl. Poult. Res.* 19: 432–443.

Hoogenboom, L. A., Kan, C. A., Bovee, T. F., van der Weg, G., Onstenk, C. and Traag, W. A. 2004. Residues of dioxins and PCBs in fat of growing pigs and broilers fed contaminated feed. *Chemosphere*. 57 (1): 35–42.

Hoogenboom, L. A., Kan, C. A., Zeilmaker, M. J., Van Eijkeren, J. and Traag, W. A. 2006. Carry-over of dioxins and PCBs from feed and soil to eggs at low contamination levels – influence of mycotoxin binders on the carry-over from feed to eggs. *Food. Addit. Contam.* 23 (5): 518–527.

Huff, W. E. and Hamilton, P. B. 1975. Decreased plasma carotenoids during ochratoxicosis. *Poult. Sci.* 54: 1308–1310.

Huff, W. E. and Doerr, J. 1981. Synergism between aflatoxin and ochratoxin A in broiler chickens. *Poult. Sci.* 60: 550–555.

Huff, W. E., Kubena, L. F., Harvey, R. B., Hagler Jr., W. M., Swanson, S. P., Phillips, T. D. and Creger, C. R. 1986. Individual and combined effects of aflatoxin and deoxynivalenol (DON, vomitoxin) in broiler chickens. *Poult. Sci.* 65: 1291–1298.

Huwe, J., Pagan-Rodriguez, D., Abdelmajid, N., Clinch, N., Gordon, D., Holterman, J., Zaki, E., Lorentzsen, M. and Dearfield, K. 2009. Survey of polychlorinated dibenzo-p-dioxins, polychlorinated dibenzofurans, and non-ortho-polychlorinated biphenyls in U.S. meat and poultry, 2007–2008: effect of new toxic equivalency factors on toxic equivalency levels, patterns, and temporal trends. *J. Agric. Food Chem.* 57: 11194–1200.

Iben, C., Böhm, J., Tausch, H., Leibetseder, J. and Luf, W. 2003. Dioxin residues in the edible tissue of broiler chicken. *J. Anim. Physiol. Anim. Nutr.* 87: 142–148.

Jones, F. T. and Richardson, K. E. 2004. *Salmonella* in commercially manufactured feeds. *Poult. Sci.* 83 (3): 384–391.

Kabir, S. M. 2010. Avian colibacillosis and salmonellosis: a closer look at epidemiology, pathogenesis, diagnosis, control and public health concerns. *Int. J. Environ. Res. Public Health*. 7 (1): 89–114.

Keyburn, A. L., Boyce, J. D., Vaz, P., Bannam, T. L., Ford, M. E. and Parker, D. 2008. NetB, a new toxin that is associated with avian necrotic enteritis caused by Clostridium perfringens. *PLoS Pathog.* 4: e26.

Kubena, L. F., Huff, W. E., Harvey, R. B., Corrier, D. E., Phillips, T. D. and Creger, C. R. 1988. Influence of ochratoxin a and deoxynivalenol (DON) on growing broiler chicks. *Poult. Sci.* 67: 253–260.

Kubena, L. F., Huff, W. E., Harvey, R. B., Phillips, T. D. and Rottinghaus, G. E. 1989. Individual and combined toxicity of deoxynivalenol and T-2 toxin in broiler chicks. *Poult. Sci.* 68: 622–626.

Kulkarni, R. R., Parreira, V. R., Sharif, S. and Prescott, J. F. 2007. Immunization of broiler chickens against *Clostridium perfringens*-induced necrotic enteritis. *Clin. Vaccine Immunol.* 14: 1070–1077.

Leeson, S., Diaz, G. J. and Summers, J. D. 1995. *Poultry Metabolic Disorders and Mycotoxins*. Guelph, ON, Canada: University Books.

Liu, J. D., Doupovec, B., Schatzmayr, D., Murugesan, G. R., Bortoluzzi, C., Villegas, A. M. and Applegate, T. J. 2020. The impact of deoxynivalenol, fumonisins, and their combination on performance, nutrient, and energy digestibility in broiler chickens. *Poult. Sci.* 99 (1): 272–279.

Maciorowski, K. G., Herrera, P., Jones, F. T., Pillai, S. D. and Ricke, S. C. 2007. Effects on poultry and livestock of feed contamination with bacteria and fungi. *Anim. Feed Sci. Technol.* 133: 109–136.

Magan, N., Medina, A. and Aldred, D. 2011. Possible climate-change effects on mycotoxin contamination of food crops pre- and postharvest. *Plant Pathol.* 60: 150–163.

Marquardt, R. R. and Frohlich, A. A. 1992. A review of recent advances in understanding ochratoxicosis. *J. Anim. Sci.* 70: 3968–3988.

Masumura, T., Sugahara, M., Noguchi, T., Mori, K. and Naito, H. 1985. The effect gizzerosine, a recently discovered compound in overheated fish meal, on the gastric acid secretion in chicken. *Poult. Sci.* 64: 356–336.

Miller, J. D., Schaafsma, A. W., Bhatnagar, D., Bondy, G., Carbone, I., Harris, L. J., Harrison, G., Munkvold, G. P., Oswald, I. P., Pestka, J. J., Sharpe, L., Sumarah, M. W., Tittlemier, S. A. and Zhou, T. 2014. Mycotoxins that affect the North American agri-food sector: State of the art and directions for the future. *World Mycotoxin J.* 7: 63–82.

Mitsch, P., Zitterl-Eglseer, K., Kohler, B., Gabler, C., Losa, R. and Zimpernik, I. 2004. The effect of two different blends of essential oil components on the proliferation of Clostridium perfringens in the intestines of broiler chickens. *Poult. Sci.* 83: 669–675.

Molnar, O., Schatzmayr, G., Fuchs, E. and Prillinger, H. 2004. Trichosporon mycotoxinivorans sp. nov., a new yeast species useful in biological detoxification of various mycotoxins. *Syst. Appl. Microbiol.* 27: 661–671.

Monson, M. S., Coulombe, R. A. and Reed, K. M. 2015. Aflatoxicosis: lessons from toxicity and responses to aflatoxin B1 in poultry. *Agriculture* 5 (3): 742–777.

Moran, E. T., Jr., Hunter, B., Ferket, P., Young, L. G. and McGirr, L. G. 1982. High tolerance of broilers to vomitoxin from corn infected with *Fusarium graminearum*. *Poult. Sci.* 61: 1828–1831.

Murugesan, G. R., Syed, B., Haldar, S. and Pender, C. M. 2015. Phytogenic feed additives as an alternative to antibiotic growth promotors in broiler chickens. *Front. Vet. Sci.* 2: 21.

Norred, W. P. 1993. Fumonisins – mycotoxins produced by Fusarium moniliforme. *J. Toxicol. Environ. Health.* 38: 309–328.

Ohimain, E. I. and Ofongo, R. T. S. 2012. The effect of probiotic and prebiotic feed supplementation on chicken health and gut microflora: a review. *Int. J. Anim. Veter. Adv.* 4: 135–143.

Osborne, D. J., Huff, W. E., Hamilton, P. B. and Burmeister, H. R. 1982. Comparison of ochratoxin, aflatoxin, and T-2 toxin for their effects on elected parameters related to digestion and evidence for specific metabolism of carotenoids in chickens. *Poult. Sci.* 61: 1646–1652.

Osselaere, A., Santos, R., Hautekiet, V., De Backer, P., Chiers, K., Ducatelle, R. and Croubels, S. 2013. Deoxynivalenol impairs hepatic and intestinal gene expression of selected oxidative stress, tight junction and inflammation proteins in broiler chickens, but addition of an adsorbing agent shifts the effects to the distal parts of the small intestine. *PLoS ONE.* 8 (7): e69014. doi:10.1371/journal.pone.0069014.

Pang, C. Y., Phillips, G. D. and Campbell, L. D. 1980. Chick edema disease syndrome in young turkey poults. *Can. Vet. J.* 21 (1): 12–16.

Paraskeuas V., Griela, E., Bouziotis, D., Fegeros, K., Antonissen, G. and Mountzouris, K. C. 2021. Effects of deoxynivalenol and fumonisins on broiler gut cytoprotective capacity. *Toxins (Basel)* 13 (10): 729.

Politis, I., Fegeros, K., Nitsch, S., Schatzmayr, G. and Kantas, D. 2005. Use of Trichosporon mycotoxinivoransto suppress the effects of ochratoxicosis on the immune system of broiler chicks. *Brit. Poult. Sci.* 46: 58–65.

Raju, M. and Devegowda, G. 2000. Influence of esterified-glucomannan on performance and organ morphology, serum biochemistry and haematology in broilers exposed to individual and combined mycotoxicosis (aflatoxin, ochratoxin and T-2 toxin). *Brit. Poult. Sci.* 41: 640–650.

Ricke, S. C. 2003. Perspectives on the use of organic acids and short chain fatty acids as antimicrobials. *Poult. Sci.* 82: 632–639.

Ruhnau, D., Hess, C., Grenier, B., Doupovec, B., Schatzmayr, D., Hess, M. and Awad, W. A. 2020. The mycotoxin deoxynivalenol (DON) promotes campylobacter jejuni multiplication in the intestine of broiler chickens with consequences on bacterial translocation and gut integrity. *Front. Vet. Sci.* 7: 573894.

Salminen, S., Collado, M. C., Endo, A., Hill, C. Lebeer, S., Quigley, E. M. M., Sanders, M. E., Shamir, R., Swann, J. R., Szajewska, H. and Vinderola, G. 2021. The International Scientific Association of Probiotics and Prebiotics (ISAPP) consensus statement on the definition and scope of postbiotics. *Nat. Rev. Gastroenterol. Hepatol.* 18: 649–667.

Santos R., Oosterveer-van der Doelen, M. A. M., Tersteeg-Zijderveld, M. H. G., Molist, F., Mézes, M. and Gehring, R. 2021. Susceptibility of Broiler Chickens to Deoxynivalenol Exposure via Artificial or Natural Dietary Contamination. Animals. 11(4): 989.

Schaeffer, J. L., Tyczkowski, J. J. and Hamilton, P. B. 1987. Alterations in carotenoid metabolism during ochratoxicosis in young broiler chickens. *Poult. Sci.* 66: 318–324.

Schatzmayr, G., Täubel, M., Vekiru, E., Moll, D., Schatzmayr, D., Binder, E. M., Krska, R. and Loibner, A. P. 2006a. Detoxification of mycotoxins by biotransformation. In D. Barug, D. Bhatnagar, H. P. van Egmond, J. W. van der Kamp, W. A. van Osenbruggen, and A. Visconti (eds.), *The Mycotoxin Factbook, Food and Feed Topics*, pp. 363–75. Wageningen, The Netherlands: Wageningen Academic Publishers.

Schatzmayr, G., Zehner, F., Täubel, M., Schatzmayr, D., Klimitsch, A., Loibner, A. P. and Binder, E. M. 2006b. Microbiologicals for deactivating mycotoxins. *Mol. Nutr. Food Res.* 50: 543–551.

Sousadias, M. G. and Smith, T. K. 1995. Toxicity and growth-promoting potential of spermine when fed to chicks. *J. Anim. Sci.* 73: 2375–2381.

Tamim, N. M. and Doerr, J. A. 2003. Effect of putrefaction of poultry carcasses prior to rendering on biogenic amine production. *J. Appl. Poult. Res.* 12: 456–460.

Thachil, A. J., McComb, B., Early, M. M., Heeder, C. and Nagaraja, K. V. 2012. Clostridium perfringens and Clostridium septicum toxoid to control cellulitis in turkeys. *J. Appl. Poult. Res.* 21 (2): 358–366.

Tran, S. T., Auvergne, A., Benard, G., Bailly, J. D., Tardieu, D., Babile, R. and Guerre, P. 2005. Chronic effects of fumonisin B_1 on ducks. *Poult. Sci.* 84: 22–28.

Veldman, A., Vahl, H. A., Borggreve, G. J. and Fuller, D. C. 1995. A survey of the incidence of *Salmonella* species and *Enterobacteriaceae* in poultry feeds and feed components. *Vet Rec.* 136 (7): 169–172.

Wang, E., Norred, W. P., Bacon, C. W., Riley, R. T. and Merrill, Jr. A. H. 1991. Inhibition of sphingolipid biosynthesis by fumonisins: implications for diseases associated with *Fusarium moniliforme*. *J. Biol. Chem*. 266: 14486–14490.

World Health Organization. 2016. Dioxins and their effects on human health. http://www.who.int/mediacentre/factsheets/fs225/en/.

World Health Organization. 2018. Salmonella (non-typhoidal). http://www.who.int/mediacentre/factsheets/fs139/en/.

Wu, F., Bhatnagar, D., Bui-Klimke, T., Carbone, I., Hellmich, R., Munkvold, G., Paul, P., Payne, G. and Takle, E. 2011. Climate change impacts on mycotoxin risks in US maize. *World Mycotoxin J*. 4: 79–93.

Wu, S. B., Stanley, D., Rodgers, N., Swick, R. A. and Moore, R. J. 2015. Two necrotic enteritis predisposing factors, dietary fishmeal and *Eimeria* infection, induce large changes in the caecal microbiota of broiler chickens. *Vet. Microbiol*. 169 (3–4): 188–197.

Zhai S, Zhu, Y., Feng, P., Li, M., Wang, W., Yang, L. and Yang, Y. 2021. Ochratoxin A: its impact on poultry gut health and microbiota, an overview. *Poult Sci*. 100 (5): 101037.

Chapter 8

Poultry feedstuff management and feed formulation utilizing big data

Bob Brill, Brilliant Alternatives Inc., USA; Matthew Clark, FeedGuys Resources Pte Ltd., Malaysia; and Dr Mirza Meghji, Salama Systems Inc., USA

1 Introduction

This chapter discusses how combining feed formulation with 'big data' will drastically change the way animal feeds are formulated. The result of these changes will be a more sustainable agriculture. Feed formulation is the 'gatekeeper' to an animal feed. One must know how to prepare information about new products or topics so that data can be included in the process of determining a new feed. Existing formulation systems do a great job generating new least cost formulas using traditional feed ingredients like corn, soybeans, etc. and nutritional properties such as protein, fat and fiber.

This chapter will review how feed formulation software is used. Future formulas will include many more variables to address new goals in addition to feed cost and animal performance. New goals include climate footprints and a more sustainable agriculture. With proper data exchange, many more participants will be able to share their knowledge so that a new feed formula

http://dx.doi.org/10.19103/AS.2024.0143.11

will reflect new priorities. The chapter will also include suggestions on how to create a 'better formula' by using a programmable optimizer or via an alternative to existing linear programming as well as the role of artificial intelligence (AI).

2 Definitions

To begin, let's define some key terms. Almost everyone that produces animal feeds uses a **least-cost formulation program**. Each feed formulation system consists of three components:

- **Database** – this has all the information about ingredients, prices and requirements for each formula;
- **Optimizer** – a linear program that uses a version of the Simplex method developed in the 1970s; and
- **Report writer** – that provides the tools needed to create reports that show least cost solutions and related information.

The database contains an ingredient matrix that has ingredients that could be selected to be included in a formula. An ingredient is defined by its nutrient properties (protein, fat, fiber, etc.) as well as cost. For each formula you wish to produce, you have requirements that must be met (such as minimum protein level, minimum energy, etc.). The ingredients that can be used in a formula plus the nutritional requirements are contained in a formula specification. In each formulation program, an optimizer (which is usually a linear program) uses the ingredient Information, cost and nutrient requirements to arrive at the lowest cost of the ingredients to produce a formula. This formula is called the least cost solution since there is no other combination of ingredients that will meet the feed requirements at a lower cost.

A **linear program** is a mathematical technique which expresses variables in terms of linear relationships that then make it possible to find an optimal point (e.g. the least cost outcome). In practical terms, this means that if an ingredient like corn has 8% protein or .08 units of protein per unit of corn, then, 2 units of corn will generate 2 × .08 or 0.16 units of protein. Since feed formulation programs have always used linear programming as the optimization technique, all ingredients had to be expressed in terms of how many units of a nutrient are in a unit of an ingredient.

Today, due to the many enzymes, phytates, amino acids and other additives that have effects on other ingredients in the formula, one cannot easily define the measurable units of each nutrient that are contained in a unit of the additive. In addition, data on these additives may reside in the additive manufacturing companies rather than feed manufacturers, complicating issues of data supply. We must estimate the effect of the additive on the other ingredients included in the feed. I mention this complication so that everyone understands there are exceptions to an ingredient being linear. With the growth of big data, it is likely some of the new properties of additives will also not be linear. However, they can be included in a formula if the value of the various nutrients provided by one unit of the additive is enough to be included in the least cost solution.

A definition of **big data** is provided by INVESTOPEDIA (https://www.investopedia.com/terms/b/big-data.asp) which defines the term by three V's:

- Greater Variety;
- Increasing Volumes; and
- More Velocity.

For the purposes of this chapter, I'll just say that big data means we have much more data from many more sources and changing much more often than in the past. The challenge we have is to be able to make use of all of the potential data that might make a better feed.

What is a **better feed**? 'Better' is a different solution than the initial solution generated by an existing feed formulation program. How or why this solution is generated will be determined by the user. The user may be willing to adopt a new feed even though it is not derived directly from the least-cost formulation program and, possibly, may be a higher cost, if it delivers other benefits.

3 Current feed formulation software

Affordable least cost formulas have traditionally been tied to the cost of a computer processor and data storage needed, and the measurable qualities of each ingredient. The development of feed formulation programs can be traced back to 1970 and Bob Brill's first formulation program – written to work on an IBM computer that was accessed by dialing the computer and using the large IBM computer as the processor. The user was able to sequentially solve many formulas but with a maximum of 50 Ingredients and 60 nutrients per ingredient. Subsequent developments have included:

- 1980: Mini-Computer – DEC PDP-11 requiring a – $40,000 Investment for feed mill or premix companies able to process a maximum of 100 ingredients and 120 nutrients.
- 1986: Personal Computer (PC) – $4,000 investment with a maximum of 200 ingredients and 120 nutrients. The combination of a PC with appropriate formulation Software was the real game changer in how formulas were done.
- 1988: Multi-blend on PC – $2,000 investment with a maximum of 200 Ingredients and 500 nutrients, combined with pioneer linear programming

developed by Bob Brill and programmer William Orchard-Hays, allowing optimization of complex problems.

- 1994: Multi-blend/Multi-Time period on PC, local area network, etc., allowing unlimited ingredients and a maximum of 999 nutrients.
- Today: All software capabilities on a PC, in the Cloud, etc. with unlimited Ingredients and 999 nutrients which could easily be expanded. Or a smaller problem can be optimized using Microsoft's Excel Solver.

Today one can solve very large problems on most formulation software programs, i.e. problems that includes many feed plants, many formulas and many time periods. If, for example, one had 50 plants, 4 time periods and 60 formulas per plant, that would be 12,000 formulas in one solution.

As Fig. 1 shows, computing costs and 'problem size limitations' in the formulation software are no longer the main limitations. Current software available for feed formulation has no practical limits on data such as ingredients, nutrients, formulas, locations, etc. Current challenges are more about how to accommodate a much wider range of data within the limitations of linear programming. More detail is provided later in the chapter.

The current technologies include the following:

- **Database** – most feed formulation databases use a standard database such as Access or SQL which can be read by most programs. This is a huge improvement from the past. Automatically updating the nutrient values of key ingredients is now possible.
- **Optimizer** – although most optimizers are based on a linear program using a version of the Simplex method developed in the 1970s, a user can supply their own Optimizer.
- **Report writer** – today output can be written to other files including Excel.
- **Data exchange tools** – There are now computer products that enable a user to get data from other sources and automatically move that data into the formulation database for use in feed formulation (typically using Excel).

In summary, there is nothing to prevent a formulator of animal feeds from evaluating any new product or nutrient.

	Hardware	Cost		Software	Capability	Ingredient:	Nutrients	Formulas	Locations
1970	IBM Call 360	$20.00	per hour	Bob Brill	Single Blend	50	40	50	10
1980	DEC PDP 11	$50,000.00	purchase	Bob Brill	Single Blend	200	60	100	10
1986	IBM PC Floppy Dis	$2,500.00	purchase	Brill Corporation	Single Blend	200	60	100	10
1988	IBM PC Hard disk	$4,500.00	purchase	Basic Optimizatio	Multiblend	1000	100	Unlimited	Unlimited
1994	Local Area Networ	$15,000.00	purchase	Ketron	Multi-Plant MB	Unlimited	200	Unlimited	Unlimited
2020	Cloud, Server, PC	$2,000.00	purchase	Ketron	Everything	Unlimited	999	Unlimited	Unlimited

Figure 1 Historical cost of computer hardware needed to generate a least cost formula.

4 Optimizing a formula using Microsoft Solver

This section has been provided by Matthew Clark. One can formulate diets using Excel and its Solver (which is free). This means that anyone can solve a pretty good-sized formula using Solver. Most of the same information is available from a solution provided by Solver as found in a feed formulation program. The purpose of this section in showing the capabilities of the Solver is to help people who do not have access to a feed formulation program to use Excel Solver, e.g. to answer questions about an ingredient that could be used in a feed formula, try different scenarios or to evaluate carbon footprint options. A complete PowerPoint presentation is provided later in the chapter which shows you how to setup and use the Solver to do a least cost formulation. An example is shown in Fig. 2.).

4.1 Adding an ingredient that is not in the existing ingredient database

In order to determine the value of an ingredient that is not already in the ingredient database, one must add the ingredient code to the database and then enter the nutrient values for that ingredient. Once the ingredient information is in the database, one can add this ingredient to any formula where the ingredient can be used. Each formula that is optimized with this new Ingredient will tell you the price at which this ingredient will enter that formula. After a price is entered for this ingredient, and the formula is optimized, the solution will show how much of the ingredient will be used in each of the optimized formula, or the price at which it will enter the solution.

To add the ingredient to the ingredient matrix, one can either import the ingredient information using Excel or some other data exchange method or manually enter all the nutrient values by using the ingredient maintenance module (see later in this chapter for instructions on the use of the import or export capabilities of the feed formulation program).

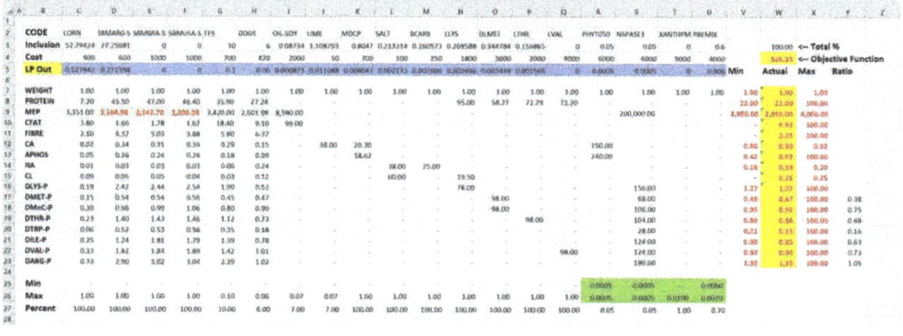

Figure 2 Optimizing a Formula using Microsoft's Solver.

In the following example, we have a feasible solution for feed 51 which is a broiler feed. Since canola has already been entered into the ingredient database, one simply adds the ingredient to the list of useable ingredients in formula 51, and then optimizes the formula. The feed without canola is shown in Fig. 3a. The cost of this feed is $148.0582.

In Fig. 3a, we will add canola to the formula specification. Since canola has already been entered into the ingredient database, we simply look it up in the database and add ingredient 175 (canola). I decided to limit the amount of canola in the feed to a maximum of 15%.

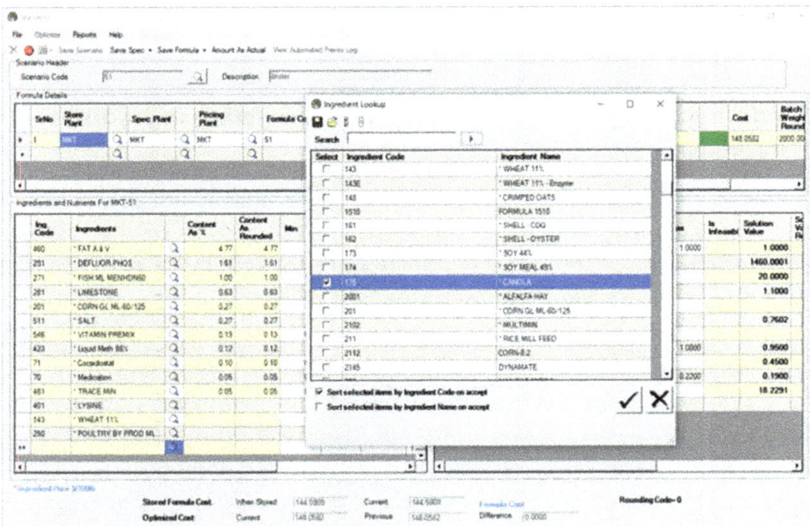

Figure 3a Formula spec with canola now added for consideration.

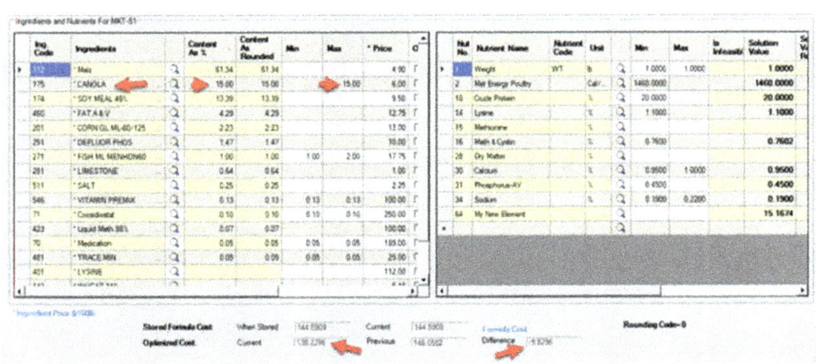

Figure 3b

The new solution shown in has canola included at the maximum of 15%. The cost is $138.2286 which is a reduction of $9.8296 when compared to the formula without the canola.

4.2 Price change example

Whenever the price of an ingredient changes, there may be a change in the least-cost solution for that formula. Each formula must be looked at separately since they have different requirements and different available ingredients. Figure 4a and Fig. 4b shows the changed in price of from 4.90 to 5.30 when wheat replaced corn/maize in the solution.

4.3 Change in nutrient values of an ingredient such as corn

Using the ingredient maintenance module, we can manually change the nutrient properties of an ingredient. In this case (Fig. 5), I have changed several nutrient values in corn (maize). I am reducing the values of Met (metabolizable) energy, crude protein, arginine, lysine and methionine. I changed the energy in corn to 1560, 1520, protein from 8.7 to 8.2, arginine from 0.4747 to 0.4000 and lysine from 0.1933 to 0.18.

When one reviews the nutrient content of the formula that was previously saved using the previous nutrient values, one can see that the calculated nutrient values are no longer meeting the minimum requirements. After nutrient values of the ingredients are changed, one must re-optimize the formula in order to make sure the formula meets requirements such as 20% crude protein, and 1460 met energy. Please see Fig. 6 to know the reduced nutrient levels.

After reoptimizing the formula (Fig. 7), the new formula now meet the requirements. The cost has increased due to the reduction in the nutrient values of corn from $144.5909 to $148.0582 per US Ton. This is a very important lesson about the need to use data that is current so that formulas actually meet animal nutrient requirements. As demonstrated, the old formula no longer meets the requirement.

4.4 New nutrients (elements) in a formula

In order to determine the amount of a nutrient (element) that is contained in a formula, we need to know how much of that element is in each of the ingredients that are included in the formula. For example, to determine how much protein is in a formula, there must be a protein value (0, if none) for corn (8.5%), soya (48.5%), wheat (11%), salt (0.0), etc. The total amount of protein in a formula is calculated by taking the quantity of each ingredient times the protein amount in that ingredient.

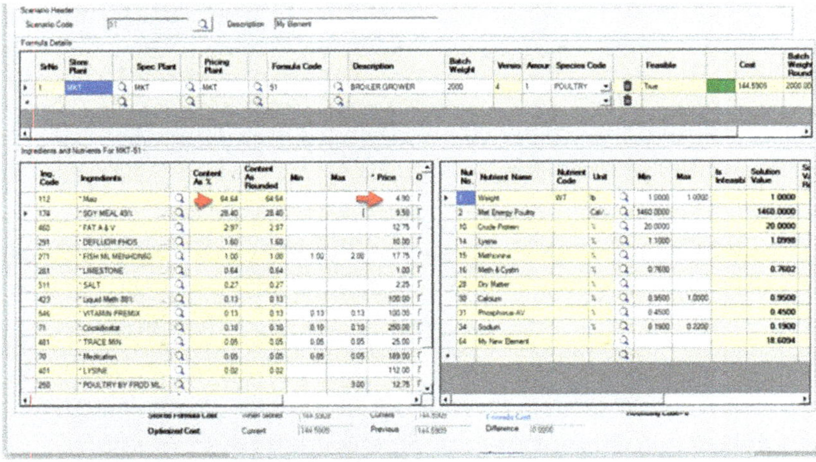

Figure 4a Shows the changing price of Maize from 4.90 to 5.30,

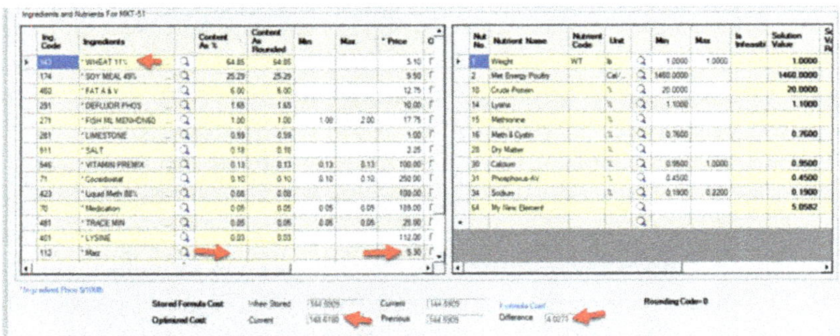

Figure 4b The new formula removes corn and adds Wheat.

If you want to calculate a new nutrient (element) in a feed formula, you will need to make sure that all ingredients offered in the formula have a value for the nutrient (element) that is being evaluated. If you are the provider of the data, then you can define (create) the new measure. Once the data for the new nutrient is in your formulation database, you can change the values to whatever you believe they should be. When you formulate the feed, the program uses your changed values.

The new information about the nutrient can be used for any formula, and the resulting optimized solution will show how much of that nutrient is in the formula, and, if there is any change in the cost of the formula as a result of adding the nutrient to the formula specification. The cost will only be affected if there is a minimum or maximum restriction on this nutrient.

Please see later in this chapter on how one enters a new nutrient into an existing formulation database so that one can see how much of that

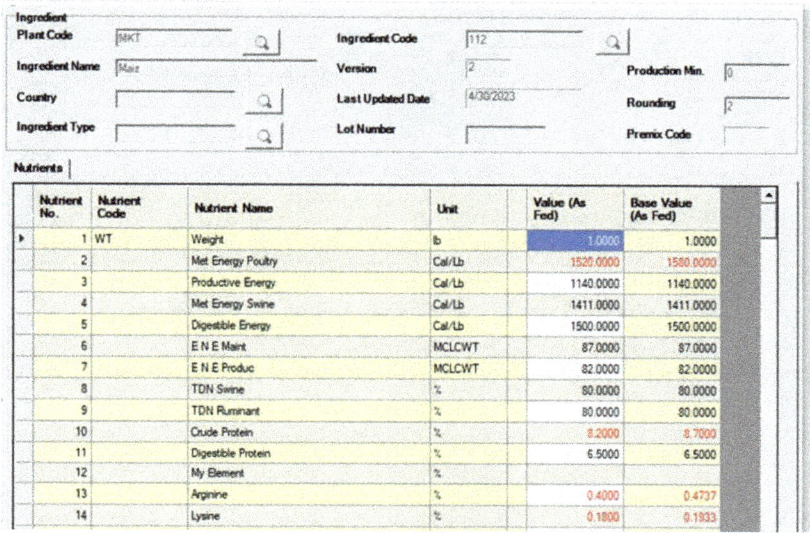

Figure 5 Change nutrient content of ME and Protein, Arginine and Lysine

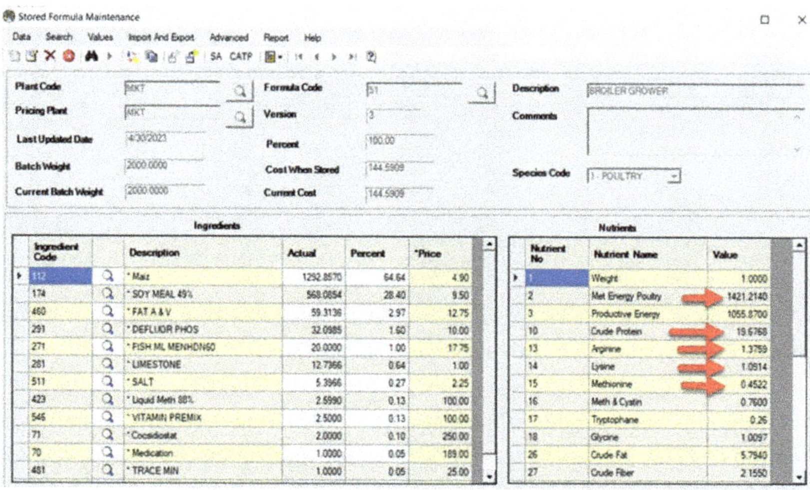

Figure 6 Nutrient levels of Formula do not meet the Formula Specifications.

nutrient is included in the feed. In the example (Fig. 8), a user creates a new nutrient number 64, enters a value for nutrient 64 and then has the value of nutrient 64 calculated for the formula. In the example, the user wants to get a 'score' for Nutrient 64 (Fig. 9). The user could have also placed a minimum or maximum restriction in the formula, but we did not for this example. For various ingredients, one can see the value of nutrient 64. Some have a value of 10, 20, 5 or 15.

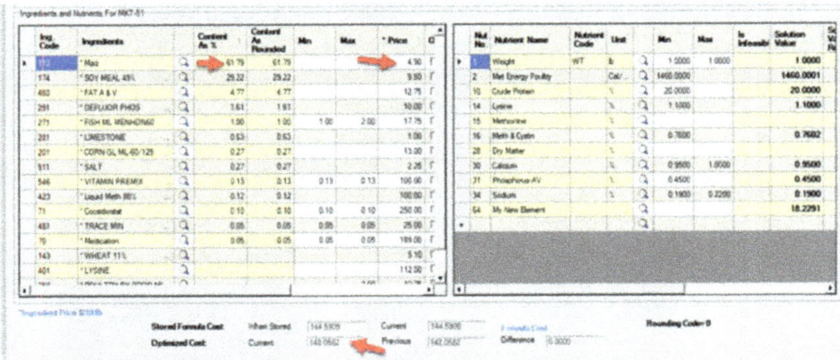

Figure 7 After reoptimizing the formula, all the requirements are met and the new cost is higher.

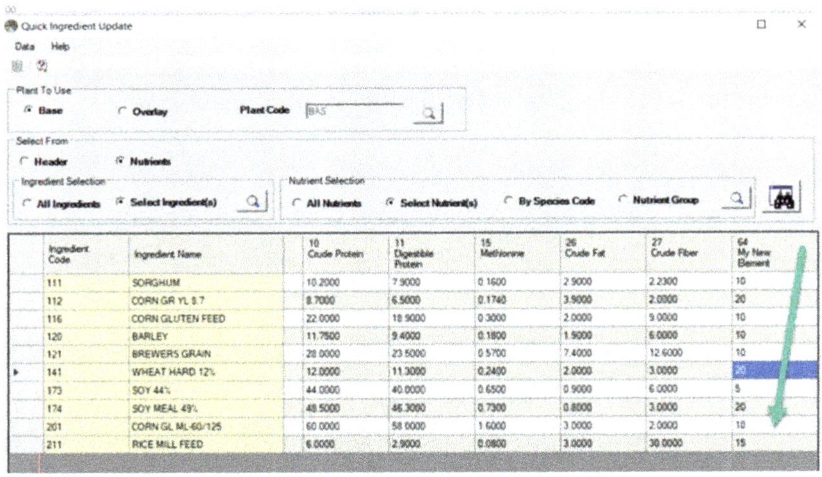

Figure 8 The new Nutrient has a value for each Ingredient.

If I want to have a minimum score of 17 for the new element, I can place a minimum of 17 as the restriction for nutrient 64, and optimize again. Likewise, I could place a maximum limit on nutrient 64. I will run two scenarios. A minimum of 17 and then a maximum of 15. There is a cost for these limits. The optimizer has generated a new least-cost formula with a minimum restriction on nutrient 64. The increased cost to increase nutrient 64 from 15.1674 to 17.0 is 5.8836. (Fig. 9). By supplying values to all the ingredients for nutrient 64, we were able to calculate a 'score' for nutrient 64 in this formula: 15.1674. We were then able to set a minimum of 17, and we were able to see how much more we would have to pay in order to get a minimum of 17 (Fig. 10).

Next, we will show what happens when we place a maximum of 15 on the nutrient 64. The new solution and the cost of that solution are shown in Fig. 11.

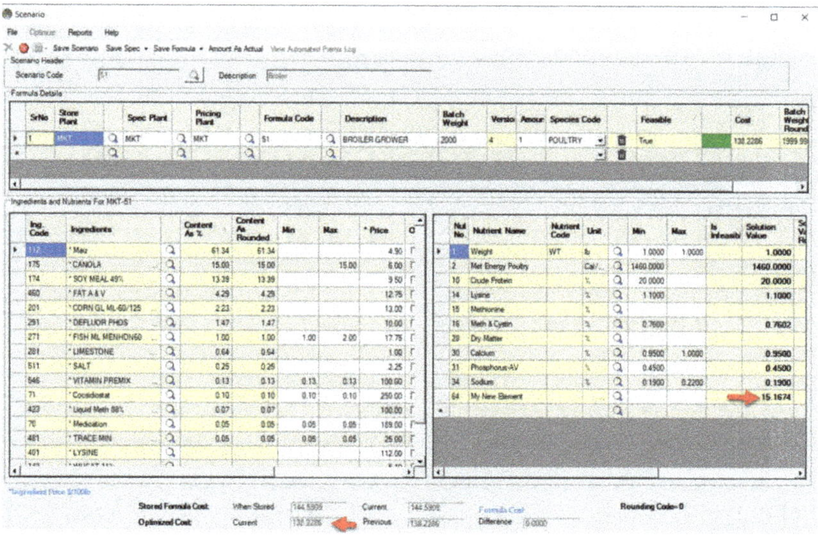

Figure 9 As a part of the Least Cost Report, the amount of nutrient 64 is shown to be 15.1674.

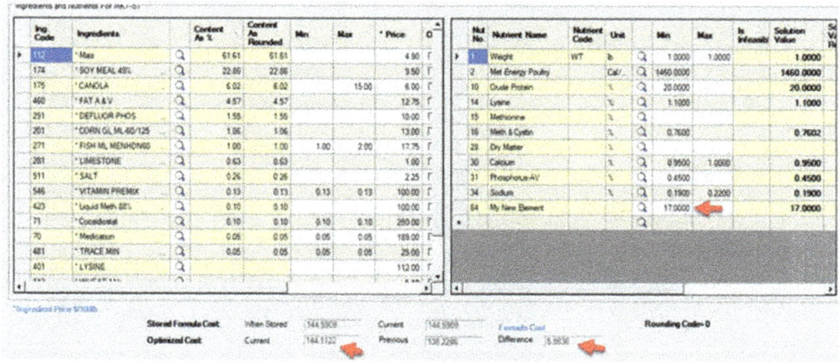

Figure 10 The solution shows a 17.00 minimum, and an increase in the cost of 5.8836.

If the right information is supplied to the feed formulation system, the least cost formula can identify the costs related to the new element (nutrient).

4.5 *Is the least-cost solution really the least cost after changes to production rules?*

The least-cost solution generates a mix of ingredients that satisfy the requirements of the formula specification at the lowest cost. Once this formula is generated by the optimizer, adjustments are made to the amount of each ingredient due to production system rules, normally by rounding each ingredient. This is done

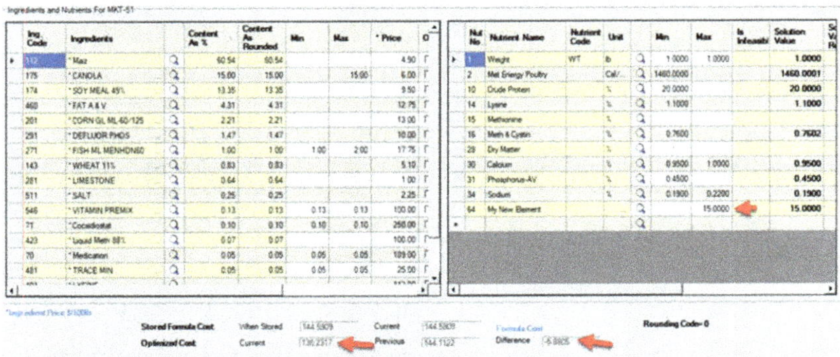

Ingredients and Nutrients For MKT-51

Ing. Code	Ingredients	Content As %	Content As Rounded	Min	Max	* Price
112	* Maiz	60.54	60.54			4.90
175	* CANOLA	15.00	15.00		15.00	6.00
174	* SOY MEAL 49%	13.35	13.35			9.50
460	* FAT A & V	4.31	4.31			12.75
201	* CORN GL ML-60/125	2.21	2.21			13.00
291	* DEFLUOR PHOS	1.47	1.47			10.00
271	* FISH ML MENHDN60	1.00	1.00	1.00	2.00	17.75
143	* WHEAT 11%	0.83	0.83			5.10
281	* LIMESTONE	0.64	0.64			1.00
511	* SALT	0.25	0.25			2.25
546	* VITAMIN PREMIX	0.13	0.13	0.13	0.13	100.00
71	* Cocsidiostat	0.10	0.10	0.10	0.10	250.00
423	* Liquid Meth 88%	0.07	0.07			100.00
70	* Medication	0.05	0.05	0.05	0.05	189.00
481	* TRACE MIN	0.05	0.05	0.05	0.05	25.00

Nut No.	Nutrient Name	Nutrient Code	Unit	Min	Max	Is Infeasibl	Solution Value
1	Weight	WT	lb	1.0000	1.0000		1.0000
2	Met Energy Poultry		Cal/...	1460.0000			1460.0001
10	Crude Protein		%	20.0000			20.0000
14	Lysine		%	1.1000			1.1000
15	Methionine		%				
16	Meth & Cystin		%	0.7600			0.7602
28	Dry Matter		%				
30	Calcium		%	0.9500	1.0000		0.9500
31	Phosphorus-AV		%	0.4500			0.4500
34	Sodium		%	0.1900	0.2200		0.1900
64	My New Element				15.0000		15.0000

*Ingredient Price $/100lb

Stored Formula Cost — When Stored: 144.5309 — Current: 144.5309 — Formula Cost
Optimized Cost — Current: 138.2317 — Previous: 144.1122 — Difference: -5.8805
Rounding Code= 0

Figure 11 New solution with a maximum of 15 placed on nutrient 64.

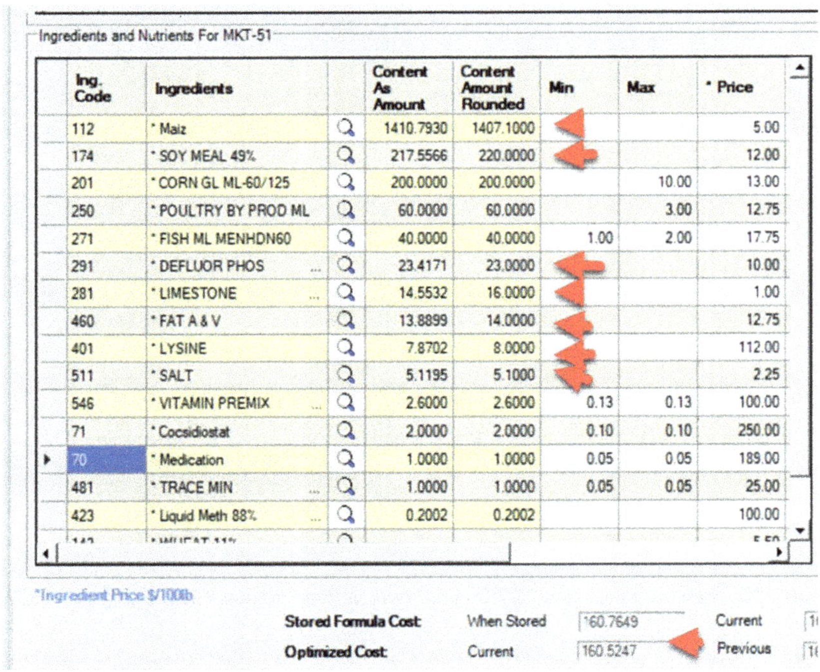

Ingredients and Nutrients For MKT-51

Ing. Code	Ingredients	Content As Amount	Content Amount Rounded	Min	Max	* Price
112	* Maiz	1410.7930	1407.1000			5.00
174	* SOY MEAL 49%	217.5566	220.0000			12.00
201	* CORN GL ML-60/125	200.0000	200.0000		10.00	13.00
250	* POULTRY BY PROD ML	60.0000	60.0000		3.00	12.75
271	* FISH ML MENHDN60	40.0000	40.0000	1.00	2.00	17.75
291	* DEFLUOR PHOS	23.4171	23.0000			10.00
281	* LIMESTONE	14.5532	16.0000			1.00
460	* FAT A & V	13.8899	14.0000			12.75
401	* LYSINE	7.8702	8.0000			112.00
511	* SALT	5.1195	5.1000			2.25
546	* VITAMIN PREMIX	2.6000	2.6000	0.13	0.13	100.00
71	* Cocsidiostat	2.0000	2.0000	0.10	0.10	250.00
70	* Medication	1.0000	1.0000	0.05	0.05	189.00
481	* TRACE MIN	1.0000	1.0000	0.05	0.05	25.00
423	* Liquid Meth 88%	0.2002	0.2002			100.00

*Ingredient Price $/100lb

Stored Formula Cost — When Stored: 160.7649 — Current
Optimized Cost — Current: 160.5247 — Previous

Figure 12 Amount of each ingredient before and after the rounding is done for production

to adjust each quantity to an amount that can be measured by scales or to account for other factors during mixing or manufacture of the feed.

In the example below (Fig. 12), the quantity for eight (8) of the ingredients in the least cost solution was modified by 'rounding" of the formula. The difference in the actual cost of the least cost formula and the rounded formula was $160.7649 less $160.5247 = $0.2602. In this case, the least cost was based upon 'unrounded' formulas which are then rounded. Small changes in ingredients or formula costs of this kind are normal. This makes it easier to experiment with new formulas. Since the currently defined least cost is always adjusted to meet feed mill production systems, it is possible to explore ways to generate a better formula as long as the change in cost does not exceed some acceptable amount.

5 Formulating feeds to address sustainability and other non-nutritive goals

Many more factors are being included in the formulation of feeds. In addition to the cost and nutritional values of ingredients, other types of data can now be included the formulation process where it can, e.g. be used as a criterion that needs to be met. This includes data related to the environmental impact of manufacturing and using a particular feed formulation. There is now a significant body of data about the environmental as well as financial costs of feed manufacturing and delivery as well as of animal waste products resulting from use of a particular feed. Gert van Duinkerken, Manager of Livestock Research at Wageningen University & Research has commented that 'More and more animal nutrition will shift its focus from feed efficiency to agro-ecological system efficiency' (https://www.allaboutfeed.net/market/market-trends/emissions-and-circular-feed-highlighted-at-all-about-feed-forum/).

At least one feed formulation program is being adapted to take account of these wider issues. One of the leading providers of feed formulation software, Bestmix, is collaborating with BASF to incorporate sustainability data. This has been explained in a very good video presentation by Karl Veraet, a senior product specialist of Bestmix, part of the 'Feed Mill of the Future' series by Watt Global Media (https://www.feedmillofthefuture.com/videos/video/15546143/bestmix-software-inc-how-feed-formulation-impacts-downstream-sustainability-goals-video).

The task of formulating a feed with new variables such as these is the same as accounting for existing nutrient values. We need to have a value of the new measure for each ingredient in the formula. Since most feed systems use linear programming, we must assume that two units of the ingredient will contribute 2 units of each new measure. An example of official types of non-nutritive data is the European Product Environmental (PEF) and Organization Environmental Footprint (EF) PEFCR Feed Guidelines (https://fefac.eu/priorities/sustainability/pefcr-feed/). An example of an unofficial type of measure is the data that additive companies generate.

Examples of companies providing tools to address the carbon footprint of feed products include:

- BASF (https://www.basf.com/global/en/media/news-releases/2023/09/p-23-289.html);

- DSM-Fermenich (https://www.feednavigator.com/Article/2023/07/10/Environmental-footprint-of-feed-dsm-firmenich-looks-to-support-industry-on-Scope-3-reporting); and
- Trouw Nutrition (https://www.trouwnutrition.com/en/programme-lister/myfeedprint-602845/).
- Trouw Nutrition, e.g. has produced a scoring system tied to the European PEFCR Feed Guidelines.

6 Data exchange with the feed formulation database

The idea for automatic data exchange with the feed formulation database is not new. Before 2000, most feed formulation programs were standalone, i.e. everything used to generate a new formulation was contained within the formulation program or its own database. Fields in the formulation database and the type of database (Btrieve, Paradox, Access, etc.) were determined by the original software developer and could not be changed without rewriting the original program.

Over time, developers of feed formulation programs began to use commonly used programs like Excel to import, manage and export data, effectively creating a standard format for potential data exchange. Formulation programs also now have the option to translate the nutrient, ingredient, and formula codes from one set of codes to another. This allows users to establish a set of translation codes for each source or recipient of the data to be transferred.

In practice, when one wants to share data with a formulation program, the easiest way to do that is to place the data into an Excel file (Figs. 13 and

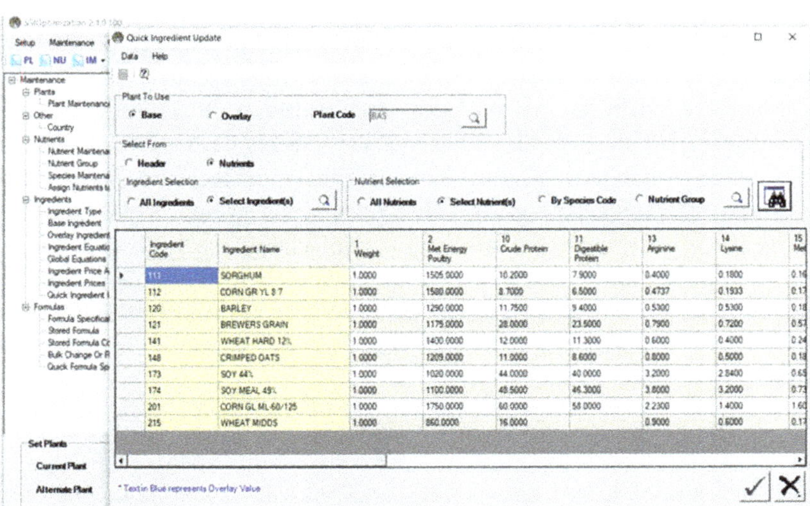

Figure 13 Ingredient information in the formulation database.

14). Should the data outside of the formulation database change, one simply updates the Excel file and imports the data from the Excel file (Fig. 15). Anyone can generate the data in the Excel file and then share the latest data with those who use feed formulation software. As you can see in Fig. 16, the data in the Excel file ends up in the formulation database.

The Excel import software was written so that one can define the data layout of the Excel file and then import the data into the feed formulation data matrix. This enables the formulation program to import data that has been made available by anyone who wishes to share data. For researchers or others outside of the feed industry, one can easily share data generated outside of the formulation program with those in the feed industry simply by making it available in an Excel file. Data about carbon footprints, sustainability information, trucking costs, energy for pelleting, etc. can now be updated into the formulation database if one wants to either formulate on that parameter, or just know what amount is included in a formula. As you can see in Fig. 17, one can define the format of the Excel file as well as translation codes for ingredient and nutrient codes from one user's codes to another user's codes.

A	B	C	D	E	F	G	H	I	J	K	L	M	N	O	P	Q
		Nutrient N	1	2	10	11	13	14	15	16	17	26	27	28	30	
Ingredient	Ingredient	Nutrient N	Weight	Met Energy	Crude Prot	Digestible I	Arginine	Lysine	Methionin	Meth & Cy	Tryptophar	Crude Fat	Crude Fibe	Dry Matter	Calcium	Phc
111	SORGHUM		1.0000	1505.0000	10.2000	7.9000	0.4000	0.1800	0.1600	0.3100	0.12	2.9000	2.2300	90.0000	0.0300	
112	CORN GR YL 8.7		1.0000	1580.0000	8.7000	6.5000	0.4737	0.1933	0.1740	0.3287	0.10	3.9000	2.0000	88.0000	0.0200	
120	BARLEY		1.0000	1290.0000	11.7500	9.4000	0.5300	0.5300	0.1800	0.3800	0.18	1.9000	6.0000	88.0000	0.0800	
121	BREWERS GRAIN		1.0000	1175.0000	28.0000	23.5000	0.7900	0.7200	0.5700	0.9600	0.24	7.4000	12.6000	93.0000	0.3000	
141	WHEAT HARD 12%		1.0000	1400.0000	12.0000	11.3000	0.6000	0.4000	0.2400	0.5000	0.10	2.0000	3.0000	89.0000	0.0500	
148	CRIMPED OATS		1.0000	1209.0000	11.0000	8.6000	0.8000	0.5000	0.1800	0.4000	0.16	4.6000	11.0000	90.0000	0.0900	
173	SOY 44%		1.0000	1020.0000	44.0000	40.0000	3.2000	2.8400	0.6500	1.3200	0.60	0.9000	6.0000	89.0000	0.3200	
174	SOY MEAL 49%		1.0000	1100.0000	48.5000	46.3000	3.8000	3.2000	0.7300	1.4500	0.65	0.8000	3.0000	89.0000	0.4000	
201	CORN GL ML-60/125		1.0000	1750.0000	60.0000	58.0000	2.2300	1.4000	1.6000	2.6800	0.30	3.0000	2.0000	88.0000	0.1000	
215	WHEAT MIDDS		1.0000	860.0000	16.0000		0.9000	0.6000	0.1700	0.3700	0.23	4.0000	7.0000	88.0000	0.1000	

Figure 14 Ingredient information in the Excel file.

A	B	C	D	E	F	G	H	I	J	K	L	M	N	O	P	Q
		Nutrient N	1	2	10	11	13	14	15	16	17	26	27	28	30	
Ingredient	Ingredient	Nutrient N	Weight	Met Energy	Crude Prot	Digestible I	Arginine	Lysine	Methionin	Meth & Cy	Tryptophar	Crude Fat	Crude Fibe	Dry Matter	Calcium	Phosp
111	SORGHUM		1.0000	1505.0000	10.2000	7.9000	0.4000	0.1800	0.1600	0.3100	0.12	2.9000	2.2300	90.0000	0.0300	0.
112	CORN GR YL 8.7		1.0000	1580.0000	8.2000	6.5000	0.4737	0.1933	0.1740	0.3287	0.10	3.9000	2.0000	88.0000	0.0200	0.
120	BARLEY		1.0000	1290.0000	11.7500	9.4000	0.5300	0.5300	0.1800	0.3800	0.18	1.9000	6.0000	88.0000	0.0800	0.
121	BREWERS GRAIN		1.0000	1175.0000	12.0000	23.5000	0.7900	0.7200	0.5700	0.9600	0.24	7.4000	12.6000	93.0000	0.3000	0.
141	WHEAT HARD 12%		1.0000	1400.0000	11.5000	11.3000	0.6000	0.4000	0.2400	0.5000	0.10	2.0000	3.0000	89.0000	0.0500	0.
148	CRIMPED OATS		1.0000	1209.0000	11.0000	8.6000	0.8000	0.5000	0.1800	0.4000	0.16	4.6000	11.0000	90.0000	0.0900	0.
173	SOY 44%		1.0000	1020.0000	44.0000	40.0000	3.2000	2.8400	0.6500	1.3200	0.60	0.9000	6.0000	89.0000	0.3200	0.
174	SOY MEAL 49%		1.0000	1100.0000	48.5000	46.3000	3.8000	3.2000	0.7300	1.4500	0.65	0.8000	3.0000	89.0000	0.4000	0.
201	CORN GL ML-60/125		1.0000	1750.0000	60.0000	58.0000	2.2300	1.4000	1.6000	2.6800	0.30	3.0000	2.0000	88.0000	0.1000	0.
215	WHEAT MIDDS		1.0000	860.0000	16.0000		0.9000	0.6000	0.1700	0.3700	0.23	4.0000	7.0000	88.0000	0.1000	0.

Figure 15 Change the nutrient values in the Excel spreadsheet so that the data can be imported to formulation database.

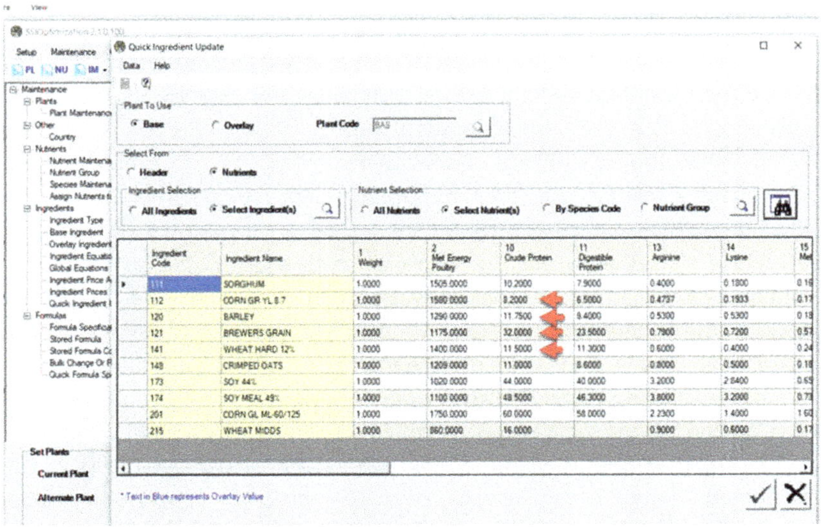

Figure 16 Data from Excel Spreadsheet is now in the formulation database.

A few examples of local data that is being automatically exchanged are as follows:

- Prices of ingredients sent from the accounting;
- New production formulas are sent from the formulation database to accounting or feed mixers;
- Nutritional properties of ingredients just received are sent directly from analytical instrumentation (e.g. near-infrared (NIR) spectroscopy) or from an Excel file generated in the quality control (QC) laboratory;
- Inventory of an ingredient in storage bins can be determined by sensors and provided to the formulation database for planning purposes; and
- The production schedule is sent to the formulation system so that future inventory can be supplied to produce future batches of feed.

Automated data exchange has reduced the costs and errors of manual inputting. In addition, this process makes full use of one of the biggest contributions to feed formulation over the past 50 years: the ability of NIR systems to generate immediate and accurate nutrient values for each ingredient sampled. If we want to formulate the least cost feed (lowest cost of ingredients to meet the requirements of the feed), we need to use the latest information about the quality of the ingredients that are included in the feed. This means we need to manage quality control (NIR and other lab) data so that the feed formulation database reflects the most recent sample information for the ingredients received.

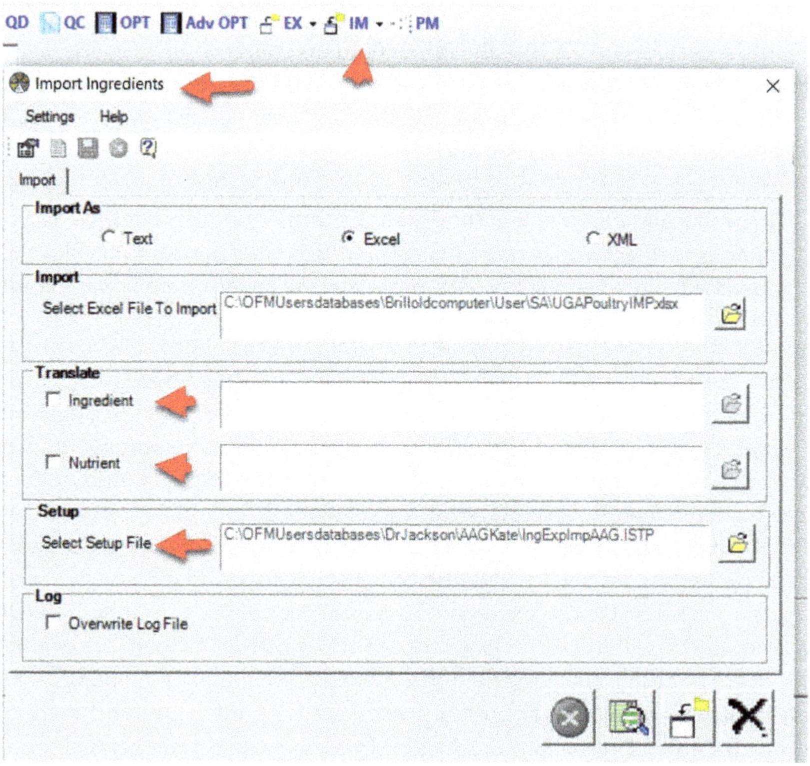

Figure 17 Defining the Excel File so the data can be imported into the Formulation Program.

The normal calculation of the safety margin for nutrient content in an ingredient (e.g. to ensure that neither too much nor too little ends up in the final formulation) is done by calculating the mean value of several samples for each nutrient in an ingredient and then reducing the mean value by 0.5% of the standard deviation (SD) for each nutrient. The resulting value is the number that is used in the feed formulation program. Calculation of the mean, SD and new nutrient value is now automatic, as is updating of new values. This has many advantages, including the following:

- Buyers of competing ingredients such as US or Brazilian soybean meal need to know the mean and SD for both products. Otherwise, the purchase will be based upon the mean or the guaranteed value, and not the value when the feed formulation program generates the least cost feed;
- If one has an NIR system, it is possible to increase the number of samples taken from incoming ingredients which will allow the formulation system to be provided with the latest data about ingredients being used. Supplying

current data should result in a much lower SD which will result in a much lower cost for the 0.5% reduction of nutrient content;

- In order for NIR data to have any effect on the least-cost formulation, the mean and SDs need to be recalculated and seamlessly transferred to the feed formulation database; and
- Many ingredient or premix suppliers offer a NIR service that will automatically analyze and send you a complete nutrient profile of your sample. In effect, you can use their expertise so that the nutrient values you are using in the formulation database are current, accurate and complete.

The additional safety margin may be the reason that the continued use of old nutrient values in feed formulation programs has not been harmful to animal performance, since these nutrient values have a lot more margin of safety built in, but at a very high cost. My co-author Matthew Clark's document 'The Cost Margins of Safety' goes into great detail to show the added cost if one continues to calculate the margin of safety as the mean minus 0.5 SD. Matthew goes on to point out that we have much more information about nutrients in ingredients, so a huge potential cost saving could be obtained by focusing on how to change from using the 0.5 SD from the mean. To quote Matthew's document: However, it is currently possible to run many more samples using NIR than was possible when wet chemistry analyses were the method of choice. We should be in a better state of knowledge about our ingredient nutrient values and therefore we may be able to formulate production formulas without the safety margins and instead rely on current analysis. The margin of safety method remains a good tool for evaluating the effect of variability in the supply chain and is still a good tool for evaluation of the cost of variability in that it is predictive of a range of possible nutrient values of future deliveries.' The document is included later in the chapter.

7 How will big data change the way we formulate feeds?

Big Data will change the way we formulate feeds in many ways, including:

- The amount of new data becoming available to help answer questions relating to sustainability and environmental impact;
- Providing data related to using additives and their effects on associated nutrients;
- The frequency of updating data; and
- The impact of artificial intelligence (AI) in data analysis.

The immediate availability of so many measurable factors in addition to the current nutrient content per ingredient will drive the changes in how we formulate. We are no longer looking at only the Ingredients and their nutrient properties. We will be looking at data from NIR systems, animal waste composition, enzymes, mold inhibitors, electricity consumption, green scores, product environmental footprints (PEF), etc. Many of these properties may not be linear or even quantitative. Potential solutions include:

- Keep using the same linear optimizer to generate formulas as we have done since 1970;
- Write a program that uses the linear optimizer, but with the program adjusting solutions according to other knowledge. A simple alternative is to try 3 or 5 alternatives and keep the best one since the speed to solve another formula is not the problem;
- Develop a new mathematical programming system that can handle non-linear situations in today's formulas; and
- Develop a new program that has access to AI during the optimization process.

As noted earlier, feed formulation software has always been a 'standalone application'. All the data needed when one formulates a diet is contained in the formulation database (silo). In the past, there have only been a few interfaces required to move data into or out of the formulation database. The developer of the formulation software usually provided the interfaces to handle what was needed. Real-time sharing of data in any 'silo' with other 'silos' was not very certain. As one looks ahead, it appears that much more data will need to be shared with the formulation program. The source of the data, the frequency of updates, etc. will probably dictate that a more automatic system be provided that allows for data from one source to be shared with the formulation database and likewise from the formulation database.

One new concept of data sharing is from a company named Team Central. I'm sure there are other companies doing similar things but, for the purposes of this chapter, let's look at what one can do with the services offered by Team Central. Team Central founder Andy Park says:

> Fortunately there is a wave of innovative tech emerging that could help solve many of these interoperability and data management challenges: No/Low-Code Integration Platform as a Service (iPaaS). Imagine connecting these siloed systems to seamlessly stream data across them without having to upgrade them or go through a disruptive major re-platform. The future is now with iPaaS. Without any custom code development, No/Low-Code iPaaS solutions are able to connect to multiple siloed data sources and automate data synching across them simultaneously and in near real time. This integration and data management automation can be achieved in hybrid environments where companies have both on-premise and cloud applications deployed as is often the case. iPaaS is a game changer because companies can leverage existing infrastructure and purpose built applications

> without having to invest in expensive and disruptive digital transformation initiatives. These no/low code integration platforms typically include pre-built connectors, business rule engines and data transform capabilities as well. Whether connecting systems or pushing data to data marts for analytics or third party consumption, iPaaS solves many types of complex data management use cases at the fraction of the cost of traditional API point to point custom integration (https://www.teamcentral.ai/about-us/).

Technologies such as iPaaS allow automatic data transfer of data from one 'silo' (such as the accounting system) can be automatically moved to the formulation database, and vice versa. They also allow data flows from new sources such as the range of data needed for new requirements such as life cycle assessment (LCA).

8 Using a callable optimizer to improve data input and analysis

Another option is to use a callable optimizer (which allows more flexible formatting of data) included in an existing program to generate a better data feed. A program with a callable optimizer offers rules for reviewing and changing solutions so the program automatically looks at many alternatives from which a better feed can be chosen.

It is likely that new data sources will provide some Excel export capability which allows the use of existing Excel import/export tools in feed formulation software. Improved tools for handling new data will make it possible to continuously update information in the formulation database. Some new 'automatic' updates of the formulation database could be the following:

- Expected nutrient profile based on analyzing crops in the field;
- Nutritional properties from suppliers (e.g. dried distiller's grains from ethanol plants);
- Chemical analysis of animal waste materials; and
- Environmental Scoring of new formulas as they are being formulated.

Existing feed formulation systems are able to get a range of data into the formulation database and then solve formulas by combining new and old data.

Let us look at a comparison of old and new ways of generating a better formula. A simple outline of using a feed formulation program is shown in Figs. 18 and 19, which can be compared with one using a callable optimizer (Fig. 20). As you can see, the second option offers many more options that

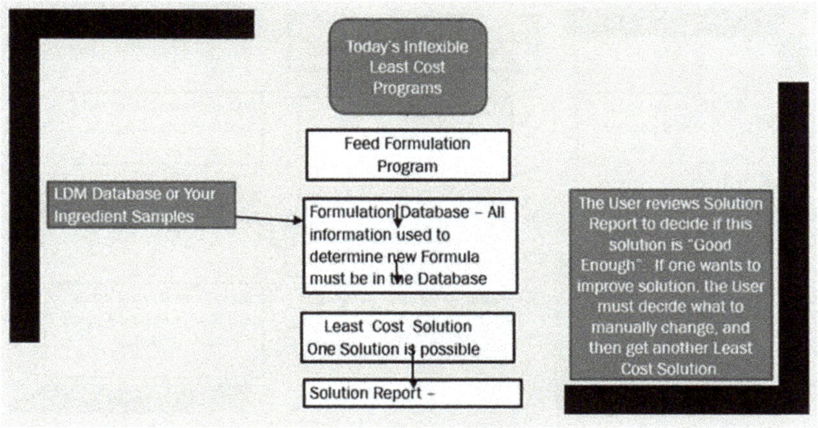

Figure 18 Current Formulation Programs

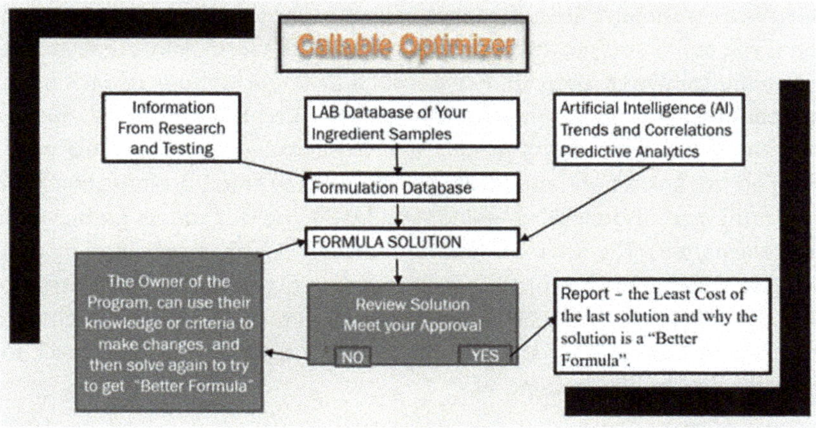

Figure 19 Callable Optimizer connected to other Applications to generate a "Better Formula"

we can use to get a better formula. This includes data from enzyme and other additive suppliers though, in the future, these guidelines will be built into the program to speed up the process. Since the use of enzymes is closely related to the nutrient composition of ingredients, there is increased benefit if enzyme-related data is used to dynamically update the formulation database.

If you use a feed additive with non-linear properties, or a new measure such as a carbon footprint, you can search for an optimizer that handles nutritional properties better than some linear programs being used today. It is possible to use a different optimizer while continuing to use an existing database and report writer. If your optimizer does a better job formulating animal feeds, you

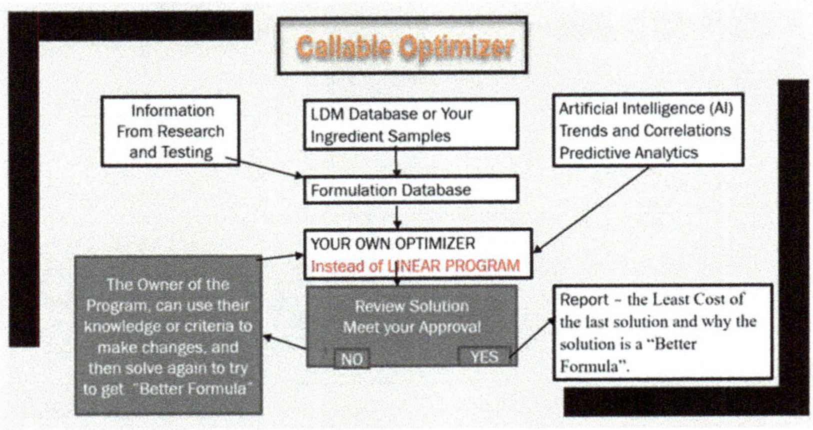

Figure 20 Callable Optimizer using your optimizer instead of Formulation's Linear Optimizer

will have a competitive advantage over your competitors. If the new optimizer is good, you could also provide it to your clients or the formulators that you wish to benefit from your research. Experts such as Roger Roberts of McKinsey & Company (in a recent presentation entitled 'The Rise of Generative AI: Business and Ethical Considerations') predict that existing software programs will be modified to link to AI applications (https://www.emorybusiness.com/2023/05/26/the-rise-of-generative-ai-industry-leaders-and-students-gather-at-6th-annual-summit/). The future of feed formulation will be determined by those willing to invest in tools to provide easy access to big data and in finding better solutions. These will probably come about by introducing new optimizers or programs that can review solutions to decide if there can be any additional improvements as well as using AI.

9 Adding a nutrient/element to the formulation database

Step 1 Find an available nutrient number (element) (Fig. 21).
Step 2 Define this nutrient and give it a name – I called this nutrient 'My New Element' (Fig. 22)
Step 3 Display a group of Ingredients and the value of nutrient 64 for each selected Ingredient. As you can see below, the nutrient value for nutrient 64 is blank or zero (Fig. 23).
Step 4 Enter your value for nutrient 64 for each of the Ingredients (Fig. 24).

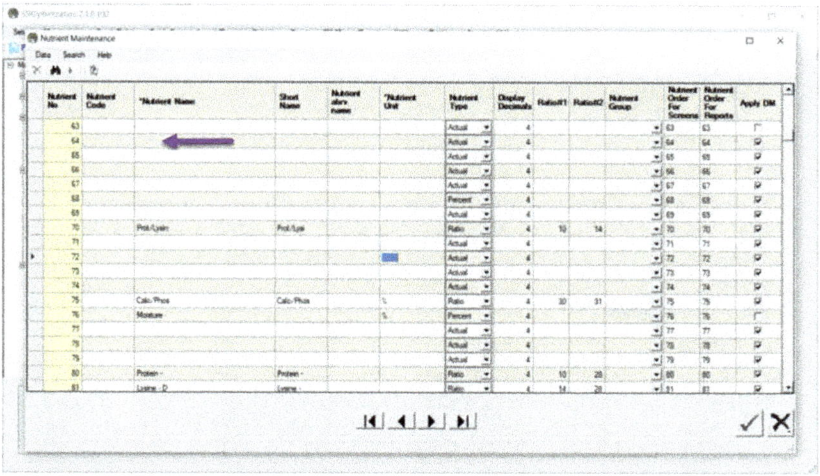

Figure 21 Choosing an unused nutrient number.

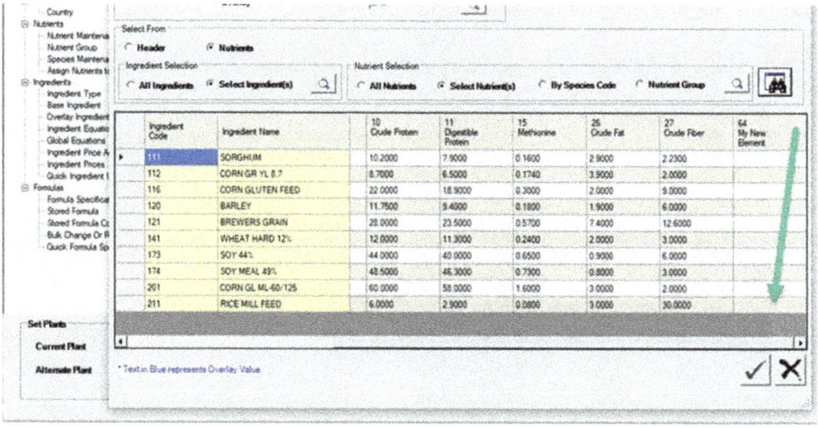

Figure 22 Giving the new nutrient a name.

Step 5 The Least Cost Solution for Formula 51 has a nutrient value of 18.6094 for nutrient 64. The person who knows what nutrient (element) 64 is, will now know that the Broiler Grower formula that we just generated has a value of 18.6094. We could put a requirement for nutrient 64 to be less than 15, 18, etc. which would make nutrient 64 a necessary nutrient in the Formula (Fig. 25).

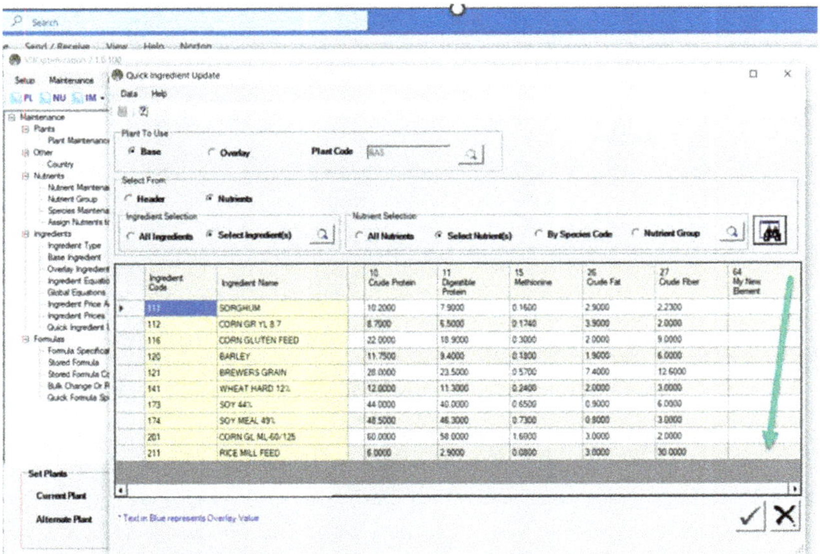

Figure 23 Starting out with no values in nutrient 64.

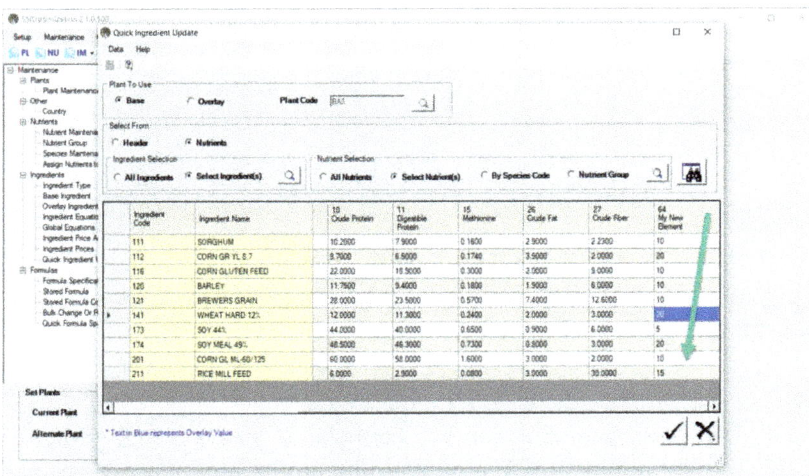

Figure 24 Enter values for nutrient 64 for each Ingredient.

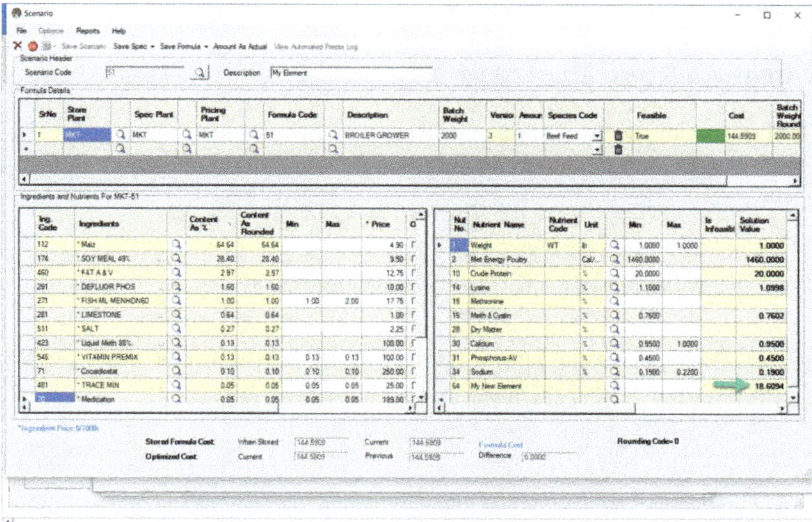

Figure 25 The amount of nutrient 64 is calculated during the formulation of the diet.

10 Using Excel Solver to generate the least cost

A formula can be generated by commercially available feed formulation software. The PowerPoint was developed by Matthew Clark of Feedguys Resources Pte Ltd.

Step 1

Step 2

Formula Spec Exercise 1

- Use the Ingredient Matrix in Poultry Formulation v1.XLSX

- Add the calculations for the formula specifications
- Install the Excel Solver
- Add the parameters for the diet optimisation

- Run the following options using Arg SBM (price out other SBM)
  - Base diet 1.32% SID Lysine, Crude protein 22%
  - 1.26% SID Lysine, Crude Protein 22%

  - Base diet 1.32% SID Lysine, No crude protein limit
  - 1.26% SID Lysine, No crude protein limit

Step 3

Check that the Solver Add-in is active

- File, More, Options, Add-ins,
- Click on Go to manage Excel Add-ins

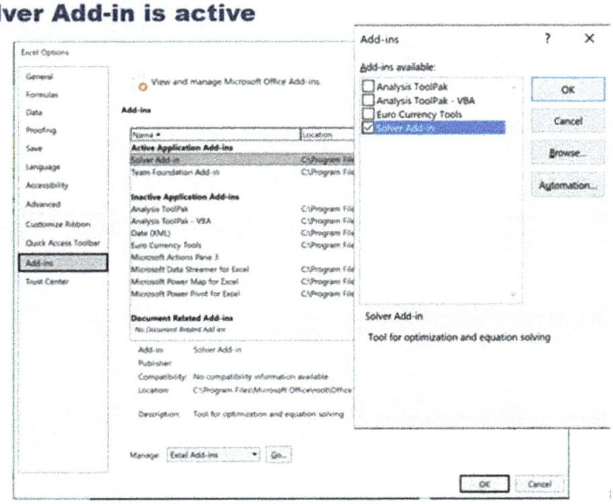

Step 4

Formula Specs Starting

CODE	CORN	SBMARG-S	SBMBRA-S	SBMUSA-S	FFS	DDGS	OIL-SOY	LIME	MDCP	SALT	BCARB	LLYS	DLMET	LTHR	LVAL	PHYTO50	NSPASE1	XANTHPMXPREMIX					
Inclusion	60	40	0	0	0	0	0	0	0	0	0	0	0	0	0	0	0	0	0		100		
Cost	386	602	602	602	772	420.74	2093	40	770	103	355	1750	2975	1860	3900	6000	4000	9300	4000		472.4		
LP Out	0.6	0.4																		Min		Max	
WEIGHT	1.00	1.00	1.00	1.00	1.00	1.00	1.00	1.00	1.00	1.00	1.00	1.00	1.00	1.00	1.00	1.00	1.00	1.00	1.00		1.00		
PROTEIN	7.20	45.50	47.00	46.40	35.90	27.24	-	-	-	-	-	95.00	58.27	72.79	73.20	-	-	-	-		22.52		
MEP	3,351.00	2,260.54	2,302.23	2,295.87	3,420.00	2,601.96	8,980.00	-	-	-	-	-	-	-	-	-	200,000.00	-	-		2,918.02		
CFAT	3.80	1.46	1.78	1.67	18.40	9.10	99.00	-	-	-	-	-	-	-	-	-	-	-	-		2.94		
FIBRE	2.10	4.32	5.03	3.88	5.80	6.37	-	-	-	-	-	-	-	-	-	-	-	-	-		2.99		
CA	0.02	0.34	0.31	0.39	0.29	0.15	-	38.00	20.30	-	-	-	-	-	-	150.00	-	-	-		0.15		
APHOS	0.05	0.26	0.24	0.26	0.18	0.59	-	-	18.62	-	-	-	-	-	-	240.00	-	-	-		0.13		
NA	0.01	0.03	0.02	0.03	0.06	0.24	-	-	-	38.00	25.00	-	-	-	-	-	-	-	-		0.02		
CL	0.09	0.05	0.05	0.04	0.03	0.12	-	-	-	60.00	-	19.50	-	-	-	-	-	-	-		0.07		
DLYS-P	0.18	2.42	2.44	2.34	1.90	0.53	-	-	-	-	-	78.00	-	-	-	-	136.00	-	-		1.08	100.00	
DMET-P	0.15	0.54	0.54	0.56	0.45	0.47	-	-	-	-	-	-	98.00	-	-	-	68.00	-	-		0.30	100.00	0.38
DM+C-P	0.30	0.96	0.99	1.06	0.80	0.90	-	-	-	-	-	-	98.00	-	-	-	106.00	-	-		0.57	100.00	0.75
DTHR-P	0.23	1.40	1.43	1.46	1.12	0.73	-	-	-	-	-	-	-	98.00	-	-	104.00	-	-		0.70	100.00	0.68
DTRP-P	0.06	0.52	0.53	0.56	0.35	0.18	-	-	-	-	-	-	-	-	-	-	28.00	-	-		0.24	100.00	0.16
DILE-P	0.25	1.74	1.81	1.79	1.39	0.78	-	-	-	-	-	-	-	-	-	-	124.00	-	-		0.85	100.00	0.63
DVAL-P	0.33	1.82	1.84	1.89	1.42	1.01	-	-	-	-	-	-	-	-	98.00	-	124.00	-	-		0.93	100.00	0.73
DARG-P	0.33	2.90	3.02	3.04	2.29	1.02	-	-	-	-	-	-	-	-	-	-	180.00	-	-		1.38	100.00	1.05

- Add the min and max limits
- Use your own mineral limits
- Start with the highest spec
- Note Cost Calculation uses =SUMPRODUCT(C5:U5,$C$6:$U$6)
- Nutrient Calculation uses =SUMPRODUCT(C9:U9,$C$6:$U$6)

Step 5

Formula Specs Completed

CODE	CORN	SBMARG-S	SBMBRA-S	SBMUSA-S	FFS	DDGS	OIL-SOY	LIME	MDCP	SALT	BCARB	LLYS	DLMET	LTHR	LVAL	PHYTO50	NSPASE1	XANTHPMXPREMIX					
Inclusion	60	40	0	0	0	0	0	0	0	0	0	0	0	0	0	0	0	0	0		100		
Cost	386	602	602	602	772	420.74	2093	40	770	103	355	1750	2975	1860	3900	6000	4000	9300	4000		472.4		
LP Out	0.6	0.4																		Min		Max	
WEIGHT	1.00	1.00	1.00	1.00	1.00	1.00	1.00	1.00	1.00	1.00	1.00	1.00	1.00	1.00	1.00	1.00	1.00	1.00	1.00		1.00		
PROTEIN	7.20	45.50	47.00	46.40	35.90	27.24	-	-	-	-	-	95.00	58.27	72.79	73.20	-	-	-	-	22.00	22.52	100.00	
MEP	3,351.00	2,260.54	2,302.23	2,295.87	3,420.00	2,601.96	8,980.00	-	-	-	-	-	-	-	-	-	200,000.00	-	-	2,950.00	2,918.02	4,000.00	
CFAT	3.80	1.46	1.78	1.67	18.40	9.10	99.00	-	-	-	-	-	-	-	-	-	-	-	-	-	2.94	7.00	
FIBRE	2.10	4.32	5.03	3.88	5.80	6.37	-	-	-	-	-	-	-	-	-	-	-	-	-	-	2.99	7.00	
CA	0.02	0.34	0.31	0.39	0.29	0.15	-	38.00	20.30	-	-	-	-	-	-	150.00	-	-	-	0.88	0.15	0.95	
APHOS	0.05	0.26	0.24	0.26	0.18	0.59	-	-	18.62	-	-	-	-	-	-	240.00	-	-	-	0.40	0.13	100.00	
NA	0.01	0.03	0.02	0.03	0.06	0.24	-	-	-	38.00	25.00	-	-	-	-	-	-	-	-	0.18	0.02	0.20	
CL	0.09	0.05	0.05	0.04	0.03	0.12	-	-	-	60.00	-	19.50	-	-	-	-	-	-	-	-	0.07	0.24	
DLYS-P	0.18	2.42	2.44	2.34	1.90	0.53	-	-	-	-	-	78.00	-	-	-	-	136.00	-	-	1.32	1.08	100.00	
DMET-P	0.15	0.54	0.54	0.56	0.45	0.47	-	-	-	-	-	-	98.00	-	-	-	68.00	-	-	0.50	0.30	100.00	0.38
DM+C-P	0.30	0.96	0.99	1.06	0.80	0.90	-	-	-	-	-	-	98.00	-	-	-	106.00	-	-	0.96	0.57	100.00	0.75
DTHR-P	0.23	1.40	1.43	1.46	1.12	0.73	-	-	-	-	-	-	-	98.00	-	-	104.00	-	-	0.90	0.70	100.00	0.68
DTRP-P	0.06	0.52	0.53	0.56	0.35	0.18	-	-	-	-	-	-	-	-	-	-	28.00	-	-	0.22	0.24	100.00	0.16
DILE-P	0.25	1.74	1.81	1.79	1.39	0.78	-	-	-	-	-	-	-	-	-	-	124.00	-	-	0.85	0.85	100.00	0.63
DVAL-P	0.33	1.82	1.84	1.89	1.42	1.01	-	-	-	-	-	-	-	-	98.00	-	124.00	-	-	0.96	0.93	100.00	0.73
DARG-P	0.33	2.90	3.02	3.04	2.29	1.02	-	-	-	-	-	-	-	-	-	-	180.00	-	-	1.38	1.36	100.00	1.05

- Set up the Solver Limits
- Set Arg SBM Price to 602 and BRA & USA to 1000 (Price out)
- Data, Solver… Then follow the setup in the next slides

Step 6

Solver Parameters

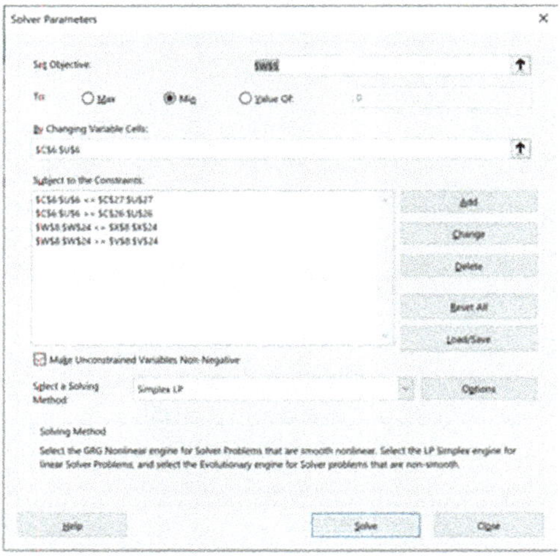

- Set the objective – the formula cost calculation
- Set the solver to minimum
- Set the min and max limits
- Set solver to Simplex
- Click on Solve when ready

Step 7

Formula Specs Completed

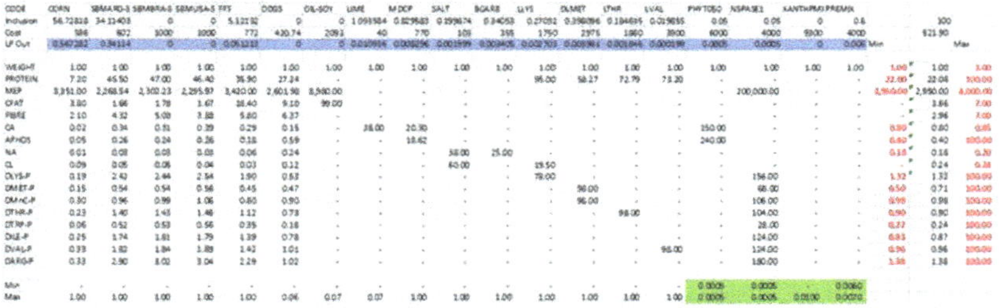

- Base solution costs 521.90

11 The high cost of margins of safety in animal feed

It has been a common practice to include safety margins when handling ingredient data. The main reason behind this practice is to ensure that the nutrient minima are adequately met. There is the additional concern that manufacturers are required to conform to their tag declarations for key nutrients, particularly protein.

This safety approach can be applied not only to the tag values for legal reasons but also sometimes to the key nutrients such as digestible amino acids to be more sure of reaching adequate nutrient levels. However, with increasing commodity prices over the last 5 years, the practice of adding safety margins (typically of 0.5 SDs) for each nutrient has led to increases in the cost of safety in formulation.

A simple exercise using adjustments to the nutrients in soybean meal will show the economic effect. In the following examples, the nutrients in SBM have been adjusted by deducting 0.5 × SDs to derive 'safe' versions of available SBM samples. These adjusted values are referred to as margin of safety (MOS) values:

Crude protein	Argentine SBM	Brazilian SBM	US SBM	India SBM
Mean	45.50	47.00	46.40	46.30
Standard deviation	**1.08**	**1.65**	**0.50**	0.98
Adjustment	**−0.54**	**−0.83**	**−0.25**	**−0.49**
'Safe' value	44.96	46.18	46.15	45.81

The starting amino acid levels and SDs of those observations in the different meals were as follows:

Amino acid	Argentine SBM	Brazilian SBM	US SBM	India SBM
Lysine	2.82	2.88	2.89	2.83
Methionine	0.62	0.63	0.64	0.61
Cysteine	0.66	0.67	0.68	0.63
Threonine	1.77	1.80	1.80	1.75
Tryptophan	0.63	0.64	0.65	0.64
Isoleucine	2.07	2.15	2.10	2.12
Valine	2.19	2.25	2.22	2.19
Arginine	3.30	3.43	3.38	3.40
Standard deviations*	Argentine SBM	Brazilian SBM	US SBM	India SBM
Lysine	0.19	0.25	0.20	0.21
Methionine	0.04	0.07	0.02	0.02
Cysteine	0.02	0.04	0.03	0.03
Threonine	0.08	0.09	0.04	0.07
Tryptophan	0.00	0.00	0.00	0.00
Isoleucine	0.09	0.16	0.10	0.11
Valine	0.16	0.23	0.11	0.19
Arginine	0.05	0.16	0.14	0.15

*Standard deviation data from Ravindran 2014.

The new ingredient values are tabulated below:

Nutrient		ARG SBM	BRA SBM	USA SBM	IND SBM
Crude Protein	Mean	45.50	47.00	46.40	46.30
	MOS	44.96	46.18	46.15	45.81
Lysine	Mean	2.82	2.88	2.89	2.83
	MOS	2.72	2.75	2.79	2.73
Methionine	Mean	0.62	0.63	0.64	0.61
	MOS	0.60	0.59	0.63	0.60
Cysteine	Mean	0.66	0.67	0.68	0.63
	MOS	0.65	0.65	0.67	0.61
Threonine	Mean	1.77	1.80	1.80	1.75
	MOS	1.73	1.76	1.78	1.72
Tryptophan	Mean	0.63	0.64	0.65	0.64
	MOS	0.63	0.64	0.65	0.64
Isoleucine	Mean	2.07	2.15	2.10	2.12
	MOS	2.03	2.07	2.05	2.07
Valine	Mean	2.19	2.25	2.22	2.19
	MOS	2.11	2.13	2.16	2.10
Arginine	Mean	3.30	3.43	3.38	3.40
	MOS	3.27	3.35	3.31	3.32

The impact of using the MOS values can be seen from the shadow price data in a typical formula:

Shadow prices	Mean values	MOS values	Difference	Percent
SBM Argentina	691.304	682.212	−9.092	−1.3%
SBM Brazil	714.667	700.738	−13.929	−1.9%
SBM USA	719.171	714.905	−4.266	−0.6%
SBM India	688.209	680.032	−8.177	−1.2%

Applying the MOS values for candidate ingredients does lower the shadow price, but it can be seen that not all ingredients are affected equally. Brazilian SBM has a 1.9% lower shadow price compared to US SBM which is lowered by 0.6%. Thus, it can be seen that the MOS application does assist the purchasing operation to further differentiate the penalty cost inherent in the variability of supply. The formula cost will also be impacted by using the MOS values:

Shadow prices	Mean values	MOS values	Difference	Percent
SBM Argentina	540.443	543.255	2.812	0.5%
SBM Brazil	534.262	539.598	5.336	1.0%
SBM USA	532.308	533.635	1.327	0.2%
SBM India	544.541	549.119	4.578	0.8%

It can be seen that the MOS deductions are a potential tool for differentiating the economic impact of the variability within different commodity sources with cost penalties ranging between 0.2% and 1%. MOS is commonly used to evaluate the cost impact of variability in ingredient sourcing.

A similar analysis on comparing mean nutrient values for corn against formulas with MOS values shows formula cost impacts (which would be cumulative) of US$0.80 to US$ 1.00 per tonne.

While there is some increase in feed cost, the bias in purchasing the less variable ingredients will have a relatively low cost impact in percentage terms and is still a viable method of providing cost effective safety in formulation.

However, it is currently possible to run many more samples using NIR than was possible when wet chemistry analyses were the method of choice. We should be in a better state of knowledge about our ingredient nutrient values and therefore we may be able to formulate production formulas without the safety margins and instead rely on current analysis. The Margin of Safety method remains a good tool for evaluating the effect of variability in the supply chain and is still a good tool for evaluation of the cost of variability in that it is predictive of a range of possible nutrient values of future deliveries

It is therefore important to make good use of the QC database in relation to the feed formulation system to fine-tune both the purchasing decisions (in supplier evaluation) and current formulation with accurate current ingredient values.

Applying linear programming to feed formulation can not only to minimize cost but also minimize waste. This can be achieved in a number of ways:

- Reduction in safety margins by reducing the standard deviation penalties as described above. With a better state of knowledge of our ingredients, we should be able to minimize the margins of safety on ingredients can capture real current values. The trend in the industry is to pair QC systems with formulation not just to lower cost, but to minimize excess nutrition; and
- The margin of safety practice is geared to regulation. In particular, the crude protein limits are seen as a legal necessity, but this may not be an environmentally sustainable procedure. Crude protein is not a nutrient for monogastric animals and excess crude protein is a pollutant.

- There are concerns for the integrity of the water supplies when nitrates accumulate due to excess nitrogen in the feed and manure.
- Excess non-protein nitrogen in the form of ammonia leads to the production of nitrous oxide as an intermediate which is a potent greenhouse gas.

This point is well demonstrated by Peter Chrystal (Asian Agribusiness 2022) in his research on reduction of crude protein in layer feeds:

	Crude protein (g/kg)		
Parameters	140.0	155.0	170.0
Feed (g/day)	116.0	115.4	113.9
Lysine (mg/day)	940.9	936.0	924.1
Production (%)	77.22	76.02	75.24
Egg weight (g)	66.63	66.82	66.67
Egg output (g/day)	51.45	50.82	50.18
Yolk (%)	24.94	24.39	24.19
Manure crude protein (g/kg)	338.8	385.4	464.9

The data shows that a reduction in crude protein from 17% to 14% has no impact on production. The key difference is excretion of nitrogen in the manure is reduced from 464.90 g/kg to 338.80 g/kg. This has important implications:

- Environmental ammonia is a health hazard to poultry and also to workers causing conjunctivitis and also respiratory damage.
- The excreted nitrogen is a water pollutant as well as a source of greenhouse gasses.

Reduction in crude protein is made possible by utilizing the growing range of supplemental amino acids and LP plays an important role in formulating the more complex diets with greater accuracy enabling nitrogen-sparing diets. However, the process does not end there. Recommendations for crude protein are still a feature of nutritional guidelines. Research and education are still needed in this area.

QC data processing plays an important role here also. If only one of the amino acids is over estimated with the effect of under supplying the amino acid in the formulation, the other amino acids will be out of balance and in excess thus causing a nitrogen over supply. Thus, the importance of the role of quantitative QC coupled to formulation cannot be overstated and this coupling is a growing trend in the industry.

In the following example, a layer feed is formulated to 16.5% crude protein and then reformulated without the limit:

Nutrient	Min	Actual	Max		Previous
Weight	1.00	1.00	1.00	MAX	1.00
Crude Protein	0.00	15.06	24.00		16.50
AMEn Poultry	2750.00	2750.00	2850.00	MIN	2750.00
AMEn Broiler	0.00	2715.45	3950.00		2704.27
Crude Fat	3.00	6.29	11.00		6.91
Crude Fiber	0.00	3.74	7.00		3.83
Xanthophyll	14.00	14.00	30.00	MIN	14.00
Calcium	4.10	4.10	4.20	MIN	4.10
Av Phos (Poultry)	0.40	0.40	100.00	MIN	0.40
Sodium	0.18	0.18	0.20	MIN	0.18
Chloride	0.00	0.23	0.23	MAX	0.23
Digestible Lys Poultry	0.76	0.76	100.00	MIN	0.76
Digestible Met Poultry	0.38	0.47	100.00		0.46
Digestible M&C Poultry	0.69	0.69	100.00	MIN	0.69
Digestible Thr Poultry	0.53	0.53	100.00	MIN	0.53
Digestible Trp Poultry	0.16	0.16	100.00	MIN	0.18
Digestible Ile Poultry	0.60	0.60	100.00	MIN	0.61
Digestible Val Poultry	0.66	0.66	100.00	MIN	0.71
Digestible Arg Poultry	0.78	0.88	100.00		0.99

The cost falls by $11.14 per tonne. But from the conservation point of view, there is a more important implication that there are savings in the consumption of soybean meal:

Ingredient	Min	Actual	Max		Price	Previous
Yellow Corn	0.00	42.07	100.00		376.09	37.27
Base SBM	0.00	13.65	100.00		691.30	18.07
DDGS BASE	0.00	10.00	10.00	MAX	428.26	10.00
Wheat Pollard	0.00	6.00	6.00	MAX	270.00	6.00
Rice Bran	0.00	15.00	15.00	MAX	237.00	15.00
Oil	0.00	1.23	100.00		878.84	1.97
Limestone	0.00	9.85	100.00		45.00	9.84
MDCP	0.00	0.65	100.00		1050.00	0.60
Salt	0.00	0.21	100.00		100.00	0.25
Bicarb	0.00	0.24	100.00		350.00	0.17

Ingredient	Min	Actual	Max		Price	Previous
L-Lysine HCl	0.00	0.16	100.00		1030.04	0.03
DL-Methionine	0.00	0.22	100.00		3090.13	0.19
L-Threonine	0.00	0.05	100.00		1802.58	0.00
L-Isoleucine	0.00	0.05	100.00		7510.73	0.00
L-Valine	0.00	0.02	100.00		3648.07	0.00
Phytase 500 g 1000 FTU (STD)	0.05	0.05	0.05	MAX	6000.00	0.05
NSP Enzyme Generic 500 g	0.05	0.05	0.05	MAX	4000.00	0.05
Xanthophyll Premix	0.00	0.00	1.00		9300.00	0.01
Premix Ingredients	0.50	0.50	0.50		4000.00	0.50

The usage of Soybean meal has fallen from 18.7% to 13.65% thus sparing an important resource.

Linear Programming plays a significant role in this as the speed of calculation enables this saving to be discovered in a matter of a few milliseconds without which we would not have the time in a large enterprise to explore this level of accuracy.

There is a role too at the smallholder level. Farm feeds tend to be formulated with a higher level of safety than industrial feeds. Is this really necessary? Indeed, we should roll out the same level of QC investigation in small scale farming as well as larger feed mills not only to save cost but to optimize the use of scarce resources. This is indeed the classical contribution of LP in scarce resource allocation and given the current scope of IT infrastructure, LP could be utilized more widely.

Thus, we can see that LP plays a significant role in the calculation, but we should also be aware that LP can only answer the questions posed. The future contribution of LP is within the realm of Big Data to make better-informed decisions against better regulation and without the luxury of excess nutrient targets and raw material nutrient penalties. Our better state of knowledge should enable us to formulate with better confidence and less attendant 'insurance' which serves only to raise the cost if our raw material values are precise.

Chapter 9

Advances in broiler breeder nutrition

Dr. R.A. (Rick) van Emous, Wageningen Livestock Research, The Netherlands

1 Introduction

Poultry meat is the most important protein source in the human diet. In 2020, in order to meet the increasing global demand, poultry production increased annually with approximately 3% to more than 100 million tons (USDA, 2021). The majority of poultry meat is produced by more than 70 billion broilers originating from approximately 600 million broiler breeders. This means that a relatively small number of broiler breeders (or parent stock) have a substantial impact on the poultry meat chain. Broiler breeders need to produce first-class hatching eggs and healthy chicks. Therefore, appropriate management is essential to produce a maximum number of fertile eggs and high-quality broiler chickens (van Emous, 2015). This is becoming increasingly challenging because of the continuing increase in the genetic growth potential of the offspring and the parent stock as well (Zuidhof et al., 2014). During the last six decades, broilers have been successfully selected for maximum growth and feed efficiency (Havenstein et al., 2003a,b; Zuidhof et al., 2014). Modern fast-growing broilers reach a body weight (BW) of 2.0–2.5 kg in 35-45 days (Augère-Granier, 2019). The high growth rate and efficiency are advantageous in broilers, but it may lead to welfare and reproduction problems in parent stock during the rearing and laying period. Feeding breeder pullets *ad libitum*, as is standard practice in broilers, results in a very high BW with negative effects on reproductive performance, mortality and health (Katanbaf et al., 1989; Savory et al., 1993;

http://dx.doi.org/10.19103/AS.2024.0143.12

Heck et al., 2004). In general, broiler breeders produce until 60–65 weeks of age (WOA), meaning that high mortality, poor reproduction and reduced efficiency have a major impact on bird welfare and economic profitability (Decuypere et al., 2010). To prevent these negative effects on reproduction and health, feed restriction is necessary to promote breeder welfare and health reasons, whereas feed restriction simultaneously creates another welfare problem (D'Eath et al., 2009). Breeders show abnormal behaviours like stereotypical pecking behaviour towards objects such as walls, empty feed systems, empty water systems and other parts of the house, sometimes in combination with excessive water intake (spoiling) (Hocking et al., 1996, 2001; Savory and Kostal, 1996; de Jong et al., 2002). There are (mostly indirect) indications that these abnormal behaviours can be indicative of hunger and frustration in broiler breeders (Mench, 2002). The discrepancy between a healthy and reproductive life and impaired behaviour and welfare is known as 'the broiler breeder paradox' (Decuypere et al., 2010). In other words, feeding broiler breeders *ad libitum* has negative consequences for health, welfare and production, but on the other hand, feed restriction has negative effects on behaviour and welfare. EFSA (2010), however, states that insufficient research has been done in this field to conclude whether feed restriction impairs the welfare of broiler breeders.

Due to the issues mentioned, the management of broiler breeders is very challenging, especially due to the continuous genetically changes in broilers and their parent stock. State-of-the-art management for housing, equipment and health is necessary to optimize the reproduction of breeders for a maximum number of healthy offspring. Besides these mentioned factors, diet composition and feeding strategies are important factors to reach this goal. This chapter describes the state-of-the-art scientific knowledge on nutrition in broiler breeders.

2 Diet composition females

2.1 Diluted diets

Due to the aforementioned welfare issues caused by feed restriction, different dietary solutions have been developed over the last three decades. Adjusting diet composition is a possibility that is frequently investigated as a method to reduce the negative effects of feed restriction. For example, by offering more feed with lower levels (diluted diets) of all nutrients, feed restriction level could be limited. Adjusting the energy to protein ratio in the feed is another possibility of diet dilution (Sandilands et al., 2006, van Emous et al., 2014, 2015a,b). A study by van Emous et al. (2014) examined the effects of three different dietary protein concentrations (15.3%, 14.2% and 13.2%) and different BW profiles (2400 g versus 2200 g at 20 WOA) on indicators of welfare during the rearing and production phase. Partly due to the higher amount of feed, a positive effect on welfare indicators was found for the low-protein diets during the rearing period. Stereotypical object pecking in the rearing phase was reduced in the low protein diet pullets compared to the high protein diet pullets.

The conclusions within the scientific community about alternative diets are contradictory to date and more research will have to show what the actual influences of these strategies are on bird welfare (D'Eath et al., 2009). Some works suggest that alternative feeding strategies result in improved bird welfare by responding to more natural feeding behaviour and/or a feeling of fullness (Zuidhof et al., 1995; de Jong et al., 2005; Sandilands et al., 2006). Others, however, conclude that this method does not eliminate 'metabolic hunger' (de Jong et al., 2005; D'Eath et al., 2009). In addition, there are studies that found no or limited positive effects on behaviour and welfare as a result of adjusted diets (Savory and Lariviere, 2000; Hocking et al., 2001; Tahamtani et al., 2020; Mens et al., 2022). For example, Decuypere et al. (2006) concluded that the provision of feed with high levels of crude fibre, resulting in a higher feed allowance with less nutritional value, gave negligible results on bird welfare. Mens et al. (2022) used a 2 × 2 factorial completely randomized block design, where pullets were fed with two diet densities and two feeding frequencies during rearing. Between 3 WOA and 23 WOA, pullets received either a standard control diet (CON) or a 16% diluted diet (DIL) containing oat hulls. These diets were provided either once (FO) or twice (FT) a day. After 23 WOA, all pullets received the same standard layer diet once a day. They concluded that feeding diluted diets and/or feeding twice a day did not have any obvious benefits to BB welfare. The study also showed that diet dilution with 10–20% oat hulls and/or increasing the feeding frequency to twice per day resulted only in minor effects on the welfare of broiler breeders in rearing, since behavioural indicators of stress, fear, motivation and frustration were not affected or showed minor changes. Increasing the feeding frequency affected behavioural patterns during the day which could indicate anticipation for the second meal. Lastly, the nutritional treatments applied in rearing had only slight effects on home pen behaviour during the laying period which aligned with the results found during the rearing period.

The use of appetite suppressants (e.g. calcium propionate) in combination with *ad libitum* oat hulls as an alternative to feed restriction seems to decrease stereotypical pecking behaviour and thus welfare of parent stock (Tolkamp et al., 2005; Sandilands et al., 2005, 2006; Morrissey et al., 2014; Arrazola et al., 2019). This alternative feeding strategy ensures that the desired growth pattern is followed despite *ad libitum* feeding. Object pecking is almost non-existent (<1%) in diets including calcium propionate compared to standard diets (about 50%). However, high inclusion rates of dietary diluents and appetite suppressants can result in negative consequences in the development and function of the GIT and welfare (Arrazola et al., 2019). In addition, other behavioural and physiological parameters (such as plasma corticosterone levels) did not differ compared to restricted birds (Sandilands et al., 2006), indicating there are no effects on the birds' stress levels.

It has been shown that pullets fed diluted diets need a higher feed allowance to compensate for lower nutrient concentration to obtain the same BW target at end of rear (Enting et al., 2007; Zuidhof et al., 2015; de los Mozos et al., 2017; van Emous et al., 2014, 2015a,b, 2021; Lesuisse et al., 2017; Heijmans et al., 2021). Breeders fed low-density diets during the laying phase showed increased egg production and egg weight as compared to control diets (Enting et al., 2007). In a recent 2 × 2 factorial study, pullets were fed once or twice a day on standard or 16% diluted diets (van Emous et al., 2021). The pullets fed the diluted diets received a higher feed allowance in combination with similar water intake which resulted in a lower water/feed ratio compared to the control pullets and thus in a higher DM content of the litter and improved litter quality. The higher feed allowance resulted in a 20% higher total manure production at 20 WOA.

2.2 Body composition

The offspring (broilers) of broiler breeders have changed enormously for production performance and body composition during the last six decades (Havenstein et al., 2003a,b; Zuidhof et al., 2014). This genetic selection has changed the body composition of broiler breeder hens (Eitan et al., 2014). In the last 30 years, the body fat mass in broiler breeder hens has decreased by approximately 50% (Eitan et al., 2014; Zuidhof, 2018). Van Emous (2015) was the first to suggest that the body composition of broiler breeder females could play an important role in egg production, incubation traits and offspring performance, and that this could be affected by a higher energy-to-protein ratio and thus lower daily protein intake. Several authors observed that a lower daily protein intake during the rearing phase led to a linear increase in abdominal fat pad and thus a higher body fat mass (van Emous et al., 2013, 2015b; Lesuisse et al., 2017, 2018; Zuidhof, 2018; Salas et al., 2019). More studies carried out on this topic found that body composition plays a major role in sexual maturation (Bédécarrats et al., 2016; Zuidhof, 2018; Hadinia et al., 2020), egg production (van der Klein et al., 2018), egg composition (Salas et al., 2017) and egg production persistency (van Emous et al., 2015). Feeding a lower daily protein intake (and higher energy daily intake) during the rearing phase resulted in a higher abdominal fat and lower breast muscle content at 20 WOA compared to a higher daily protein intake (and lower daily energy intake) (van Emous et al., 2013, 2015). In the study of van Emous et al. (2015b), a higher daily energy intake during the rearing phase showed an improved egg production persistency between 45 WOA and 60 WOA (Fig. 1). They speculated that the higher abdominal fat content was beneficial for the laying persistency due to the fact that breeders with a higher body fat content are probably more able to mobilize energy reserves in periods of a negative energy balance which prevent them for molting. The higher persistency due to the different body composition was, however, not found in the studies of Lesuisse et al. (2017, 2018), Salas et al. (2019), and Heijmans et al. (2021).

2.3 Low dietary protein laying phase

Feeding broiler breeders a 16% CP concentrations diets during the entire laying period resulted in decreased fertility (Lopez and Leeson, 1995a; Ekmay et al., 2013). Contrary, feeding breeders a lower daily protein intake second the second phase of the laying increased hatchability of fertile eggs (van Emous et al., 2015b). This was, however, not observed in a study of van Emous et al. (2018) who used a three-phase feeding program with isocaloric diets, and calculated CP content of the low CP diets was 15 g/kg lower than the standard CP diets during all phases. The higher hatchability in the study of van Emous et al. (2015b) aligns with earlier research from Pearson and Herron (1982), Whitehead et al. (1985) and Lopez and Leeson (1995b). The differences in hatchability of the fertile eggs were caused by differences in embryonic mortality. A higher or lower embryonic mortality leads to a lower or higher hatchability of fertile eggs, respectively. This observation supports the earlier work of Pearson and Herron (1982) who found that lowering daily protein intake (27.0 vs. 21.3 g/bird) resulted in a decreased mortality and malformation of embryos. Ekmay et al. (2013) observed, in a dose response trial, that increasing levels of especially lysine and isoleucine reduced fertility. They suggested that the differences in fertility may be partly explained by differences in the quality of the sperm cells after a period of storage in the hen's sperm storage tubules what is maybe caused by a changed microenvironment (e.g. pH level).

The effect of an improved hatchability and decreased embryonic mortality, while feeding a low daily protein intake, was underlined by the decreased proportion of second grade chicks in the study of van Emous et al. (2015a). The relationship between low embryonic mortality and fewer second grade chicks was previously found in studies of Reijrink et al. (2010) and Molenaar et al. (2011). An additional benefit of feeding breeders a low-CP diet is the decreased nitrogen excretion due to the lower nitrogen intake, higher nitrogen retention (Lopez and Leeson, 1995c) and reducing the ecological footprint (Lesuisse et al., 2018).

Egg weight can be affected by daily dietary protein and amino acids intake (e.g. Lopez and Leeson, 1995a). More specifically, a higher egg weight is caused by a higher daily intake of sulfur amino acids (affects albumen and

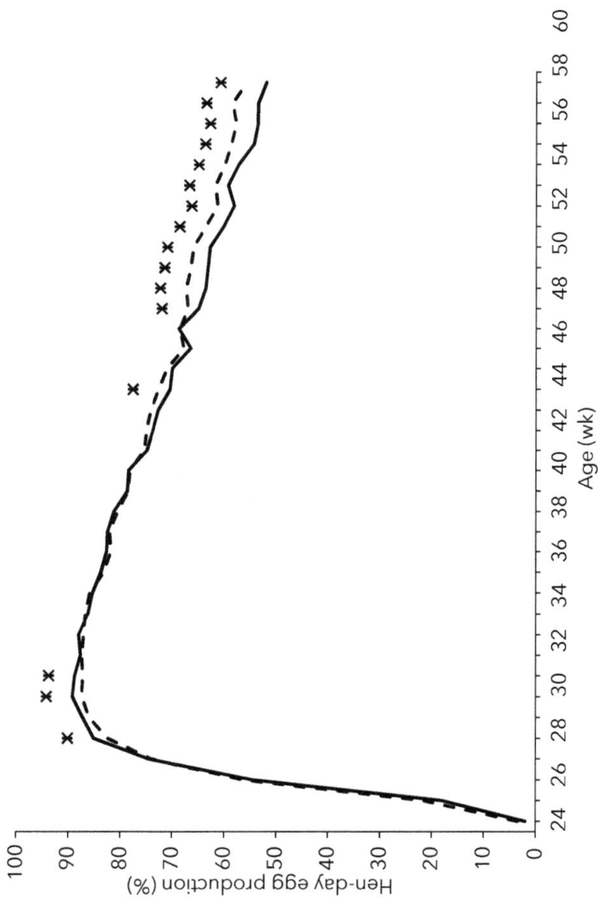

Figure 1 Carry-over effect of dietary protein intake during the rearing phase on hen-day egg production (%) (van Emous et al., 2015b). Solid line = pullets fed low daily energy intake (196 kcal/d) and dashed line = pullets fed high daily energy intake (222 kcal/d). Asterisks indicate significant differences ($P \leq 0.05$) between treatments.

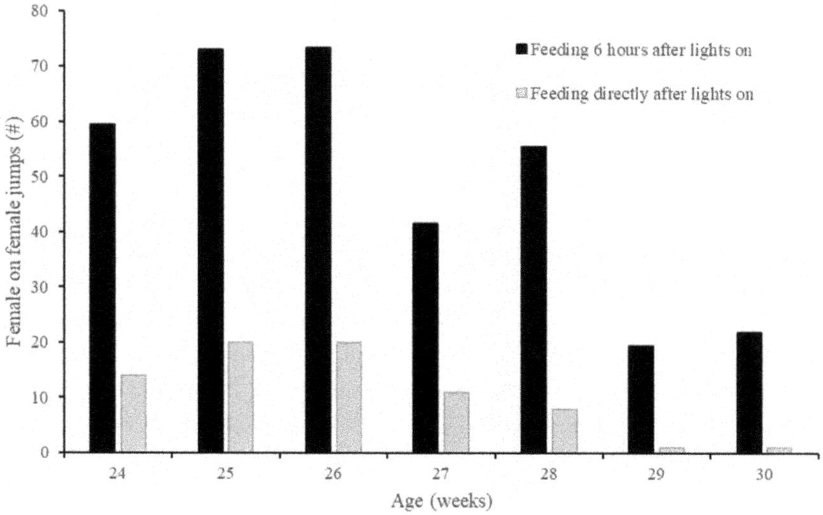

Figure 2 Number of female on female jumps around the feeding systems during the first 5 min after start of the feeding system between 24 WOA and 30 WOA (R.A. van Emous, unpublished data).

yolk) and/or higher daily intake of linoleic acid (affects on yolk). Contrary to above, higher daily protein intake in other experiments did not show higher egg weight during initial lay (Pearson and Herron, 1981; Joseph et al., 2002; Mohiti-Asli et al., 2012; van Emous et al., 2015b).

2.4 Transgenerational effects

First class (or grade A in the US) and healthy day-old chicks (DOC) is the foundation for optimal production performance, low mortality and high profitability of the flock. Previously, the majority of the research focussed on improving chick quality by egg handling, egg storage, incubation temperature and humidity. More recently, breeder nutrition has become of more interest to influence DOC quality via the transgenerational approach (van Emous et al., 2015c; Moraes et al., 2014, 2019; Lesuisse et al., 2017, 2018; Heijmans et al., 2022). This increasing interest is partly based on studies on the transgenerational effects of the Dutch famine during World War II on human health (Roseboom et al., 2011). This study showed that maternal undernutrition during gestation had long-lasting effects in later life dependent on the timing of the malnutrition during pregnancy. In the studies of van Emous et al. (2015b,c, 2018), no effect from daily protein intake in breeders on DOC weight was observed. By contrast, in the experiment of Lesuisse et al. (2017), a lower daily protein intake resulted in lower DOC weight. DOC weight was 9% lower, respectively, when breeders receive a 23% lower daily CP intake during rearing and production (Lesuisse et al., 2017). This was confirmed in the study of Heijmans et al. (2022) who applied four different dietary energy-to-protein ratios.

Several studies found an effect from breeder dietary energy-to-protein ratio on progeny performance (Moraes et al., 2014, 2019; Lesuisse et al., 2017, 2018; Heijmans et al., 2022). In the study of Heijmans et al. (2022), DOCs had a higher BW from breeders fed the highest daily CP intake, whereas they had the lowest BW at slaughter age (32 d). DOC's BW decreased linearly with decreasing daily CP intake, whereas BW at slaughter age increased linearly with decreased daily CP intake. Broilers originating from breeders fed a lower daily CP intake had a higher daily BW growth, a higher feed intake and lower FCR, than broilers originating from breeders fed a higher daily CP intake. They concluded that the observed results are indicative that maternal diet influences offspring performance which could be caused by transgenerational epigenetic programming (Berghof et al., 2013). Modification in gene expression in specific tissues could be the mode of action to these phenotypic changes in the offspring (Rao et al., 2009). In the study of Heijmans et al. (2022), breeders were fed a lower daily protein intake. They speculated that broilers originating from these breeders were programmed for CP poor circumstances and thereby were more efficient in using dietary CP resulting in an improved FCR and higher growth of the offspring. In line with this hypothesis, Lesuisse et al. (2018) observed an enhanced nitrogen retention in broilers originating from breeders fed a low CP diet, compared to a high CP diet.

2.5 Vitamins and minerals

Studies on the direct effects of vitamins and trace minerals in breeder nutrition on the birds themselves are very limited. However, much can be gleaned from knowledge of vitamin and mineral transfer efficiency by hens into table eggs (Naber, 1993). Most of the studies focussed on the transgenerational effects of vitamins and trace minerals on the offspring. There is substantial evidence that confirms the negative effects of deficiencies of vitamins and trace minerals in breeder diets on chick quality and offspring performance (Kidd, 2003; Calini and Sirri, 2007; Hocking, 2007). Vitamins and trace minerals with the most documented impact on chick quality, chick viability and skeletal development are the fat-soluble vitamins D and E and the trace elements selenium, zinc and manganese (Chang et al., 2016).

Sufficient daily vitamin D3 intake in breeders is very important for the health of the breeders but also for optimal offspring development, viability and carcass development (Sunde et al., 1978). Incidence of tibial dyschondroplasia in offspring was lower in young (30 WOA and 37 WOA) but not in older (45

WOA and 57 WOA) broiler breeders receiving between 2000 IU and 4000 IU vitamin D3/kg (Atencio et al., 2005a). Moreover, feeding breeders a better biological value of vitamin D3 (25-OH-D3) resulted in lower embryonic mortality and higher bone ash in the offspring (Atencio et al., 2005b). Vitamin E can enhance immunity in offspring by improving adaptive antibody transfer from the breeder to the offspring (Boa-Amponsem et al., 2001). An improved immune system in offspring was also found in a study by de Beer (2009) who supplemented vitamins B, E and K 20% above Aviagen breeder recommendations resulting in a 2.2% lower mortality at slaughter age in the offspring.

Breeders were fed in a 2 × 2 factorial design with two vitamin E concentrations (30 mg/kg and 120 mg/kg) and two selenium sources (sodium selenite and zinc-L-selenomethionine) (Urso et al., 2015). They concluded that dietary supplementation of zinc-L-selenomethionine and vitamin E (120 mg/kg feed) could be used to improve chick quality expressed in lower embryonic mortality and higher chick weight. The combination of increased vitamin E and selenium concentrations was also studied in an experiment of Surai (2000). Broiler breeder diets were supplemented with vitamin E (40–200 mg/kg α-tocopherol) and selenium (200 μg/kg and 400 μg/kg) and they found that the liver antioxidant system in 1- and 5-day-old chicks was improved with potential benefits for chick health. Adding 100 IU/kg vitamin E in breeder diets is recommended by the breeder companies to support optimal chick health and performance (Chang et al., 2016).

Selenium (Se) plays an important role in good semen quality and as an antioxidant defence of the embryo for optimal flock fertility and hatchability (Surai and Fisinin, 2014). The organic form of Se is identified as a potential important antioxidant and the studies showed that supplementing Se to breeders diets increased Se concentrations in the eggs and offspring (Surai, 2000; Kemp et al., 2001; Pappas et al., 2005, 2006; Jlali et al., 2013; Couloigner et al., 2015). Feeding breeders 500 μg/kg of organic Se resulted in an increased antioxidant status of eggs and a 4% higher hatchability, due to a lower embryonic mortality (Rajashree et al., 2014).

It was concluded by Kidd (2003) that higher zinc concentrations improved cellular immune function and early survival of the offspring. Supplementation of a combination of organic zinc and manganese has been observed to improve offspring liveability and some immune parameters and measures of cardiac function from day old to d 34 (Virden et al., 2003, 2004). Moreover, there is also evidence that substantially higher maternal dietary concentrations of Se, zinc and manganese than recommended may improve immune function and liveability of offspring (Hocking, 2007).

3 Feeding strategies females

3.1 Every day feeding during rear

Broiler breeders have been selected for five to six decades for production of fertile eggs and for growth rate and growth efficiency in their offspring (de Jong and van Emous, 2017). Broiler breeders are the parent stock of broilers and they have genetically a comparable high growth potential to broilers (Havenstein et al., 2003a, 2003b; Zuidhof et al., 2014). Zuidhof et al. (2014) showed that over a 56-day growth period, a 2005 broiler strain grew 4.6 times the rate of a 1957 strain. Due to the continuous genetically improvement, it is conceivable that modern strains from the last years reach an even higher BW and lower FCR. A high BW growth rate may lead to problems in broiler breeders during their entire lifecycle. *Ad libitum* feeding of breeder pullets resulted in a two and half times higher BW at onset of lay, with adverse effects on reproductive performance (Katanbaf et al., 1989; Savory et al., 1993; Heck et al., 2004). Heck et al. (2004) applied *ad libitum* feeding during the rearing period and found a five times higher mortality, 34% lower egg production and a 74% higher feed conversion ratio between 32 WOA and 40 WOA in comparison to restricted fed birds. To prevent these negative effects, feed restriction during the pullet and production phase is necessary with the magnitude of restriction necessary increasing with each subsequent generation (Mench, 2002; D'Eath et al., 2009).

Since the last decade, feeding programmes with feed-less days during the rearing period are prohibited in North West Europe, and most farmers apply daily feeding systems. Contrary, in many regions in the world (e.g. North America, Latin America, Asia), feeding programmes with feed-less days are still applied (EFSA, 2010; de Jong and van Emous, 2017). Different feeding programmes as skip-a-day (feed day followed by feed less day), 6/1, 5/2, or 4/3 feeding programmes (1, 2, or 3 days without feed each week and a larger portion feed on the feeding days) are applied in practice. The main reason for using feeding programmes was (and is) to obtain a high BW uniformity in feed-restricted pullets (Leeson and Summers, 2009). Feed restriction or, more precisely, energy restriction during the rearing phase has been estimated at 47% (van Emous, 2021). Contrary, previous research mentioned feed restriction levels between 67% and 75% (Savory et al., 1993; Savory and Kostal, 1996; de Jong et al., 2002). It is hypothesized by van Emous (2021) that feed restriction level is overestimated for many years due to the comparison at the same age and not at the same BW. Already in the 1990s, the age comparison was questioned by Savory et al. (1993) because of the (large) BW differences in restricted and *ad libitum* fed pullets at the same age. Nutrient and feed requirement is strongly dependent on BW since the energy requirement for maintenance of heavier birds is much higher than for smaller birds. Savory et al. (1993) found that feed allowance of restricted birds was 44% at 2 WOA and around 25% between 7 WOA and 15 WOA and 51% at 20 WOA compared to *ad libitum* fed birds. When Savory et al. (1993) used a comparison at the same BW (and thus different age), they found a feed intake of 40–48% for the restricted pullets when compared to the *ad libitum* fed pullets resulting in a 52–60% feed restriction level. It is very interesting that the mentioned nuance in the study of Savory et al. (1993) is ignored in recent studies performed on behaviour and welfare in broiler breeders.

The most severe restriction period is between 7–8 and 15–16 WOA, resulting in a feed clean-up time in daily-fed birds between 15 min and 30 min resulting in worse BW uniformity (de Jong and Jones, 2006) leading to increased competition around feeding time (de Jong et al., 2002). This is typically for pullet houses with insufficient feeder space and slow feed distribution (de Jong and Swalander, 2013). Nowadays, pullet houses are equipped with enough feeder space and fast chain feeders (up to 30 m per minute) or pan feeders that guarantee an equal feed distribution within 3 min in the entire rearing house (Aviagen, 2018; Cobb, 2018). State-of-the-art pullet houses are equipped with elevated feeding systems who are filled without bird access and/or in darkness to improve feed distribution (Abad, 2020).

Everyday fed pullets are shown to experience less stress than skip-a-day fed pullets, evidenced by lower corticosterone and insulin-like factor plasma concentrations (Ekmay et al., 2010). Additional, daily feeding is 5–10% more efficient in terms of feed intake to obtain target BW at the end of rear compared to feeding programmes (Vaughters et al., 1987; Bennett and Leeson, 1989; de Beer and Coon, 2007; Montiel, 2016). The higher efficiency in daily fed pullets is because they don't experience the cycle of storing nutrients (fat and protein) on feed days and utilizing nutrients for maintenance and growth on the feed less days (de Beer and Coon, 2007; Leeson and Summers, 2009). This physiological process of deposition and re-utilization of nutrients is not 100% efficient resulting in a decreased efficiency and higher feed requirement to obtain the same BW. The first authors compared four different feeding programmes (skip-a-day, 4/3, 5/2 and every day) and observed a 8% lower feed conversion ratio in every day fed birds, to obtain the same bodyweight with comparable BW uniformity at 21 WOA (de Beer and Coon, 2007). Improved feed efficiency and comparable uniformity in daily fed pullets was also found by Zuidhof et al. (2015). Moreover, de Beer and Coon (2007) observed a carry-over effect in the laying period with earlier peak egg production and higher settable egg production in everyday fed pullets compared to feeding programmes. This was confirmed in a recent study where every day fed birds achieved first egg at 166 d of age while the skip-a-day fed birds achieved first egg at 173 d of age (Sweeney et al., 2022). More recently a 9% decreased feed allowance during rear was found when applying everyday feeding compared to skip-a-day feeding (Montiel, 2016). In the study of Sweeney et al. (2022) pullets fed daily produce eggs with better eggshell quality and higher egg weight than pullets fed skip-a-day. The latter authors also observed deeper crypt depth and higher villus height in the jejunum at different ages in pullets fed daily suggestion improved gut health and digestibility. This improved gut health could also a part of the explanation for the higher efficiency in every fed pullets.

Montiel (2016) postulated that feeding programmes (with off-feed days) have a negative impact on innate immune responses. This negative effect on the immune system could be due to fasting during the off-feed days in feed programmes. Contrary of feeding programmes (with feed-less days), van Emous et al. (2021) applied twice a day feeding in combination with control and diluted diets. The twice-a-day fed pullets showed a lower BW CV (is a higher BW uniformity) at 10 WOA, however, no effect was found at 20 WOA. The pullets fed twice a day showed surprisingly a lower BW CV (better BW uniformity) at 10 WOA, however, no effect was found at 20 WOA. They hypothesized that the improved BW CV at 10 WOA was caused by physical factors. When feed is provided once a day, especially at the start of the rearing period between 2 weeks and 5 weeks of age, heavier pullets typically have a physical advantage (due to a bigger crop) over lighter birds (Wehner and Harrold, 1982), resulting in higher feed consumption. When feed is provided twice a day, the smaller pullets have the option to feed continuously before feed runs out. During the laying period, the pullets fed twice a day tended to have earlier onset of lay, higher total egg production at 30 WOA and better fertility.

Although body composition was not measured in the study of van Emous et al. (2021), feeding twice a day during the rearing period appeared to distribute nutrients in a more balanced manner throughout the day, as described by Backhouse and Gous (2006). This, in turn, may result in altered body composition (more fat and less protein) at the end of the rearing period and improved production performance (van Emous et al., 2015b).

Finally, it is hypothesized that feeding twice-a-day is more in line with natural feeding and activity behaviour, what is observed in the ancestors of broiler breeders. In the study of Collias and Collias (1967), the red jungle fowls were observed in natural habitat of North-Central India. They found that these birds were active during dawn (eating, drinking, and egg oviposition) and dusk (eating, drinking, and mating) and sitting in the resting tree during the hottest part of the day.

3.2 Feeding on the litter during rear

Under practical circumstances, feed is distributed to the pullets via pan feeders, chain feeders or the so-called spin-feeders (van Emous, 2022). Pullet feed is normally provided in mash or crumb form where pellets are used in combination with spin feeders (de Jong and van Emous, 2017). To obtain a good BW uniformity a good and rapid distribution of the total amount of feed over all birds is of utmost importance (Abad, 2020). It is recently postulated by van Emous (2022) that, due to the short clean-up time, the rule of thumb to feeding pullets within 3 min is not more valid. The novel guideline is: 'every individual pullet must receive her portion of feed within three seconds'. Distribution of feed on the litter (spin-feeder system) has been adopted in several European countries (e.g. Sweden, Denmark, UK and the Netherlands) as a more equal and natural feed distribution method (de Jong et al., 2005; Zuidhof et al., 2015; Montiel, 2016). Early studies in the 1970s showed that pullets fed on litter (twice a day) had less abdominal fat at 26 WOA to 27 WOA and produced heavier eggs than pullets fed via pan feeders (Chu, 1979). A longer shank length (13.1 vs. 12.5 cm) was observed at 20 WOA when male pullets were fed via a spin-feeder compared to male pullets fed via feeding pans, suggesting that these birds had better-developed legs due to scratching activity (van Emous et al., 2004).

Montiel (2016) observed an improved innate immune response in broiler breeder pullets when they were fed on the floor compared to pans. Everyday fed pullets on the litter cleared *E. coli* faster and exhibited lower *Salmonella enteritis* colonization rates, than birds that were fed skip-a-day and everyday fed in feeder pans, towards the end of the restriction period (17–22 WOA). These findings suggest that feeding pullets in pans instead on litter have a positive impact on innate immune responses. They postulated that this was caused by the constant contact of the pullets to the litter. A disadvantage of using spin-feeders during rear, appears after transferring pullets to a production house with a trough or pan feeder. This sometimes causes adaption problems to the new feeder system, which can have negative effects on flock start-up and uniformity. In general, however, a spin-feeder system results in more uniform BW in pullets, less mortality, fewer leg problems, improved litter condition and good performance results during the laying phase (Aviagen, 2018).

3.3 Twice-a-day feeding during lay

Despite additional calcium source provision (oyster shell or large limestone) and/or higher calcium levels in breeder diets, eggshell quality sometimes decreased while aging (Leeson and Summers, 2009). Providing a limited amount of feed in the morning, which is consumed within 2–6 h is the major cause of this phenomenon (Roland and Farmer, 1984; Backhouse and Gous, 2005). Birds fed in this way learn to satisfy their feed requirements in a short period of intense feeding activity (Lepkovsky et al., 1960). This feeding practice does not optimally support the breeder female nutrient requirements for eggshell development (e.g. Backhouse and Gous, 2006).

Approxmately 90% of all eggs are produced in the morning between 2 h and 7 h after lights-on (Zakaria et al., 2005; Zakaria and Omar, 2013; Akil and Zakaria, 2015). Following ovulation, albumen formation occurs over approximately 6 h primarily involving protein synthesis. This is followed by eggshell formation during the next 18 h (evening and night), predominantly necessitating calcium (Leeson and Summers, 2005). It was earlier hypothesized by Backhouse and Gous (2006) that twice a day feeding can improve the availability of nutrients for egg albumen and eggshell formation, maybe resulting in improved eggshell quality. Moreover, several studies demonstrate that broiler breeder hens fed twice a day produced more eggs compared to breeders fed once a day (Spradley et al., 2008; Taherkhani et al., 2010; Moradi et al., 2013; Soltanmoradi et al., 2014). Other authors, however, did not observe such an effect (de Avila et al., 2003; Backhouse and Gous, 2005; Londero et al., 2015, 2016).

In the last decade, more and more layers have been fed specially formulated morning and afternoon diets (so called 'split-feeding') which can help to provide the different nutritional requirements during egg albumen and eggshell formation (de Los Mozos and Sanchez, 2014; Molnár et al., 2018). More recently, this feeding management is also applied in broiler breeders (Anon, 2021; van Emous and Mens, 2021; van Emous, 2023). De Los Mozos and Sanchez (2014) investigated split-feeding in old laying hens (between 95 WOA and 98 WOA) and found higher eggshell weight, thicker eggshell and 30% less cracked and shell-less eggs. A study with breeders fed via split-feeding showed higher chick production (+1.9 DOC/hen housed), 9% lower feed costs and improved feather cover due to less pecking (Anon, 2021). In the study of van Emous and Mens (2021), it was concluded that twice a day feeding (the same diets) improved behaviour while split feeding' improved both egg production between 51 WOA and 55 WOA, and behaviour in broiler breeders. However, no effects were observed on eggshell quality and incubation traits. In the latest study of van Emous (2023), it was concluded that twice-a-day feeding decreased water intake and water-to-feed ratio, whereas split-feeding tended to improve egg production in broiler breeders between 45 WOA and 65 WOA. However, no effects were observed on eggshell quality and incubation traits, whereas the behavioural patterns of the birds fed twice a day differed considerably with potential better welfare. In a field study, a higher fertility at the end of the laying phase (60 WOA) was observed when breeders were fed via the split-feeding feeding (van Emous, unpublished data). It was hypothesized that the higher fertility was due to the higher activity at the end of the day in the houses with split-feeding (only females). In that study, 35% of the daily amount of feed was provided 4 h before lights switch off what increased the activity for at least 3 h. From mating behaviour, it is known that the majority of the matings takes place at the end of the day (Harris et al., 1980; Bilcik and Estevez, 2005). The last 3–4 h of the day are biologically and physiologically the optimal moment for egg fertilization (Løvlie and Pizzari, 2007). More activity and especially mixing of the females and males in the afternoon leads to more mating activity and thus a good fertility persistence while birds age (van Emous, 2010).

3.4 Feeding time

The majority of all eggs are produced in the morning between 2 h and 7 h after lights-on (e.g. Zakaria et al., 2005). To avoid floor and/or slat eggs, broiler breeders are fed once a day, before or after the peak of egg production (Aviagen, 2015). The discussion about the best feeding time (directly after lights on or later after the peak production) has been going on already for many years. To understand more about the effect of the two different methods of feeding, an on-farm study was carried out by van Emous (unpublished data). On a breeder production farm with four houses, one house was fed directly after lights on and three houses were fed 6 h after lights on. Bird behaviour was observed and feather cover measured weekly between 20 WOA and 30 WOA. A novel parameter was introduced to measure the activity around the feeding system: 'female on female jumps'. Due to the set-up of the on-farm trial; 1 house compared to 3 houses, the data could not be analysed. The differences in 'female on female jumps', however, were very clear with an overall 4.6 higher female on female jumps during the first 5 min of observation after the start of feeding (Fig. 2). The female on female jumps decreased as hens aged which was likely caused by the increasing daily feed allowance from 123 g to 166 g between 24 WOA and 30 WOA.

4 Water management for breeders

Due to the feed restriction practice, broiler breeder pullets and breeders are also water restricted (van Emous, 2022). Unrestricted access to water can lead to over-drinking in restricted fed broiler breeders, which may result in wet litter and, in severe cases, in polydipsia (Hocking et al., 1993). Excellent litter quality is essential for optimal foot health in pullets and breeders (Kaukonen et al., 2016) and preventing nest and egg contamination (EFSA, 2010). Therefore, breeders are limited during the rearing and production phase by a fixed water-to-feed ratio or a fixed number of hours of water during the day (de Jong et al., 2016). Under commercial conditions, a water-to-feed ratio of around 1.8 is recommended, dependent on house temperature (Hubbard, 2015; Aviagen-EPI, 2017; Cobb, 2018). When using the fixed-hours-of-water method, farmers closed the access to water approximately 2 h after clean-up time (van Emous, 2022). During hot days (>25°C), however, most breeder farmers give birds additional access to water (between 30 min and 1 h) in the late afternoon. With both methods of water restriction, daily crop assessment, a few hours after feeding on firmness of at least 20 birds per house during the entire lifecycle is recommended (Aviagen, 2018; Cobb, 2018).

Experts indicate that there may be a negative effect of water restriction on the welfare of breeders (de Jong et al., 2016). The only study carried out with water restriction in breeders, however, did not find any effects on welfare indicators (Hocking et al., 1993). In a study of van Emous et al. (2021), breeder pullets had 7 h of the day access to water and water-to-feed ratio was on average 2.3. Stocking density in the experiment was somewhat lower (8 vs. 10 birds/m^2) than what is common under commercial circumstances. The results of the study, however, showed that the higher water intake does not necessarily result in wet litter, which was confirmed in several studies of Aviagen at their research facilities in the Netherlands (J. de Lesuisse, 2021, personal communication).

The problem with over-drinking and litter problems in previous studies were probably caused by using open water systems (Kostal et al., 1992; Savory et al., 1993; Hocking et al., 1996, 2001) sometimes in combination with much longer access (14 h) to the water system due to the longer day length (Kostal et al., 1992; Savory et al., 1993). Over-drinking behaviour was not observed in the study of van Emous et al. (2021) and therefore they postulated that this was caused by the drinking system (nipple drinkers) and the shorter period (7 h) of water access. In the aforementioned study of van Emous et al. (2021), half of the pullets were fed control or 16% (on average) diluted diets. Against expectations, no difference in water intake was found between the two treatments resulting in lower water-to-feed ratio for pullets fed the diluted diets. The authors speculated that water intake might be maximized by the capacity of the crop (Wehner and Harrold, 1982) and not per se by the amount of feed. Overall, it is concluded that severe water restriction is not necessary when pullets have longer access to water and thereby showing calmer water intake behaviour. It is previously postulated that 'over-drinkers' are not hatched in that way, but raised (O. van Tuijl, 2018, personal communication).

Recently, a field experiment on a rearing farm was carried out with restricted and unrestricted access to water (van Emous, unpublished data). Pullets in two rearing houses had unrestricted access to water and pullets in another house had restricted access to water (Fig. 3). The figure shows that there was no difference in water-to-feed ratio between 5 WOA and 10 WOA between the pullets that had restricted and unrestricted access to water. Both treatments showed between 5 WOA and 10 WOA a high water-to-feed ratio and it is therefore hypothesized that this phenomenon is due to physiological or health (e.g. subclinical coccidiosis) issues.

The high water-to-feed ratio is thus independent of the water management program. In practice, however, many famers response to that higher water-to-feed ratio by shortening the period of access to water to prevent the litter quality from deteriorating. In reality, this will not work but will lead to intensively and fast drinking of the pullets due to the short period of access to water. In the on-farm study, the water-to-feed ratio was, from 11

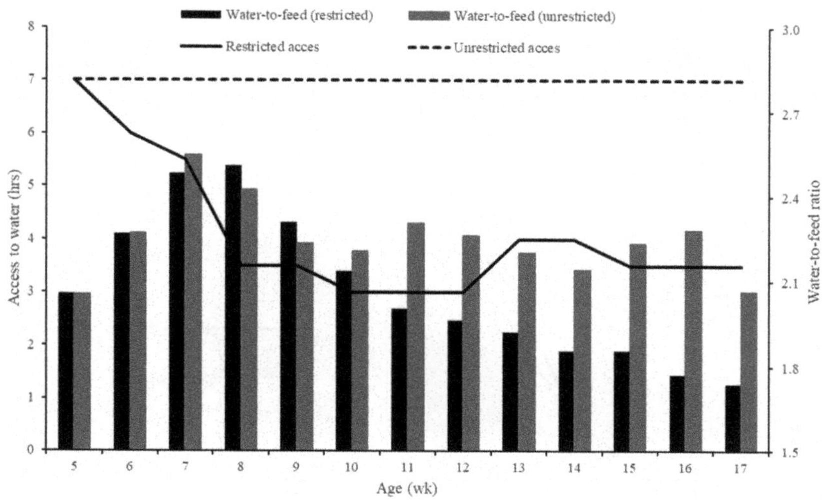

Figure 3 The effect on water-to-feed ratio in pullets with restricted or unrestricted access to water (R.A. van Emous, unpublished data).

WOA, much higher (on average 2.1) in the pullets with unrestricted access to the compared to the pullets with restricted access to water (on average 1.8). During that period, the pullets receive a relatively low amount of feed, which may cause the higher water intake due to boredom. Spoiling water could be also the case as earlier mentioned by Hocking et al. (1993) with detrimental effects on litter quality.

The trend of the cumulative water intake during the day shows that pullets that had unrestricted access to water, drank slower than pullets that had restricted access to water (Fig. 4). The unrestricted pullets reach the same total daily water intake roughly 2 h later than the restricted pullets. Because the birds had longer access to water, the total average water intake was higher which is caused to boredom and/or over-drinking.

It is important to mention that the water intake pattern changed between approximately 20 WOA and 30 WOA due to the decreased feed restriction level. Birds under practical conditions are trained during the rearing phase, to eat as quickly as possible and after feed clean-up, drinking as quickly as possible to prevent running out of feed and water, respectively. This behaviour is also observed in the production house during roughly the first five weeks, whereafter the feed allowance is increasing rapidly to approximately 90% of the maximum daily amount of feed (Aviagen-EPI, 2017). This results in a peak water intake after 1–2 h after feed distribution, which implies that the water capacity of the system must be calculated to that water peak. If not, water availability at the rear of the breeder house will be insufficient with an unequal water distribution and egg production over the flock as a result.

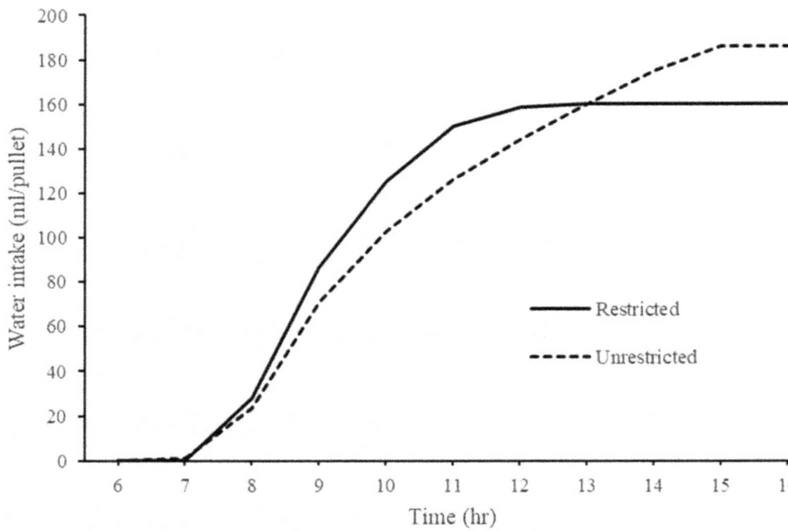

Figure 4 Cumulative water intake (ml/pullet) during the day in pullets with restricted or unrestricted access to water.

5 Male diets

Special male diets are globally still not common practice, which means that males receive more nutrients (especially calcium and protein) than they require (Leeson and Summers, 2005, 2009). Experiments, however, showed that feeding males low protein diets during the rearing and laying phase has positive effects on semen characteristics and thereby fertility. Males fed low protein diets (120 vs. 170 g/kg CP) during the rearing phase, resulted in a higher fertility in hatching eggs during the production phase (Romero-Sanchez et al, 2007). The first studies on lower dietary protein for males were performed by McDaniel and Wilson (1986), and he found that semen quality and fertility were improved. Subsequently, several papers have investigated the effect of crude protein on semen production and fertility in broiler breeder males (Wilson et al., 1987a,b; Hocking, 1989; Hocking and Bernard, 1997; Zhang et al., 1999; Borges et al., 2006; Silveira et al., 2014). In the study of Borges et al. (2006), breeder males were fed five different daily amounts of crude protein: 12.0, 14.2, 16.4, 18.6 and 20.8 g. They observed quadratic effects of daily crude protein intake on semen volume, spermatic concentration, motility, spermatozoa vigour and fertility. They concluded that 16.5 g crude protein per male per day (=−12.7% CP) meets the requirements for optimal semen quality and fertility. In a more recent on-farm study of Silveira et al. (2014) with 26 houses with each approximately having 8000 breeders (females and males) were fed a standard breeder diet (2830 kcal ME/kg and 15.0% CP) or a specially male diet (2750 kcal ME/kg and

13.5% CP). BW of males at 65 WOA in houses with the special male diet was 6% lower, and egg fertility and hatchability was 1.5 and 4.2 percentage points higher, respectively.

6 Conclusion and future trends

Poultry meat is the most important protein source in human diet, and global demand increased annually. Broiler breeders are the parents of broiler and has a substantial impact on the poultry meat chain producing first class hatching eggs with healthy chicks. This is becoming increasingly challenging, because the continuing increase in the genetic growth potential of the offspring and the parent stock as well. State-of-the-art management for housing, equipment and health is necessary to optimize reproduction of breeders for a maximum number of healthy offspring. Besides these mentioned factors, diet composition and feeding strategies are important factors to reach this goal. *Ad libitum* fed broiler breeders support fast growth and causes too high BWs with detrimental effects on health and reproduction. Therefore, broiler breeders are feed restricted which in turn causes abnormal behaviour towards equipment or other birds, which can be indicative of frustration, boredom and hunger. To minimize the negative effects of feed restriction on pullet behaviour, different feeding strategies as higher BW profile, diluted diets, twice a day feeding and scattering feeding are studied. The majority of these strategies, or combinations of more strategies, had only small effects on abnormal behaviour and physiological parameters. Body composition can be affected by a lower protein and higher energy intake during the rearing phase and plays a major role in sexual maturation, egg production, egg composition and laying persistency. Feeding low daily CP intake during, especially during the second phase of lay, improved fertility and hatchability. Several studies found an effect of breeder dietary energy-to-protein ratio on progeny performance; however, the results are inconclusive. Most of the studies on the effects of vitamins and trace minerals in breeder nutrition are focussed on the offspring. Vitamins D3 and E play a major role in DOC quality, health and immunity where selenium is important as antioxidant protection of the embryo increasing fertility and hatchability. Due to feed restriction, breeder pullets are often fed via feeding programmes with feed-less days. More and more evidence is available that every day feeding has different benefits on higher feed efficiency (5–10%), improved gut health, improved immunity and better BW uniformity. Against expectations, feeding breeder pullets twice-a-day can improve BW uniformity, however a novel system for an optimal feed distribution has to be developed. Feeding on the litter is often applied in rearing farms in several EU countries due to longer clean-up time, improved litter, fewer leg problems and higher BW uniformity. Feeding twice-a-day or split-feeding (especially morning and afternoon diet) during the laying phase can improve egg production, behaviour and egg fertility. Water is the most important nutrient, and unrestricted access to water as an important factor for behaviour, welfare, reproduction, and health. Especially male diets with lower levels crude protein and calcium can improve fertility, hatchability and health.

Future trends in broiler breeder nutrition:

- More research is necessary to diminish the effects of feed restriction, and the most promising directions to a solution should be the combination of different feeding methods and strategies. Supplementing novel ingredients can be a solution to increase the satiety level due to the higher water capacity building.
- An innovative feeding system must be developed to feed pullets twice-a-day with an even distribution within 3 s over the total flock.
- Transgenerational effects are an important field of study in broiler breeders, because a small number of birds have a major effect on billions broilers.
- Unrestricted access to water must focus on the rearing phase to train the birds to drink slower and to prevent worse litter due to higher water intake and spoiling of the water.

7 Where to look for further information

The following books contain interesting scientific and practical applicable knowledge about broiler breeder nutrition:

- *Feeding Programs for Broiler Breeders in Commercial Poultry Nutrition*, 3rd edition by S. Leeson and J.D. Summers.
- *Broiler Breeder Production* by S. Leeson and J.D. Summers.
- *Breeder Management and Nutrition. Moving the Industry Forward.* Novus International Inc. by J. Buyse, S. Duroux, and S. Peris.
- *Breeder Signals, a Practical Guide to Optimal Management of Broiler Breeders* by R.A. van Emous, J. Holleman, and T. van Schie.

8 References

Abad, J. C. (2020). Management of males in broiler breeder operations. Available at: https://avicultura.info/en/management-of-males-in-broiler-breeder-operations/.

Akil, R. and Zakaria, A. H. (2015). Egg laying characteristics, egg weight, embryo development, hatching weight and post-hatch growth in relation to oviposition time of broiler breeders, *Anim. Reprod. Sci.* 156: 103–110.

Anon. (2021). Innovative split-feeding system for broiler breeders, *Int. Hatchery Pract.* 35: 7–9.

Arrazola, A., Mosco, E., Widowski, T. M., Guerin, M. T., Kiarie, E. G. and Torrey, S. (2019). The effect of alternative feeding strategies for broiler breeder pullets: 1. Welfare and performance during rearing, *Poult. Sci.* 98(9): 3377–3390.

Atencio, A., Edwards, H. M. Jr. and Pesti, G. M. (2005a). Effects of vitamin D3 dietary supplementation of broiler breeder hens on the performance and bone abnormalities of the offspring, *Poult. Sci.* 84(7): 1058–1068.

Atencio, A., Pesti, G. M. and Edwards, H. M. Jr. (2005b). Twenty-five hydroxycholecalciferol as a cholecalciferol substitute in broiler breeder hen diets and its effect on the performance and general health of the offspring, *Poult. Sci.* 84(8): 1277–1285.

Augère-Granier, M. L. (2019). The EU poultry meat and egg sector. Available at: https://www.europarl.europa.eu/thinktank/en/document.html?reference=EPRS_IDA(2019)644195.

Aviagen (2015). Best practice in the breeder house. Preventing floor eggs. Available at: http://en.aviagen.com/assets/Tech_Center/Ross_Tech_Articles/Ross-BestPractice-Floor-Eggs-2015-EN.pdf.

Aviagen (2018). Ross parent stock management handbook. Available at: https://en.aviagen.com/assets/Tech_Center/Ross_PS/RossPSHandBook2018.pdf.

Aviagen-EPI (2017). Management guide for Ross broiler breeders (in Dutch). Roermond, The Netherlands.

Backhouse, D. and Gous, R. M. (2005). The effect of feeding time on shell quality and oviposition time in broiler breeders, *Br. Poult. Sci.* 46(2): 255–259.

Backhouse, D. and Gous, R. M. (2006). Responses of adult broiler breeders to feeding time, *Worlds Poult. Sci. J.* 62(2): 269–281.

Bédécarrats, G. Y., Baxter, M. and Sparling, B. (2016). An updated model to describe the neuroendocrine control of reproduction in chickens, *Gen. Comp. Endocrinol.* 227: 58–63.

Bennett, C. D. and Leeson, S. (1989). Research note: Growth of broiler breeder pullets with skip-a-day versus daily feeding, *Poult. Sci.* 68(6): 836–838.

Berghof, T. V. L., Parmentier, H. K. and Lammers, A. (2013). Transgenerational epigenetic effects on innate immunity in broilers: An underestimated field to be explored?, *Poult. Sci.* 92(11): 2904–2913.

Bilcik, B. and Estevez, I. (2005). Impact of male-male competition and morphological traits on mating strategies and reproductive success in broiler breeders, *Appl. Anim. Behav. Sci.* 92(4): 307–323.

Boa-Amponsem, K., Price, S. E. H., Dunnington, E. A. and Siegel, P. B. (2001). Effect of route of inoculation on humoral immune response of White Leghorn chickens selected for high or low antibody response to sheep red blood cells, *Poult. Sci.* 80(8): 1073–1078.

Borges, C. A. Q., Rostagno, H. S., da Silva, J. H. V., Albino, L. F. T., Torres, C. A. A., Filho, J. J. and Ribeiro, M. L. G. (2006). Protein requirements and carcass composition of male broiler breeders from 27 to 61 weeks of age (in Portuguese), *Rev. Bras. Zootec.* 35(5): 1971–1977.

Calini, F. and Sirri, F. (2007). Breeder nutrition and offspring performance, *Rev. Bras. Cienc. Avic.* 9(2): 77–83.

Chang, A., Halley, J. and Silva, M. (2016). Can feeding the broiler breeder improve chick quality and offspring performance?, *Anim. Prod. Sci.* 56(8): 1254–1262.

Chu, S. S. W. (1979). The effect of feeding management during the growing and laying periods on productivity and the characteristics of adipose tissue of the broiler breeders. Master thesis, University of British Columbia, Canada.

Cobb (2018). Cobb breeder management guide. Cobb-Vantress, Siloam Springs, AR.
Collias, N. E. and Collias, E. C. (1967). A field study of the Red jungle fowl in North-Central India, *Condor* 69(4): 360–386.
Couloigner, F., Rouffineau, F., Geraert, P. A. and Mercier, Y. (2015). Dietary selenium supplementation of breeders influences growth and selenium status of offspring, *Australian Poultry Science Symposium* 26: 80–83.
de Avila, V. S., Penz Jr., A. M., de Brum, P. A. R., Rosa, P. S., Guidoni, A. L. and de Figueiredo, É. A. P. (2003). Performance of female broiler breeders submitted to different feeding schedules, *Rev. Bras. Cienc. Avic.* 5(3): 197–201.
de Beer, M. (2009). Current approaches to feeding broiler breeders. In *Proceedings of the 17th European Symposium of Poultry Nutrition, Edinburgh, Scotland, World's Poult. Sci. Assoc.*, Attleborough, pp. 104–114.
de Beer, M. and Coon, C. N. (2007). The effect of different feed restriction programs on reproductive performance, efficiency, frame size, and uniformity in broiler breeder hens, *Poult. Sci.* 86(9): 1927–1939.
D'Eath, R. B., Tolkamp, B. J., Kyriazakis, I. and Lawrence, A. B. (2009). 'Freedom from hunger' and preventing obesity: The animal welfare implications of reducing foot quantity or food quality, *Anim. Behav.* 77(2): 275–288.
Decuypere, E., Hocking, P. M., Tona, K., Onagbesan, O., Bruggeman, V., Jones, E. K. M., Cassy, S., Rideau, N., Metayer, S., Jego, Y., Putterflam, J., Tesseraud, S., Collin, A., Duclos, M., Trevidy, J. J. and Williams, J. (2006). Broiler breeder paradox: A project report, *Worlds Poult. Sci. J.* 62(3): 443–453.
Decuypere, E., Bruggeman, V., Everaert, N., Li, Y., Boonen, R., De Tavernier, J., Janssens, S. and Buys, N. (2010). The Broiler Breeder Paradox: Ethical, genetic and physiological perspectives, and suggestions for solutions, *Br. Poult. Sci.* 51(5): 569–579.
de Jong, I. C. and Jones, B. (2006). Feed restriction and welfare in domestic birds. In V. Bels (Ed.), *Feeding in Domestic Vertebrates. From Structure to Behavior*. CABI Publishing, Wallingford, pp. 120–135.
de Jong, I. C. and Swalander, M. (2013). Housing and management of broiler breeders and turkey breeders. In Sandilands, V. and Hocking, P. M. (Eds.), *Alternative Systems for Poultry: Health, Welfare and Productivity*. CABI Publishing, Wallingford, pp. 225–249.
de Los Mozos, J. and Sanchez, F. (2014). Split feeding meets layers' specific needs. Available at: http://www.thepoultrysite.com/articles/3091/split-feeding-system-for-more-sustainable-egg-production.
de Jong, I. C. and van Emous, R. A. (2017). Broiler breeding flocks: Management and animal welfare. In Applegate, T. (Ed.), *Achieving Sustainable Production of Poultry Meat*. Burleigh Dodds Science Publishing Limited, Cambrigde, UK, pp. 211–230.
de Jong, I. C., van Voorst, S., Ehlhardt, D. A. and Blokhuis, H. J. (2002). Effects of restricted feeding on physiological stress parameters in growing broiler breeders, *Br. Poult. Sci.* 43(2): 157–168.
de Jong, I. C., Fillerup, M. and Blokhuis, H. J. (2005). Effect of scattered feeding and feeding twice a day during rearing on parameters of hunger and frustration in broiler breeders, *Appl. Anim. Behav. Sci.* 92(1–2): 61–76.
de Jong, I. C., van Harn, J., Koene, P., Ellen, H., van Emous, R. A., Rommers, J. M. and van den Brand, H. (2016). Risk assessment of water provision for broilers and broiler breeders (in Dutch). Wageningen Livestock Research, report 989, Wageningen, the Netherlands.

de los Mozos, J., García-Ruiz, A. I., den Hartog, L. A. and Villamide, M. J. (2017). Growth curve and diet density affect eating motivation, behavior, and body composition of broiler breeders during rearing, *Poult. Sci.* 96(8): 2708–2717.

EFSA (2010). Scientific Opinion on welfare aspects of the management and housing of the grandparent and parent stocks raised and kept for breeding purposes, *EFSA J.* 8: 81.

Eitan, Y., Lipkin, E. and Soller, M. (2014). Body composition and reproductive performance at entry into lay of anno 1980 versus anno 2000 broiler breeder females under fast and slow release from feed restriction, *Poult. Sci.* 93(5): 1227–1235.

Ekmay, R. D., de Beer, M., Rosebrough, R. W., Richards, M. P., McMurtry, J. P. and Coon, C. N. (2010). The role of feeding regimens in regulating metabolism of sexually mature broiler breeders, *Poult. Sci.* 89(6): 1171–1181.

Ekmay, R. D., de Beer, M., Mei, S. J., Manangi, M. and Coon, C. N. (2013). Amino acid requirements of broiler breeders at peak production for egg mass, body weight, and fertility, *Poult. Sci.* 92(4): 992–1006.

Enting, H., Boersma, W. J. A., Cornelissen, J. B., van Winden, S. C. L., Verstegen, M. W. A. and van der Aar, P. J. (2007). The effect of low-density broiler breeder diets on performance and immune status of their offspring, *Poult. Sci.* 86(2): 282–290.

Hadinia, S. H., Carneiro, P. R. O., Fitzsimmons, C. J., Bédécarrats, G. Y. and Zuidhof, M. J. (2020). Post-photostimulation energy intake accelerated pubertal development in broiler breeder pullets, *Poult. Sci.* 99(4): 2215–2229.

Harris, G. C. Jr., Benson, J., Sellers, R. and Cole, T. A. (1980). The mating activity of broiler breeder cockerels, *Arkansas Farm Res.* 29: 15.

Havenstein, G. B., Ferket, P. R. and Qureshi, M. A. (2003a). Growth, livability, and feed conversion of 1957 versus 2001 broilers when fed representative 1957 and 2001 broiler diets, *Poult. Sci.* 82(10): 1500–1508.

Havenstein, G. B., Ferket, P. R. and Qureshi, M. A. (2003b). Carcass composition and yield of 1957 versus 2001 broilers when fed representative 1957 and 2001 broiler diets, *Poult. Sci.* 82(10): 1509–1518.

Heck, A., Onagbesan, O., Tona, K., Metayer, S., Puterflam, J., Jego, Y., Trevidy, J. J., Decuypere, E., Williams, J., Picard, M. and Bruggeman, V. (2004). Effects of ad libitum feeding on performance of different strains of broiler breeders, *Br. Poult. Sci.* 45(5): 695–703.

Heijmans, J., Duijster, M., Gerrits, W. J. J., Kemp, B., Kwakkel, R. P. and van den Brand, H. (2021). Impact of growth curve and dietary energy-to-protein ratio on productive performance of broiler breeders, *Poult. Sci.* 100(7): 101131.

Heijmans, J., Duijster, M., Gerrits, W. J. J., Kemp, B., Kwakkel, R. P. and van den Brand, H. (2022). Impact of growth curve and dietary energy-to-protein ratio of broiler breeders on offspring quality and performance, *Poult. Sci.* 101(11): 102071.

Hocking, P. M. (1989). Effect of dietary crude protein concentration on semen yield and quality in male broiler breeder fowls, *Br. Poult. Sci.* 30(4): 935–945.

Hocking, P. M. (2007). Optimum feed composition of broiler breeder diets to maximise offspring performance. In *Proceedings of the 16th European Symposium of Poultry Nutrition, World Poultry Science Association*, Strasbourg, pp. 101–108.

Hocking, P. M. and Bernard, R. (1997). Effects of dietary crude protein content and food intake on the production of semen in two lines of broiler breeder males, *Br. Poult. Sci.* 38(2): 199–202.

Hocking, P. M., Maxwell, M. H. and Mitchell, M. A. (1993). Welfare assessment of broiler breeder and layer females subjected to food restriction and limited access to water during rearing, *Br. Poult. Sci.* 34(3): 443–458.

Hocking, P. M., Maxwell, M. H. and Mitchell, M. A. (1996). Relationships between the degree of food restriction and welfare indices in broiler breeder females, *Br. Poult. Sci.* 37(2): 263–278.

Hocking, P. M., Maxwell, M. H., Robertson, G. W. and Mitchell, M. A. (2001). Welfare assessment of modified rearing programs for broiler breeders, *Br. Poult. Sci.* 42(4): 424–432.

Hubbard (2015). Breeder management manual. Quintin, France.

Jlali, M., Briens, M., Rouffineau, F., Mercerand, F., Geraert, P. A. and Mercier, Y. (2013). Effect of 2-hydroxy-4-methylselenobutanoic acid as a dietary selenium supplement to improve the selenium concentration of table eggs, *J. Anim. Sci.* 91(4): 1745–1752.

Joseph, N. S., Dulaney, A. A. J., Robinson, F. E., Renema, R. A. and Zuidhof, M. J. (2002). The effects of age at photostimulation and dietary protein intake on reproductive efficiency in three strains of broiler breeders varying in breast yield, *Poult. Sci.* 81(5): 597–607.

Katanbaf, M. N., Dunnington, E. A. and Siegel, P. B. (1989). Restricted feeding in early and late-feathering chickens. 1. Growth and physiological responses, *Poult. Sci.* 68(3): 344–351.

Kaukonen, E., Norring, M. and Valros, A. (2016). Effect of litter quality on foot pad dermatitis, hock burns and breast blisters in broiler breeders during the production period, *Avian Pathol.* 45(6): 667–673.

Kemp, C., Wylie, L. and Fisher, C. (2001). Broiler breeder nutrition, nutrient transfer and broiler performance. In *13th European Symposium on Poultry Nutrition, World Poultry Science Association*, Blankenberge, Belgium, pp. 61–67.

Kidd, M. T. (2003). A treatise on chicken dam nutrition that impacts on progeny, *Worlds Poult. Sci. J.* 59(4): 475–494.

Kostal, L., Savory, C. J. and Hughes, B. O. (1992). Diurnal and individual variation in behaviour of restricted-fed broiler breeders, *Appl. Anim. Behav. Sci.* 32(4): 361–374.

Leeson, S. and Summers, J. D. (2005). *Commercial Poultry Nutrition*. University Books, Guelph.

Leeson, S. and Summers, J. D. (2009). *Broiler Breeder Production*. University Books, Guelph.

Lepkovsky, S., Chari-Bitron, A., Lemmon, R. M., Ostwald, R. C. and Dimick, M. K. (1960). Metabolic and anatomic adaptation in chickens "trained" to eat their daily food in two hours, *Poult. Sci.* 39(2): 385–389.

Lesuisse, J., Li, C., Schallier, S., Leblois, J., Everaert, N. and Buyse, J. (2017). Feeding broiler breeders a reduced balanced protein diet during the rearing and laying period impairs reproductive performance but enhances broiler offspring performance, *Poult. Sci.* 96(11): 3949–3959.

Lesuisse, J., Li, C., Schallier, S., Clímaco, W. L. S., Bautil, A., Everaert, N. and Buyse, J. (2018). Multigenerational effects of a reduced balanced protein diet during the rearing and laying period of broiler breeders. 1. Performance of the F1 breeder generation, *Poult. Sci.* 97(5): 1651–1665.

Londero, A., Rosa, A. P., Giacomini, C. B. S., Vivas, C. E. B., Orso, C., de Freitas, H. M., Gressler, L. T. and Vargas, A. C. (2015). Effect of different feeding schedules on reproductive parameters and egg quality of broiler breeders, *Anim. Feed Sci. Technol.* 210: 165–171.

Londero, A., Rosa, A. P., Vivas, C. E. B., Orso, C., Fernandes, M. O., Paixão, S. J., Giacomini, C. B. S., Andrade, C. M. and Palma, H. E. (2016). The effect of different feeding schedules on egg quality, blood and bone parameters in broiler breeders, *Anim. Reprod.* 13(1): 14–20.

Lopez, G. and Leeson, S. (1995a). Response of broiler breeders to low-protein diets. 1. Adult breeder performance, *Poult. Sci.* 74(4): 685–695.

Lopez, G. and Leeson, S. (1995b). Response of broiler breeders to low-protein diets. 2. Offspring performance, *Poult. Sci.* 74(4): 696–701.

Løvlie, H. and Pizzari, T. (2007). Sex in the morning or in the evening? Females adjust daily mating patterns to the intensity of sexual harassment, *Am. Nat.* 170(1): E1–13.

McDaniel, G. R. and Wilson, J. C. (1986). Feeding breeder males. In *Proceedings of the Carolina Nutrition Conference*, Charlotte, NC, pp. 66–74.

Mench, J. A. (2002). Broiler breeders: Feed restriction and welfare, *Worlds Poult. Sci. J.* 58(1): 23–29.

Mens, A. J. W., de Jong, I. C., van Riel, J. W., Gunnink, H., van Hattum, T. and van Emous, R. A. (2022). Diet dilution and feeding frequency have only minor effects on the behaviour of broiler breeder pullets, *Appl. Anim. Behav. Sci.* 253. https://doi.org/10.1016/j.applanim.2022.105678.

Mohiti-Asli, M., Shivazad, M., Zaghari, M., Rezaian, M., Aminzadeh, S. and Mateos, G. G. (2012). Effects of feeding regimen, fiber inclusion, and crude protein content of the diet on performance and egg quality and hatchability of eggs of broiler breeder hens, *Poult. Sci.* 91(12): 3097–3106.

Molenaar, R., Hulet, R., Meijerhof , R., Maatjens, C. M., Kemp, B. and van den Brand, H. (2011). High eggshell temperatures during incubation decrease growth performance and increase the incidence of ascites in broiler chickens, *Poult. Sci.* 90(3): 624–632.

Molnár, A., Kempen, I., Sleeckx, N., Zoons, J., Maertens, L., Ampe, B., Buyse, J. and Delezie, E. (2018). Effects of split feeding on performance, egg quality, and bone strength in brown laying hens in aviary system, *J. Appl. Poult. Res.* 27(3): 401–415.

Montiel, E. R. (2016). Influence of feeding programs on innate and adaptive immunity in broiler breeders. Thesis University of Georgia, Athens, GA.

Moradi, S., Zaghari, M., Shivazad, M., Osfoori, R. and Mardi, M. (2013). The effect of increasing feeding frequency on performance, plasma hormones and metabolites, and hepatic lipid metabolism of broiler breeder hens, *Poult. Sci.* 92(5): 1227–1237.

Moraes, T. G. V., Pishnamazi, A., Mba, E. T., Wenger, I. I., Renema, R. A. and Zuidhof, M. J. (2014). Effect of maternal dietary energy and protein on live performance and yield dynamics of broiler progeny from young breeders, *Poult. Sci.* 93(11): 2818–2826.

Moraes, T. G. V., Pishnamazi, A., Wenger, I. I., Renema, R. A. and Zuidhof, M. J. (2019). Energy and protein dilution in broiler breeder pullet diets reduced offspring body weight and yield, *Poult. Sci.* 98(6): 2555–2561.

Morrissey, K. L. H., Widowski, T., Leeson, S., Sandilands, V., Arnone, A. and Torrey, S. (2014). The effect of dietary alterations during rearing on growth, productivity, and behaviour in broiler breeder females, *Poult. Sci.* 93(2): 285–295.

Naber, E. C. (1993). Modifying vitamin composition of eggs: A review, *J. Appl. Poult. Res.* 2(4): 385–393.

Pappas, A. C., Karadas, F., Surai, P. F. and Speake, B. K. (2005). The selenium intake of the female chicken influences the selenium status of her offspring, *Comp. Biochem. Physiol., B Biochem. Mol. Biol.* 142(4): 465–474.

Pappas, A. C., Acamovic, T., Surai, P. F. and McDevitt, R. M. (2006). Maternal organo-selenium compounds and polyunsaturated fatty acids affect offspring performance

and levels of selenium and docosahexaenoic acid in the chick tissues, *Poult. Sci.* 85(9): 1610–1620.
Pearson, R. A. and Herron, K. M. (1981). Effects of energy and protein allowances during lay on the reproductive performance of broiler breeder hens, *Br. Poult. Sci.* 22(3): 227–239.
Pearson, R. A. and Herron, K. M. (1982). Effects of maternal energy and protein intakes on the incidence of malformations and malpositions of the embryo and time of death during incubation, *Br. Poult. Sci.* 23(1): 71–77.
Rajashree, K., Muthukumar, T. and Karthikeyan, N. (2014). Comparative study of the effects of organic selenium on hen performance and productivity of broiler breeders, *Br. Poult. Sci.* 55(3): 367–374.
Rao, K., Xie, J., Yang, X., Chen, L., Grossmann, R. and Zhao, R. (2009). Maternal low-protein diet programmes offspring growth in association with alterations in yolk leptin deposition and gene expression in yolk-sac membrane, hypothalamus and muscle of developing Langshan chicken embryos, *Br. J. Nutr.* 102(6): 848–857.
Reijrink, I. A. M., Berghmans, D., Meijerhof, R., Kemp, B. and van den Brand, H. (2010). Influence of egg storage time and preincubation warming profile on embryonic development, hatchability, and chick quality, *Poult. Sci.* 89(6): 1225–1238.
Roland, D. A. and Farmer, M. (1984). Egg shell quality II: Importance of time of calcium intake with emphasis on broiler breeders, *Worlds Poult. Sci. J.* 40(3): 255–260.
Romero-Sanchez, H., Plumstead, P. W. and Brake, J. (2007). Feeding broiler breeder males. 1. Effect of feeding program and dietary crude protein during rearing on body weight and fertility of broiler breeder males, *Poult. Sci.* 86(1): 168–174.
Roseboom, T. J., Painter, R. C., van Abeelen, A. F. M., Veenendaal, M. V. E. and de Rooij, S. R. (2011). Hungry in the womb: What are the consequences? Lessons from the Dutch famine, *Maturitas* 70(2): 141–145.
Salas, C., Ekmay, R. D., England, J. A., Cerrate, S. and Coon, C. N. (2017). Mechanisms of lipid mobilization towards egg formation in broiler breeder hens using stable isotopes, *Poult. Sci.* 96(2): 383–387.
Salas, C., Ekmay, R. D., England, J., Cerrate, S. and Coon, C. N. (2019). Effect of body weight and energy intake on body composition analysis of broiler breeder hens, *Poult. Sci.* 98(2): 796–802.
Sandilands, V., Tolkamp, B. J. and Kyriazakis, I. (2005). Behaviour of food restricted broilers during rearing and lay – Effects of an alternative feeding method, *Physiol. Behav.* 85(2): 115–123.
Sandilands, V., Tolkamp, B. J., Savory, C. J. and Kyriazakis, I. (2006). Behaviour and welfare of broiler breeders fed qualitatively restricted diets during rearing: Are there viable alternatives to quantitative restriction?, *Appl. Anim. Behav. Sci.* 96(1–2): 53–67.
Savory, C. J. and Maros, K. (1993). Influence of degree of food restriction, age and time of day on behaviour of broiler breeder chickens, *Behav. Processes* 29(3): 179–189.
Savory, C. J. and Kostal, L. (1996). Temporal patterning of oral stereotypies in restricted-fed fowls: 1. investigations with a single daily meal, *Inter. J. Comp. Psychol.* 9(3): 117–139.
Savory, C. J. and Lariviere, J. M. (2000). Effects of qualitative and quantitative food restriction treatments on feeding motivational state and general activity level of growing broiler breeders, *Appl. Anim. Behav. Sci.* 69(2): 135–147.
Silveira, M. M., de Freitas, A. G., Moraes, C. A., Gomes, F. S., Litz, F. H., Martins, J. M. S., Fagundes, N. S. and Fernandes, E. A. (2014). Feeding management strategy for male

broiler breeders and its effects on body weight, hatchability and fertility, *Brazil. J. Poult. Sci.* 16(4): 397–402.
Soltanmoradi, M. G., Seidavi, A., Dadashbeiki, M., Laudadio, V., Centoducati, G. and Tufarelli, V. (2014). Influence of feeding frequency and timetable on egg parameters and reproductive performance in broiler breeder hens, *Avian Biol. Res.* 7(3): 153–159.
Spradley, J. M., Freeman, M. E., Wilson, J. L. and Davis, A. J. (2008). The influence of a twice-a-day feeding regimen after photostimulation on the reproductive performance of broiler breeder hens, *Poult. Sci.* 87(3): 561–568.
Sunde, M., Turk, C. M. and DeLuca, H. F. (1978). The essentiality of vitamin D metabolites for chick development, *Sci.* 200(4345): 1067–1069.
Surai, P. F. (2000). Effect of selenium and vitamin E content of the maternal diet on the antioxidant system of the yolk and the developing chick, *Br. Poult. Sci.* 41(2): 235–243.
Surai, P. F. and Fisinin, V. I. (2014). Selenium in poultry breeder nutrition: An update, *Anim. Feed Sci. Technol.* 191: 1–15.
Sweeney, K. M., Aranibar, C. D., Kim, W. K., Williams, S. M., Avila, L. P., Starkey, J. D., Starkey, C. W. and Wilson, J. L. (2022). Impact of every-day versus skip-a-day feeding of broiler breeder pullets during rearing on body weight uniformity and reproductive performance, *Poult. Sci.* 101(8): 101959. https://doi.org/10.1016/j.psj.2022.101959.
Tahamtani, F. M., Moradi, H. and Riber, A. B. (2020). Effect of qualitative feed restriction in broiler breeder pullets on stress and clinical welfare indicators, *Front. Vet. Sci.* 7: 316.
Taherkhani, R., Zaghari, M., Shivazad, M. and Zare Shahneh, A. (2010). A twice-a-day feeding regimen optimizes performance in broiler breeder hens, *Poult. Sci.* 89(8): 1692–1702.
Tolkamp, B. J., Sandilands, V. and Kyriazakis, I. (2005). Effects of qualitative feed restriction during rearing on the performance of broiler breeders during rearing and lay, *Poult. Sci.* 84(8): 1286–1293.
Urso, U. R. A., Dahlke, F., Mairoka, A., Bueno, I. J. M., Schneider, A. F., Surek, D. and Rocha, C. (2015). Vitamin E and selenium in broiler breeder diets: Effect on live performance, hatching process, and chick quality, *Poult. Sci.* 94(5): 976–983.
USDA (2021). Livestock and poultry: World markets and trade, United States Department of Agriculture Foreign Agricultural Service, January 12, 2021. Available at: https://apps.fas.usda.gov/psdonline/circulars/livestock_poultry.pdf.
Vaughters, P. D., Pesti, G. M. and Howarth, B. J. (1987). Effects of feed consumption and feeding schedule on growth and development of broiler breeder males, *Poult. Sci.* 66(1): 134–146.
van der Klein, S. A. S., Bédécarrats, G. Y. and Zuidhof, M. J. (2018). The effect of rearing photoperiod on broiler breeder reproductive performance depended on body weight, *Poult. Sci.* 97(9): 3286–3294.
van Emous, R. A. (2010). Innovative housing in broiler breeder production. In *Proceedings of the 2nd International Symposium: Highlights in Nutrition and Welfare in Poultry Production, Wageningen, the Netherlands*. Animal Science Group, Lelystad, The Netherlands, pp. 37–44.
van Emous, R. A. (2015). Body composition and reproduction in broiler breeders: Impact of feeding strategies. PhD Thesis. Wageningen University, Wageningen.
van Emous, R. A. (2021). Literature study controlled fed broiler breeders (in Dutch), Report 1291. Wageningen Livestock Research, Wageningen, the Netherlands.

van Emous, R. A. (2022). Housing, equipment and management of broiler breeders. In Buyse, J., Duroux, S. and Peris, S. (Eds.), *Breeder Management and Nutrition: Moving the Industry Forward*. Novus International Inc., Saint Charles, pp. 11–139.

van Emous, R. A. (2023). Effects of feeding strategies during lay on broiler breeder production performance, eggshell quality, incubation traits, and behavior, *Poult. Sci.* 102(6): 102630.

van Emous, R. A. and Mens, A. J. W. (2021). Effects of twice-a-day feeding and split feeding during lay on broiler breeder production performance, eggshell quality, incubation traits, and behavior, *Poult. Sci.* 100(11): 101419.

van Emous, R. A., Lourens, A. and van Harn, J. (2004). Vitality of broiler breeders and ammonia measurements (in Dutch with English summary), Report Poultry 13. Animal Sciences Group, Praktijkonderzoek Veehouderij, Lelystad.

van Emous, R. A., Kwakkel, R. P., van Krimpen, M. M. and Hendriks, W. H. (2013). Effects of growth patterns and dietary crude protein levels during the rearing on body composition and performance in broiler breeder females during the rearing and laying period, *Poult. Sci.* 92(8): 2091–2100.

van Emous, R. A., Kwakkel, R. P., van Krimpen, M. M. and Hendriks, W. H. (2014). Effects of growth pattern and dietary protein level during rearing on feed intake, eating time, eating rate, behavior, plasma corticosterone concentration, and feather cover in broiler breeder females during the rearing and laying period, *Appl. Anim. Behav. Sci.* 150: 44–54.

van Emous, R. A., Kwakkel, R. P., van Krimpen, M. M. and Hendriks, W. H. (2015a). Effects of different dietary protein levels during rearing and different dietary energy levels during lay on behaviour and feather cover in broiler breeder females, *Appl. Anim. Behav. Sci.* 168: 45–55.

van Emous, R. A., Kwakkel, R. P., van Krimpen, M. M. and Hendriks, W. H. (2015b). Effects of dietary protein levels during rearing and different dietary energy levels during lay on body composition and reproduction in broiler breeder females, *Poult. Sci.* 94(5): 1030–1042.

van Emous, R. A., Kwakkel, R. P., van Krimpen, M. M., van den Brand, H. and Hendriks, W. H. (2015c). Effects of growth patterns and dietary protein levels during rearing of broiler breeders on fertility, hatchability, embryonic mortality, and offspring performance, *Poult. Sci.* 94(4): 681–691.

van Emous, R. A., de La Cruz, C. E. and Naranjo, V. D. (2018). Effects of dietary protein level and age at photo stimulation on reproduction traits of broiler breeders and progeny performance, *Poult. Sci.* 97(6): 1968–1979.

van Emous, R. A., Mens, A. J. W. and Winkel, A. (2021). Effects of diet density and feeding frequency during the rearing period on broiler breeder performance, *Br. Poult. Sci.* 62(5): 686–694.

Virden, W. S., Yeatman, J. B., Barber, S. J., Zumwalt, C. D., Ward, T. L., Johnson, A. B. and Kidd, M. T. (2003). Hen mineral nutrition impacts offspring livability, *J. Appl. Poult. Res.* 12(4): 411–416.

Virden, W. S., Yeatman, J. B., Barber, S. J., Willeford, K. O., Ward, T. L., Fakle, T. M., Wideman, R. F. Jr. and Kidd, M. T. (2004). Immune system and cardiac functions of offspring chicks from dams fed diets differing in zinc and manganese level and source, *Poult. Sci.* 83(3): 344–351.

Wehner, G. R. and Harrold, R. L. (1982). Crop volume of chickens as affected by body size, sex, and breed, *Poult. Sci.* 61(3): 598–600.

Whitehead, C. C., Pearson, R. A. and Herron, K. M. (1985). Protein requirements of broiler breeders fed diets of different protein content and effect of insufficient protein on the viability of progeny, *Br. Poult. Sci.* 26(1): 73–82.

Wilson, J. L., McDaniel, C. D. and Sutton, C. D. (1987a). Dietary protein levels for broiler breeder males, *Poult. Sci.* 66(2): 237–242.

Wilson, J. L., McDaniel, G. R., Sutton, C. D. and Renden, J. A. (1987b). Semen and carcass evaluation of broiler breeder males fed low protein diets, *Poult. Sci.* 66(9): 1535–1540.

Zakaria, A. H., Plumstead, P. W., Romero-Sanchez, H., Leksrisompong, N., Osborne, J. and Brake, J. (2005). Oviposition pattern, egg weight, fertility, and hatchability of young and old broiler breeders, *Poult. Sci.* 84(9): 1505–1509.

Zakaria, A. H. and Omar, O. H. (2013). Egg laying pattern, egg weight, body weight at hatch and sex ratio bias relative to oviposition time of young- and mid-age broiler breeders, *Anim. Reprod. Sci.* 141(1–2): 80–85.

Zhang, X., Berry, W. D., McDaniel, G. R., Roland, D. A., Liu, P., Calvert, C. and Wilhites, R. (1999). Body weight and semen production of broiler breeder males as influenced by crude protein levels and feeding regimes during rearing, *Poult. Sci.* 78(2): 190–196.

Zuidhof, M. J. (2018). Lifetime productivity of conventionally and precision-fed broiler breeders, *Poult. Sci.* 97(11): 3921–3937.

Zuidhof, M. J., Robinson, F. E., Feddes, J. J. R., Hardin, R. T., Wilson, J. L., McKay, R. I. and Newcombe, M. (1995). The effects of nutrient dilution on the wellbeing and performance of female broiler breeders, *Poult. Sci.* 74(3): 441–456.

Zuidhof, M. J., Schneider, B. L., Carney, V. L., Korver, D. R. and Robinson, F. E. (2014). Growth, efficiency, and yield of commercial broilers from 1957, 1978, and 2005, *Poult. Sci.* 93(12): 2970–2982.

Zuidhof, M. J., Holm, D. E., Renema, R. A., Jalal, M. A. and Robinson, F. E. (2015). Effects of broiler breeder management on pullet body weight and carcass uniformity, *Poult. Sci.* 94(6): 1389–1397.

Part 3

Individual nutrients/additives

Chapter 10

Advances in understanding and improving the use of amino acids in poultry nutrition

M. T. Kidd, University of Arkansas, USA; S. Y. Liu and P. Selle, The University of Sydney, Australia

1 Introduction

The benefits of incorporating amino acids in poultry feed formulations, such as improved broiler and layer feed efficiency, health, and performance, are well established (Dozier and Tillman, 2017; Lee et al., 2023). Developing robust feed formulations with the use of feed-grade amino acids in broilers requires knowledge of formulation minimums of the less limiting amino acids (in optimizing protein synthesis). Such minimums do not necessitate strict constraints in least-cost formulation at present but rather provide an estimate that can then be assessed and subsequently tested for use by integrators in the poultry supply chain. Research insights into formulating reduced protein diets for broilers have been recently reviewed (Kidd et al., 2021). Developing low-protein diets for broilers (with the aim of optimizing production efficiency and sustainability) requires discussion of least-cost formulation ingredients (e.g. L -form amino acids, soybean meal, and algae meal) and nutrient matrices (e.g. threonine, electrolyte balance, and non-essential amino acids) (Kidd et al., 2021). Other issues requiring attention include non-essential amino acid

http://dx.doi.org/10.19103/AS.2024.0143.13

levels, total nitrogen supply, and the ratio of cereals to oilseeds, which dictates increased feed-grade amino acids and cereal starch.

Branched-chain amino acids are also pivotal in minimizing diet nitrogen excretion and have been recently reviewed (Kidd, 2021). In a previous review, it was determined that two of the branched-chain amino acids (i.e. valine and isoleucine) are fourth limiting (in terms of protein synthesis) when the two most abundant ingredients (i.e. corn and soybean meal) are used in broiler diets (Kidd and Hackenhaar, 2005). However, when arginine was formulated to a 1.05 ratio to lysine, it became the fourth limiting in a broiler diet simulation model when sorghum grain replaced corn (Kidd and Hackenhaar, 2005). It must be pointed out that arginine (as discussed in a subsequent section) has numerous physiological functions associated with the production of nitic oxide and health-related needs. This may necessitate higher nutrient minimums, which further limits arginine in broiler diets.

In the least-cost formulation for broiler diets, the first limiting amino acid (that is not present in a feed grade amino acid supplement form) sets the protein level of diet, assuming no crude protein (CP) constraint minimum is imposed. Thus, understanding the need for less limiting amino acids is paramount to minimizing nitrogen excesses and diet costs. In this chapter, we have assessed arginine, glycine and serine, histidine, and phenylalanine and tyrosine research, and outlined nutrient implications for future considerations in diet formulation for broilers.

2 Arginine

As poultry production is poised for exponential growth expansion in the coming decades, oilseed demand is projected to be high, with supply remaining tight and the availability of protein meals varying globally by region. In addition to increased demand for protein meals, heightened demand for oil for biofuels will further decrease regional supply availability due to the increase in soybean processing for oil rather than whole beans destined for export markets. This is important because the arginine-to-lysine constraint in least-cost formulation for broiler diets is primarily met with soybean meal or other oilseeds used in lesser amounts (e.g. peanut or canola meal). Indeed, the arginine minimum in formulation (i.e. estimated requirement) is subject to much debate. The objective of this section is to highlight research in broilers that supports variation in arginine needs, to aid nutritionists in establishing arginine minimums, and to help researchers identify areas for arginine research.

Arginine, an indispensable amino acid in poultry, can easily become the critical pressure point in a least-cost formulation based on its specification in the nutrient matrix, which can range from levels approximating 1.0 to 1.2 in practice. For example, the work by Brake et al. (1998) demonstrated that the arginine-to-lysine ratio varies in broilers subjected to different environmental temperatures. Moreover, when comparing the intestinal epithelial uptake of equal arginine and lysine concentrations *in vitro*, cells from heat-stressed broilers had significantly reduced uptake of arginine compared to lysine (Brake et al., 1998). Hence, the discrepancy between lysine and arginine may be associated with arginine's gastrointestinal roles.

Work conducted by Kessler and Thomas (1976) titrated arginine from 0.74% to 1.28% of the diet for male and female broilers and estimated a live performance need no greater than 1.1% of the diet. Although the work was conducted with total amino acids, the ratio to lysine approximates 1.0–1.05. Cuca and Jensen (1990) conducted three experiments with male broilers to assess arginine needs in the starting period and concluded that arginine needs are around 1.25% of diet, also approximately near a 1.0 ratio to lysine, or lower. Hurwitz et al. (1998) conducted research with varying amounts of arginine, lysine, and CP and determined that the sensitivity of arginine supplementation was more effective in low CP diets and that breast muscle yield responded favorably to increasing arginine. Corzo et al. (2003) conducted an arginine dose response in 42-to-56-day-old broilers by feeding graduations of arginine from 0.80% to 1.25% of the total diet. Results indicated that arginine needs approximated 1.0%, which agreed with NRC (1994), resulting in an arginine-to-lysine requirement of 1.15. However, this study also assessed carcass skin scratches and determined that broiler arginine needs for immunological or mechanical challenges to connective skin are higher than 1.0 (Corzo et al. 2003).

In a study by Kidd (2004), feeding broilers three levels of an arginine:lysine ratio (0.9, 1.1, and 1.3) from 1-to-42 days of age, when birds were challenged with coccidiosis at day 28, resulted in reduced feed conversion being improved at 1.1, but that reduced mortality required an arginine:lysine ratio of 1.3. Although the former represented a practical trial with an *Eimeria* challenge isolated in practice, the ability of arginine to improve health during an enteric challenge has gained interest over the past two decades. For example, Perez-Carbajal et al. (2010) evaluated immunity in broilers after the *Eimeria* challenge via heterophil and monocyte oxygen burst and found improvements when basal diets were supplemented with 0.3% arginine, but not 0.6%, to a basal diet meeting arginine NRC (1994) needs.

Recently, there has been significant work from the laboratory of Professor Woo Kim at the University of Georgia on the arginine needs of broilers when exposed to coccidiosis. Castro et al. (2020) challenged Cobb 500 male chicks with a sporulated combination of *Eimeria* consisting of E. *arcervulina*, *maxima*, and *tenella*. Dietary arginine levels fed ranged from 1.04% to 1.44% in 0.1% increments. Results indicated that arginine levels between 1.24% and 1.44% of the diet were warranted for best growth and recovery to *Eimeria* (Castro et al., 2020). Teng et al. (2021) fed low CP diets to *Eimeria*-challenged Cobb 500 male broilers and supplemented 0.75% additional arginine, glutamine, methionine, and threonine both separately and in combination. Arginine, and the independent addition of glutamine or threonine, showed promise in improving intestinal health in broilers after coccidiosis infection. The level of arginine in the control diet relative to lysine was 1.05, which resulted in the supplemental level approximating a 1.8 arginine-to-lysine ratio, showing intestinal morphology benefits (Teng et al., 2021). Liu et al. (2023) evaluated the effects of a 50% supplemental excess of both branched-chain amino acids and arginine in reduced-CP diets in Cobb male broilers. In addition to arginine, branched-chain amino acids play a critical role in mucin integrity (Wise et al., 2021). Birds fed both amino acid supplement combinations had improved immunity after *Eimeria* infection, but those fed higher arginine showed a better intestinal immune response (Liu et al., (2023).

In the United States of America, the majority of broiler feeds no longer contain antibiotics given consumer attitudes. Coccidiostats (e.g. ionophores) are considered antibiotics in the United States of America. Continued broiler research with arginine above levels considered adequate for live performance on immunity and recovery from *Eimeria* is warranted. Globally, coccidiosis is one of the most economically important infections in broilers, and an understanding of an arginine need in formulation for an *Eimeria* infection will aid integrators in making decisions.

3 Glycine and serine

Glycine and serine have been described as the first limiting non-essential amino acids in broiler chickens (Ospina-Rojas et al., 2012). Both are well-known 'building-blocks' of protein, as glycine constitutes 115 mg/g and serine 45 mg/g, of whole-body protein in broiler chickens (Wu, 2014). The connection between glycine and serine may have been first suggested in broiler chickens by Wixom et al. (1955) and has been considered by other researchers (Baker et al., 1968; Featherston, 1976). Using radioactively labelled 2-^{14}C glycine, Sugahara and Kandatsu (1976) concluded that glycine is directly converted to serine in roosters based on similar ratios (0.7–1.5) of 2-C serine to 2-C glycine in tissue proteins. The metabolic interconversion of glycine and serine is assumed to be continuous and has led to the expression of their dietary requirements as glycine equivalents where serine concentrations are corrected by a factor of 0.7143 to adjust for molecular weights (Siegert and Rodehutscord, 2019).

Increasing attention has been paid to glycine and serine since Dean et al. (2006) reported that the addition of glycine to reduced-CP diets can support 'optimal' growth performance in broiler chickens. Ideal amino acid profiles, where amino acid requirements are expressed relative to lysine (100), are not frequently proposed for non-essential amino acids. However, Wu (2014) recommended 176 glycine and 69 serine (225 glycine equivalents), and more conservatively, Maharjan et al. (2021) recommended 61 glycine and 75 serine (115 glycine equivalents). These differences in glycine incongruity suggest that considerably more research is needed in this area, which may largely stem from the complexities of glycine and serine biosynthesis.

Glycine metabolism in mammalian species has been extensively reviewed by Wang et al. (2013), indicating that glycine biosynthesis may not be sufficient to meet metabolic needs. Serine is converted to glycine by serine hydroxymethyltransferase. Additional substrates for glycine biosynthesis include threonine, choline, and glyoxylate. However, the extent to which threonine metabolism generates glycine in poultry is still unknown. In theory, threonine may be metabolized in reactions catalyzed by threonine dehydrogenase, threonine dehydratase, or threonine aldolase, but threonine dehydrogenase is the dominant enzyme (88%) in poultry (Akagi et al., 2004). However, free threonine plasma concentrations typically spike in birds offered reduced-CP diets. This was first reported by Fancher and Jensen (1989) and has been frequently observed since. This is totally inconsistent with the conversion of threonine to glycine. Threonine spikes may potentially be attributed to the downregulation of threonine dehydrogenase activity (Selle et al., 2021a). This theory was based on the possibility that additional acetyl CoA (derived from higher starch concentrations in reduced-CP diets) inhibits threonine dehydrogenase activity via a negative feedback mechanism (Geurranti et al., 2001), as threonine dehydrogenase converts threonine to acetyl CoA and glycine (Davis and Austic, 1982). It was subsequently suggested that free threonine plasma concentrations could serve as an indicative biomarker for the precision with which reduced-CP broiler diets are formulated (Macelline et al., 2021).

Numerous papers have examined the impact of glycine equivalents in reduced-CP diets since the publication by Dean et al. (2006), including Yuan et al. (2012) who concluded that glycine may be critical in reduced-CP broiler diets. Nevertheless, a clear picture has yet to emerge. However, a series of four similar studies (Chrystal et al., 2020a, 2020b, 2020c, 2021) provides some insight. Collectively, CP levels of corn-based diets were reduced from 20.6% to 16.5% and, without fortification, analyzed glycine equivalents declined from 1.53% to 1.05% in the first three studies. Weight gains declined by an average of 2.40% (1913 versus 1960 g/bird), and feed conversion was compromised by 4.92% (1.600 versus 1.525) from 14 to 35 days post-hatch. In the fourth study, dietary CP was reduced from 22.2% to 16.5% CP, but analyzed glycine equivalents were maintained at 1.68% and 1.61%, respectively, by the inclusion of non-bound glycine and serine in the reduced-CP diet. Birds offered the 16.5% CP diet enjoyed a significant weight gain advantage of 7.05% (2370 versus

2214 g/bird) over their standard CP counterparts, although feed conversion was numerically compromised by 1.38% (1.473 versus 1.453). Maintaining dietary glycine equivalent concentrations in the order of 1.6% in reduced-CP diets was thus clearly advantageous.

Nevertheless, the 'ideal' glycine equivalent concentration in reduced-CP diets remains a challenge, as a broad range from 1.1% to 2.0% was proposed by Siegert and Rodehutscord (2019). More recently, Aguihe et al. (2021, 2022) addressed this issue. In their first study, they suggested minimum levels of 1.79–1.94% glycine equivalents, depending on dietary methionine plus cysteine concentrations for optimum growth. In their second study, they proposed a minimum concentration of 1.55% glycine equivalents in 17.4% CP diets containing 0.93% digestible threonine, but this could increase to over 2.09% in diets containing lower threonine levels.

One pivotal factor promoting these variations in appropriate glycine equivalent recommendations could well be the involvement of glycine in ammonia (NH_3) detoxification. NH_3 is inherently toxic (Stern and Mozdziak, 2019), and high NH_3 plasma concentrations have been shown to depress broiler growth performance (Namroud et al., 2008; Ospina-Rojas et al., 2013, 2014). Following detoxification, NH_3-N is ultimately excreted as uric acid-N. However, every mole of uric acid excreted requires an input of one mole of glycine into the Krebs uric acid cycle in avian species (Salway, 2018). It was retrospectively estimated (Selle et al., 2021b) that in the Chrystal et al. (2021) study, an average of 49.2% of dietary glycine entered the Krebs uric acid in the NH_3 detoxification pathway, but this proportion ranged from 25.0% to 80.9% across all dietary treatments. This calculation was based on the total excreta collection period (35–37 days post hatch) when an average of 6.8 g uric acid-N was excreted per replicate cage as opposed to an 18.3 g dietary glycine intake. This estimate does not account for any glycine biosynthesis from serine, threonine, or other substrates. This may be relatively limited, but the extent of glycine biosynthesis may be the cause of the variability. Nevertheless, it is evident that the appropriate amount of dietary glycine is variable and largely dependent on the volume of uric acid that is synthesized and excreted. This is essentially dictated by the quantities of surplus, or unbalanced, amino acids that are being catabolized, triggering NH_3 generation (Selle et al., 2020).

However, additional factors are complicating the identification of ideal glycine equivalent concentrations in reduced-CP broiler diets. Both serine and methionine are involved in the biosynthesis of cysteine (Belloir et al., 2015); however, serine and cysteine constitute 13.7 and 8.2 mol/100 mol, respectively, of protein in feathers of broiler chickens (Greenhalgh et al., 2020). Given the real possibility that amino acids are preferentially partitioned toward feathering in poultry (Wylie et al., 2003), this may limit glycine biosynthesis and glycine *per se* is incorporated into feathers at 11.4 mol/100mol.

The effects of additional glycine in reduced-CP broiler diets with varying methionine plus cysteine levels were investigated by Powell et al. (2011). Glycine enhanced feed conversion in diets with suboptimal methionine plus cysteine concentrations but not when cysteine levels met requirements. This outcome illustrates the relationship between cysteine and glycine equivalent requirements in broiler chickens. Threonine was shown to reduce glycine requirements in reduced-CP broiler diets by Lambert et al. (2015); however, as discussed, the extent to which threonine does serve as a glycine precursor requires clarification.

The interactive effects of dietary concentrations of glycine equivalents (0.9–2.1%), cysteine (0.2–0.5%), and choline (0.05–0.17%) in broiler chickens were investigated by Hofmann et al. (2020). However, the interactive effects of these three dietary factors were limited, which may have been due to a lack of variation in efficiency of nitrogen utilization across treatments. In an earlier study (Siegert et al., 2015), choline was shown to influence the intakes of glycine equivalents and threonine necessary to achieve responses in weight gain and feed conversion. However, the researchers conceded that more consideration of these interactive effects is required.

In addition to being substrates for protein synthesis, amino acids act as modulators of signal transduction pathways that control metabolism and cell functions (Tesseraud et al., 2011), which have been considered specifically in relation to the nutritional regulation of feathering in poultry (Chen et al., 2020). This illustrates the challenge that identifying viable recommendations for glycine equivalent inclusions in broiler diets represents to enhance chicken-meat production. However, the likelihood is that, if appropriate ideal amino acid ratios for reduced-CP diets can be developed, glycine requirements will become more consistent and conservative. This could stem from a better balance of dietary amino acids that will reduce the need for glycine's involvement in the elimination of uric acid. Finally, maintaining glycine plus serine concentrations at 2.3% permitted a 4.5% CP reduction in corn-based broiler diets from 8 to 42 days post-hatch without compromising growth performance or carcass composition (Wang et al., 2020). While the glycine plus serine fortification was not able to overcome the deleterious impacts of a 6.0% CP dietary reduction, this study vividly demonstrates the crucial role of glycine equivalents in reduced-CP broiler diets.

4 Histidine

The composition of histidine in poultry carcasses is 3.7%, but the histidine composition of breast tissue is 4.7% (Conde-Aguilera et al., 2013). The former discrepancy is thought to be due to lower histidine-containing molecules in connective tissue and higher concentrations in IIB-type muscle fibers (Brosnan and Brosnan, 2020). The histidine content of broiler breast and thigh muscles was analyzed as 4.34% and 3.53%, respectively (Hamm, 1981). Chickens fed a histidine-free diet produce breast meat with lower than normal carnosine (Leveille et al. 1960). Thus, when assessing histidine breast muscle or whole-body status, other compounds linked to histidine metabolism should be considered.

Enzyme functionality is largely dependent on histidine as approximately 5% of all enzyme active sites contain histidine and react with other amino acids or divalent cations (Schneider, 1978, Liao et al., 2013). In addition to protein synthesis and enzyme activity, histidine can be decarboxylated to histamine and is a precursor for synthesis of carnosine (β-alanyl-histidine) and its methylated analogs: anserine (β-alanyl-Nπ-methyl-histidine) and ophidine (β-alanyl-Nτ-methyl-histidine)(Mora et al., 2007). The metabolism of methionine and histidine are linked. Hence, carnosine synthase is thought to synthesize anserine (Drozak et al., 2010) or carnosine methylation may produce anserine (Drozak et al., 2013). Anserine methylation requires S-adenosylmethionine (Drozak et al., 2013). It has also been demonstrated that broilers fed methionine-deficient diets had decreased breast tissue anserine and ophidine, but increased breast tissue carnosine when compared to broilers fed adequate dietary methionine (Conde-Aguilera et al., 2015).

In terms of breast muscle development, carnosine and anserine may contribute up to 40% of the buffering capacity of type IIB muscle fibers (Davey, 1960). Qi et al. (2021) fed four levels of supplemental histidine (0, 0.07, 0.13, and 0.20) and two dietary β-alanine levels (0% and 0.12%) to male broilers and found that both supplements improved breast meat redox balance and quality, increased breast muscle carnosine and anserine content, and increased overall carnosine synthesis via upregulating associated genes. It must be pointed out that increases in L-histidine reduced average daily body wight gain (Qi et al. (2021). However, it must be kept in mind that carnosine represents an endogenous reserve of histidine (Lin, 1981), which may necessitate feeding a histidine-deficient diet prior to the initiation of research conducted to assess dietary histidine needs.

Several key experiments to assess histidine were conducted in Professor David Baker's laboratory in the 1970s and 1980s. Robbins et al. (1977) conducted three experiments to assess histidine needs in starting and growing chickens fed cornstarch and crystalline amino acid-based diets. The unique aspect of this work was that all three experiments initialized the dose response of zero histidines. Experiments 1 and 2 consisted of dose responses of L-histidine from 0% to 0.44% of the diet, and experiment 3 consisted of dose responses from 0% to 0.57% of the diet. Plasma histidine in birds was low until dietary histidine reached 0.41% and was unaffected by dietary L-carnosine supplementation. Growth rate was the primary variable to assess histidine requirements. These studies demonstrated that total histidine needs for growth approximated levels around 0.32%, and total dietary lysine from L -lysine hydrochloride was 1.14%. It must be pointed out, however, that this research also suggests that muscle carnosine is metabolized to histidine and β-alanine, as plasma histidine did not reach its maximum until peak weight gain and muscle carnosine levels were reached (Robbins et al., 1977).

Additional histidine studies were conducted by Han et al. (1991), who fed histidine-deficient diets based on corn, feather meal, and soybean meal to chicks and quantified a digestible histidine need of 0.31% of the diet for growth and feed conversion for the 8-to-22-day period. Confirming the basic work of Robbins et al. (1977) in practical broiler feeds, Haug et al. (2008) fed 0.1% histidine to broilers and noted a breast muscle increase of 64% for carnosine and 10% for anserine. The research by Robbins et al. was conducted in diets primarily based on wheat and soy extract, and treatments consisted of 0%, 0.1%, 0.2%, and 0.3% histidine supplementation. It has been suggested that future work should assess the effects of dietay histidine on breast muscle storage attributes (e.g. appearance, shelf life, and health benefits) based on increased concentrations of carnosine and anserine, which could result in an improved human product, poultry products with improved shelf life, or even poultry products serving as functional foods (Kralik et al., 2018). In addition, improving broiler myopathies with dietary histidine may also be a possibility, as addressed in the next section.

Lackner et al. (2022) fed Ross 308 male broilers with graduated amounts of histidine relative to lysine from 0.41 to 0.57 from hatching to slaughter (days 38–41). Although histidine levels from the initial dose were fed above that considered adequate, the primary objective of this work was to assess histidine needs on muscle myopathies. Dietary increases in histidine above 0.41 increased pale–soft–exudative breast meat but decreased woody breast syndrome (Lackner et al., 2022).

Poultry breast tissue is rich in the dipeptides carnosine and anserine, both of which have been reported to have high antioxidant potential (Kohen et al., 1988). To assess breast muscle carnosine and anserine, Kai et al. (2015) fed female broilers three levels of histidine relative to the 1994 Nutrient Requirements of Poultry (67%, 100%, and 200% of NRC) from 14 to 24 days of age. Results obtained indicate that the low and high histidine diets fed to broilers resulted in reduced and increased, respectively, carnosine and anserine in breast muscle (Kai et al., 2015).

As the occurrence of breast tissue myopathies has increased in commercial broilers in the past decade, dietary histidine needs should not only be assessed for live performance and carcass traits but also skeletal tissue health. Franco et al. (2017) fed graduated amounts of dietary histidine expressed relative to lysine (0.28–0.40) to Cobb 500 males from 8 to 17 days of age and determined that a 0.38 histidine-to-lysine ratio supports live performance needs. Clearly, studies assessing histidine well in excess of known needs to support the growth of broilers may identify higher histidine needs to support skeletal muscle health.

5 Phenylalanine and tyrosine

Phenylalanine and tyrosine are two proteinogenic amino acids that represent 4.15% and 4.07%, respectively, of breast-meat protein in broiler chickens (Hamm, 1981). Notionally, phenylalanine is an essential amino acid, whereas tyrosine is considered non-essential, and both are aromatic amino acids, similar to tryptophan. However, phenylalanine is often overlooked by nutritionists as it is seldom limiting in conventional broiler diets. Despite this, Chrystal et al. (2020d) have highlighted the need for more research into the ideal ratio of phenylalanine plus tyrosine to lysine to support optimal growth performance in broilers offered reduced-CP diets. The ideal phenylalanine inclusion in broiler diets is 66 (range: 60–75), and tyrosine 52 (range: 45–59) relative to lysine (100), based on the average recommendations of Roth et al. (2001), Wu (2014), and Maharjan et al. (2021). Graded digestible phenylalanine plus tyrosine ratios from 94 to 118 have been found to generate quadratic responses in weight gain and linear responses in breast weights (Franco et al., 2017). This study concluded that the ideal phenylalanine plus tyrosine ratio was 112 in relation to digestible lysine. This total ratio of 112 falls within the range of 105–134 for the total ratio suggested by previous studies.

The relationship between phenylalanine and tyrosine in broiler chickens was initially studied by Grau (1947), although this relationship had been investigated earlier in rats by Womack and Rose (1934) and others. Subsequently, Sasse and Baker (1972) proposed that dietary concentrations of phenylalanine and tyrosine for maximal growth in broiler chickens were 0.5% and 0.37%, respectively, and that tyrosine was capable of furnishing 42.5% of combined requirements. The potential basis of this relationship is that phenylalanine may serve as a precursor of tyrosine when 1.000 phenylalanine is converted to 1.097 tyrosine on a mass basis (Creek, 1968).

The apparent distal ileal digestibility coefficients of phenylalanine (0.781) and tyrosine (0.763) were similar in broilers offered standard sorghum-based diets in Liu et al. (2013). The two amino acids share common Na^+-dependent and Na^+-independent transport systems (Hyde et al., 2003). However, the digestion rate constant of protein-bound phenylalanine (2.39×10^{-2}/min) was noticeably more rapid than tyrosine (1.85×10^{-2}/min), as reflected in the proximal jejunal digestibility coefficients where phenylalanine amounted to 20.8% (0.413 versus 0.342). Indeed, tyrosine had the slowest digestion rate constant of the 16 amino acids assessed. The more rapid and proximal intestinal uptakes of phenylalanine versus tyrosine may impact on post-enteral catabolism of phenylalanine in tyrosine provision.

Following digestion and absorption, the post-enteral fate of phenylalanine and tyrosine is, in part, net protein synthesis, with whole-body protein of broiler chickens composed of 3.48% phenylalanine and 2.66% tyrosine (Wu, 2014). Additionally, phenylalanine may be converted to tyrosine via hydroxylation by phenylalanine hydroxylase, which is the sole catabolic pathway for phenylalanine. This conversion requires dihydropteridine reductase and the coenzyme tetrahydrobiopterin and is allosterically activated by phenylalanine (Kaufman, 1987; Hufton et al., 1995). Therefore, phenylalanine that is not deposited as protein is catabolized to generate tyrosine. Tyrosine, on the other hand, may be catabolized by either tyrosine hydroxylase or hepatic tyrosine aminotransferase (Hufton et al., 1995; Tabiri et al., 2002).

The activity as well as expression of phenylalanine hydroxylase in broiler chickens subjected to phenylalanine imbalances or toxicity was investigated by Lartey and Austic (2009). These researchers found that either a deficiency or an excess of phenylalanine impairs growth performance, which is reflected in lower or higher free phenylalanine plasma concentrations. The addition of a blend of indispensable amino acids lacking phenylalanine to a phenylalanine-deficient diet or to a diet containing excess phenylalanine results in substantial reductions in phenylalanine plasma concentrations.

The kinetics of phenylalanine and tyrosine in humans has been reviewed by Matthews (2007). In the postabsorptive state, the flux of phenylalanine in the body is derived from phenylalanine released from protein breakdown as there is no entry of amino acids from dietary sources. The input of phenylalanine is matched by phenylalanine removal via protein synthesis and via metabolic disposal by conversion to tyrosine. Phenylalanine levels in the postabsorptive state are indicative of the rate of whole-body proteolysis. For this reason, consideration will be given to free plasma concentrations of phenylalanine and tyrosine in broiler chickens later in this section.

Decades ago, the amino acid composition of protein and its dietary concentration was shown to influence thyroid activity in poultry (March et al., 1964). More specifically, phenylalanine and tyrosine have been shown to impact circulating thyroid hormone levels in broiler chickens, with tyrosine directly involved (Elkin et al., 1980). These researchers concluded that phenylalanine alone could not supply sufficient tyrosine when dietary amino acid concentrations were suboptimal. When birds were offered diets at 85% of amino acid requirements, the addition of 0.44% phenylalanine plus 0.40% tyrosine increased 14-day weight gain by 79.6% (169.7 versus 94.5 g/bird) and improved feed conversion by 16.5% (1.67 versus 2.00) compared to 0.81% phenylalanine alone. Comparable responses were not observed in birds offered diets at 100% and 115% of total amino acid requirements.

Interestingly, free tyrosine plasma concentrations in broilers exposed to heat (32°C versus 22°C) were depressed by 37.4% (18.9 versus 30.2 mg/mL), but phenylalanine concentrations remained unaltered (22.5 versus 24.5 mg/mL) (Geraert et al., 1996). Phenylalanine was one of the few amino acids whose plasma concentrations were not influenced by chronic heat exposure. The effects of chronic heat stress and dietary tyrosine levels on hepatic tyrosine aminotransferase activity in broiler chickens were investigated by Tabiri et al. (2002). Tyrosine aminotransferase activity significantly increased in chickens exposed to 36°C for 14 days, whereas hepatic phenylalanine hydroxylase was not increased. These outcomes are consistent with the findings of Geraert et al. (1996).

Importantly, tyrosine hydroxylase initiates the conversion of tyrosine to dopamine (Hufton et al., 1995). Tyrosine and phenylalanine are involved in the synthesis of catecholamines, including dopamine, in cerebral tissue (Fernstrom and Fernstrom, 2007). A deficiency in the neurotransmitters, dopamine and serotonin, which are derived from tryptophan, can lead to aggressive behavior in poultry, especially in layer hens (Birkl et al., 2017). However, feather-pecking and poor feathering were observed in broiler chickens in a reduced-CP feeding study reported by Greenhalgh et al. (2020). In this study, dietary crude protein concentrations of wheat-based diets were reduced from 21.5% to 19.8%, 18.0%, and 16.3%, which substantially compromised growth performance from 7 to 35 days post-hatch. Greenhalgh et al. (2020) suggested that feather-pecking and, to some extent, extremely poor growth performances observed may have been due to inadequacies of dopamine and serotonin, triggered by dietary deficiencies in tyrosine, phenylalanine, and tryptophan in birds offered reduced-CP diets, which corresponds with other research (de Hass and van der Eijk, 2018).

The impacts of reducing CP concentrations of sorghum-based diets from 22.0 to 19.0 and 16.0% and L-carnitine inclusions were investigated by Greenhalgh et al. (2022), in which all diets were formulated to 1.12% digestible lysine with an energy density of 12.95 MJ/kg. The experimental diets contained up to ten non-bound amino acids. Combined amounts increased from 1.52% to 2.93% and 5.1% as CP levels decreased, but non-bound phenylalanine and tyrosine were not included in any diets. Consequently, relative to 1.12% lysine, total phenylalanine and tyrosine concentrations declined from 12.9% to 9.5% and 6.3%, as dietary levels were reduced from 22.0% to 16.0% CP. The total ratios in the reduced-CP diets were thus clearly inadequate in comparison to the ideal ratio of 112 found by Franco et al. (2017). Only the control diets without L-carnitine are considered in the following discussion.

Dietary CP reductions linearly compromised 7–33-day growth performance in the study by Greenhalgh et al. (2022). For example, feed conversion was progressively increased from 1.343 to 1.385 and 1.615, where 1.491 is the 2019

Aviagen performance objective. Dietary CP reduction from 22.0% to 16.0% compromised feed conversion by 20.3% (1.615 versus 1.343) due to multiple factors. Declining apparent distal ileal disappearance rates (intestinal uptake rates) of both phenylalanine ($r = 0.993$; $P < 0.0001$) and tyrosine ($r = 0.993$; $P < 0.0001$) decreased feed conversion quadratically. Regression equations suggest that a phenylalanine disappearance rate of 0.901 g/bird/day would support a minimum feed conversion of 1.340, and that a tyrosine reduction rate of 0.343 g/bird/day would support a minimum feed conversion of 1.329. Given comparable rates of digestion, this suggests that a dietary ratio of 0.72 phenylalanine to 0.28 tyrosine would be appropriate.

The net portal flux of phenylalanine and tyrosine (as a proportion of ileal digestion) is 94.5% and 85.6%, respectively, in pigs (Van der Meulen et al., 1997). It appears that substantially more phenylalanine enters the systemic pool than tyrosine, but broadly similar amounts are incorporated into whole-body protein, which is composed of 3.48% phenylalanine and 2.66% tyrosine. Interestingly, intestinal uptakes of phenylalanine were not related ($r = 0.007$; $P = 0.980$) to its free plasma concentrations; however, there was a significant relationship ($r = 0.577$; $P = 0.019$) for tyrosine in the Greenhalgh et al. (2022) study. Moreover, intestinal uptakes of phenylalanine were quadratically related ($r = 0.645$; $P = 0.030$) to free tyrosine plasma concentrations as shown in Fig. 1. It may be deduced from the regression equation that tyrosine plasma concentrations increase when phenylalanine intestinal uptakes exceed a threshold of 0.533 g/bird/day. The implication is that phenylalanine is converted to tyrosine by phenylalanine

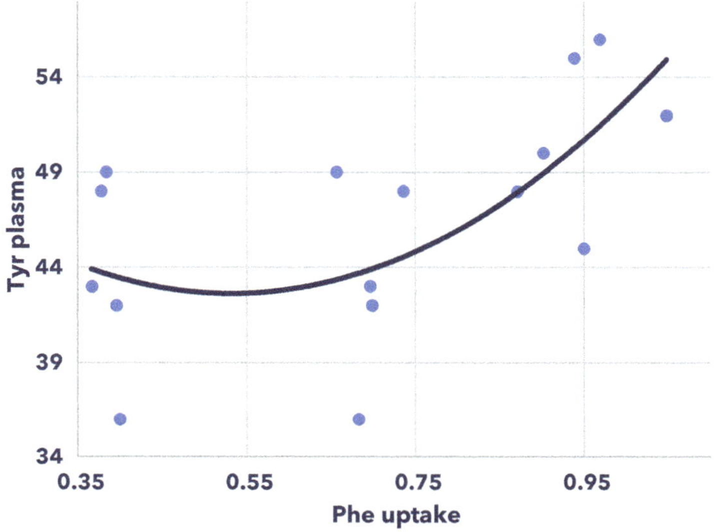

Figure 1 Quadratic relationship ($r = 0.645$; $P = 0.030$) between phenylalanine uptake rates at distal ileum (g/bird/day) and free tyrosine concentrations in systemic plasma (mg/mL), where $y_{(tyr)} = 55.691 - 49.104 \times phe + 46.145 \times phe^2$.

hydroxylase activity given adequate systemic quantities of phenylalanine, but if they are not satisfied, preference is given to protein accretion for phenylalanine.

Dietary CP reduction from 22.0% to 16.0% compromised weight gains by 38.2% (1374 versus 2225 g/bird) in the Greenhalgh et al. (2022) study, which is completely unacceptable. Again, multiple factors would have been responsible but, clearly, inadequate dietary phenylalanine and tyrosine concentrations contributed. Broiler growth performance was severely depressed when dietary phenylalanine plus tyrosine concentrations were reduced to 60% of requirements as demonstrated by Sugahara et al. (1969). Indeed, their deleterious impacts were the most pronounced of all amino acids assessed. Thus, careful attention should be paid to phenylalanine and tyrosine in the formulation of reduced-CP diets, and it would be prudent not to assume that phenylalanine alone can meet tyrosine requirements. Phenylalanine hydroxylase activity may not be present in cerebral tissue (Tourian et al., 1969), so a deficiency of tyrosine *per se* could limit dopamine synthesis and tyrosine may not be provided by phenylalanine catabolism at a sufficiently rapid rate, especially when dietary phenylalanine concentrations are relatively low, as can be the case in reduced-CP diets.

6 Conclusion

Feed receiving and feed mill quality control areas must maintain a high-quality ingredient analysis program that can provide data to commercial formulators quickly in order to optimize precision feeding. Advances in the analysis of incoming ingredients will continue to be the most important approach to balancing dietary amino acids. However, precision formulation also relies on meeting, rather than exceeding, amino acid minimums in the nutrient matrix. From the research reviewed, more consideration of the nutrient minimums for arginine, glycine and serine, histidine, and phenylalanine and tyrosine can contribute to dietary nutrient minimum evaluation.

7 Where to look for further information

Arginine

- Bortoluzzi, C., Rochell, S. J. and Applegate, T. J. 2018. Threonine, arginine, and glutamine: Influences on intestinal physiology, immunology, and microbiology in broilers. *Poultry Science* 97, 937–945.
- Castro, F. L. S. and Kim, W. K. 2020. Secondary functions of arginine and sulfur amino acids in poultry health: review. *Animals (Basel)* 10, 2106.

Glycine and serine:

- Siegert, W. and Rodehutscord, M. 2019. The relevance of glycine and serine in poultry nutrition: a review. *British Poultry Science* 60, 579–588.

Histidine

- Moro, J., Tomé, D., Schmidely, P., Demersay, T.-C. and Azzout, M. 2020. Histidine: a systematic review on metabolism and physiological effects in human and different animal species. *Nutrients* 12, 1414.

8 References

Aguihe, P. C., Hirata, K. A., Ospina-Rojas, C. I., dos Santos, T. C., Pozza, P. C., Iyayi, E. A. and Murakami, A. E. 2022. Effect of glycine equivalent levels in low protein diet containing different SID threonine concentrations on performance, serum metabolites and muscle creatine of broiler chickens. *Italian Journal of Animal Science* 21(1), 1000–1009. https://doi.org/10.1080/1828051X.2022.2080593

Aguihe, P. C., Ospina-Rojas, I. C., Sakamoto, M. I., Pozza, P. C., Iyayi, E. A. and Murakami, A. E. 2021. Dietary glycine equivalent and standardized ileal digestible methionine + cysteine levels for male broiler chicken fed low-crude-protein diets. *Canadian Journal of Animal Science* 102(1), 19–29. https://doi.org/10.1139/cjas-2021-0009

Akagi, S., Sato, K. and Ohmori, S. 2004. Threonine metabolism in Japanese quail liver. *Amino Acids* 26(3), 235–242. https://doi.org/10.1007/s00726-004-0074-8

Baker, D. H., Sugahara, M. and Scott, H. M. 1968. The glycine-serine interrelationship in chick nutrition. *Poultry Science* 47(4), 1376–1377. https://doi.org/10.3382/ps.0471376

Belloir, P., Lessire, M., Berri, C., Lambert, W., Corrent, E. and Tesseraud, S. 2015. *Revisiting amino acid nutrition*. Proceedings, 20th European Symposium on Poultry Nutrition, Prague, Czech Republic.

Birkl, P., Franke, L., Bas Rodenburg, T. B., Ellen, E. and Harlander-Matauschek, A. 2017. A role for plasma aromatic amino acids in injurious pecking behavior in laying hens. *Physiology and Behavior* 175, 88–96. https://doi.org/10.1016/j.physbeh.2017.03.041

Brake, J., Balnave, D. and Dibner, J. J. 1998. Optimum dietary arginine:lysine ratio for broiler chickens is altered during heat stress in association with changes in intestinal uptake and dietary sodium chloride. *British Poultry Science* 39(5), 639–647. https://doi.org/10.1080/00071669888511

Brosnan, M. E. and Brosnan, J. T. 2020. Histidine metabolism and function. *Journal of Nutrition* 150(Suppl. 1), 2570S–2575S. https://doi.org/10.1093/jn/nxaa079

Castro, F. L. S., Teng, P. Y., Yadav, S., Gould, R. L., Craig, S., Pazdro, R. and Kim, W. K. 2020. The effects of L-arginine supplementation on growth performance and intestinal health of broiler chickens challenged with Eimeria spp. *Poultry Science* 99(11), 5844–5857. https://doi.org/10.1016/j.psj.2020.08.017

Chen, M.-J., Xie, W.-Y., Jiang, S.-G., Wang, X.-Q., Yan, H.-C. and Gao, C.-Q. 2020. Molecular signaling and nutritional regulation in the context of poultry feather growth and regeneration. *Frontiers in Physiology* 10, 1609. https://doi.org/10.3389/fphys.2019.01609

Chrystal, P. V., Greenhalgh, S., McInerney, B. V., McQuade, L. R., Akter, Y., de Paula Dorigam, J. C., Selle, P. H. and Liu, S. Y. 2021. Maize-based diets are more conducive to crude protein reductions than wheat-based diets for broiler chickens. *Animal Feed Science and Technology* 275, 114867. https://doi.org/10.1016/j.anifeedsci.2021.114867

Chrystal, P. V., Greenhalgh, S., Selle, P. H. and Liu, S. Y. 2020d. Facilitating the acceptance of tangibly reduced-crude protein diets for chicken-meat production. *Animal Nutrition* 6(3), 247–257. https://doi.org/10.1016/j.aninu.2020.06.001

Chrystal, P. V., Moss, A. F., Khoddami, A., Naranjo, V. D., Selle, P. H. and Liu, S. Y. 2020b. Effects of reduced crude protein levels, dietary electrolyte balance and energy density on the performance of broiler chickens offered maize-based diets with evaluations of starch, protein and amino acid metabolism. *Poultry Science* 99, 421.

Chrystal, P. V., Moss, A. F., Khoddami, A., Naranjo, V. D., Selle, P. H. and Liu, S. Y. 2020a. Impacts of reduced-crude protein diets on key parameters in male broiler chickens offered maize-based diets. *Poultry Science* 99(1), 505–516. https://doi.org/10.3382/ps/pez573

Chrystal, P. V., Moss, A. F., Yin, D., Khoddami, A., Naranjo, V. D., Selle, P. H. and Liu, S. Y. 2020c. Glycine equivalent and threonine inclusions in reduced-crude protein, maize-based diets impact on growth performance, fat deposition starch-protein digestive dynamics and amino acid metabolism in broiler chickens. *Animal Feed Science and Technology* 261, 114387. https://doi.org/10.1016/j.anifeedsci.2019.114387

Conde-Aguilera, J. A., Cobo-Ortega, C., Tesseraud, S., Lessire, M., Mercier, Y. and van Milgen, J. 2013. Changes in body composition in broilers by a sulfur amino acid deficiency during growth. *Poultry Science* 92(5), 1266–1275. https://doi.org/10.3382/ps.2012-02796

Conde-Aguilera, J. A., Tesseraud, S., Lessire, M., Geraert, P. A., Mercier, Y. and Van Milgen, J. 2015. *A deficient supply of methionine changes the composition and metabolism of the pectoralis major muscle in growing broilers*. Proceedings of the 26th Australian poultry science symposium, Sydney, Australia.

Corzo, A., Moran, Jr., E. T. and Hoehler, D. 2003. Arginine need of heavy broiler males: applying the ideal protein concept. *Poultry Science* 82(3), 402–407. https://doi.org/10.1093/ps/82.3.402

Creek, R. D. 1968. Non equivalence in mass in the conversion of phenylalanine to tyrosine and methionine to cystine. *Poultry Science* 47(4), 1385–1386. https://doi.org/10.3382/ps.0471385

Cuca, M. and Jensen, L. S. 1990. Arginine requirement of starting broiler chicks. *Poultry Science* 69(8), 1377–1382. https://doi.org/10.3382/ps.0691377

Davey, C. L. 1960. The significance of carnosine and anserine in striated skeletal muscle. *Archives of Biochemistry and Biophysics* 89, 303–308. https://doi.org/10.1016/0003-9861(60)90059-x

Davis, A. T. and Austic, R. E. 1982. Threonine-degrading enzymes in the chicken. *Poultry Science* 61(10), 2107–2111. https://doi.org/10.3382/ps.0612107

de Haas, E. N. and van der Eijk, J. A. J. 2018. Where in the serotonergic system does it go wrong? Unravelling the route by which the serotonergic system affects feather

pecking in chickens. *Neuroscience and Biobehavioral Reviews* 95, 170–188. https://doi.org/10.1016/j.neubiorev.2018.07.007
Dean, D. W., Bidner, T. D. and Southern, L. L. 2006. Glycine supplementation to low protein, amino acid-supplemented diets supports optimal performance of broiler chicks. *Poultry Science* 85(2), 288–296. https://doi.org/10.1093/ps/85.2.288
Dozier, W. A. and Tillman, P. B. 2017. Advances in understanding and improving the role of amino acids in poultry nutrition. In T. Applegate (Ed.), *Achieving sustainable production of poultry meat volume 2: breeding and nutrition* (pp. 113–126). Cambridge, UK: Burleigh Dodds Science Publishing. https://doi.org/10.19103/AS.2016.0011.07
Drozak, J., Chrobok, L., Poleszak, O., Jagielski, A. K. and Derlacz, R. 2013. Molecular identification of carnosine N-methyltransferase as chicken histamine N-methyltransferase-like protein (HNMT-Like). *Plos One* 8(5), e64805. https://doi.org/10.1371/journal.pone.0064805
Drozak, J., Veiga-da-Cunha, M., Vertommen, D., Stroobant, V. and Van Schaftingen, E. 2010. Molecular identification of carnosine synthase as ATP-grasp domain-containing protein 1 (AGTPD1). *Journal of Biological Chemistry* 285(13), 9346–9356. https://doi.org/10.1074/jbc.M109.095505
Elkin, R. G., Featherston, W. R. and Rogler, J. C. 1980. Effects of dietary phenylalanine and tyrosine on circulating thyroid hormone levels and growth in the chick. *Journal of Nutrition* 110(1), 130–138. https://doi.org/10.1093/jn/110.1.130
Fancher, B. I. and Jensen, L. S. 1989. Dietary protein levels and essential amino acid content: influence upon female broiler performance during the growing period. *Poultry Science* 68(7), 897–908. https://doi.org/10.3382/ps.0680897
Featherston, W. R. 1976. Glycine-serine interrelations in the chick. *Federation Proceedings* 35(8), 1910–1913.
Fernstrom, J. D. and Fernstrom, M. H. 2007. Tyrosine, phenylalanine, and catecholamine synthesis and function in the brain. *Journal of Nutrition* 137(6), 1539S–1548S. https://doi.org/10.1093/jn/137.6.1539S
Franco, S. M., Tavernari, F. C., Maia, R. C., Barros, V. R. S. M., Albino, L. F. T., Rostagno, H. S., Lelis, G. R., Calderano, A. A. and Dilger, R. N. 2017. Estimation of optimal ratios of digestible phenylalanine + tyrosine, histidine, and leucine to digestible lysine for performance and breast yield in broilers. *Poultry Science* 96(4), 829–837. https://doi.org/10.3382/ps/pew305
Geraert, P. A., Padilha, J. C. F. and Guillaumin, S. 1996. Metabolic and endocrine changes induced by chronic heat exposure in broiler chickens: biological and endocrinological variables. *British Journal of Nutrition* 75(2), 205–216. https://doi.org/10.1079/bjn19960125
Grau, C. R. 1947. Interrelations of phenylalanine and tyrosine in the chick. *Journal of Biological Chemistry* 170(2), 661–669. https://doi.org/10.1016/S0021-9258(17)30847-5
Greenhalgh, S., Hamilton, E. J., Macelline, S. P., Toghyani, M., Chrystal, P. V., Liu, S. Y. and Selle, P. H. 2022. Dietary crude protein concentrations and L-carnitine inclusions interactively influence performance parameters of grower broiler chickens offered sorghum-based diets. *Animal Feed Science and Technology* 291, 115392. https://doi.org/10.1016/j.anifeedsci.2022.115392
Greenhalgh, S., McInerney, B. V., McQuade, L. R., Chrystal, P. V., Khoddami, A., Zhuang, M. A. M., Liu, S. Y. and Selle, P. H. 2020. Capping dietary starch: protein ratios in

moderately reduced crude protein, wheat-based diets showed promise but further reductions generated inferior growth performance in broiler chickens. *Animal Nutrition* 6(2), 168–178. https://doi.org/10.1016/j.aninu.2020.01.002

Guerranti, R., Pagani, R., Neri, S., Errico, S. V., Leoncini, R. and Marinello, E. 2001. Inhibition and regulation of rat liver L-threonine dehydrogenase by different fatty acids and their derivatives. *Biochimica et Biophysica Acta* 1568(1), 45–52. https://doi.org/10.1016/s0304-4165(01)00197-0

Hamm, D. 1981. Amino acid composition of breast and thigh meat from broilers produced in four locations of the United States. *Journal of Food Science* 46(4), 1122–1124. https://doi.org/10.1111/j.1365-2621.1981.tb03005.x

Han, Y. M., Suzuki, H. and Baker, D. H. 1991. Histidine and tryptophan requirement of growing chicks. *Poultry Science* 70(10), 2148–2153. https://doi.org/10.3382/ps.0702148

Haug, A., Rødbotten, R., Mydland, L. T. and Christophersen, O. A. 2008. Increased broiler muscle carnosine and anserine following histidine supplementation of commercial broiler feed concentrate. *Acta Agriculturae Scandinavica, Section A – Animal Science* 58(2), 71–77. https://doi.org/10.1080/09064700802213545

Hofmann, P., Siegert, W., Ahmadi, H., Krieg, J., Novotny, M., Naranjo, V. D. and Rodehutscord, M. 2020. Interactive effects of glycine equivalent, cysteine, and choline on growth performance, nitrogen excretion characteristics, and plasma metabolites of broiler chickens using neural networks optimized with genetic algorithms. *Animals: an Open Access Journal from MDPI* 10(8), 1392. https://doi.org/10.3390/ani10081392

Hufton, S. E., Jennings, I. G. and Cotton, R. G. 1995. Structure and function of the aromatic amino acid hydroxylases. *Biochemical Journal* 311(2), 353–366. https://doi.org/10.1042/bj3110353

Hurwitz, S., Sklan, D., Talpaz, H. and Plavnik, I. 1998. The effect of dietary protein level on the lysine and arginine requirements of growth chickens. *Poultry Science* 77(5), 689–696. https://doi.org/10.1093/ps/77.5.689

Hyde, R., Taylor, P. M. and Hundal, H. S. 2003. Amino acid transporters: roles in amino acid sensing and signaling in animal cells. *Biochemical Journal* 373(1), 1–18. https://doi.org/10.1042/bj20030405

Kai, S., Watanabe, G., Kubota, M., Kadowaki, M. and Fujimura, S. 2015. Effect of dietary histidine on contents of carnosine and anserine in muscles of broilers. *Animal Science Journal* 86(5), 541–546. https://doi.org/10.1111/asj.12322

Kaufman, S. 1987. Enzymology of the phenylalanine-hydroxylating system. *Enzyme* 38(1–4), 286–295. https://doi.org/10.1159/000469218

Kessler, J. W. and Thomas, O. P. 1976. The arginine requirement of the 4–7 week old broiler. *Poultry Science* 55(6), 2379–2382. https://doi.org/10.3382/ps.0552379

Kidd, M. T. 2004. Nutritional modulation of immune function in broilers. *Poultry Science* 83(4), 650–657. https://doi.org/10.1093/ps/83.4.650

Kidd, M. T. 2021. *Branched-chain amino acids. Metabolism, benefits, and role in disease* (pp. 1–332). Hauppauge: Nova Science Publishers.

Kidd, M. T. and Hackenhaar, L. 2005. Dietary threonine for broilers: dietary interactions and feed additive supplement use. *CABI Reviews* 1, 1–6. https://doi.org/10.1079/PAVSNNR20061005

Kidd, M. T., Maynard, C. W. and Mullenix, G. J. 2021. Progress of amino acid nutrition for diet protein reduction in poultry. *Journal of Animal Science and Biotechnology* 12(1), 45. https://doi.org/10.1186/s40104-021-00568-0

Kohen, R., Yamamoto, Y., Cundy, K. C. and Ames, B. N. 1988. Anti-oxidant activity of carnosine, homocarnosine, aand anserine present in muscle and brain. *Proceedings of the national Avademy of Sciences of the United States of America* 85, 3175–3179.

Kralik, G., Sak-Bosnar, M., Grčević, M. and Kralik, Z. 2018. Effect of amino acids on growth performance, carcass characteristics, meat quality, and carnosine concentration in broiler chickens. *Journal of Poultry Science* 55(4), 239–248. https://doi.org/10.2141/jpsa.0170083

Lackner, J., Hess, V., Stef, L. and Sauerwein, H. 2022. Effects of feeding histidine to lysine ratios on performance, meat quality, and the occurrence of breast myopathies in broiler chickens. *Poultry Science* 101(2), 101568. https://doi.org/10.1016/j.psj.2021.101568

Lambert, W., Rovers, M., Ensink, J., Tesseraud, S., Corrent, E., De Lange, L. and Star, L. 2015. *Interaction between threonine and glycine at low dietary crude protein and the effect on production performance, meat quality and plasma metabolites in broilers*. Proceedings, 20th European Symposium on Poultry Nutrition, Prague, Czech Republic.

Lartey, F. M. and Austic, R. E. 2009. Phenylalanine hydroxylase activity and expression in chicks subjected to phenylalanine imbalance or phenylalanine toxicity. *Poultry Science* 88(4), 774–783. https://doi.org/10.3382/ps.2008-00370

Lee, J. T., Rochell, S. J., Kriseldi, R., Kim, W. K. and Mitchell, R. D. 2023. Functional properties of amino acids: improving health status and sustainability. *Poultry Science* 102(1), 102288. https://doi.org/10.1016/j.psj.2022.102288

Leveille, G. A., Shapiro, R. and Fisher, H. 1960. Amino acid requirements for maintenance in the adult rooster. IV. The requirements for methionine, cystine, phenylalanine, tyrosine, and tryptophan; th adequacy of the determined requirements. *Journal of Nutrition* 72, 8–15. https://doi.org/10.1093/jn/72.1.8

Liao, S. M., Du, Q. S., Meng, J. Z., Pang, Z. W. and Huang, R. B. 2013. The multiple roles of histidine in protein interactions. *Chemistry Central Journal* 7(1), 44. https://doi.org/10.1186/1752-153X-7-44

Lin, C. Y. 1981. Relationship between increased body weight and fat deposition in broilers. *World's Poultry Science Journal* 37(2), 106–110. https://doi.org/10.1079/WPS19810010

Liu, G., Ajao, A. M., Shanmugasundaram, R., Taylor, J., Ball, E., Applegate, T. J., Selvaraj, R., Kyriazakis, I.; Olukosi, O. A. and Kim, W. K. 2023. The effects of arginine and branched-chain amino acid supplementation to reduced-protein diet on intestinal health, cecal short-chain fatty acid profiles, and immune response in broiler chickens challenged with Eimeria spp. *Poultry Science* 102(7), 102773. https://doi.org/10.1016/j.psj.2023.102773

Liu, S. Y., Selle, P. H., Court, S. G. and Cowieson, A. J. 2013. Protease supplementation of sorghum-based broiler diets enhances amino acid digestibility coefficients in four small intestinal sites and accelerates their rates of digestion. *Animal Feed Science and Technology* 183(3–4), 175–183. https://doi.org/10.1016/j.anifeedsci.2013.05.006

Macelline, S. P., Chrystal, P. V., Liu, S. Y. and Selle, P. H. 2021. Implications of elevated threonine plasma concentrations in the development of reduced-crude protein diets for broiler chickens. *Animal Production Science* 61(14), 1442–1448. https://doi.org/10.1071/AN20554

Maharjan, P., Martinez, D. A., Weil, J., Suesuttajit, N., Umberson, C., Mullenix, G., Hilton, K. M., Beitia, A. and Coon, C. N. 2021. Review: physiological growth trend of current meat broilers and dietary protein and energy management approaches for sustainable broiler production. *Animal* 15(Suppl. 1), 100284. https://doi.org/10.1016/j.animal.2021.100284

March, B. E., Biely, J. and Pastro, K. R. 1964. The effect of protein level and amino acid balance upon thyroid activity in the chick. *Canadian Journal of Biochemistry* 42, 341–344. https://doi.org/10.1139/o64-040

Matthews, D. E. 2007. An overview of phenylalanine and tyrosine kinetics in humans. *Journal of Nutrition* 137(6), 1549S–1573S. https://doi.org/10.1093/jn/137.6.1549S

Mora, L., Sentandreu, M. A. and Toldrá, F. 2007. Hydrophilic chromatographic determination of carnosine, anserine, creatine, and creatinine. *Journal of Agricultural and Food Chemistry* 55(12), 4664–4669. https://doi.org/10.1021/jf0703809

Namroud, N. F., Shivazad, M. and Zaghari, M. 2008. Effects of fortifying low crude protein diet with crystalline amino acids on performance, blood ammonia level, and excreta characteristics of broiler chicks. *Poultry Science* 87(11), 2250–2258. https://doi.org/10.3382/ps.2007-00499

NRC. 1994. *Nutrient requirements of poultry* (9th ed.). Washington, DC: National Academy Press.

Ospina-Rojas, I. C., Murakami, A. E., Duarte, C. R., Eyng, C., Oliveira, C. A. and Janeiro, V. 2014. Valine, isoleucine, arginine and glycine supplementation of low-protein diets for broiler chickens during the starter and grower phases. *British Poultry Science* 55(6), 766–773. https://doi.org/10.1080/00071668.2014.970125

Ospina-Rojas, I. C., Murakami, A. E., Eyng, C., Nunes, R. V., Duarte, C. R. A. and Vargas, M. D. 2012. Commercially available amino acid supplementation of low-protein diets for broiler chickens with different ratios of digestible glycine+serine:lysine. *Poultry Science* 91(12), 3148–3155. https://doi.org/10.3382/ps.2012-02470

Ospina-Rojas, I. C., Murakami, A. E., Moreira, I., Picoli, K. P., Rodrigueiro, R. J. B. and Furlan, A. C. 2013. Dietary glycine + serine responses of male broilers given low-protein diets with different concentrations of threonine. *British Poultry Science* 54(4), 486–493. https://doi.org/10.1080/00071668.2013.794257

Perez-Carbajal, C., Caldwell, D., Farnell, M., Stringfellow, K., Pohl, S., Casco, G., Pro-Martinez, A. and Ruiz-Feria, C. A. 2010. Immune response of broiler chickens fed different levels of arginine and vitamin E to a coccidiosis vaccine and Eimeria challenge. *Poultry Science* 89(9), 1870–1877. https://doi.org/10.3382/ps.2010-00753

Powell, S., Bidner, T. D. and Southern, L. L. 2011. Effects of glycine supplementation at varying levels of methionine and cystine on the growth performance of broilers fed reduced crude protein diets. *Poultry Science* 90(5), 1023–1027. https://doi.org/10.3382/ps.2010-01247

Qi, B., Wang, J., Hu, M., Ma, Y., Wu, S., Qi, G., Qiu, K. and Zhang, H. 2021. Influences of beta-alanine and L-histidine supplementation on growth performance, meat quality, carnosine content, and mRNA expression of carnosine-related enzymes in broilers. *Animals: an Open Access Journal from MDPI* 11(8), 2265. https://doi.org/10.3390/ani11082265

Robbins, K. R., Baker, D. H. and Norton, H. W. 1977. Histidine status in the chick as measured by growth rate, plasma free histidine and breast muscle carnosine. *Journal of Nutrition* 107(11), 2055–2061. https://doi.org/10.1093/jn/107.11.2055

Roth, F. X., Gruber, K. and Kirchgessner, M. 2001. The ideal dietary amino acid pattern for broiler-chicks of age 7 to 28 days. *Archiv für Gelflügelkunde* 65, 199–206.

Salway, J. G. 2018. The Krebs uric acid cycle: a forgotten Krebs cycle. *Trends in Biochemical Sciences* 43(11), 847–849. https://doi.org/10.1016/j.tibs.2018.04.012

Sasse, C. E. and Baker, D. H. 1972. The phenylalanine and tyrosine requirements and their interrelationship for the young chick. *Poultry Science* 51(5), 1531–1536. https://doi.org/10.3382/ps.0511531

Schneider, F. 1978. Histidine in enzyme active centers. *Angewandte Chemie* 17(8), 583–592. https://doi.org/10.1002/anie.197805831

Selle, P. H., Cantor, D. I., McQuade, L. R., McInerney, B. V., de Paula Dorigam, J. Cde, P., Macelline, S. P., Chrystal, P. V. and Liu, S. Y. 2021b. Implications of excreta uric acid concentrations in broilers offered reduced-crude protein diets and dietary glycine requirements for uric acid synthesis. *Animal Nutrition* 7(4), 939–946. https://doi.org/10.1016/j.aninu.2021.03.011

Selle, P. H., Chrystal, P. V. and Liu, S. Y. 2020. The cost of deamination in reduced-crude protein broiler diets. *Proceedings, Australian Poultry Science Symposium* 31, 63–66.

Selle, P. H., Macelline, S. P., Chrystal, P. V. and Liu, S. Y. 2021a. A rationale for elevated free threonine plasma concentrations in chickens offered reduced-crude protein diets. *Proceedings, Australian Poultry Science Symposium* 32, 121–124.

Siegert, W., Ahmadi, H., Helmbrecht, A. and Rodehutscord, M. 2015. A quantitative study of the interactive effects of glycine and serine with threonine and choline on growth performance in broilers. *Poultry Science* 94(7), 1557–1568. https://doi.org/10.3382/ps/pev109

Siegert, W. and Rodehutscord, M. 2019. The relevance of glycine and serine in poultry nutrition: a review. *British Poultry Science* 60(5), 579–588. https://doi.org/10.1080/00071668.2019.1622081

Stern, R. A. and Mozdziak, P. E. 2019. Differential ammonia metabolism and toxicity between avian and mammalian species, and effect of ammonia on skeletal muscle: a comparative review. *Journal of Animal Physiology and Animal Nutrition* 103(3), 774–785. https://doi.org/10.1111/jpn.13080

Sugahara, M., Baker, D. H. and Scott, H. M. 1969. Effect of different patterns of excess amino acids on performance of chicks fed amino acid-deficient diets. *Journal of Nutrition* 97(1), 29–32. https://doi.org/10.1093/jn/97.1.29

Sugahara, M. and Kandatsu, M. 1976. Glycine serine interconversion in the rooster. *Agricultural and Biological Chemistry* 40(5), 833–837. https://doi.org/10.1080/00021369.1976.10862153

Tabiri, H. Y., Sato, K., Takahashi, K., Toyomizu, M. and Akiba, Y. 2002. Hepatic tyrosine aminotransferase activity is affected by chronic heat stress and dietary tyrosine in broiler chickens. *British Poultry Science* 43(4), 629–634. https://doi.org/10.1080/0007166022000004561

Teng, P.-Y., Choi, J., Yadav, S., Tompkins, Y. H. and Kim, W. K. 2021. Effects of low-crude protein diets supplemented with arginine, glutamine, threonine, and methionine on regulating nutrient absorption, intestinal health, and growth performance of Eimeria-infected chickens. *Poultry Science* 100(11), 101427. https://doi.org/10.1016/j.psj.2021.101427

Tesseraud, S., Everaert, N., Boussaid-Om Ezzine, S., Collin, A., Métayer-Coustard, S. and Berri, C. 2011. Manipulating tissue metabolism by amino acids. *World's Poultry Science Journal* 67(2), 243–252. https://doi.org/10.1017/S0043933911000274

Tourian, A., Goddard, J. and Puck, T. T. 1969. Phenylalanine hydroxylase activity in mammalian cells. *Journal of Cellular Physiology* 73(2), 159–170. https://doi.org/10.1002/jcp.1040730210

Van der Meulen, J., Bakker, J. G. M., Smits, B. and De Visser, H. 1997. Effect of source of starch on net portal flux of glucose, lactate, volatile fatty acids and amino acids in the pig. *British Journal of Nutrition* 78(4), 533–544. https://doi.org/10.1079/bjn19970173

Wang, W., Wu, Z., Dai, Z., Yang, Y., Wang, J. and Wu, G. 2013. Glycine metabolism in animals and humans: implications for nutrition and health. *Amino Acids* 45(3), 463–477. https://doi.org/10.1007/s00726-013-1493-1

Wang, W.-W., Wang, J., Wu, S.-G., Zhang, H.-J. and Qi, G.-H. 2020. Response of broilers to gradual dietary protein reduction with or without an adequate glycine plus serine level. *Italian Journal of Animal Science* 19(1), 127–136. https://doi.org/10.1080/1828051X.2019.1704634

Wise, T. L., Kidd, M. T. and Dozier III, W. A. 2021), Chapter 9. Digestion, intestinal functionality, and metabolism of isoleucine and valine with emphasis in poultry. In M. T. Kidd (Ed.), *Branched-chain amino acids. Metabolism, benefits, and role in disease* (pp. 249–274). New York: Nova Science Publishers, Inc.

Wixom, R. L., Pipkin, G. E. and Day, P. L. 1955. Interrelationship of serine and glycine for chick growth. *Journal of Nutrition* 56(3), 409–422. https://doi.org/10.1093/jn/56.3.409

Womack, M. and Rose, W. C. 1934. Feeding experiments with mixtures of highly purified amino acids. *Journal of Biological Chemistry* 107(2), 449–458. https://doi.org/10.1016/S0021-9258(18)75363-5

Wu, G. 2014. Dietary requirements of synthesizable amino acids by animals: a paradigm shift in protein nutrition. *Journal of Animal Science and Biotechnology* 5(1), 34. https://doi.org/10.1186/2049-1891-5-34

Wylie, L. M., Robertson, G. W. and Hocking, P. M. 2003. Effects of dietary protein concentration and specific amino acids on body weight, body composition and feather growth in young turkeys. *British Poultry Science* 44(1), 75–87. https://doi.org/10.1080/0007166031000085346

Yuan, J., Karimi, A., Zornes, S., Goodgame, S., Mussini, F., Lu, C. and Waldroup, P. W. 2012. Evaluation of the role of glycine in low-protein amino acid-supplemented diets. *Journal of Applied Poultry Research* 21(4), 726–737. https://doi.org/10.3382/japr.2011-00388

Chapter 11

Advances in understanding and improving the use of enzymes in broiler nutrition

Tanner L. Wise, Jung Yeol Sung, and Olayiwola Adeola, Purdue University, USA

1 Introduction

The use of exogenous enzymes in the diets of non-ruminant animals has grown significantly from the beginning of the twenty-first century. Many of the enzymes in common use today have been developed over many decades and arose out of a necessity to increase livestock feed efficiency (Jensen et al., 1957; Willingham et al., 1961). Through constant development, there is now an increased ability to modify microorganisms to produce such enzymes that has made their production both easier and cheaper (Ravindran, 2013). This means that these enzymes can be included in commercial diets of animals such as broilers and growing pigs. Although there is now a wide array of enzyme products and manufacturers, the understanding of the function and efficacy

http://dx.doi.org/10.19103/AS.2024.0143.15

of many enzymes is still in its early stages. New enzymes are developed and released every year, making it imperative to continue to research and evaluate these products to understand how they can be used to more efficiently feed livestock.

Some of the most common enzymes in commercial use are phytases, carbohydrases, and proteases. These enzymes can help to lower feed costs. Phytase for example can reduce the need for expensive inorganic phosphorus (P) sources as it is able to liberate phytate-bound P from organic sources that are already in the diet. Phytase has also been shown to positively impact the digestibility of many other nutrients, including cationic minerals and amino acids (AAs). Carbohydrases can increase the energy availability of diets by breaking down carbohydrates that are not readily degradable by endogenous enzymes, allow inclusion of less digestible ingredients, and reduce the need for oil sources in the diet. Proteases may increase the AA digestibility of some feed ingredients and therefore reduce the amount of expensive protein substrates that are required in the diet.

Beyond the potential for reduced feed costs, many of these products can increase the production efficiency of animals. Exogenous enzymes can stimulate greater release of endogenous enzymes, increase the secretion of protective mucins in the small intestine, cause shifts in gut microbes, and prevent the flow of nutrients to pathogens in the large intestine. The enzymes discussed previously have the potential to positively affect the production and efficiency of non-ruminant animals. However, continued research is key to understanding the benefits and limitations of these products. This chapter reviews research relating to better understanding the mechanisms of action and enabling more efficient use of these top three classes of enzymes being used in monogastric production.

2 Phytases

Phytase is perhaps the most widely used and available exogenous enzyme for non-ruminant animal production. Its primary purpose is to systematically degrade phytate molecules from plant origin to release P that is available to the animal. Most animals do not produce adequate amounts of endogenous phytase, so the P bound in phytate is largely not utilized by the animal but excreted instead. This is not only wasteful for the producer but can also contribute to negative effects on the environment. Exogenous phytases have been shown to be very effective in the liberation of P from organic sources and allow for the reduced use of expensive and limited inorganic rock P sources. This not only makes diets more sustainable but can also reduce dietary costs. The effects of phytases on improving P digestibility and utilization are well studied and the observed effects are very consistent.

However, phytase has many other benefits to the diets deemed, 'extra-phosphoric effects'. These effects include promoting growth performance and feed intake, improving calcium (Ca) and other mineral digestibility, increasing dietary apparent metabolizable energy (AME), and improving the digestibility of AAs. Many of these effects stem primarily from reducing the anti-nutritional properties of phytate. At physiological pH, phytate molecules have a strong negative charge and are able to undergo chelate formation with positively charged substrates in the stomach and intestinal lumen. This reduces the availability of substrates such as starches, minerals, and proteins to their respective endogenous enzymes and subsequent absorption. Degradation of phytate in the gut as a result of phytase supplementation can have a number of positive impacts on nutrient digestibility and broiler performance. These extra-phosphoric effects have garnered much research attention in recent years and are becoming better understood. In fact, many phytase enzymes are supplemented at the expense of some concentration of dietary Ca, energy and AAs.

However, some areas of phytase efficacy are still not well understood. Factors like the dietary concentration of energy or phytate and the underlying digestibility of AA have been shown to have some relationship with the efficacy of phytase. Additionally, there is speculation that some of the extra-phosphoric effects may be driven by the effects of inositol liberated by phytase and its effects on feed intake and muscle accretion, more so than its effects on nutrient digestibility. The following sections will focus primarily on exploring these areas of research for phytase enzymes.

3 Factors affecting phytase efficacy

Phytases are well known to have very high efficacy in improving digestibility and utilization of P in broiler diets. Additionally, phytases have been shown to produce a host of other benefits including increasing growth performance and the digestibility of minerals, energy, and AA. While all of these effects are relatively consistent, there are a number of factors, both dietary and otherwise, that can affect the relative efficacy to degrade phytate (IP6) and enact beneficial effects. This includes factors like:

- concentrations of Ca and inorganic P (P_i) in the diet;
- dietary concentration of IP6 and what it is chelated with; and
- inclusion of other dietary additives.

All of these factors interact with phytase efficacy and impact its ability to degrade its target substrate.

3.1 Concentrations of calcium and inorganic phosphorus in the diet

It has been long understood that there is a relationship between the dietary concentration of Ca and the digestibility of P with research showing that high Ca inclusion reduced the ileal digestibility of P_i, IP6, and total P for broilers (Tamin et al., 2004). This observation has been made for P digestibility many times and has been suggested as the primary factor reducing phytase efficacy (Liu et al., 1998; Kornegay, 2001; Angel et al., 2002). Ca readily chelates with IP6 and forms a complex that is very difficult to degrade both by phytase or endogenous actions. While Ca does not have the greatest affinity of all cations to IP6, it is often the most abundant in broiler diets, making it a primary factor (Liu et al., 1998).

Beyond the abundance and high incidence of these IP6–Ca compounds, there is also a belief that these compounds cause greater anti-nutritional effects compared to other chelated compounds like IP6–Na or IP6–protein (Tran et al., 2011). This has been an important area of research, and researchers have found a number of possible ways to combat the antagonistic effects of Ca. Dersjant-Li et al. (2015) observed that, in the total absence of dietary Ca, P had very high digestibility and phytase showed limited overall improvement due to this high digestibility. This is not, of course, a practical solution; however, it does demonstrate that taking Ca away can be a solution. This is exactly what many researchers have observed for P digestibility with phytase (Lei et al., 1994; Liu et al., 1998; Lei and Stahl, 2000; Kornegay, 2001) and without phytase (Liu et al., 1998; Angel et al., 2002; Tamin et al., 2004; Walk et al., 2012). It has manifested by reducing the Ca:P ratios in diets, regardless of phytase inclusion. The relative consensus of this research is that appropriate total Ca:P ratios are around 1.2:1 for maximizing P digestibility and phytase efficacy. This means that it is critical that the Ca content of diets is well managed, and care is given to evaluating feed ingredients. Even with this heightened awareness, Walk et al. (2012) observed that the average diet being used to evaluate phytase efficacy was supplemented with approximately 0.2% excess Ca, regardless of the matrix value assigned to the phytase.

In addition to excess Ca in the diet, excess dietary P_i has been shown to reduce the efficacy of phytase enzymes (Greiner et al., 1993; Ravindran et al., 2000; Kornegay, 2001). It has been reported that the use of a phytase enzyme with high activity at low pH (as in the proventriculus and gizzard) can help combat these negative impacts, as it creates lower IP esters early in digestion that have a reduced ability to chelate nutrients (Walk et al., 2012; Derjant-Li et al., 2015). It has also been reported that the negative effects of increased Ca can be partially ameliorated by increasing the dietary concentration of Vitamin D (cholecalciferol), which can help to facilitate greater Ca absorption

(Lei et al., 1994; Lei and Stahl, 2000). There has also been research on the effect of limestone particle size in reducing the effects of high levels of Ca (Manangi and Coon, 2007).

3.2 Inclusion of other dietary additives

The ingredient composition of diets has also been shown to influence the relative efficacy of phytase to increase P and AA digestibility. Morales et al. (2012a, 2012b) reported a significant effect of the primary plant protein source on the effects of phytase, with the largest differences coming from soy and canola meals. This response was also observed by Newkirk and Classen (1998) where IP6–canola protein compounds proved more difficult to degrade than IP6–soy protein compounds, an effect that has also been observed by Tran et al. (2011), though IP6–protein compounds are as difficult to hydrolyze as IP6–mineral compounds (Angel et al., 2002). This could suggest that the source of phytate can affect phytase efficacy. However, there are multiple reports that observed a relatively consistent effect of phytase across diets with differing inclusions of ingredients and nutrient densities but with consistent IP6 concentrations (Leske and Coon, 1999; Babatunde et al., 2021; Kriseldi et al., 2021d). This may suggest that it is not the source but rather the concentration of IP6 that can affect phytase efficacy.

This interaction has been investigated several times with somewhat inconsistent results. Ravindran et al. (2000) reported no interaction between dietary IP6 concentration and phytase on phytase-related improvements in P and AA digestibility, though the main effect of phytase was significant at all levels of IP6. Similar results were obtained by Dilger et al. (2004), who observed no significant interaction; however, the digestibility of P and AA were numerically improved for the high IP6 diets compared to the control. This is mirrored by another study that found no significant interaction but observed numerically greater phytase-related responses of P digestibility with higher IP6 concentrations (Babatunde et al., 2021). More recently, Wise et al. (Unpublished data) observed greater phytase-related responses of P and crude protein digestibility in diets with elevated IP6 concentrations compared to a control; however, this study also did not have a significant interaction.

It has been suggested, however, that dietary IP6 concentration is a primary driver of relative phytase efficacy (Kornegay, 2001). This is supported by the findings of Sands et al. (2007) who observed a significant improvement of relative phytase efficacy in diets with increased IP6 concentrations, even when IP6 inclusions did not elicit significant anti-nutritive effects. The relationship between IP6 and phytase is likely based on the negative impact of IP6 on the basal endogenous flow of AA and other nutrients. Cowieson and Ravindran (2007) reported that the presence of undegraded IP6 in the diet can increase the endogenous flow of AA by 25%, which also has a negative impact on the endogenous losses of many other nutrients. However, phytase inclusion was found to nearly fully ameliorate these negative effects, an observation that has been made in other studies (Cowieson et al., 2004, 2008).

There is another class of feed additive that has been shown to increase the efficacy of phytases and in many cases produce synergistic effects: organic acids. These compounds, primarily lactic and citric acids, are added to diets to act as energy sources and help in modulation of the mucosal layer and microbial growth, but they have also been observed to increase the relative efficiency of phytases (Lei and Stahl, 2000). This has been observed with cocktails of organic acids, as well as individually with lactic acid (Kemme et al., 1999) and citric acid (Phromkunthong et al., 2010) in broiler diets.

However, not all feed additives interact positively with phytases. The addition of pharmacological levels of Cu and Zn as alternatives to antibiotics has been shown to be antagonistic to the effects of phytases (Augspurger et al., 2004). This is likely due to the strong positive charge of these minerals and their relatively high affinity for IP6, causing increased chelation. Additionally, it has been suggested that additions of other exogenous enzymes have the potential to impact the relative efficacy of phytases, both synergistically and antagonistically (Cowieson and Bedford, 2009), a topic that will be discussed in greater detail later in the chapter.

4 Inositol-related effects of phytase

Phytase degrades the phytate molecule in a stepwise fashion by removing one phosphate group at a time, resulting in what are referred to as lower phytate esters (Selle and Ravindran, 2007). These molecules can have anywhere from one to five (IP1 to IP5) remaining phosphate groups; however, the majority of partially degraded molecules are IP3 or IP4 (Walk et al., 2014). Ultimately, if all of the phosphate groups are removed, only the central inositol molecule will remain (Selle and Ravindran, 2007).

Myo-inositol is a bioactive molecule that is very highly digested in the small intestine of broilers (potentially as high as 99.8% digestibility) (Croze and Soulage, 2013). Some have suggested that inositol may play a significant role in the extra-phosphoric effects of phytase (Beeson et al., 2017) and others have argued that inositol gives some level of growth-promoting effect (Cowieson et al., 2013; Walk et al., 2014). This may be accomplished through increased lean tissue development, as inositol can be converted into phosphatidylinositol after absorption, which plays a role in skeletal muscle growth (Lee and Bedford, 2016).

It has also been postulated that the increases in growth performance are more directly related to increased feed intake as opposed to other effects of phytase such as effects on nutrient digestibility (Walk and Olukosi, 2019). These authors reported a positive correlation between increased intake of AA stimulated by phytase supplementation and improved growth performance. Kriseldi et al. (2019) reported a similar relationship between increased intake from phytase supplementation and improved growth and muscle accretion. If both AA intake and pre-cecal digestibility are improved, there could be potential for substantial improvements in muscle growth.

It has also been hypothesized that increases in feed intake with phytase supplementation are directly associated with the liberation and uptake of inositol or some other nutrient liberated by phytase (Dridi, 2017). It has been suggested that inositol may act as an indicator or precursor to alter the gene expression and production of appetite hormones and catecholamines and ultimately stimulate greater feed intake. These hypotheses rely, however, on phytase effectively liberating inositol fully from phytase.

Bedford and Walk (2016) reported that lower inclusions of phytase (e.g. 500 FTU/kg) struggle to achieve full degradation of phytase to inositol. This is supported by other reports that there is commonly an accumulation of IP3 and IP4 esters in the gut of broilers (Walk et al., 2014; Beeson et al., 2017). While there are still positive effects observed from this level of degradation, particularly for P, this still presents an issue. This incomplete breakdown does not facilitate maximum inositol liberation, and lower IP esters are still capable of enacting significant anti-nutritional effects, though lesser than phytate (Yu et al., 2012). This would therefore necessitate greater inclusions of phytase in order to evaluate the possible effects of inositol. Kriseldi et al. (2021a) investigated the ability of low (400 FTU/kg) and moderate (1200 FTU/kg) concentrations of phytase to degrade phytase and increase inositol in the gut and subsequently in the plasma. These authors found that, in the gizzard, only 1200 FTU/kg significantly reduced phytate concentrations and increased inositol concentrations. Additionally, total IP esters were only reduced by the higher dose, though there was no difference between 1200 and 400 FTU/kg in the IP3 and IP4 concentrations. The results were similar for ileal digesta. However, the results for inositol liberation were much stronger with 1200 FTU/kg producing a more than 300% increase over 400 FTU/kg. As inositol is very highly absorbed, this translated into increases in plasma inositol concentrations being greater for the 1200 FTU/kg treatment. Although there was substantial inositol liberation in this study, there were still significant concentrations of the lower IP esters in the ileum, indicating incomplete breakdown of much of the phytate (Kriseldi et al., 2021a).

Subsequent work by Kriseldi et al. (2021c) evaluated the ability of greater concentrations (up to 40,500 FTU/kg) to completely degrade phytate (Fig. 1). These authors reported a nearly complete degradation of dietary phytate in the gizzard and ileum with 40,500 FTU/kg of phytase; however, similar results were

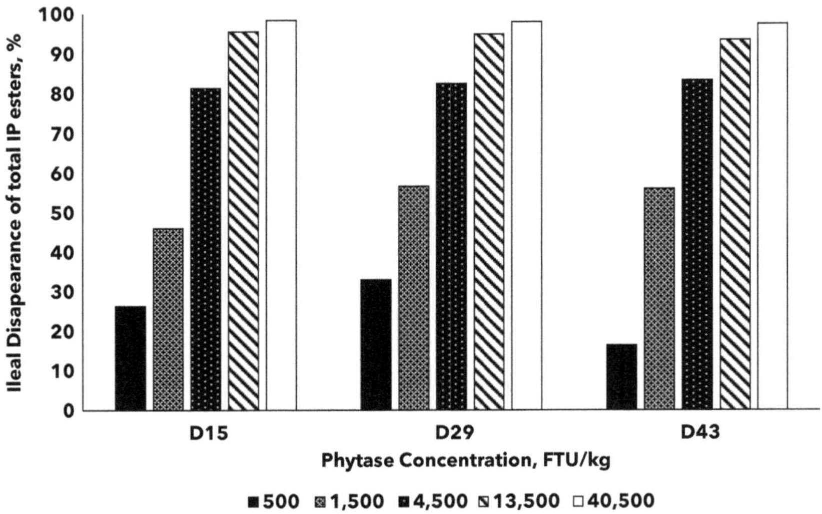

Figure 1. Ileal disappearance of total inositol phosphate (IP) esters at 500, 1500, 4500, 13,500, and 40,500 FTU/kg supplemented phytase on d 15, 29, and 43 for broiler chickens. Source: Data adapted from: Kriseldi et al. (2021c).

obtained with 4500 FTU/kg. This degradation resulted in a nearly complete liberation of inositol and left very limited lower IP esters in the gut. As in the previous study, this resulted in increased plasma concentrations of inositol, but there were diminishing returns above 4500 FTU/kg (Kriseldi et al., 2021c). Interestingly, these authors also included a diet where inositol concentrations equal to that released by total phytate degradation were fed independently from phytase and this diet showed a less efficient uptake of inositol from the gut to the plasma, indicating that there may be some other effect of phytase aiding in the uptake of inositol.

This data shows that inositol from phytate degradation can be efficiently released and absorbed by broilers; however, this is only meaningful if it can be translated into growth performance. Sommerfeld et al. (2018) reported that increased plasma inositol concentrations from feeding 3000 FTU/kg were correlated with increases in feed intake along with increased ileal digestibility of P, Ca, and AA. This is an interesting result; however, it does not tie these responses directly to increased inositol concentrations.

Kriseldi et al. (2021b) investigated this relationship between growth performance and carcass characteristics for broilers. They fed broilers diets with increasing phytase concentrations as well as a diet supplemented with a concentration of purified inositol that would result from total phytate degradation. The authors observed linear and quadratic effects of increased phytase for growth, feed intake, and efficiency of broilers, with optimum growth performance coming from birds supplemented with 4500 FTU/kg phytase.

However, the diets supplemented with purified inositol did not match the performance of the birds fed with differing concentrations of phytase and, in some cases, broilers showed worse performance than the control (Kriseldi et al., 2021b). These authors also observed positive effects on the carcass characteristics of broilers with phytase supplementation, with 4500 FTU/kg providing the optimum response. Similarly to growth performance, the diet supplemented with inositol did not match the performance of any diet supplemented with phytase. However, the birds fed inositol did have numerically improved carcass and breast meat weights compared to the control, indicating a possible effect of inositol on lean tissue development (Kriseldi et al., 2021b). This study demonstrated that, even though high concentrations of phytase can translate into high concentrations of plasma inositol, inositol alone is not likely to improve the growth performance of broilers, though understanding this exact relationship would be a good area for additional study.

While the research discussed so far did not uncover a direct relationship between inositol and growth performance, it has been reported that a primary driver of phytase-related effects on growth performance is strongly related to increased feed intake (Walk and Olukosi, 2019). Phytase has been shown to stimulate feed intake both dependently and independently of large increases in growth, and this likely represents some stimulatory impact of phytase on broilers' drive to eat. As mentioned earlier, Dridi (2017) hypothesized that inositol, or some other metabolite of phytate degradation, may act as a precursor or stimulant to the increased production of appetite hormones or catecholamines and that this could be responsible for the phytase-related intake responses.

Kriseldi et al. (2021d) evaluated the effects of supplementing 4500 FTU/kg phytase (which had been previously shown to produce near complete phytate degradation) to broilers on the production of appetite hormones and catecholamines. These authors reported no effect of phytase supplementation on the production of ghrelin or neuropeptide Y; however, there was an observed increase in proopiomelanocortin, a hormone related to satiety. There was also an observed increase in dopamine production with phytase inclusion, which could be tied to increased drive to eat. While this result cannot be tied directly to inositol, Gonzales-Uarquin et al. (2020) reported a relationship between plasma inositol concentrations and increased dopamine production. It should be noted that the effect on intake observed by Kriseldi et al. (2021d) was clear if modest. These results may indicate some relationship between phytase and increased hypothalamic intake stimulation, but this is an area that requires additional research. While the relationship between inositol and feed intake is not clear, this research does make it clear that plasma inositol concentrations are a consistent indicator of phytase efficacy in total phytate degradation, and this could be a useful tool in further evaluations of exogenous phytases.

5 Proteases

Proteases are adept at the hydrolysis of large, three-dimensional-structured polypeptides-proteins (AAs residues joined together by peptide bonds connecting the carboxyl group of one AA to the amino group of the next AA in the chain). Dietary supplementation of exogenous proteases is a widely accepted strategy to improve the digestibility of proteins found in common feed ingredients (Erdaw et al., 2017). These enzymes have historically been incorporated into multi-enzyme blends in order to improve the general performance of monogastric animals. The specific role of proteases in these enzyme cocktails was not well understood. However, with the rising cost of supplying dietary protein, and greater investigation into their implicit value, mono-component proteases have become popular tools in dietary formulation.

Some have employed formulation techniques similar to those for phytases, where a protease is assigned matrix values for AA and there is a subsequent reduction in dietary AA density via a reduction in expensive protein concentrates (Romero et al., 2014). However, this can be very difficult to accomplish for proteases. Unlike phytases that have one substrate, proteases can accommodate a wide range of proteins and peptides of various sizes and complexity. This can make the efficacy of these products quite variable, meaning relatively large variations in appropriate matrix values. However, regardless of these inconsistencies in effect, proteases have been shown to produce relatively consistent results in the overall performance of broilers, even with reduced dietary protein. This section will review the current literature on the effects of mono-component proteases on broiler performance and nutrient digestibility, along with current methods of employing these enzymes.

6 Protease effects on growth performance

The use of exogenous proteases began due to their positive effects on growth performance, and this remains a primary draw for their use in commercial diets. Often proteases will be evaluated in diets where one or more primary nutrients are reduced in the expectation that the enzyme will produce consistent performance with a control. This is often done by assigning a matrix value of nutrients to the protease and making a subsequent reduction of these nutrients in the diet.

Lee et al. (2023) evaluated a mono-component protease (ProAct 360™) in the diet of broilers based on the full recommended matrix value from the manufacturer. These authors completed a three-phase feeding of three diets: a PC, NC [making the deduction for the matrix values without enzyme addition; 25 kcal/kg AME and 0.721% crude protein (CP)], and the NC + 50 mg/kg protease. The authors did not observe any significant changes in the growth or efficiency of the broilers during any period. However, due to the cost reduction of the NC diet, the addition of the protease increased the profit margin by about 5%. Amerah et al. (2017) evaluated a protease (*Bacillus subtilis*) under similar dietary conditions (reduction of approximately 50 kcal/kg and 0.3% CP) and found that 4000 units/kg of protease produced similar results to the control for day (d) 1 to 21 weight gain and feed conversion ratio (FCR) and improved d 1 to 42 FCR over the NC. These results are similar to that of Angel et al. (2011) who found that even with greater CP reduction (2.0%) weight gain and FCR could be returned to that of the PC with 7500 and 15,000 units/kg of protease (*Bacillus licheniformis*), respectively. McCafferty et al. (2022) also evaluated a protease (Ronozyme ProAct CT™) from d 1 to 35 in diets with three levels of AME and found that the protease increased the d 1 to 29 weight gain and decreased the d 1 to 15 FCR independently of the dietary energy level. The protease did not appear to have any effect on the growth performance from d 29 to 35. However, the performance of the protease at 15,000 units/kg was greater than that of the control in the starter and grower phases.

These studies support protease having value in marginally deficient corn-soybean meal-based diets. However, one of the primary difficulties in predicting protease performance is variability in dietary ingredients. Simbaya et al. (1996) observed that protease elicited improvements in weight gain and FCR from broilers being fed semi-purified canola meal-based diets, subsequently confirmed in more practical diets with canola meal and soybean meal (Hessing et al., 1996). More recently, Toghyani et al. (2017) evaluated a protease (Ronozyme ProAct™) in diets with canola meal and soybean meal each as primary protein sources. These authors found that there was no difference in the action of protease between the diet types and observed significant improvements in the d 35 body weight and the d 10 to 35 FCR compared to the controls. This is contrary to the findings of Siegert et al. (2019) who found that diets with soybean meal as the primary protein source saw improvements due to protease inclusion, while diets with a high proportion of canola meal were unaffected. Additionally, diets with high concentrations of sunflower meal produced similar results to the soybean diets. These protease-related improvements were in feed efficiency while all diet types were unaffected by protease inclusion with regards to weight gain.

Yan et al. (2017) evaluated another protease (CIBENZA DP100™) in rye–wheat–soybean meal-based diets. These authors observed improvements in the d 1 to 21 weight gain and d 1 to 7 FCR, similarly to what was observed by McCafferty et al. (2022) with corn–soy-based diets. There may be some evidence for variable effects of proteases based on the ingredient composition of the diets. However, it is likely that these differences would be more clearly demonstrated using nutrient digestibility analyses.

Proteases have also been evaluated in complete diets that do not have nutrient deficits. Aderibigbe et al. (2020a) evaluated Ronozyme ProAct™ at

15,000 unit/kg in broilers from d 1 to 21 and observed improvements of 5% in weight gain and 4% in feed efficiency in corn–soybean meal-based diets. However, these observations were considerably better than other studies with similar designs. Kalmendal and Tauson (2012) observed improvements in the d 1 to 34 FCR but no effect on weight gain. Borda-Molina et al. (2019) also observed moderate responses for feed efficiency and average daily gain from two of the three proteases evaluated. However, these responses were variable and were not consistent across multiple inclusions of the enzymes. Even in evaluations where consistent results were obtained, the results were not observed at the same magnitude as for Aderibigbe et al. (2020a), and positive results were also primarily found in the starter and grower phases (Simbaya et al., 1996; Yan et al., 2017; McCafferty et al., 2022).

Simbaya et al. (1996) evaluated five different mono-component proteases (but did not identify individual enzymes) and found quite variable results. In the authors' first evaluation with semi-purified diets, two of the enzymes produced improvements in growth in canola-meal-based diets. In a second experiment with practical diets, it was found that only one of the enzymes promoted growth. However, the response was dependent on both the protein level of the diets and ingredient composition with positive responses being for diets with lower protein and soybean meal as the primary protein substrate (Simbaya et al., 1996). These results suggest that there are still many factors that require consideration when evaluating protease-related effects and that effects of individual enzymes are a paramount factor to be considered. In considering the wide range of dietary factors contributing to the effects of protease, Lee et al. (2018a) performed a meta-analysis of 67 published studies. These authors observed the average effects of protease as a 1.4% increase in weight gain and about a 1% decrease in FCR. These values take into account a wide variety of dietary formulations, enzymes, and conditions and can be seen as representative of what the expectations should be for protease in practical commercial diets.

7 Protease effects on nutrient digestibility

As discussed earlier, the use of proteases began with a primary focus on increasing growth performance of animals, and any effects on the digestibility and utilization of dietary nutrients were secondary. However, as the interest in these enzymes has increased, considerably more research has focused on the potential benefits exogenous proteases can have on dietary nutrient digestibility. As the primary metabolic action of proteases is to degrade proteinaceous substrates, it would seem evident that they would have the potential to increase protein and AA digestibility. This assumption has proven to be generally true. However, there are several factors that can affect protease efficacy and multiple modes of action for their improvement.

As with other enzymes like phytases, proteases can also have additional benefits beyond their target substrates. Proteases have been shown to increase the digestibility of energy sources (Kalmendal and Tauson, 2012; Olukosi et al., 2015; Amerah et al., 2017), increase mineral digestibility (Olukosi et al., 2015) and produce improvements in intestinal physiology and mucosal production (Kalmendal and Tauson, 2012; Yan et al., 2017; Lee et al., 2023). This section will focus on the proteinaceous and extra-proteinaceous effects of proteases for broilers and how they may affect their use.

As in the case of growth performance, proteases are often utilized to produce a net neutral response to a test diet against a control with one or more nutrient deficits. This is also true for evaluations of protein and AA digestibility. However, in many cases, the addition of a protease to a NC will result in elevated performance relative to both the NC and the PC. This was true in the evaluation performed by Lee et al. (2023) where a NC (marginally reduced CP and AME) supplemented with 50 mg/kg protease produced 4.3% greater digestibility of CP compared to the NC and a 1% greater response than the PC. This response was evenly distributed between dispensable and indispensable AA with an average response of about +3.0% for each. Both responses were significantly greater than both the NC and PC (Lee et al., 2023).

A similar response was observed by Angel et al. (2011) where protease elevated the digestibility of CP and many AA in a protein-deficient diet. The responses of CP, Lys, and most other affected AA were greater than both the NC and PC at all levels of protease supplementation. The digestibility of Met was only elevated over the PC at the highest two supplementations but showed greater responses than the NC at all levels. Threonine was also the most affected AA in this study. Amerah et al. (2017) also observed similar results where protease increased the ileal and total tract digestibility of CP over the NC (CP and energy deficient) and the PC. The response was greater for ileal digestibility than for total tract digestibility, which is contrary to what was observed by Olukosi et al. (2015); however, these test diets were not at a nutrient deficit. Olukosi et al. (2015) observed increases in CP ileal digestibility of 3.3–6.7% with protease doses from 3000 to 10,000 units/kg and increases in total tract digestibility from 4.0% to 17.7% over the same range of supplementation. Kalmendal and Tauson (2012) also evaluated total tract digestibility of CP and found an average increase of 7.7% over a nutritionally adequate control with protease supplementation. This is significantly greater than the average response for ileal digestibility reported in other studies (Cowieson and Roos, 2014; Lee et al., 2018a). This may indicate that proteases have a greater response for total tract digestibility relative to ileal digestibility; however, this effect is not clear.

Many of the previously discussed results suggest that the nutrient density of the diet may have an effect on the efficacy of proteases. However, a meta-analysis performed by Cowieson and Roos (2014) found a mean protease-related increase in AA digestibility of 3.74% that was independent of both species effect and dietary composition. There were no significant effects of ingredient composition or AA density observed in this analysis, but there may have been effects from the underlying digestibility of AA. Further analysis of a similar data set by Cowieson and Roos (2016) found a strong correlation between the underlying digestibility of AA and the relative effects of proteases in increasing digestibility, an effect also observed by Aderibigbe et al. (2020a; Fig. 2). In fact, underlying AA digestibility accounted for 49% of the variability in the responses of proteases with the lowest digestibility producing the greatest effects. Lee et al. (2018a) observed a similar effect in a meta-analysis of protease-related effects on AA digestibility. These authors found an average increase of 1.6% in the ileal digestibility of AA, and they observed that the underlying digestibility of AA exerted a significant effect on the efficacy of proteases. Lee et al. (2018a) also observed that the efficacy of proteases was significantly impacted by the supplementation of other enzymes in the diet, which will be discussed in more detail later in the chapter.

Beyond the effects of AA density and digestibility in the diet, it has been observed that energy content of diets may influence the protease-related effects on AA digestibility. This was observed in multiple studies, where diets marginally deficient in energy produced greater relative effect on AA digestibility (Amerah et al., 2017; Lee et al., 2023). However, it was unclear if the energy deficit was the main factor as these diets also were deficient in CP. McCafferty et al. (2022) performed a factorial analysis of dietary AME

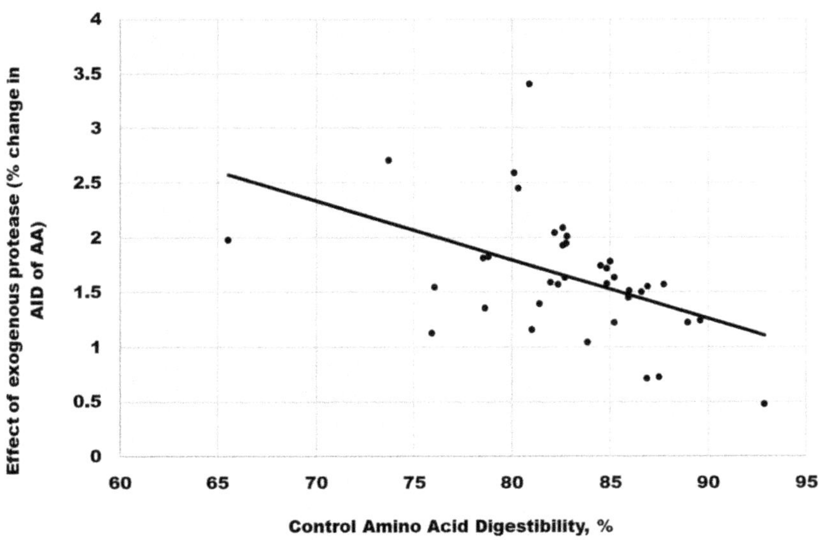

Figure 2. Relationship between the control amino acid digestibility on the relative effects of protease to increase average apparent ileal digestibility (AID) of amino acids (AAs) for broiler chickens. Source: Data adapted from Aderibigbe et al. (2021).

concentration and protease in order to evaluate this relationship. They did not find protease increased ileal or jejunal digestibility of CP or energy; however, they did observe some interactions of energy content and protease. At the lowest AME concentration, jejunal digestibility of CP was improved by protease; however, at moderate and high energy densities, protease created marginal decreases in CP digestibility. Similarly, the digestibility of energy was increased in both the ileum and jejunum when protease was supplemented at low AME concentration, but there was no observed effect at moderate and high levels (McCafferty et al., 2022). These data suggest that there are a number of dietary factors that have the potential to affect the efficacy of proteases.

Proteases are also utilized in nutritionally adequate diets and also have the potential to improve CP and AA digestibility under these conditions. Aderibigbe et al. (2020a) observed a 2.5% increase in the ileal digestibility of CP as well as significant increases in all AA except for Met and Cys. Similarly, Siegert et al. (2019) found that protease supplementation elicited a 2.6% increase in CP digestibility, while Toghyani et al. (2017) observed a nearly identical increase of 2.5%. Both of these studies identified Thr as the AA most affected with about a 3% increase. These reports indicate a relatively consistent response in nutritionally adequate diets of approximately 2.5%. However, Cowieson et al. (2020a, 2020b) observed greater average responses. These authors reported average increases in ileal digestibility for all AA from 3.9% to 4.1%. Cowieson et al. (2020a) found that the response was greater for dispensable AA relative to indispensable AA with average responses of 5.5% and 4.0%, respectively. Both of these studies also found Thr to be the individual AA that increased most with a peak response of 8.2%.

None of the previous studies identified any significant effects of dietary ingredient composition on protease-related effects on CP or AA digestibility. These included variations of both cereal grains as well as protein concentrates in both practical and semi-purified diets (Toghyani et al., 2017; Siegert et al., 2019; Cowieson et al., 2020a, 2020b). This may indicate protease has relatively consistent action to liberate AA across different ingredients. However, there were some interactions observed between primary protein concentrates for growth performance, which suggests greater understanding is required (Siegert et al., 2019).

Proteases have also been shown to have a number of effects beyond their improvements on growth performance and in degrading proteinaceous substrates. They have been shown to increase digestibility of other nutrients including minerals and energy. However, this is a relatively new area of research. Olukosi et al. (2015) observed that multiple inclusions of protease increased the total tract digestibility of both Ca and P as well as increasing the AME of the diets by as much as 177 kcal/kg. However, these improvements in mineral digestibility have not yet been widely studied elsewhere. Aderibigbe et al. (2021) observed protease-related increases in the ileal digestibility of copper and potassium, though these were in semi-purified diets. Amerah et al. (2017) observed increases in both the ileal and total tract digestibility of gross energy in both nutritionally adequate diets and diets with marginal energy deficiencies. Similar responses were observed by Kalmendal and Tauson (2012), who found a 114 kcal/kg increase in nitrogen-corrected AME with protease supplementation, and Cowieson et al. (2015) who observed increases of 77 and 131 kcal/kg for ileal digestible energy and AME, respectively. It is likely these effects are directly related to increases in AA digestibility.

There have also been observed improvements in the digestibility of specific energy substrates. The ileal and total tract digestibility of starch have been observed to increase with protease supplementation with an average increase of approximately 2% (Wang et al., 2008; Kalmendal and Tauson, 2012; Amerah et al., 2017). The digestibility and retention of fat have also been shown to increase with protease supplementation and have been observed to increase from 2.0% to 4.6% (Fru-Nji et al., 2011; Kalmendal and Tauson, 2012). The increase in fat digestibility may also be seen in the decrease in jejunal taurine in birds fed diets supplemented with protease (Huang et al., 2014; Cowieson et al., 2016). Intestinal taurine is known to be correlated with the luminal concentration of pancreatic bile and may be indicative of reduced bile production as a response to increased fat digestibility (Yuan and Wang, 2010). As mentioned, the increase of energy digestibility has been linked to greater protein digestibility; however, this alone is unlikely to cause such a significant response. These observed increases in starch and fat digestibility certainly seem to be involved in the protease-related increases in energy digestibility.

It should be noted, however, that proteases have not resulted in increased energy digestibility in all evaluations. McCafferty et al. (2022) observed no protease-related effects on the total tract digestibility of energy in diets at three densities of dietary energy. Toghyani et al. (2017) also reported protease having no effects on the ileal digestibility of energy in multiple diet types. These authors did not report any negative impacts of protease inclusion. As this area of protease research is still developing, much more work is required to understand these effects.

The methods by which proteases improve the digestibility of nutrients are not clear, though a number of pathways have been proposed. It has been suggested that protease-related effects on digestibility are a consequence of improved gut health and morphology. Cowieson et al. (2015) reported an overall increase in the tensile strength of the jejunum of broilers fed protease concurrently with a decrease in overall epithelial thickness. The authors argued, perhaps correctly, that this translated into a direct increase in intestinal integrity. This may be supported by Kalmendal and Tauson (2012) who observed a protease-related decrease in the relative empty weights of the ileum and jejunum along with a decrease in the length of each organ. This could suggest an overall thinning of the epithelia, which is thought to be a measure of gut health and absorptive efficiency. Cowieson et al. (2015) reported an upregulation in Claudin 1 expression as well as an upregulation of several epithelial AA transporters. This might suggest both an increase in the tight junction integrity of the gut (Claudin 1) and a potentially heightened absorptive efficiency (AA transporters).

Increased absorptive efficiency is a popular theory for the increase in nutrient digestibility, and there are a number of studies that support this view. Increases in intestinal villus height as well as decreases in crypt depth have been reported in association with increases in the digestibility of CP and AA (Wang et al., 2008; Yan et al., 2017; Aderdigbe et al., 2020a; Aderibigbe et al., 2021; Lee et al., 2023). These responses resulted in an increased villus height:crypt depth ratio, which is highly associated with an increase in the digestive efficacy of the gut. Along with these results, protease has also been reported to decrease the viscosity of jejunal and ileal digesta (Yan et al., 2017) and improve the fecal scores of broilers (Lee et al., 2023). Decreases in digesta viscosity are associated with increased nutrient digestibility linked to augmented passage rates and increased access of endogenous enzymes to their substrates.

There is also some evidence to suggest that proteases have an effect on the endogenous flows of AA in the gut, ultimately leading to increased digestibility. As reported earlier, in nearly all evaluations of the protease-related effects on AA digestibility, Thr is the most affected AA. This may indicate that proteases can affect the production of secretory mucins in the small intestine, as Thr is a primary constituent of these glycoproteins (Moran, 2016). This response may be due to this process or could be a byproduct of overall increased digestibility of nutrients in the small intestine. This idea is supported by the reduction in the expression of MUC-2 observed by Aderibigbe et al. (2020a) in birds fed diets supplemented with a protease. This suggests a reduction in the overall production of mucins. However, Aderibigbe et al. (2021) reported that protease supplementation had no significant effect on the production of crude mucin. These authors also did not observe any protease-related effects on endogenous flows of any AA. This contradiction does not have a readily available explanation. However, it could be related to the observed increases in the excretion of pancreatic proteases. In both studies by Aderibigbe et al. (2020a, 2021), exogenous protease supplementation resulted in the increased production of endogenous trypsin and chymotrypsin, and perhaps this was poorly recovered, leading to a lack of effect on endogenous AA flows.

One other suggested mode of action for exogenous proteases is their interactions with dietary anti-nutrients such as trypsin inhibitors (TI). It has been suggested that these exogenous proteases will either bind to anti-nutrients and lower their ability to interact with endogenous proteases or they will directly degrade them. Castanon and Marquardt (1989) reported that proteases improved performance with soybean-based diets but not with those based on other protein concentrates. This led to further work by Huo et al. (1993) who observed that acid-proteases could degrade TI and lectins *in vitro* and reported some evidence of this in a limited *in vivo* trial. Further *in vitro* work provided more evidence that proteases had some ability to bind or degrade TI and reduce the total antigenicity of soybean meal (Hessing et al., 1996; Rooke et al., 1998). Ghazi et al. (2002) performed an *in vivo* trial where they observed protease-related reduction in the soybean-specific antibodies in the serum of broilers; however, this study did not link this response to improvements in digestibility. More recently, Aderibigbe et al. (2020) performed a 2 × 2 factorial trial of protease supplementation and purified TI addition to investigate this relationship in practical diets. These authors found TI to be a potent anti-nutrient, observing reductions in growth performance, nutrient digestibility, and gut morphology, with protease supplementation unable to ameliorate the negative effects of the TI. A subsequent study was performed by Aderibigbe et al. (2021) to investigate a similar model for effects on endogenous AA flows. The results for this study were similar, with TI reducing animal performance and protease having positive effects being unable to counteract negative effects of TI. These data provide solid evidence that there is no strong relationship between exogenous proteases and TI, despite positive early evaluations. It is much more likely that the primary roles of proteases to improve nutrient digestibility are related to improvements in gut health and morphology and the degradation of feed and exogenous proteins.

8 Practical applications of proteases

Proteases are one of the most commonly used exogenous enzymes for non-ruminant animal production. From the beginning, they have been used to promote increased growth and improved feed efficiency in several diet types. This response has been shown to be relatively robust, especially for diets with reduced nutrient density or ingredient quality. This also remains the most common way proteases are used, either as a mono-component supplement or as a part of an enzyme blend. This allows the use of some lower-quality ingredients in commercial diets without significant risk to overall animal production. The assignment of matrix values to protease can be quite advantageous to dietary formulations, as it can allow for a reduction in the use of very expensive protein concentrates as well as a marginal decrease in dietary energy. Nutritionists must be very careful, however, because the effects of proteases can be variable, depending on dietary conditions. This means that a given protease must be thoroughly investigated in the diet to be used in assigning matrix values and, even then, it is advisable to utilize conservative values. The other enzymes included in the diet must also be considered as they are likely to have an effect on the ultimate action of the protease, a topic that will be explored later in the chapter.

Cowieson and Roos (2016) argued that proteases are still a relatively new area of research with many areas that require further investigation. This is certainly true for understanding the roles proteases may play in improving nutrient digestibility, gut morphology, microbial communities, and many other critical portions of animal production. A deeper understanding of the extra-proteinaceous effects of proteases will help nutritionists to better understand how to implement and utilize their effects in a variety of practical diets. This is particularly true for understanding better how proteases work to improve the absorptive capacity of the gut and the digestibility of energy substrates. Additionally, understanding the direct and indirect roles proteases have on the degradation and absorption of AA, as well as AA endogenous flow and mucin production, will help nutritionists to implement them in increasingly efficient ways. Exogenous proteases are a useful and promising tool to improve nutritional and animal production efficiency.

9 Carbohydrases: xylanases

The objective of supplementing carbohydrases in broiler chicken diets is to enhance the breakdown of polymeric carbohydrates, ultimately reducing their molecular weight (Adeola and Cowieson, 2011). Non-starch-degrading enzymes such as xylanases are commonly employed for this purpose. Non-starch polysaccharides present in poultry diets are considered anti-nutritional factors because they are not efficiently digested by the host enzymes. As a result, the utilization of non-starch polysaccharides is low. Moreover, the presence of non-starch polysaccharides also impedes the digestion of other nutrients (Choi et al., 2020).

Another widely used type of carbohydrase is amylase. Compared to non-starch polysaccharide degrading enzymes, pancreatic amylase secretion is significant. However, supplementing exogenous amylase in poultry diets can significantly enhance growth performance, particularly during critical stages such as early post-hatching and rapid growth phases. This is because there might be a limitation in fully degrading and digesting starch in broiler chickens during those stages (Svihus, 2014).

While there are many available exogenous mono-component carbohydrases, the following discussion will focus primarily on xylanases and amylase as they are widely researched and have shown consistent efficacy for broilers. This section discusses xylanases.

Xylan is a type of hemicellulose and belongs to the group of non-starch polysaccharides. The specific type of xylan is determined according to side chains attached to the backbone of the xylan. Arabinoxylan is the most common form of xylan, which is found in wheat, sorghum, corn, and soybean meal (Moita and Kim, 2022). Xylan is considered a type of soluble non-starch polysaccharide, and its presence generally has a negative impact on nutrient digestibility due to increased digesta viscosity (Moita et al., 2022). Moreover, soluble non-starch polysaccharides can have adverse effects on intestinal health and gut microbiota (Kim and Duarte, 2021). In the case of broiler chicken diets, the presence of arabinoxylan has the potential to promote the growth of pathogenic bacteria, trigger inflammatory responses and oxidative stress, and compromise the integrity of the intestinal barrier.

The addition of exogenous xylanase as a supplement is considered a potential solution to mitigate the negative impacts of xylan in broiler chicken diets. Xylanase is an enzyme that hydrolyzes the 1,4-beta-D-xylopyranosyl linkages in xylan. There are two main types of xylanases (Juturu and Wu, 2014). Endo-xylanase is primarily utilized in swine and poultry diets due to its easier production and its specific ability to hydrolyze the xylosidic bonds within the internal structure of the xylan backbone. In contrast, exo-xylanase predominantly targets the terminal bonds of xylan but is less commonly employed in poultry diets.

The primary function of exogenous xylanase in broiler chicken diets is to reduce digesta viscosity. It accomplishes this by hydrolyzing xylopyranosyl bonds, which aids in the release of trapped nutrients. This promotes effective digestion of nutrients by enzymes. Based on a review of peer-reviewed papers published after 2000, the supplementation of exogenous xylanase in broiler chicken diets resulted in an average decrease in jejunal digesta viscosity of approximately 37% compared to diets without xylanase (Moita and Kim, 2022). The reduction in digesta viscosity is one factor in increased nutrient and energy digestibility. Previous studies have reported that supplementing broiler chickens with exogenous xylanase improves the digestibility of energy and nitrogen (Stefanello et al., 2023; Vasanthakumari et al., 2023; Table 1). This improvement in nutrient and energy digestibility is potentially linked to enhanced growth performance. Based on their review of previously published papers, Moita et al. (2022) reported that the supplementation of exogenous xylanase led to an average increase in body weight gain of broiler chickens of 8.1%. However, a reduction in digesta viscosity does not always result in improvements in nutrient digestibility or growth performance (McCafferty et al., 2019).

An additional benefit of exogenous xylanase supplementation is the potential to reduce oxidative stress and inflammation while increasing antioxidant capacity (Moita and Kim, 2022). The exact mechanism of action is not yet fully understood. However, the release of fermentable non-starch polysaccharides such as xylo-oligosaccharides or ferulic acid

Table 1 Effects of xylanase on broiler chickens in recent publications

Feeding duration	Diet or ingredient type	Unit or inclusion rate	Effects	Reference
d 14 to 21	Corn (waxy, semi-dent, or semi-flint)	100 FXU/kg	Increased ileal digestibility of nitrogen and energy and jejunal digestibility of total and resistant starch	Stefanello et al. (2023)
d 0 to 21	Corn-SBM diet	0, 45,000, or 90,000 U/kg	Increased feed efficiency and ileal digestibility of energy, starch, and nitrogen	Vasanthakumari et al. (2023)
d 0 to 28	Pelleted wheat-SBM diet	0, 30,000, 45,000, or 90,000 U/kg	Linearly increased or tended to increase weight gain, feed efficiency, and metabolizable energyQuadratically decreased ileal digesta viscosity	Gorenz et al. (2022)
d 0 to 16	Corn-SBM diet	Not available	No effect of weight gain, ileal soluble xylose, and total soluble non-starch polysaccharide regardless of necrotic enteritis challenge	Kim et al. (2022)
d 0 to 16	Wheat-SBM diet	Not available	Increased weight gain regardless of necrotic enteritis challengeDecreased ileal soluble xylose and total soluble non-starch polysaccharide concentration regardless of necrotic enteritis challenge	Kim et al. (2022)
d 0 to 35	Corn-SBM diet	0, 30,000, 45,000, or 90,000 U/kg	Increased weight gain, feed efficiency, the count of *Lactobacillus*, concentration of butyrate, and total volatile fatty acid concentrations in cecal digestaDecreased the count of *E. coli* in cecal digesta	Van Hoeck et al. (2021)
d 1 to 32	Corn-SBM diet	0, 3750, 7500, or 11,250 XU/kg	Linearly increased weight gain, feed efficiency, ileal digestibility of nitrogen and soluble dietary fiber, metabolizability of energy, and villus height of the duodenum and jejunumLinearly decreased ileum digesta viscosity	Lee et al. (2020)

(Continued)

Table 1 (*Continued*)

Feeding duration	Diet or ingredient type	Unit or inclusion rate	Effects	Reference
d 0 to 28	Wheat-SBM diet	0 or 16,000 BXU/kg	Increased feed efficiency and ileal digestibility of dry matter and nitrogen	Olukosi et al. (2020)
d 1 to 42	Corn-rye-SBM diet	0 or 200 mg/kg of feed	Increased feed efficiencyDecreased the viscosity of the jejunum and ileum digesta	Arczewska-Wlosek et al. (2019)
d 0 to 42	Wheat-SBM diet	0 or 16,000 BXU/kg	Increased feed efficiencyDecreased the relative weight of the crop	González-Ortiz et al. (2019)
d 1 to 40	Corn-SBM diet or wheat-SBM diet	0, 12,000, or 16,000 BXU/kg	No effect	McCafferty et al. (2019)
d 1 to 35	Wheat-SBM diet	0 or 1875, 3750, or 5625 XU/kg	Linearly increased weight gain, feed efficiency, ileal digestibility of energy and amino acids, villus height of the small intestine, and *Lactobacillus* count in cecal digestaLinearly decreased ileal digesta viscosity and *E. coli* count in cecal digesta	Liu and Kim (2017)
d 1 to 21	Corn-SBM diet	0 or 2000 U/kg	Increased weight gain and ileal digestibility of energy, starch, and nitrogenDecreased soluble non-starch polysaccharide concentration in ileal digesta	Amerah et al. (2017)

SBM, soybean meal.

from the hydrolysis of xylan structures may be a factor (Liu and Kim, 2017). Oligosaccharides released can also undergo fermentation by beneficial bacteria, such as *Bifidobacteria* and *Lactobacillus*, resulting in the production of volatile fatty acids. These volatile fatty acids have several important functions. Firstly, they can lower the pH of the gut, inhibiting the growth of potentially harmful bacteria. Some volatile fatty acids, such as butyrate, can also serve as an energy source for intestinal cells. Van Hoeck et al. (2021) reported that the addition of exogenous xylanase to a corn–soybean meal-based diet, ranging from 0 to 90,000 U/kg, resulted in an increase in *Lactobacillus* count and a decrease in *E. coli* count in cecal digesta. Xylanase supplementation also increases the concentration of propionate, butyrate, and total volatile fatty acids in cecal digesta.

In general, the impact of xylanase supplementation is more pronounced when birds are fed diets based on wheat or diets that contain fibrous ingredients, compared to corn-based diets, because of the larger quantity of soluble non-starch polysaccharides. For this reason, viscosity may not be a significant concern, and reduction in digesta viscosity may not play a crucial role in improving growth when birds are fed corn-based diets (Lee et al., 2017). The mechanism by which xylanase enhances nutrient digestibility and growth performance is still not fully understood. One possible explanation is that xylanase hydrolyzes the insoluble non-starch polysaccharide fraction found in the cell wall of corn, leading to the release of nutrients (Meng and Slominski, 2005). However, there is uncertainty about whether exogenous xylanase can sufficiently degrade corn cell walls within the small intestine by the end of the digestion process. For this reason, the mechanism of exogenous xylanase in corn-based diets may operate indirectly.

Another potential mechanism is the stimulation of peptide YY (PYY) production by volatile fatty acid production through xylanase (Lee et al., 2017). As PYY (a neuropeptide that is released by L-cells) has the ability to modulate the hypothalamus and delay gastric emptying, the passage rate of digesta in the gizzard and proventriculus decreases. This, in turn, leads to a reduction in particle size and an improvement in protein digestion, respectively. Furthermore, PYY can increase the active transport of glucose in the small intestine by modulating sodium-glucose cotransporter (SGLT1) or glucose transporter 2 (GLUT2). To test this hypothesis, corn-soybean meal-based diets were supplemented with 18,100 BXU/kg of exogenous xylanase on d 0, 7, 14, 21, or 28 post hatching (Lee et al., 2017). The control group was fed a diet without xylanase supplementation. The concentration of blood PYY on d 35 was greater when xylanase was supplemented on d 14, 21, or 28 compared to the control, indicating a beneficial effect on young birds exposed to xylanase. However, there was no significant difference in mRNA expression of SGLT1 or GLUT2.

When birds are challenged with necrotic enteritis, wheat-based diets tend to worsen the negative effects of the condition compared to corn-based diets because of greater concentration of soluble non-starch polysaccharides, which increases the viscosity of the gut's digesta. The increased viscosity prolongs the passage rate of the digesta and creates an environment conducive to the proliferation of pathogenic bacteria such as *C. perfringens* (Annett et al., 2002). According to Kim et al. (2022), the supplementation of xylanase to the diets of 16-d-old broiler chickens resulted in increased weight gain and reduced concentrations of ileal soluble xylose and soluble non-starch polysaccharides, regardless of the necrotic enteritis challenge. However, it is worth noting that the positive impact of xylanase was not observed when the birds were fed corn-based diets. Moreover, the weight gain observed in challenged birds fed wheat-based diets was not significantly different from that of the healthy birds. This suggests that the impact of exogenous xylanase may remain consistent regardless of the health status of the birds.

10 Carbohydrases: amylase

Starch is quantitatively the largest energy source in poultry diets. While poultry have the ability to adapt to starch digestion after hatching through the secretion of endogenous amylase, their capacity to fully degrade and digest starch is limited (Svihus, 2014). This limitation may limit their growth performance. Young broiler chickens may have insufficient pancreatic amylase secretion in their gastrointestinal tract, which can impede the digestion of starch. Moreover, endogenous pancreatic amylase secretion in broiler chickens around 28–35 d post hatching may prove inadequate for starch digestion (Sklan and Noy, 2003). This is primarily due to their high feed intake, which accelerates the passage rate of digesta. Consequently, the rapid movement of digesta limits the opportunities for endogenous amylase to effectively access and break down starch.

The supplementation of exogenous α-amylase can enhance starch digestibility in poultry, leading to improved energy digestibility and growth performance. Furthermore, application of exogenous α-amylase decreases endogenous α-amylase secretion, reducing the energy needed for production of α-amylase in the pancreas (Perz et al., 2023a). Supplementation of exogenous α-amylase can also potentially increase jejunal Na^+-K^+-ATPase activity, which is beneficial for nutrient absorption (Wang et al., 2020).

Aderibigbe et al. (2020b) found a linear effect from amylase supplementation (0, 80, or 160 KNU/kg) of a corn–soybean meal-based diet on weight gain and gain-to-feed ratio of broiler chickens from day 0 to 21 (Table 2). These responses were not significant from d 0 to 14, suggesting that the amylase had the greatest effect over the final 7 d. There was a positive

Table 2 Effects of amylase on broiler chickens in recent publications

Feeding duration	Diet or ingredient type	Unit or inclusion rate	Effects	Reference
d 0 to 35	Corn–SBM–faba bean diet or corn–SBM–field pea diet	0 or 80 KNU/kg	Increased weight gain, feed efficiency, and ileal digestibility of dry matter and amino acids. Decreased sialic acid	Perz et al. (2023a)
d 1 to 42	Corn–SBM–field pea diet	0 or 80 KNU/kg	Increased ileal digestibility of dry matter, starch, and amino acids	Perz et al. (2023b)
d 1 to 16	Corn–SBM–faba bean diet	0 or 80 KNU/kg	Increased ileal digestiblity of starch and amino acids, and nitrogen-corrected metabolizable energy	Perz et al. (2022)
d 0 to 40	Corn–SBM diet	0 or 80 KNU/kg	Increased nitrogen-corrected metabolizable energy, weight gain, and feed efficiency	Córdova-Noboa et al. (2021)
d 1 to 41	Corn–SBM diet	0, 400, 600, 800, or 1,200 U/kg	Linearly decreased starch digestibility in the proximal jejunum and jejunal amylase activityLinearly increased serum insulin and glucagon concentration in the fasting and postprandial state	Zhou et al. (2021)
d 0 to 21	Corn–SBM diet	0 , 80, or 160 KNU/kg	Increased weight gain, feed efficiency, starch, and energy digestibility, and jejunal villus heightDecreased viscosity of jejunal digesta	Aderibigbe et al. (2020a)
d 0 to 56	Corn–SBM diet	0 or 80 KNU/kg	Increased growth, feed efficiency, jejunal villus height, duodenal amylase activity, and ileal digestibility of starch and energyDecreased jejunal digesta viscosity and pancreatic amylase activity	Aderibigbe et al. (2020b)
d 0 to 42	Corn–SBM diet	0, 80, or 160 KNU/kg	Increased or tended to increase weight gain, feed efficiency, relative intestinal length to weight, and ileal digestibility of dry matter, protein, and starch	Córdova-Noboa et al. (2020)
d 1 to 28	Corn–SBM diet containing 6 different batches of corn	0 or 80 KNU/kg	Increased feed efficiency and ileal digestible energyReduced variability of ileal digestible energy	Ma et al. (2020)
d 21 to 63	Corn–SBM diet	0, 300 or 500 mg/kg of feed	Increased the activity of jejunal Na+-K+-ATPaseIncreased relative abundance of Lactobacillus in cecal digesta (only by 300 mg/kg)	Wang et al. (2020)
d 14 to 25	Corn–SBM diet	0, 40, 80, 120, or 160 KNU/kg	Increased weight gain, ileal digestible energy, and metabolizable energy	Stefanello et al., (2019)
d 14 to 25	Corn	0, 40, 80, 120, or 160 KNU/kg	Increased ileal digestible energy and metabolizable energy	Stefanello et al., (2019)

SBM, soybean meal.

linear response to amylase in the digestibility of starch in the posterior jejunum, anterior ileum, posterior ileum, and the excreta, with amylase supplementation causing increased disappearance of starch from the gut.

Aderibigbe et al. (2020c) evaluated the effects of amylase supplementation on broilers in multiple growth stages. This study was a 2 × 4 factorial trial of amylase supplementation (0 or 80 KNU/kg) and growth period (d 0 to 11, d 11 to 21, d 21 to 42, and d 42 to 56) for broilers. Amylase also produced an improvement in weight gain and gain-to-feed ratio for broilers. However, body weight and gain-to-feed ratio responses were lower in the last two stages of growth (1.8% vs. 4.3% and 1.1% vs. 5.3%, respectively). There were improvements in ileal digestibility of starch and energy with amylase supplementation, but with d 11 to 21 phase having much lower improvements than the other stages. This effect did not transfer to the metabolizability of starch and energy, and they were improved by amylase in each phase. The observed effect for the second growth phase and the ileal digestibility of starch may be related to amylase activity in the gut. Supplemental dietary amylase increased the duodenal amylase activity for all phases. However, the increase was only 4.1% in the second phase compared to an average of 35.4% for the other three phases. This is also paired with an increase in the pancreatic amylase activity in this phase when it was decreased in all others. There were also increases in the jejunal villus height for each growth phase; however, increases were greater in the latter two stages.

There is clear evidence from these studies that amylase has the potential to improve the growth performance and nutrient digestibility for broilers. However, these studies also demonstrate that there may be differing effects based on growth phase, intestinal site, and possibly amylase concentration. The effects of amylase tend to be relatively consistent compared to other carbohydrases which often show potential but tend to produce inconsistent results in trials. This may be due to in part to differences in the proportion and activities of the individual enzymes in a certain blend. This may necessitate greater analysis of these enzymes individually in order to ascertain the correct mixtures and proportions for optimum performance in pigs and broilers.

The effect of α-amylase on corn–soybean meal-based diets is not consistent, due to the relatively easy degradation of starch in corn and soybean meal by endogenous amylase in poultry (Perz et al., 2023a). However, the degradation of resistant starch is more challenging than normal starch. For this reason, it may be beneficial to supplement α-amylase when incorporating feed ingredients with high levels of resistant starch, such as faba beans and peas. In the study of Perz et al. (2022), supplementation of 80 KNU/kg of α-amylase increased nitrogen-corrected AME and ileal digestibility of starch, protein, and most indispensable AAs in broiler chickens fed diets containing 50% of faba bean. The supplementation of 80 KNU/kg of α-amylase also increased body weight gain, feed efficiency, and ileal starch digestibility when broiler chickens were fed corn–soybean meal diets containing faba bean or pea (Perz et al., 2023a).

Generally, feeding fibrous diets increases the production of mucin, which leads to increased endogenous losses. Because the characteristics of resistant starch are highly similar to those of dietary fiber, resistant starch may lead to overproduction of mucin, which can cover intestinal surfaces and potentially decrease digestibility (Cowieson et al., 2003). Perz et al. (2023a) reported that the supplementation of α-amylase decreased total sialic acid excretion, a marker for endogenous losses and mucin production. However, the reduction of sialic acid by exogenous α-amylase was not consistent in other studies (Perz et al., 2022, 2023a, 2023b). Even in diets with lower concentrations of dietary fiber or resistant starch, amylase has the potential to reduce the viscosity of digesta. This decrease in viscosity can enhance nutrient digestibility, as the presence of a viscous gel in the gastrointestinal tract hinders the digestion and absorption of nutrients. According to Aderibigbe et al. (2020b, 2020c), the addition of exogenous amylase to a diet based on corn–soybean meal resulted in a reduction in the viscosity of jejunal digesta.

Supplementing with exogenous substances may also have beneficial effects on gut health, although the mode of action is not clear. In the study conducted by Córdova-Noboa et al. (2020), the supplementation of 80 KNU/kg of α-amylase resulted in an increase in the length of the intestine relative to the body weight of broiler chickens, which might have contributed to increased ileal nutrient digestibility and growth performance. Aderibigbe et al. (2020b) also observed a linear increase from amylase supplementation in villus height in the jejunum which can contribute to the increased digestibility of other nutrients.

The effect of α-amylase is influenced by several factors, including kernel hardness, particle size of grains, and the ratio of amylopectin to amylose in starch. The endosperm hardness of grains plays a crucial role as it directly impacts the overall structure of starch. In the study by Córdova-Noboa et al. (2021), supplementation of 80 KNU/kg of α-amylase in broiler chicken diets increased body weight gain and feed efficiency of broiler chickens, regardless of corn kernel hardness. However, the effect of α-amylase was more significant in broiler chickens fed diets containing corn with normal kernel hardness. The effectiveness of exogenous amylase may be also influenced by the particle size of corn. Addition of 80 KNU/kg of exogenous amylase increased ileal digestibility of dry matter and protein, regardless of corn particle size (Córdova-Noboa et al., 2020). However, the degree of increase attributed to exogenous amylase was greater in broiler chickens fed diets with coarse corn compared to those with fine corn.

The main components of starch are amylopectin and amylose. Amylopectin is a highly branched molecule, while amylose consists of a linear chain structure with fewer branching points. Due to its branched nature, amylopectin is more readily digested compared to amylose, which contains more intramolecular hydrogen bonds (Yin et al., 2010). This suggests that increase in starch digestibility using amylase would be more pronounced in diets with high levels of amylose as compared to diets with high levels of amylopectin.

Another advantage of using exogenous α-amylase is its potential to reduce the variability of dietary energy concentration. In a study conducted by Liu et al. (2020), the coefficient of variation in ileal digestible energy decreased from 5.2% to 0.7% in six diets containing different batches of corn. However, increasing exogenous α-amylase up to 1200 U/kg linearly decreased glucose transporter (GLUT2 and SGLT1) in the jejunum and ileum and starch digestibility in the proximal jejunum while increasing glucose levels in the fasting and post-prandial state in broiler chickens, indicating that high dose of supplementation of α-amylase may inhibit starch digestion (Zhou et al., 2021).

11 Enzyme combinations

It is common practice to use various combinations of enzymes in broiler chicken diets rather than using a single enzyme. Each enzyme is designed to target specific substrates and, when multiple enzymes are employed together, there is a potential for synergistic or additive effects. The effects of enzyme combinations are influenced by factors such as substrate quantity, substrate diversity, enzyme types and the growth stage of birds (Aftab and Bedford, 2018). There has been a growing body of research that investigates the impact of enzyme combinations that will be explored in the following sections.

11.1 Xylanases in enzyme combinations

Xylanase is one of the most commonly used carbohydrases, often used in combination with other enzymes in broiler chicken diets. One reason for using xylanase in combination with other enzymes is to enhance the accessibility of these enzymes to substrates by reducing the viscosity of the digesta, particularly in broiler chicken diets with high concentrations of soluble non-starch polysaccharides. Another mechanism involves the breakdown of the indigestible fraction of diets (Cowieson, 2005). Xylanase plays a crucial role in enhancing the accessibility of endogenous or exogenous enzymes to proteins and starches within the endosperm cell. This is achieved by hydrolyzing the insoluble arabinoxylan present in the cell wall.

In a study conducted by dos Santos et al. (2017), the effect of different enzyme combinations on sorghum–soybean meal-based diets was investigated in broiler chickens using a 2 × 2 × 2 factorial treatment arrangement. The enzymes evaluated included xylanase (0 or 16,000 BXU/kg), protease (0 or 7500 PROT/kg), and phytase (500 or 1500 FTU/kg). No significant three-way interaction was observed in the study. However, it is worth noting that the inclusion of 1500 FTU/kg of phytase in a diet supplemented with xylanase resulted in an increased transit time of digesta. When phytase was added to a diet supplemented with protease, this effect on increased transit time of digesta was not significant. The reason for this remains unclear. However, it is important to note that longer transit time generally allows for extended nutrient digestion and absorption, indicating that combination of xylanase and super-dosing phytase potentially enhances nutrient digestibility.

In a study by Lee et al. (2018b), broiler chickens were fed wheat–soybean meal-based diets arranged in a 2 × 2 factorial treatment arrangement. The diets included either 0 or 16,000 BXU/kg of xylanase and 500 or 2500 FTU/kg of phytase. An interaction between xylanase and phytase was observed in terms of weight gain from d 0 to 42. The addition of xylanase to a diet containing 500 FTU/kg of phytase, or increasing phytase concentration up to 2500 FTU/kg, increased weight gain of birds. However, no synergetic effect was observed when combining xylanase with 2500 FTU/kg of phytase, indicating that phytase and xylanase may have non-additive effects. Karimi et al. (2013) also reported no synergistic effect of combining phytase and xylanase on broiler chickens fed corn–sorghum-based diets. In contrast, Kühn et al. (2013) reported a synergistic effect when using 1500 FTU/kg of phytase and 16,000 BXU/kg of xylanase, leading to improved growth performance in 35-d-old broiler chickens fed wheat-based diets. Although the results of Lee et al. (2018b) and Kühn et al. (2013) are not consistent, it is expected that a synergistic interaction between xylanase and phytase is more likely to occur in wheat-based diets because of the high concentration of phytate storage in the aleurone layer of wheat, which primarily consists of arabinoxylan (O'Dell et al., 1972; Burton and Fincher, 2014). This is because xylanase can degrade the aleurone layer, allowing phytase to gain access to phytate for effective hydrolysis.

Woyengo et al. (2019) also reported a synergistic effect when combining phytase (2500 FTU/kg) with carbohydrase cocktails consisting of 203 IU/kg of xylanase, 60 IU/kg of cellulase, and 53 IU/kg of beta-glucanase, resulting in improved ileal digestibility of energy and protein. Similarly, a synergistic interaction was observed between phytase and amylase (80 KNU/kg), as well as between amylase and carbohydrase cocktails. However, no three-way interaction was observed among these enzymes.

Amerah et al. (2017) investigated the separate effects of xylanase (2000 U/kg), amylase (200 U/kg), and protease (4000 U/kg), as well as their combined effects, on broiler chickens fed corn–soybean meal-based diets. Birds fed a diet supplemented with a combination of the three enzymes showed greater feed efficiency from d 0 to 42 compared to birds fed diets with individual enzyme supplementation. When the three enzymes were used together, nitrogen-corrected AME in diets was significantly higher compared to using xylanase or phytase individually, indicating that there might be synergistic interactions among different enzymes. It has been suggested that protease potentially assists in the action of xylanase, specifically in breaking down cell wall structures (Żyła et al., 1999). Furthermore, both amylase and protease are able to disrupt starch–protein interactions within cells, which potentially contributes to synergistic interactions among different enzymes (Amerah, 2015).

11.2 Proteases in enzyme combinations

The use of mono-component proteases as individual feed additives has increased over the past decade. However, the main use of these enzymes is in combination with other enzymes like carbohydrases and phytase. One commercially popular combination is the use of xylanase (~2000 units/kg), amylase (~200 units/kg), and protease (~4000 units/kg) in one supplement (XAP). This combination has been shown to produce increases in apparent ileal digestibility (AID) of ash, nitrogen, gross energy, AA, and starch as well as improving apparent total tract utilization (ATTU) of gross energy, nitrogen, and starch while increasing dietary AME (Singh et al., 2019). These authors also reported a decrease in total NSP flow in the ileum with even greater response in the excreta. While Singh et al. (2019) did not report on growth performance, it has been shown in many other evaluations that XAP increases weight gain and decreases FCR from 1 to 42 d of age along with improvements in nutrient digestibility (Cowieson and Adeola, 2005; Olukosi et al., 2007; Amerah et al., 2017). It is clear that the XAP blend is efficacious in broiler diets, but it is unclear if this combination is superior to supplementation of individual enzymes and if the effects of the enzymes are additive or potentially synergistic.

Barekatain et al. (2013) evaluated a combination of protease and xylanase without amylase and reported increases in 1 to 35 d weight gain concentrations of short-chain fatty acids in the ceca; however, the effects on nutrient digestibility were not substantial. These authors only fed birds a combination diet so it was not clear which enzyme might be having the greater effect. Kalmendal and Tauson (2012) fed a similar combination of protease and xylanase as well as individual enzymes. These authors reported no effects of any treatment on weight gain but both individual enzymes improved FCR and the ATTU of starch, fat, and nitrogen as well as increasing diet nitrogen-corrected AME. The enzyme combination produced statistically similar results for each individual enzyme and did not produce any additive results.

Olukosi et al. (2015) evaluated the effects of an individual protease against a XAP combination and reported the protease increased the ATTU of P, Ca, gross energy, and nitrogen along with modest improvements in growth performance. However, the XAP combination produced much better results for all of these parameters as well as others, including ileal flow of NSP. A closer examination of the potential relative effects of protease in the combination shows about 5–45% of effects potentially coming from protease (with the greatest relative effect on nitrogen digestibility) in Olukosi et al (2015). This demonstrates that protease is a critical part of this enzyme combination. Amerah et al. (2017) fed birds xylanase, amylase, and protease individually as well as the XAP combination. They reported that xylanase individually and XAP improved weight gain compared to amylase and protease. XAP responses were equal to that of xylanase, suggesting that xylanase was the primary driver of this response. All individual enzymes improved FCR. XAP improved FCR by an additional 2% compared to the average of the individual enzymes, suggesting some level of additive or synergistic effect (Amerah et al., 2017).

The authors evaluated enzyme effects on AID and ATTU of nutrients. All enzymes improved the AID of starch with amylase with XAP producing the best results. All enzymes improved the AID of nitrogen; however, xylanase and the XAP produced the greatest responses with protease having the least effect. Additionally, all treatments increased the AID of gross energy, but the XAP produced the greatest response. All of the ileal digestibility results showed the XAP treatment having the greatest response in digestibility; however, there was no evidence of truly synergistic effects. This was similar to the ATTU of nitrogen where protease and the XAP produced the best responses. None of the individual enzymes improved ATTU of gross energy, while XAP increased this metric by approximately 4%. This may indicate some potential for synergistic effects. In all cases the XAP performed at least as well as the individual enzyme that was the primary driver of the response; however, there was limited evidence of any truly synergistic effects (Amerah et al., 2017). This would suggest that rather than the positive effects coming from enzymes improving each other's performance, it is the presence of the wide array of enzymes acting individually that gives a slightly additive effect.

Blends of xylanase, amylase, and protease are also commonly fed alongside a phytase enzyme in practical diets. Cowieson and Adeola (2005) evaluated the relationship of XAP and phytase at both 500 and 1000 FTU/kg. At 500 FTU/kg the combination showed synergistic effects for weight gain, with the combination performing 6.0% above the additive effects of phytase and XAP in isolation. At 1000 FTU/kg the combination did show synergistic effects. Results were partially additive with the combination producing results 71% above the average response of the individual treatments (Cowieson and Adeola, 2005). The authors did not find additive results for AID of nitrogen, Ca, and P when compared to individual treatments.

Olukosi et al. (2007) performed a similar evaluation with 1000 FTU/kg phytase and found additive results of the XAP and phytase for weight gain where there was an almost exactly 100% increase of the combination over the individual treatments. The results for nitrogen and P digestibility, however, showed no additive relationship as the combination did not perform as well as the individual treatments, though it was still better than the control. Both of these studies show the potential for a strong positive relationship between the XAP and phytase, especially for growth performance (Cowieson and Adeola, 2005; Olukosi et al., 2007).

Protease has also been shown to have some synergistic effects with phytase (Cowieson et al., 2020a). These authors reported a synergistic relationship in increasing the standardized ileal digestibility of threonine, proline, and cysteine. This response may be related to the relative changes to endogenous secretions and recovery produced by these enzymes. The authors also reported that this effect is consistent across multiple ingredient combinations.

There have been evaluations of other carbohydrase combinations with protease. Mohammadigheisar et al. (2021) evaluated an enzyme combination containing phytase (1500 FTU/kg), protease (800 units/kg), beta-glucanase (70 units/kg), and cellulase (80 units/kg) and observed increases in the 0 to 28 d weight gain as well as in the ATTU of nitrogen, fiber fractions, and Ca. The authors also observed an increase in the nitrogen-corrected AME of experimental diets. It is difficult to attribute any of these effects to individual enzymes, though this combination clearly produced a positive response. Yan et al. (2017) evaluated the effects of a combination (XBG) of xylanase (1000 units/kg), beta-glucanase (37.5 units/kg), and galactosidase (12.5 units/kg) alone and in combination with a protease. These authors reported significant individual effects of protease and the XBG to improve weight gain and FCR with the XBG have greater relative effects. There was mild additivity for the combination with responses being 28.5% above the average of the individual treatment for weight gain and FCR. Both individual treatments reduced the viscosity of ileal digesta, but there was no additive effect observed. Additionally, there was a synergistic effect in increasing concentration of plasma superoxide dismutase, potentially signaling an increased immune response (Yan et al., 2017).

Toghyani et al. (2017) reported that the combination of a proprietary blend of carbohydrase and protease produced antagonistic effects for weight gain, FCR, and the digestibility of gross energy and nitrogen, whilst the carbohydrase blend and protease individually increased these metrics. This result highlights how critical it is to evaluate individual enzyme combinations, as there is a potential for the production of antagonistic effects. Overall, these combinations (especially the XAP) appear to produce at least equal performance to the feeding of individual enzymes, with the potential for additive or synergistic effects at times. This is an area in need of significant additional investigation but has the potential to be a powerful tool for increasing broiler production.

12 Conclusion

The use of exogenous enzymes in the feed of non-ruminant animals continues to evolve as new enzymes are developed and new methods for utilizing them are devised. For this reason, it is pivotal that the research and study of enzymes continue so that these products can be best used to improve the efficiency of production. There is potential in the use of enzyme combinations which may produce a greater and more consistent effects than individual enzymes. There is much to learn about how to use these enzymes in different diet types and at what level of substitution. Additionally, greater work is needed to understand the potential of antagonistic, additive, or synergistic effects of enzymes under various dietary conditions. The precise mode of action by which protease acts to improve production and digestibility is still not well understood. However, it is clear that these products have a generally positive impact on production and can be useful in practical diets. Phytase has shown consistent utility for P and AA digestibility. However, there is still much to learn about how phytase may interact with other enzymes that are supplemented alongside it and how the many dietary factors that affect it may affect this relationship. This is of particular importance given the wide use of phytase for pigs and chickens. Animal production continues to move toward greater efficiency, and the use of exogenous enzymes is one powerful tool to further this improvement. This chapter has covered a wide array of findings for the modern use of enzymes. However, it also clearly demonstrates how much further there is still to come.

13 Where to look for further information

- Adeola, O., and A. J. Cowieson. 2011. Board-invited review: opportunities and challenges in using exogenous enzymes to improve nonruminant animal production. *J. Anim. Sci.* 89:3189–3218.
- Cowieson, A. J., and M. R. Bedford. 2009. The effect of phytase and carbohydrase on ileal amino acid digestibility in monogastric diets: complementary mode of action? *Worlds Poult. Sci. J.* 65:609–624.
- Lee, S. A., M. R. Bedford, and C. L. Walk.2018a. Meta-analysis: explicit value of mono-component proteases in monogastric diets. *Poult. Sci.* 97:2078–2085.
- Ravindran, V. 2013. Feed enzymes: the science, practice, and metabolic realities. *J. Appl. Poult. Res.* 22:628–636.

14 References

Adeola, O. and Cowieson, A. J. 2011. Board-invited review: opportunities and challenges in using exogenous enzymes to improve nonruminant animal production. *J. Anim. Sci.* 89, 3189–3218.

Aderibigbe, A. S., Cowieson, A. J., Sorbara, J. O., Pappenberger, G. and Adeola, O. 2020a. Growth performance and amino acid digestibility responses of broiler chickens fed diets containing purified soybean trypsin inhibitor and supplemented with a monocomponent protease. *Poult. Sci.* 99, 5007–5017.

Aderibigbe, A. S., Cowieson, A. J., Sorbara, J. O. and Adeola, O. 2020b. Intestinal starch and energy digestibility in broiler chickens fed diets supplemented with α-amylase. *Poult. Sci.* 99, 5907–5914.

Aderibigbe, A. S., Cowieson, A. J., Sorbara, J. O. and Adeola, O. 2020c. Growth phase and dietary α-amylase supplementation effects on nutrient digestibility and feedback secretion in broiler chickens. *Poult. Sci.* 99, 6867–6876.

Aderibigbe, A. S., Cowieson, A. J., Ajuwon, K. M. and Adeola, O. 2021. Contribution of purified soybean trypsin inhibitor and exogenous protease to endogenous amino acid losses and mineral digestibility. *Poult. Sci.* 100, 101486.

Aftab, U. and Bedford, M. R. 2018. The use of NSP enzymes in poultry nutrition: myths and realities. *Worlds Poult. Sci.* J. 74, 277–286.

Amerah, A. M. 2015. Interactions between wheat characteristics and feed enzyme supplementation in broiler diets. *Anim. Feed Sci. Technol.* 199, 1–9.

Amerah, A. M., Romero, L. F., Awati, A. and Ravindran, V. 2017. Effect of exogenous xylanase, amylase, and protease as single or combined activities on nutrient digestibility and growth performance of broilers fed corn/soy diets. *Poult. Sci.* 96, 807–816.

Angel, C. R., Tamin, N. M., Applegate, T. J., Dhandu, A. S. and Ellestad, E. L. 2002. Phytic acid chemistry: influence on phytin-phosphorus availability and phytase efficacy. *J. Appl. Poult. Res.* 11, 471–480.

Angel, C. R., Saylor, W., Vietra, S. L. and Ward, N. 2011. Effects of a moncomponent protease on performance and protein utilization in 7- to 22-day-old broiler chickens. *Poult. Sci.* 90, 2281–2280.

Annett, C. B., Viste, R. J., Chirino-Trejo, M., Classen, L. H., Middleton, M. D. and Simko, E. 2002. Necrotic enteritis: effect of barley, wheat and corn diets on proliferation of Clostridium perfringens type A. *Avian Pathol.* 31, 598–601.

Augspurger, N. R., Spencer, D. J., Webel, M. D. and Baker, H. D. 2004. Pharmacological zinc levels reduce the phosphorus-releasing efficiency of phytase in young pigs and chickens. *J. Anim. Sci.* 82, 1732–1739.

Babatunde, O. O., Bello, A., Dersjant-Li, Y. and Adeola, O. 2021. Evaluation of the responses of broiler chickens to varying concentrations of phytate phosphorus and phytase. I. Starter phase (day 1–11 post hatching). *Poult. Sci.* 100, 101369.

Barekatain, M. R., Antipatis, C., Rodgers, N., Walkden, Brown, W. S., Iji, A. P. and Choct, M. 2013. Evaluation of high dietary inclusion of distillers dried grains with solubles and supplementation of protease and xylanase in the diets of broiler chickens under necrotic enteritis challenge. *Poult. Sci.* 92, 1579–1594.

Bedford, M. R. and Walk, L. C. 2016. Reduction of phytate to tetrakisphosphate (IP_4) to trikisphosphate (IP_3), or perhaps even lower, does not remove its antinutritive properties. In L. C. Walk, I. Kuhn, H. H. Stein, T. M. Kidd and M. Rodehutscord (Eds.), *Phytate Destruction - Consequences for Precision Animal Nutrition* (pp. 45–51). Wageningen, The Netherlands: Wagenningen Academic Publishers.

Beeson, L. A., Walk, L. C., Bedford, R. M. and Adeola, O. 2017. Hydrolysis of phytase to its lower esters can influence the growth performance and nutrient utilization of broilers with regular or superdoses of phytase. *Poult. Sci.* 96, 2243–2253.

Borda-Molina, D., Zuber, T., Siegert, W., Camarinha-Silva, A., Feuerstein, D. and Rodehutscord, M. 2019. Effects of protease and phytase supplements on small intestinal microbiota and amino acid digestibility in broiler chickens. *Poult. Sci.* 98, 2906–2918.

Burton, R. A. and Fincher, B. G. 2014. Evolution and development of cell walls in cereal grains. *Front. Plant Sci.* 5, 456.

Castanon, J. I. R. and Marquardt, R. R. 1989. Effects of enzyme addition, autoclave treatment and fermenting on the nutritive value of field beans (Vicia faba L.). *Anim. Feed Sci. Technol.* 26, 71–79.

Choi, H., Sung, J. Y. and Kim, B. G. 2020. Neutral detergent fiber rather than other dietary fiber types as an independent variable increases the accuracy of prediction equation for digestible energy in feeds for growing pigs. *Asian-Australas. J. Anim. Sci.* 33, 615–622.

Córdova-Noboa, H. A., Oviedo-Rondón, E. O., Ortiz, A., Matta, Y., Hoyos, S., Buitrago, D. G., Martinez, D. J., Yanquen, J., Peñuela, L. and Sorbara, J. O. 2020. Corn drying temperature, particle size, and amylase supplementation influence growth performance, digestive tract development, and nutrient utilization of broilers. *Poult. Sci.* 99, 5681–5696.

Córdova-Noboa, H. A., Oviedo-Rondón, E. O., Matta, Y., Ortiz, A., Buitrago, D. G., Martinez, D. J., Yanquen, J., Hoyos, S., Castellanos, L. A. and Sorbara, J. O. 2021. Corn kernel hardness, drying temperature and amylase supplementation affect live performance and nutrient utilization of broilers. *Poult. Sci.* 100, 101395.

Cowieson, A. J., Acamovic, T. and Bedford, R. M. 2003. The effect of phytase and phytic acid on endogenous losses from broiler chickens. *Br. Poult. Sci.* 44, 23–24.

Cowieson, A. J., Acamovic, T. and Bedford, R. M. 2004. The effects of phytase and phytic acid on the loss of endogenous amino acids and minerals from broiler chickens. *Br. Poult. Sci.* 45, 101–108.

Cowieson, A. J. 2005. Factors that affect the nutritional value of maize for broilers. *Anim. Feed Sci. Technol.* 119, 293–305.

Cowieson, A. J. and Adeola, O. 2005. Carboydrases, proteases, and phytase have an additive beneficial effect in nutritionally marginal diets for broiler chickens. *Poult. Sci.* 84, 1860–1867.

Cowieson, A. J. and Ravindran, V. 2007. Effect of phytic acid and microbial phytase on the flow and amino acid composition of endogenous protein at the terminal ileum of growing broiler chickens. *Br. J. Nutr.* 98, 745–752.

Cowieson, A. J., Ravindran, V. and Selle, H. P. 2008. Influence of dietary phytic acid and source of microbial phytase on ileal endogenous amino acid flows in broiler chickens. *Poult. Sci.* 87, 2287–2299.

Cowieson, A. J. and Bedford, R. M. 2009. The effect of phytase and carbohydrase on ileal amino acid digestibility in monogastric diets: complementary mode of action? *Worlds Poult. Sci. J.* 65, 609–624.

Cowieson, A. J., Aptak, Mackowiak, P., Sassek, M., Pruszynska-Oszmalek, E., Zyla, K., Swiatkiewicz, S., Kaczmarek, S. and Jozefiak, D. 2013. The effect of microbial phytase and myo-inositol on performance and blood biochemistry of broiler chickens fed wheat/corn-based diets. *Poult. Sci.* 92, 2124–2134.

Cowieson, A. J. and Roos, F. F. 2014. Bioefficacy of a mono-component protease in the diets of pigs and poultry: a meta-analysis of effect on ileal amino acid digestibility. *J. Appl. Anim. Nutr.* 2, 13–21.

Cowieson, A. J., Lu, H., Ajuwon, K., Knap, I. and Adeola, O. 2015. Interactive effects of dietary protein source and exogenous protease on growth performance, immune competence and jejunal health of broiler chickens. *Anim. Prod. Sci.* 57, 252–261.

Cowieson, A. J. and Roos, F. F. 2016. Toward optimal value creation through the application of exogenous mono-component protease in the diets of non-ruminants. *Anim. Feed Sci. Technol.* 221, 331–340.

Cowieson, A. J., Bhuiyan, M. M., Sorbara, J. O. B., Pappenberger, G., Pedersen, B. M., Choct, M. 2020a. Contribution of individual broilers to variation in amino acid digestibility in soybean meal and the efficacy of an exogenous monocomponent protease. *Poult. Sci.* 99, 1075–1083.

Cowieson, A. J., Sorbara, J. O. B., Pappenberger, G., Abdollahi, R. M. and Ravindran, V. 2020b. Toward standardized amino acid matrices for exogenous phytase and protease in corn-soybean meal-based diets for broilers. *Poult. Sci.* 99, 3196–3206.

Croze, M. L. and Soulage, O. C. 2013. Potential role and therapeutic interests of myo-inositol in metabolic disease. *Biochimie* 95, 1811–1827.

Dersjant-Li, Y., Awati, A., Schulze, H. and Partridge, G. 2015. Phytase in non-ruminant: a critical review on phytase activities in the gastrointestinal tract and influencing factors. *J. Sci. Food Agric.* 95, 878–896.

Dilger, R. N., Onyango, M. E., Sands, S. J. and Adeola, O. 2004. Evaluation of microbial phytase in broiler diets. *Poult. Sci.* 83, 962–970.

Dridi, S. 2017. Understanding feed and water intake in poultry. In T. Applegate (Ed.), *Achieving Sustainable Production of Poultry Meat* (pp. 69–94). Philadelphia, PA: Burleigh Dodds Science Publishing Limited.

dos Santos, T. T., O'Neill, H. V. M., González-Ortiz, G., Camacho-Fernández, D. and López-Coello, C. 2017. Xylanase, protease and superdosing phytase interactions in broiler performance, carcass yield and digesta transit time. *Anim. Nutr.* 3, 121–126.

Erdaw, M. M., Wu, S., and Iji, A. P. 2017. Growth and physiological responses of broiler chickens to diets containing raw, full-fat soybean and supplemented with a high-impact microbial protease. *Asian-Australas. J. Anim. Sci.* 30, 1303–1313.

Fru-Nji, F., Kluenter, M. A., Fischer, M. and Pontoppidan, K. 2011. A feed serine protease improves broiler performance and increases protein and energy digestibility. *J. Poult. Sci.* 48, 239–246.

Ghazi, S., Rooke, A. J., Galbraith, H., Bedford, R. M. 2002. The potential for the improvement of the nutritive value of soya-bean meal by different proteases in broiler chickens and roiler cockerels. *Br. Poult. Sci.* 42, 70–77.

Gonzales-Uarquin, F., Kenez, A., Rodehutscord, M. and Huber, K. 2020. Dietary phytase and myo-inositol supplementation are associated with distinct plasma metabolome profile in broiler chickens. *Animal* 14, 549–559.

Gorenz, B., Iseri, V., Rubach, J. and Dilger, R. N. 2022. Xylanase supplementation of pelleted wheat-based diets increases growth efficiency and apparent metabolizable energy and decreases viscosity of intestinal contents in broilers. *Poult. Sci.* 101:102220

Greiner, R., Konietzny, U. and Jany, D. K. 1993. Purification and characterization of two phytases form Escherichia coli. *Arch. Biochem. Biophys.* 303, 107–113.

Hessing, G. C., Van Laarhoven, H., Rooke, A. J. and Morgan, A. 1996. Quality of soybean meals (SBM) and effect of microbial enzymes in degrading soya antinutritional compounds (ANC). In *2nd International Soyabean Processing and Utilization Conference* (pp. 8–13).

Huang, C., Gou, Y. and Yuan, J. 2014. Dietary yaurine impairs intestinal growth and mucosal structure of broiler chickens by increasing toxic bile acid concentrations in the intestine. *Poult. Sci.* 93, 1475–1483.

Huo, G. C., Fowler, R. V., Inborr, J. and Bedford, R. M. 1993. The use of enzymes to denature antinutritive factors in soyabean. In *2nd Workshop on Antinutritional Factors in Legume Seed* (pp. 517–521).

Jensen, L. S., Fry, E. R., Allred, B. J. and McGinnis, J. 1957. Improvement in the nutritional value of barley for chicks by enzyme supplementation. *Poult. Sci.* 36, 919–921.

Juturu, V. and Wu, J. C. 2014. Microbial exo-xylanases: a mini review. *Appl. Biochem. Biotechnol.* 174, 81–92.

Kalmendal, R. and Tauson, R. 2012. Effects of a xylanase and protease individually or in combination, and an ionophore coccidiostat on performance, nutrient utilization, and intestinal morphology in broiler chickens fed a wheat-soybean meal-based diets. *Poult. Sci.* 91, 1387–1393.

Karimi, A., Coto, C., Mussini, F., Goodgame, S., Lu, C., Yuan, J., Bedford, R. M. and Waldroup, W. P. 2013. Interactions between phytase and xylanase enzymes in male broiler chicks fed phosphorus-deficient diets from 1 to 18 days of age. *Poult. Sci.* 92, 1818–1823.

Kemme, P. A., Jongbloed, W. A., Mroz, Z., Kogut, J. and Beynen, C. A. 1999. Digestibility of nutrients in growing-finishing pigs is affected by Aspergillus niger phytase, phytase and lactic acid levels. 2. Apparent total tract digestibility of phosphorus, calcium and magnesium and ileal degradation of phytic acid. *Livest. Prod. Sci.* 58, 119–127.

Kim, S. W. and Duarte, E. M. 2021. Understanding intestinal health in nursery pigs and the relevant nutritional strategies. *Anim. Biosci.* 34, 338–344.

Kim, E., Moss, F. A., Morgan, K. N., Gharib-Naseri, K., Ader, P. and Choct, M. 2022. Ileal profile of non-starch polysaccharides and oligosaccharides in response to exogenous enzymes in broiler chickens offered wheat-or maize-based diets under subclinical necrotic enteritis challenge. *Anim. Nutr.* 10, 207–215.

Kornegay, E. T. 2001. Digestion of phosphorus and other nutrients: the role of phytases and factors influencing their activity. In M. R. Bedford and G. G. Partridge (Eds.), *Enzymes in Farm Animal Nutrition* (pp. 237–271). Walingford, UK: CAB International.

Kriseldi, R., Walk, L. C., Bedford, R. M. and Dozier, A. W. III. 2019. Effects of inositol and gradient phytase supplementation on growth performance and meat yield of broilers from 1 to 41 days of age. *Poult. Sci.* 98(Suppl. 1), 22 Abstr.

Kriseldi, R., Johnson, A. J., Walk, L. C., Bedford, R. M. and Dozier, A. W. III. 2021a. Influence of exogenous phytase supplementation on phytate, degradation, plasma inositol, alkaline phosphatase, and glucose concentrations of broilers at 28 days of age. *Poult. Sci.* 100, 224–234.

Kriseldi, R., Walk, L. C., Bedford, R. M. and Dozier, A. W. III. 2021b. Inositol and gradient phytase supplementation in broiler diets during a 6-week production period: 1. Effects on growth performance and meat yield. *Poult. Sci.* 100, 964–972.

Kriseldi, R., Walk, L. C., Bedford, R. M. and Dozier, A. W. III. 2021c. Inositol and gradient phytase supplementation in broiler diets during a 6-week production period: 2. effects on phytate degradation and inositol liberation in gizzard and ileal digesta contents. *Poult. Sci.* 100, 100899.

Kriseldi, R., Bedford, R. M., Dilger, N. R., Foradori, D. C., MacKay, L. and Dozier, A. W. III. 2021d. Effects of phytase supplementation and increased nutrient density on growth performance, carcass characteristics, and hypothalamic appetitive hormone expression and catecholamine concentrations in broilers from 1 to 43 days of age. *Poult. Sci.* 100, 101495.

Lee, S. A. and Bedford, R. M. 2016. Inositol - an affective growth promoter? *Worlds Poult. Sci. J.* 72, 743–759.

Lee, S. A., Wiseman, J., Masey, O'Neill, V. H., Scholey, V. D., Burton, J. E. and Hill, E. S. 2017. Understanding the direct and indirect mechanisms of xylanase action on starch digestion in broilers. *J. World Poult. Res.* 7, 35–47.

Lee, S. A., Bedford, R. M. and Walk, L. C. 2018a. Meta-analysis: explicit value of mono-component proteases in monogastric diets. *Poult. Sci.* 97, 2078–2085.

Lee, S. A., Dunne, J., Febery, E., Brearley, A. C., Mottram, T. and Bedford, R. M. 2018b. Exogenous phytase and xylanase exhibit opposing effects on real-time gizzard pH in broiler chickens. *Br. Poult. Sci.* 59, 568–578.

Lee, S. H., Hosseindoust, A., Laxman Ingale, S., Rathi, C. P., Yoon, Y. S., Choi, W. J. and Kim, S. J. 2020. Thermostable xylanase derived from Trichoderma citrinoviride increases growth performance and non-starch polysaccharide degradation in broiler chickens. *Br. Poult. Sci.* 61, 57–62.

Lee, J., Oh, H., Kim, Y., Song, D., An, J., Chang, S., Go, Y., Cho, H., Lee, B., Kim, W. and Cho, J. 2023. Effects of exogenous protease on performance, economic evaluation, nutrient digestibility, fecal score, intestinal morphology, blood profile, carcass trait, and meat quality in broilers fed normal diets and diets considered with matrix value. *Poult. Sci.* 102, 102565.

Lei, X. G., Ku, P. K., Miller, R. E., Yokoyama, T. M. and Ullrey, E. D. 1994. Calcium level affects the efficacy of supplemental microbial phytase in corn-soybean meal diets of weanling pigs. *J. Anim. Sci.* 72, 139–143.

Lei, X. G. and Stahl, H. C. 2000. Nutritional benefits of phytase and dietary determinants of its efficacy. *J. Appl. Anim. Res.* 17, 97–112.

Leske, K. L. and Coon, N. C. 1999. A bioassay to determine the effect of phytate phosphorus hydrolysis and total phosphorus retention of feed ingredients as determined with broilers and laying hens. *Poult. Sci.* 78, 1151–1157.

Liu, J., Bollinger, W. D., Ledoux, R. D. and Veum, L. T. 1998. Lowering the dietary calcium to total phosphorus ratio increases phosphorus utilization in low-phosphorus corn-soybean meal diets supplemented with microbial phytase for growing -finishing pigs. *J. Anim. Sci.* 76, 808–813.

Liu, W.-C. and Kim, H. I. 2017. Effects of dietary xylanase supplementation on performance and functional digestive parameters in broilers fed wheat-based diets. *Poult. Sci.* 96, 566–573.

Liu, H., Sorbara, J. O. B., Cowieson, A. J., Romero, F. L., Wang, K. S., Wu, J. L. and Kluenter, M. A. 2020. Exogenous α-amylase supplementation reduces the variability of ileal digestible energy in broiler chickens fed complete diets with maize batches of variable protein solubility. *Anim. Feed Sci. Technol.* 268, 114610.

Manangi, M. K. and Coon, N. C. 2007. The effect of calcium carbonate particle size and solubility on the utilization of phosphorus from phytase for broilers. *Int. J. Poult. Sci.* 6, 85–90.

McCafferty, K. W., Bedford, R. M., Kerr, B. and Dozier, A. W. III. 2019. Effects of cereal grain source and supplemental xylanase concentrations on broiler growth performance and cecal volatile fatty acid concentrations from 1 to 40 d of age. *Poult. Sci.* 98, 2866–2879.

McCafferty, K. W., Morgan, K. N., Cowieson, A. J., Choet, M. and Moss, F. A. 2022. Varying apparent metabolizable energy concentrations and protease supplementation affected broiler performance and jejunal and ileal nutrient digestibility from 1 to 35 d of age. *Poult. Sci.* 101, 101191.

Meng, X. and Slominski, A. B. 2005. Nutritive values of corn, soybean meal, canola meal, and peas for broiler chickens as affected by a multicarbohydrase preparation of cell wall degrading enzymes. *Poult. Sci.* 84, 1242–1251.

Mohammadigheisar, M., Shouldice, L. V., Torrey, S., Widowski, M. T., Ward, E. N. and Kiarie, G. E. 2021. Growth performance, organ attributes, nutrient and caloric utilization in broiler chickens differing in growth rates when fed a corn-soybean meal diet with multi-enzyme supplement containing phytase, protease and fiber degrading enzymes. *Poult. Sci.* 100, 101362.

Moita, V. H. C., Duarte, E. M. and Kim, W. S. 2022. Functional roles of xylanase enhancing intestinal health and growth performance of nursery pigs by reducing the digesta viscosity and modulating the mucosa-associated microbiota in the jejunum. *J. Anim. Sci.* 100:skac116.

Moita, V. H. C. and Kim, W. S. 2022. Nutritional and functional roles of phytase and xylanase enhancing the intestinal health and growth of nursery pigs and broiler chickens. *Animals* 12, 3322.

Morales, G. A., de Rodriganez, M. S., Marquez, L., Diaz, M. and Moyano, J. F. 2012a. Effect of phytase on the solubility of protein fractions in selected plant ingredients for fish feeds using SDS-PAGE image analysis: book abstract. In *15th International Symposium on Fish Nutrition and Feeding*, Molde, Norway, p. 48, 4–7 June.

Morales, G. A., de Rodriganez, M. S., Marquez, L., Diaz, M. and Moyano, J. F. 2012b. Effect of phytase on the hydrolysis of protein and phosphorus in selected plant ingredients simulating fish gastric environment: book abstract. In *15th International Symposium on Fish Nutrition and Feeding*, Molde, Norway, p. 47, 4–7 June.

Moran, E. T. 2016. Gastric digestion of protein through pancreozyme action optimizes intestinal forms for absorption, mucin formation and villus integrity. *Anim. Feed Sci. Technol.* 221, 284–303.

Newkirk, R. W. and Classen, L. H. 1998. *In vitro* analysis of phytate in canola meal with purified and pure sources of phytase. *Anim. Feed Sci. Technol.* 72, 315–327.

O'Dell, B. L., De Boland, A. R., and Koirtyohann, R. S. 1972. Distribution of phytate and nutritionally important elements among the morphological components of cereal grains. *J. Agric. Food Chem.* 20, 718–723.

Olukosi, O. A., Cowieson, A. J. and Adeola, O. 2007. Age-related influence of a cocktail of xylanase, amylase, and protease or phytase individually of in combination in broilers. *Poult. Sci.* 86, 77–86.

Olukosi, O. A., Beeson, A. L., Englyst, K. and Romero, F. L. 2015. Effects of exogenous proteases without or with carbohydrases on nutrient digestibility and disappearance of non-starch polysaccharides in broiler chickens. *Poult. Sci.* 94, 2662–2669.

Olukosi, O., G. González-Ortiz, H. Whitfield, and M.Bedford. 2020. Comparative aspects of phytase and xylanase effects on performance, mineral digestibility, and ileal phytate degradation in broilers and turkeys. *Poult. Sci.* 99:1528–1539.

Perz, K., Nowaczewski, S., Kaczmarek, A. S., Cowieson, and, Hejdysz, J. A. M. 2022. Research note: amylase supplementation improves starch and amino acids digestibility of faba bean for broilers. *Poult. Sci.* 101, 102117.

Perz, K., Kaczmarek, A. S., Nowaczewski, S., Cowieson, A. J., Ciszewski, A. and Hejdysz, M. 2023a. The effect of reduction of resistant starch content of faba bean and pea by amylase supplementation on performance, nutrient digestibility, and sialic acid excretion of broiler chickens. *Anim. Feed Sci. Technol.* 298, 115621.

Perz, K., Kaczmarek, A. S., Nowaczewski, S., Cowieson, A. J. and Hejdysz, M. 2023b. The effect of exogenous amylase supplementation on the nutritional value of pea (Pisum sativum L.) for broiler chickens. *Animals* 13, 816.

Phromkunthong, W., Nuntapong, N. and Gabaudan, J. 2010. Interaction of phytatse Ronozyme P(L) and citric acid on the utilization of phosphorus by common carp (Cyprinus carpio). *Songklanakarin J. Sci. Technol.* 32, 547–554.

Ravindran, V., Cabahug, S., Ravindran, G., Selle, H. P. and Bryden, L. W. 2000. Responses of broiler chickens to microbial phytase supplementation as influenced by dietary phytic acid and non-phytate phosphorus levels. II. Effects on apparent metabolizable energy, nutrient digestibility and nutrient retention. *Br. Poult. Sci.* 41, 193–200.

Ravindran, V. 2013. Feed enzymes: the science, practice, and metabolic realities. *J. Appl. Poult. Res.* 22, 628–636.

Romero, L. F., Sands, S. J., Indrakumar, E. S., Plumstead, W. P., Dalsgaard, S. and Ravindran, V. 2014. Contribution of protein, starch and fat to the apparent ileal digestible energy of corn- and wheat-based broiler diets in response to exogenous xylanase and amylase with or without protease. *Poult. Sci.* 93, 2501–2513.

Rooke, J. A., Slessor, M., Fraser, H. and Thomson, R. J. 1998. Growth performance and gut function of piglets weaned at four weeks of age and fed protease-treated soya-bean meal. *Anim. Feed. Sci. Technol.* 70, 175–190.

Sands, J. S., Dilger, N. R., Ragland, D. and Adeola, O. 2007. Ileal amino acid and phosphorus digestibility responses of pigs to microbial phytase supplementation of high-phytin diets. *Livest. Sci.* 109, 208–211.

Selle, P. H. and Ravindran, V. 2007. Microbial phytase in poultry nutrition. *Anim. Feed Sci. Technol.* 135, 1–41.

Siegert, W., Zuber, T., Sommerfeld, V., Krieg, J., Feuerstein, D., Kurrle, U. and Roehutscord, M. 2019. Prececal amino acid digestibility and phytate degradation in broiler chickens when using different oilseed meals, phytase and protease supplements in the feed. *Poult. Sci.* 98, 5700–5713.

Simbaya, J., Slominski, A. B., Guenter, W., Morgan, A. and Campbell, D. L. 1996. The effects of protease and carbohydrase supplementation on the nutritive value of canola meal for poultry: *In vitro* and *in vivo* studies. *Anim. Feed Sci. Technol.* 61, 219–234.

Singh, A. K., Tiwari, P. U., Berrocoso, D. J., Dersjant, Y.-Li, Awati, A. and R. Jha. 2019. Effects of a combination of xylanase, amylase, and protease, and probiotics on major nutrients including amino acids and non-starch polysaccharides utilization in broilers fed different level of fibers. *Poult. Sci.* 98, 5571–5581.

Sklan, D. and Y. Noy. 2003. Crude protein and essential amino acid requirements in chicks during the first week posthatch. *Br. Poult. Sci.* 44, 266–274.

Sommerfield, V., Schollenberger, M., Kuhn, I. and Rodehutscord, M. 2018. Interactive effects of phosphorus, calcium, and phytase supplements on products of phytate degradation in the digestive tract of broiler chickens. *Poult. Sci.* 97, 1177–1188.

Stefanello, C., S. L. Vieira, P. Soster, B. M. Dos Santos, Y. K. Dalmoro, A. Favero, and A. J. Cowieson. 2019. Utilization of corn-based diets supplemented with an exogenous α-amylase for broilers. *Poult.Sci.* 98:5862–5869.

Stefanello, C., Vieira, L. S., Rios, V. H., Soster, P., Simoes, T. C., Godoy, G. and Fascina, V. 2023. Research note: corn energy and nutrient utilization by broilers as affected by geographic areas and carbohydrases. *Poult. Sci.* 102, 102366.

Svihus, B. 2014. Starch digestion capacity of poultry. *Poult. Sci.* 93, 2394–2399.
Tamin, N. M., Angel, R. and Chrstman, M. 2004. Influence of dietary calcium and phytate phosphorus hydrolysis in broiler chickens. *Poult. Sci.* 83, 1358–1367.
Toghyani, M., Wu, S. B., Perez-Maldanado, R. A., Iji, A. P. and Swick, A. R. 2017. Performance, nutrient utilization, and energy partitioning in broiler chickens offered high canola meal diets supplemented with multicomponent carbohydrases and mono-component protease. *Poult. Sci.* 96, 3960–3972.
Tran, T. T., Hatti-Kaul, R., Dalsgaard, S. and Yu, S. 2011. A simple and fast kinetic assay for phytases using phytic acid-protein complex as substrate. *Anal. Biochem.* 410, 177–184.
Van Hoeck, V., Wu, D., Somers, I., Wealleans, A., Vasanthakumari, B., Sanchez, G. A. and Morisset, D. 2021. Xylanase impact beyond performance: a prebiotic approach in broiler chickens. *J. Appl. Poult. Res.* 30, 100193.
Vasanthakumari, B. L., Gedye, R. K., Abdollahi, R. M., Di Benedetto, M., Sanchez, G. D., Wealleans, A. and Ravindran, V. 2023. A new monocomponent xylanase improves performance, ileal digestibility of energy and nutrients, intestinal morphology, and intestinal microbiota in young broilers. *J. Appl. Poult. Res.* 32, 100301.
Walk, C. L., Addo-Chidie, E. K., Bedford, R. M. and Adeola, O. 2012. Evaluation of a highly soluble calcium source and phytase in the diets of broiler chickens. *Poult. Sci.* 91, 2255–2263.
Walk, C. L., Santos, T. T. and Bedford, R. M. 2014. Influence of superdoses of a novel microbial phytase on growth performance, tibia ash, and gizzard phytate and inositol in young broilers. *Poult. Sci.* 93, 1172–1177.
Walk, C. L. and Olukosi, A. O. 2019. Influence of graded concentrations of phytase in high-phytate diets on growth performance, apparent ileal amino acid digestibility, and phytate concentration in broilers from hatch to 28 d post-hatch. *Poult. Sci.* 98, 3884–3893.
Wang, H., Guo, Y. and Shih, J. C. H. 2008. Effects of dietary supplementation of keratinase on growth performance, nitrogen retention, and intestinal morphology of broiler chickens fed diets fed with soybean and cottonseed meals. *Anim. Feed. Sci. Technol.* 140, 376–384.
Wang, Y., Wang, Y., Lin, X., Gou, Z., Fan, Q., Ye, J., and Jiang, S. 2020. Potential effects of acidifier and amylase as substitutes for antibiotic on the growth performance, nutrient digestion and gut microbiota in yellow-feathered broilers. *Animals* 10, 1858.
Willingham, H. E., Fry, E. R., Leong, C. K., Jensen, S. L. and McGinnis, J. 1961. Studies on stability of enzyme supplements to pelleting, long storage, and other treatments. *Poult. Sci.* 40, 854–857.
Woyengo, T. A., Bogota, J. K., Noll, L. S. and Wilson, J. 2019. Enhancing nutrient utilization of broiler chickens through supplemental enzymes. *Poult. Sci.* 98, 1302–1309.
Yan, F., Dibner, J. J., Knight, D. C. and Vazquez-Anon, M. 2017. Effect of carbohydrase and protease on growth performance and gut health of young broilers fed diets containing rye, wheat, and feather meal. *Poult. Sci.* 96, 817–828.
Yin, F., Zhang, Z., Huang, J. and Yin, Y. 2010. Digestion rate of dietary starch affects systemic circulation of amino acids in weaned pigs. *Br. J. Nutr.* 103, 1404–1412.
Yu, S., Cowieson, A., Gilbert, C., Plumstead, P. and Dalsgaard, S. 2012. Interaction of phytate and myo-inositol phosphate esters ($IP_{1\text{-}5}$) including IP_5 isomers with dietary protein and iron and inhibition of pepsin. *J. Anim. Sci.* 90, 1824–1832.

Yuan, J. M. and Wang, Z. 2010. Effect of taurine on intestinal morphology and utilization of soy oil in chicken. *Br. Poult. Sci.* 51, 540–545.

Zhou, H., Wu, Y., Sun, X., Yin, D., Wang, Y., Mahmood, T. and Yuan, J. 2021. Effects of exogenous α-(1, 4)-amylase on the utilisation of corn starch and glucose metabolism in broiler chickens. *Animal* 15, 100396.

Żyła, K., Gogol, D., Koreleski, J., Świątkiewicz, S. and Ledoux, R. D. 1999. Simultaneous application of phytase and xylanase to broiler feeds based on wheat: *in vitro* measurements of phosphorus and pentose release from wheats and wheat-based feeds. *J. Sci. Food. Agric.* 79, 1832–1840.

Chapter 12

Advances in understanding and improving the use of macro-minerals in poultry nutrition: the example of phosphorus

A. Narcy[1], BOA-INRAE-Université France; F. Hervo, BOA-INRAE-Université de Tours, France and Université Laval, Canada; C. R. Angel, University of Maryland, USA; A. Bonadiman, Université Laval, Canada; M. Duclos, B. Méda and S. Mignon-Grasteau, BOA-INRAE-Université de Tours, France; M. Reis, Trouw Nutrition R&D – El Viso de San Juan, Spain; and M. P. Létourneau-Montminy, Université Laval, Canada

1 Introduction

Phosphorus (P) is the second most abundant mineral in the animal's body and is essential for birds due to its critical functions in metabolic processes, such as cellular metabolism, bone mineralization, and regulatory mechanisms (Suttle, 2010). A sufficient dietary supply of P is thus required to reach birds' optimum genetic potential in growth, feed efficiency, and skeletal integrity. However, the efficiency of P retention is rather low with diets based only on cereals, soybean meal, and minerals, i.e. approximately 50% in growing chickens (Plumstead et al., 2008; Rousseau et al., 2012). Strategies aiming to improve the efficiency

1 Corresponding author: agnes.narcy@inrae.fr.

http://dx.doi.org/10.19103/AS.2024.0143.16

of P use are required, notably because P originating from poultry manure contribute to the eutrophication of freshwater and seawater. According to Cordell et al. (2009), the world reserves of mineral phosphates are now limited to 35–85 years and should be preserved. For sustainable poultry production, P excretion and the use of non-renewable resources should thus be decreased throughout the production process (Dourmad et al., 2019).

To face these challenges, the P supply may be optimized without negatively affecting animal welfare and production performance. In this view, a nutritionist's objective is to fit the dietary P supply to the bird's requirements and to improve its retention rate. This strategy relies on the precise quantification of the feed ingredient P content and its availability to poultry, as well as a better knowledge of the requirements of birds, depending on their characteristics (including age, gender, and strain). This also includes better handling of the use of microbial phytase (PhytM) and determining its optimal application conditions. Enhancing the efficiency of animals to use minerals based on the stimulation of their adaptive capacity or by selection represents another way of improvement.

Over the last few decades, data acquired on P utilization by animals have greatly advanced our understanding of the mechanisms involved and the effects of the main factors of variation. Today, this wealth of knowledge can form the basis for new perspectives on formulation based on multi-criteria approaches to animal requirements and responses. This latter question is even more crucial, as production performance is no longer the only response criterion to be considered given the challenges of sustainability.

The objective of this chapter is to review the key research achieved for accurately evaluating the P values of feed ingredients and the P requirements of poultry, as well as to provide promising strategies to reduce P input and excretion in the environment while maintaining performance and bone health.

2 Challenges in optimizing mineral nutrition in poultry

2.1 Better characterization of the phosphorus value of feed ingredients

The precise supply of dietary P to meet the specific requirements of an animal involves reliable knowledge about the availability of P from feed ingredients. Phytic acid (PA) is the main storage form of P in plants, corresponding to an inositol esterified up to six phosphate groups: from IP1 to IP6. It accounts for 50–80% of the P content and is mainly found in the form of salts with cations, i.e. phytates. Poultry hydrolyzes phytates only partially due to their low endogenous phytase activity in the digestive tract. P from plants is thus poorly available without the use of exogenous phytases (Selle and Ravindran, 2007). PhytMs are enzymes that hydrolyze PA and release phosphate groups in the digestive tracts of birds. Non-phytic P (nPP) is obtained by subtracting the quantity of phytic P (PP) from that of the total P. As for mineral and animal nPP, the availability of the latter form is rather high (around 70%).

The total dietary P content and PP concentration in feed ingredients can easily be determined by chemical analyses. However, they do not reflect the portion used by the animals. Basically, available P (aP) represents the portion of dietary total P that, at a suboptimal level of P supply, can be utilized to cover the P requirement of the animal. However, this term has been widely used in different methodological approaches, which is confusing. Indeed, nPP (NRC, 1994) and aP are often used interchangeably, although studies have clearly demonstrated that nPP is not totally available and that PP is not totally unavailable to the animal (Angel et al., 2002; Coon et al., 2002). In the French system, aP refers to the proportion of P retained by the animal (based on bone mineralization) compared to a highly available reference source assumed to have a biological value of 100%, e.g. monocalcium phosphate (Sauvant et al., 2004), and thus is a relative value. This method has long been used but has some drawbacks related to the lack of standardization with criteria and reference sources that vary between experiments.

Digestible P for poultry has been implemented in the Dutch system in 1997 (van der Klis and Blok, 1997). The P digestibility of feedstuffs is determined through balance trials at marginal levels of P ensuring a negligible P excretion via the urine. A correction is applied to consider variations in the breakdown of inositol phosphate due the level of digestible P in the feed (CVB, 2018).

More recently, ileal digestibility (e.g. apparent, standardized, or true) has been proposed as a suitable method with a clearly described methodology to minimize variations between laboratories (WPSA, 2013; Mutucumarana et al., 2014; Babatunde et al., 2020). Apparent ileal digestibility (AID) is the proportion of nutrients that disappear up to the last half of the ileum compared with the diet concentration, determined with the help of an indigestible marker (e.g. Cr, TiO_2, and celite). True ileal digestibility (TID) and standardized ileal digestibility (SID) both consider P endogenous losses through the small intestine, either estimated by regression or measured using a P-free diet, respectively (Dilger and Adeola, 2006; Mutucumarana et al., 2014; Babatunde et al., 2020).

2.2 Factors interfering with phosphorus availability

The digestive utilization of dietary P is influenced by many factors, such as the proportions of the different P forms (PP and nPP), incorporation of PhytM, and calcium (Ca) content (Selle and Ravindran, 2007). These effects and interactions need to be precisely quantified to improve diet formulations.

2.2.1 Toward a more efficient utilization of phytic phosphorus

As stated above, 50–80% of the P content in poultry diets is in the form of phytates that are considered poorly available to monogastric animals without hydrolysis by the phytase enzyme. For optimal action, PA must be hydrolyzed by phytases upstream from the sites of P absorption, which occurs mostly in the upper small intestine (Suttle et al., 2010). There are four sources of phytase: (1) the mucosa of the small intestine; (2) microorganisms in the large intestine; (3) the phytasic activity of cereals; and (4) exogenous PhytM added to the diet. A unit of phytase activity is defined as the release of 1 μmol of inorganic P per minute in a solution containing 5.1 mmol of sodium phytate per liter at pH 5.5 and 37°C (Engelen et al., 1994).

Phytase activity in the proximal part of the gut is generally considered low in monogastrics. However, some studies have shown that poultry have effective phytase and phosphatase activity in the intestinal mucosa, allowing to hydrolyze up to 40% without exogenous phytase (Applegate et al., 2003), but this activity varies according to dietary factors, such as dietary Ca and P (McCuaig et al., 1972; Shastak et al., 2014).

As plant phytase activity is partially or completely inactivated after high temperature treatment (>70°C) such as those for pelleting (Greiner and Konietzny, 2006), it is generally considered negligible. Currently, exogenous PhytMs are commonly used in poultry diets. The first exogenous phytase was marketed in the 1990s in the Netherlands, the first country that had strict regulations intended to limit P discharge from pig and poultry farming. The characteristics and efficacy of phytases have evolved considerably over time with significant progress thanks to genetic engineering (Rizwanuddin et al., 2023).

2.2.2 Impact of calcium on phosphorus utilization

PA can chelate some divalent cations (e.g. Fe^{2+}, Zn^{2+}, Cu^{2+}, and Ca^{2+}) to form phytates (Angel et al., 2002). These complexes may avoid PP degradation by phytase by limiting the access of the enzyme to its binding site. A high pH (≥5.0) is known to favor the complexation process between Ca and PA (Nolan and Duffin, 1987). Thus, in poultry, the increase of the pH along the small intestine favors Ca–PA complexes. This complexation process strongly depends on the level of PA esterification, so it decreases from highly esterified IP6 to poorly esterified IP1 (Lutrell, 1993). In poultry diets, PA is mainly present as IP6 (Lott et al., 2000). Additionally, the binding of Ca to PA depends on the Ca-to-PA ratio (Wise, 1983). When this ratio is high, the probability of complex formation increases. In broilers, the Ca:PA ratio is 1.05, on average, while in laying hen, due to the high Ca requirement for the eggshell, the Ca:PA ratio is around 4.6.

In poultry diets, limestone is the main Ca source, and it can be supplied as fine limestone particles (FLPs) or coarse limestone particles (CLPs; Scott et al., 1971). The formation of the Ca–PA complex is greatly influenced by limestone particle size, as FLP facilitate the binding of Ca to PA. These FLPs have greater solubility compared with CLP and rapidly provide a high quantity of soluble Ca in the gizzard (Kim et al., 2019). This soluble Ca can easily bind PA, reducing phytase efficiency (Manangi and Coon, 2007) and, in turn, decreasing the AID of P and Ca in broilers (Tamim et al., 2004; Kim et al., 2018) and laying hens (Hervo et al., 2023). For instance, Li et al. (2021) found in broilers that the AID of P decreased from 87% to 61% in diets supplemented with PhytM when the same limestone was reduced in particle size from 0.80 to 0.15 mm. Similarly, in laying hens fed diets supplemented with PhytM, the AID of P decreased by 16% when limestone particle size decreased from 2.26 to 0.26 mm (Hervo et al., 2023).

2.3 Better evaluation of requirements

Poultry nutritional requirements can be evaluated using empirical, factorial, or mechanistic approaches. The empirical approach consists of determining the requirements in a series of dose-response trials. Bone mineralization is often used as a response criterion (Shastak and Rodehutscord, 2013) because the majority of body P is contained in bone tissue. The empirical method is the most common but running these trials is time consuming and costly. Moreover, the results vary widely depending on the response criterion considered, the method of interpretation, animal factors (e.g. age, gender, and strain), diet composition, and environmental conditions. Ravindran et al. (1995) noticed that the choice of the response criteria greatly influences the results of this approach.

The factorial approach has also been used to determine the P requirements of growing chickens. The overall nutrient requirements represent the sum of requirements for 'maintenance' plus requirements for 'growth' (Khaksar et al., 2017; Rostagno et al., 2017). Using this approach, the aP requirement is determined *via* the estimation of endogenous P loss for maintenance and via P retained per kilogram of body weight gain for growth, with body P composition the main component of the P requirement.

Around 60% of body P and more than 95% of body Ca are found in the chicken skeleton bound together to form hydroxyapatite ($Ca_{10}(PO_4)_6$; Narcy et al., 2015). The deposition of Ca and P into bone is therefore strictly dependent on the presence of other minerals. The reminder of body P is mainly found in muscle tissues, where it serves several vital mechanisms and has no common function with Ca. Soft tissue and bone tissue depositions therefore evolve differentially and independently depending on dietary P and Ca intakes (Narcy et al., 2015).

Mathematical models allow to replicate reality through the integration of various theories, thereby enabling their application in diverse situations (Fisher, 2015). To achieve this, the endeavor involves assembling the underlying mechanisms that drive specific responses, ultimately enabling the accurate prediction of such responses. Upon successful development, these 'mechanistic' models can be useful to nutritionists in accomplishing multiple specific objectives. These objectives may include optimizing performance traits and reducing nutrient excretion, without impairing animal welfare. Considering the complexity of P and Ca interactions, the mechanistic approach represents a robust method to establish requirements of the bird. Mathematical modeling has already been used to better understand the metabolic fate of dietary P and Ca in laying hens (Kebreab et al., 2009) and pigs (Besancon and Gueguen, 1969; Fernandez, 1995; Létourneau-Montminy et al., 2015). With the aim of developing a more precise and robust methodology to assess P requirements of broilers, a model that combines the factors that drive the fate of Ca and P in the gastrointestinal tract upon their absorption (digestible Ca and P) and the responses of broilers enabling the estimation of the required quantities of Ca and P for achieving maximum bone mineralization was developed. This innovative approach, described in Section 4, combines both empirical and mechanistical modeling.

3 Case study: Meta-analysis for assessing precisely the phosphorus value of diet

3.1 Approach

All methods described earlier to estimate the P value of feedstuffs gives a unique P and Ca value for each feedstuff regardless of interactions with other components of the diet. To tackle this challenge, it is worth noting that there has been a substantial body of research on the measurement of AID of P in broiler chickens over the past 10 years. In order to gain a deeper understanding and quantify the impact of the primary sources of variation on this response, a meta-analysis is a valuable approach (Sauvant et al., 2008). A study was therefore conducted to integrate quantitative information from the literature and develop a model for predicting the impacts and interactions of dietary Ca and P concentration, forms of P and the addition of PhytM on P digestibility.

3.2 Data collection and coding

An exhaustive literature search (2010–2021) of publications dealing with the effect of dietary P and Ca concentrations and PhytM supplementation on AID of P in fast-growing broiler chickens reared between 1 and 42 days (d) of age was performed (Khaksar et al., 2022). The database contained general information (e.g. author's name, year of publication, and type of scientific journal) as well as qualitative (e.g. strain, age, and sex) and quantitative data (e.g. PP, nPP, Ca, and PhytM concentrations). Each observation represented the average of one treatment. The complete database used for the meta-analysis included 625 treatments from 118 experiments and 74 publications. The publications covered the periods 1–21 and 22–42 d in the following proportions: 292 and 333 treatments, 55 and 63 experiments, and 40 and 49 references, respectively. Of these data, more than 75% of the treatments come from corn-based diet.

3.3 Calculation and determination of dependent and independent variables

To ensure the overall consistency of the database, total dietary Ca, Ca from plant, mineral and animal origin, total dietary P, PP, and nPP from plant, mineral, and animal origin were recalculated using ingredient nutritional composition data (Sauvant et al., 2004). In 84% of treatments, diets included phosphate (62% dicalcium, 35% monocalcium, and 3% monosodium) and 10% contained nPP in animal form. Of the 625 treatments, 273 were supplemented with PhytM. The independent variables used in the statistical models were dietary PP, total nPP (plant + mineral + animal) and Ca concentrations (g/kg diet), and phytase activity (FTU/kg diet). These were used to predict the 'digestible P'-dependent variable which is the AID coefficient multiplied by the total P (digP; g/kg diet). Data were analyzed using the GLM model (Minitab, version 20, State College, PA, USA) considering the experiment as a fixed effect.

Based on the graphical examinations of intra-experimental responses to nPP, quadratic models were fitted as follows:

$$Y_{ij} = \alpha + \alpha_i + b_1 \times nPP_{ij} + b_2 \times nPP_{ij}^2 + e_{ij}$$

Where, Y_{ij} is the value of the dependent variable Y in experiment *i* with nPP concentration *j*, α is the overall intercept; α_i is the effect of experiment *i* on the intercept α with the condition that the sum of each α_i is equal to 0; b_1 and b_2 are the linear and quadratic coefficients of the overall response to dietary nPP, respectively; and e_{ij} is the residual error. Significant ($P < 0.05$) and trending ($P < 0.10$) parameters were included in the final models.

3.4 Findings

The response of digP to nPP was linear which means that the maximum absorption rate was not reached. However, the response to nPP was influenced by many other factors (Ca × nPP × PhytM, $P < 0.01$). This three-way interaction shows that digP is optimized in the presence of PhytM and for lower Ca contents. The positive effect of lowering dietary Ca (Fig. 1) is well-known and can be

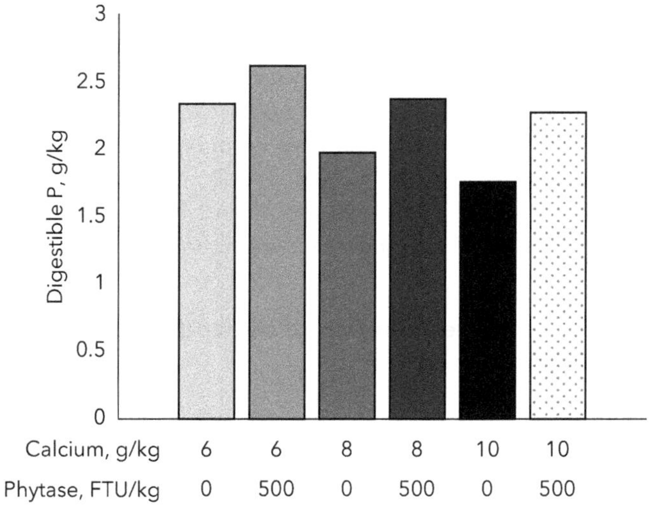

Figure 1 Impact of dietary calcium (Ca) and microbial phytase (PhytM) on digestible P; Interaction Calcium x Phytase (P=0.001) and interaction Calcium x Phytase x Phytase (P=0.03).

due to reduction of both Ca–phosphate and Ca–PA complexes in the gut, thus improving P absorption (Selle et al., 2009; Heaney, 2004). Birds have also the capacity to increase their absorption by hormonal regulation in low Ca diets (Proszkowiec-Weglarz and Angel, 2013). Additionally, low Ca also promote the activity of intestinal enzymes involved in PP degradation. McCuaig et al. (1972), for example, showed that reducing Ca concentrations from 15 to 6 g/kg increased intestinal alkaline phosphatase activity by 52% and intestinal phytase activity by 75%, as Ca competes with the active sites of these enzymes. Similar results were reported by Applegate et al. (2003), who found that dietary Ca, at concentrations commonly used in broiler diets (9 g/kg), led to a reduction in intestinal phytase activity and apparent hydrolysis of PP at the ileal level compared with a lower concentration of Ca (4 g/kg). Ca thus influences the activity of endogenous phosphatases and phytases. This may explain part of the positive effect of reducing dietary Ca on P utilization efficiency. As a result, the PhytM effect is potentialized in high Ca diet; supplying 500 FTU/kg of phytase increases digestible P by 12%, 20%, and 29% for 6, 8, and 10 g dietary Ca/kg (Fig. 1).

Furthermore, in a diet with high Ca content, the broiler chickens' response to PhytM fades as nPP increases. For example, in a diet containing 10 gCa/kg, PhytM supplementation (500 FTU/kg) improves digP by 0.54 g/kg at low nPP concentrations (1 g/kg), whereas this effect is reduced to 0.23 g/kg at higher nPP concentrations (5 g/kg). The possible mechanism is the deepened P deficiency and the stimulation of the gut's capacity to absorb P (Rousseau et al.,

2016). Also, low nPP intake could induce an increase in alkaline phosphatase activity like low Ca intake (Liu et al., 2017).

In summary, this meta-analysis shows and quantifies the negative effect of Ca on digP. The PhytM responses also depends on nPP and Ca, with higher response in high Ca and low nPP diets. This emphasizes the importance of evaluating phytase using a standardized methodology (WPSA, 2013). From a practical point of view, this meta-analysis work allowed to develop an equation that predicts digP from a complete diet. Thus, by the chemical analysis of Ca, PP, total P, and PhytM, it will be possible to obtain digestible P value of the diet. This is less time consuming than evaluating each feedstuff for their digestibility and more precise than considering a unique value per feedstuffs.

Although coming from exhaustive literature data, the equation needs external validation to assess the accuracy of predictions. A first validation trial was made, formulating eight diets to have a linear increase of digP through different concentrations of Ca, nPP, and PhytM and measuring AID of Ca, PP, and nPP in 21-day-old broilers (Narcy et al., 2023). Results showed that the equation underestimates digestibility of PP in low-P diets in accordance with results cited previously that showed the capacity of the bird to increase PP degradation in cases of deficiencies. Therefore, considering the PP digestibility measured in the experiment allowed to have an accurate prediction (RMSPE, 5.6%; R2, 90%). It should be noted that this equation has been performed in a database containing mostly corn-based diets so validation with other feedstuffs will also be necessary.

4 Case study: Mechanistic modeling

A significant portion of P and Ca is deposited alongside body constituents. Evidence demonstrates that the amounts of P and Ca present in protein and lipids in the body remain relatively constant (van der Klis et al., 1990), in contrast to their concentrations in bone tissue, which may vary based on the availability of dietary sources (Proszkowiec-Weglarz and Angel, 2013). Therefore, properly estimating the growth of body constituents, such as protein, lipid, and bone ash, becomes a crucial aspect in determining the P and Ca requirements of the animal for specific purposes. By employing this technique, we can better understand the actual requirements of a broiler, thereby avoiding either an excessive or insufficient supply. The method employed in this estimation is a factorial approach, whereby the quantities of P and Ca required for maintenance and growth are aggregated to determine the overall needs of the animals.

4.1 Model structure

The mathematical model presented herein uses the mechanisms described by Reis et al. (2023a) to simulate the fate of dietary Ca and P and estimate the metabolic requirement for maximum bone mineralization. The model is organized into three modules: (1) a digestive module, predicting the absorption of Ca and P in the diet available for tissue deposition; (2) a soft tissue module that simulates the growth of protein, lipids, ash, and water in the body, based on the Reis et al. (2023b); and (3) a body ash module that projects the accumulation of ash minerals in soft tissue and bone separately, adapted from Létourneau-Montminy et al. (2015).

4.2 Absorption of dietary calcium and phosphorus

The absorption has been simulated by using the equation developed in the previous section using meta-analysis approach. The same database was utilized to estimate dietary Ca digestibility. To estimate the intestinal absorption (in g/d) of Ca (Ca_{Abs}) and P (P_{Abs}), the endogenous losses of Ca ($Ca_{EndLossF}$) and P ($P_{EndLossF}$) in the feces were subtracted from the absorbed fraction as follows:

$$P_{Abs} = \left(-P_{EndLossF} + digP\right) \times Feed\,Intake$$

$$Ca_{Abs} = \left[\left(-Ca_{EndLossF} \times DM intake\right) + digP\right] \times Feed\,Intake$$

4.3 The simulation of the soft tissue growth

The method used herein to estimate broiler growth is the same as that used by Reis et al. (2023b). The driving force behind predicting the growth lies in understanding body protein deposition, which then enables the prediction of lipid, water, and ash through allometry equations. It is proposed that the nutritional requirements of an animal can be estimated based on its growth and body composition. Due to the variation in protein composition between the feather-free body and feathers (Stilborn et al., 1997) and the differences in growth ratios (Reis et al., 2018), it becomes necessary to calculate the growth of each compartment separately. This approach ensures a more accurate estimation of the animal's nutritional needs and growth potential, considering the specific characteristics of different body components. The ash composition of soft tissue was generated for Ross 308 from day 7 to day 42 on a weekly basis (Hamdi et al., 2017). It was assumed that soft tissue contains a constant concentration of P and Ca (Reis et al., 2023a). Therefore, the dietary intake of P and Ca is initially directed toward fulfilling soft tissue growth.

4.4 The simulation of body ash

The primary components of bone ash are Ca and phosphate, along with various other minerals (Crenshaw, 2001). Hydroxyapatite salts, with a fixed ratio of 2:1

for P and Ca deposition (Matuszewski et al., 2020), are the primary structures in which these minerals are deposited in bone. Therefore, we calculate bone ash as the sum of Ca, P, and miscellaneous minerals in bone. Potential of Ca retention in bone dictates the deposition of all minerals, maintaining the proportion of minerals in bone. The 2:1 ratio of Ca:P was used to estimate the potential amount of P in bone ash. The concentration of miscellaneous minerals in bone ash is assumed to be 48% of the total bone ash. The model is also designed to predict the mobilization of P from bones when the intake is insufficient to support soft tissue growth.

The actual deposition of Ca and P in bone is a complex process that is driven by the quantity absorbed versus the genetic potential for Ca deposition and the amount and ratio of Ca and P absorbed. To maximize bone mineralization, Ca and P absorption have to be absorbed in quantity fitting the possible deposition based on genetic potential (see previous sections) and in the right ratio. Urinary excretion of Ca and P comprises both obligatory losses and the minerals that cannot be retained in soft tissue or bone due to mineral imbalances or excess intake. The obligatory urinary losses of Ca and P are minimal and have been estimated in the model at 0.678 and 0.874 mg/kg of body weight, respectively, based on the work of Manangi et al. (2018).

4.5 Model behavior

The model showed that most part of Ca is found into bone, while P is equally divided between bone and soft tissues (Fig. 2; Reis et al., 2023a). The model also allows predicting imbalances in Ca and P for bone deposition which results in the excretion of the non-limiting mineral (Fig. 3; Reis et al., 2023a).

4.6 Prediction of calcium and phosphorus requirements

Following inversion modelling principles (Doeschl-Wilson et al., 2006) the model of Reis et al. (2023a) described in the previous sections was inversed to predict requirement. Phosphorus requirements is the sum of the maintenance requirement and the amount that should be absorbed to reach soft tissue ash and bone ash potential. It can therefore be expressed based on AID or SID in gram per kilogram diet (Fig. 4). To obtain AID requirement, only the unavoidable urinary loss is considered. To obtain SID, the basal endogenous at the digestive tract level is also considered.

It is obvious that the requirements for Ca and P are influenced by both the genetic potential and the current state of the animal. Therefore, to accurately determine the nutritional requirements of an animal, it is imperative to provide a detailed description of the specific animal under consideration. In the simulations conducted in this study, our focus was on the genetic potential of a male Ross 308 broiler, which has been extensively characterized by Gonçalves

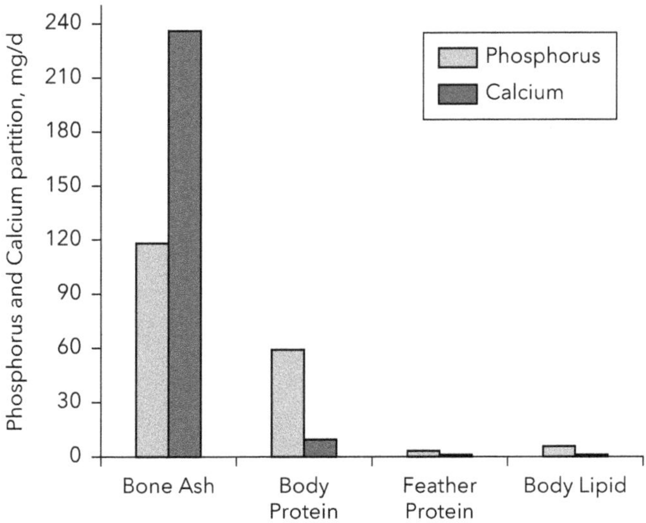

Figure 2 Body composition and proportion of phosphorus and calcium in the body of 21-day-old broiler fed recommended nutrient levels according to Rostagno et al. (2017).

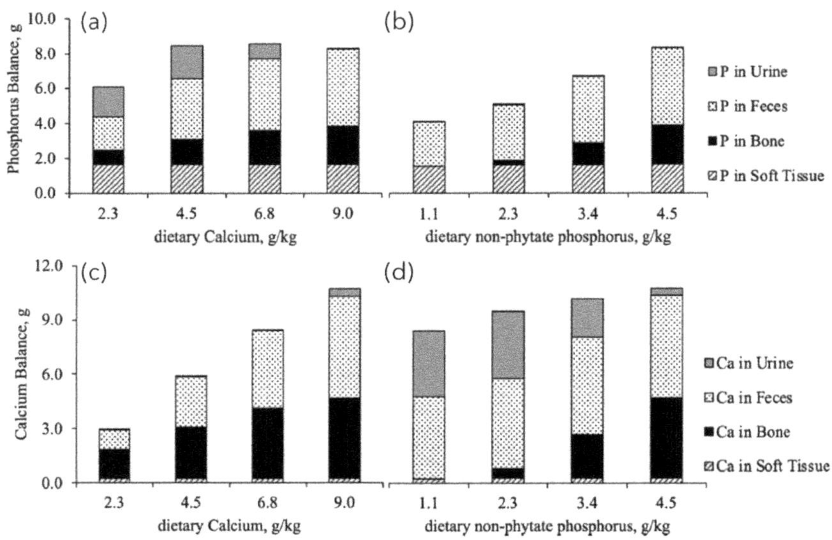

Figure 3 Metabolic fate of phosphorus (A and B) and calcium (C and D) on broilers body, where (A and C) are the simulation results of a broiler consuming increasing levels of dietary calcium (from 0.23% to 0.9%) with constant dietary available phosphorus (0.45%), and (B and D) are the simulation results of a broiler consuming increasing levels of dietary available phosphorus (from 0.11% to 0.45%) with constant dietary calcium (0.9%) (From Reis et al., 2023a).

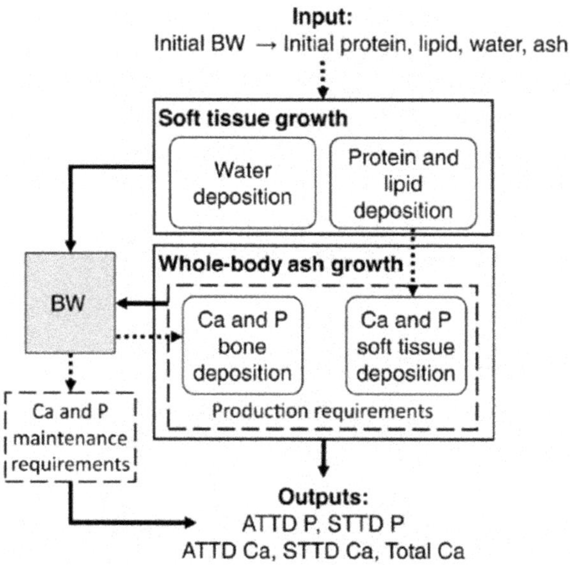

Figure 4 General layout of the inversed model predicting calcium and phosphorus requirement in apparent and standardized ileal digestibility (AID and SID) of growing broilers; Lautrou et al. (2020).

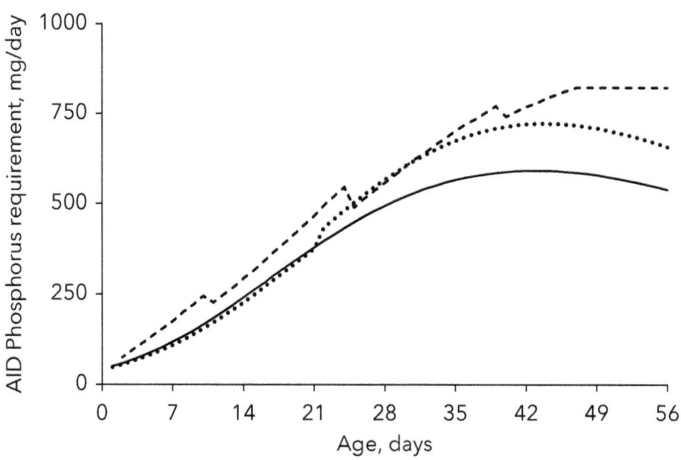

Figure 5 Apparent ileal digestible (AID) phosphorus required for maximum bone mineralization as predicted by the model of Reis et al. (2023a) (black full line) and the available phosphorus recommended by Rostagno et al. (2017; dotted line) and Aviagen (2022; dashed line).

et al. (2020) and Vargas et al. (2020). The results were then compared to the recommendations available in the Brazilian tables (Rostagno et al., 2017) and Ross 308 (Aviagen, 2022) (Fig. 5). The results showed that the current model predict requirements that are close but lower than Ross 308 and close to Rostagno et al. (2017). The model also allows the prediction of digestible Ca in apparent or standardized ileal value. However, at the moment, the digestible Ca system in not yet available although some research is ongoing (see Section 5). Using this approach, requirements can be defined based on different criteria of interest that depend on production objectives (e.g. profitability, animal welfare, and environment). The next step will be the validation of this model in both traditional phase-feeding and precision feeding (see Section 5).

5 Future trends

5.1 Toward precision feeding for sustainable broiler and layer production

In livestock production, precision feeding can be defined as providing an animal with the right amount of a given nutrient at the right time (Zuidhof, 2020) versus classic phase feeding which leads to periods of excess and deficiency (Fig. 6). Therefore, this approach requires a good knowledge of both animal requirements (and their dynamics over time) and feed characteristics (chemical composition, digestibility), and strongly relies on the development of mathematical models. However, while this concept has been widely studied in swine production for more than one decade (Gaillard et al., 2020), few studies are available for poultry.

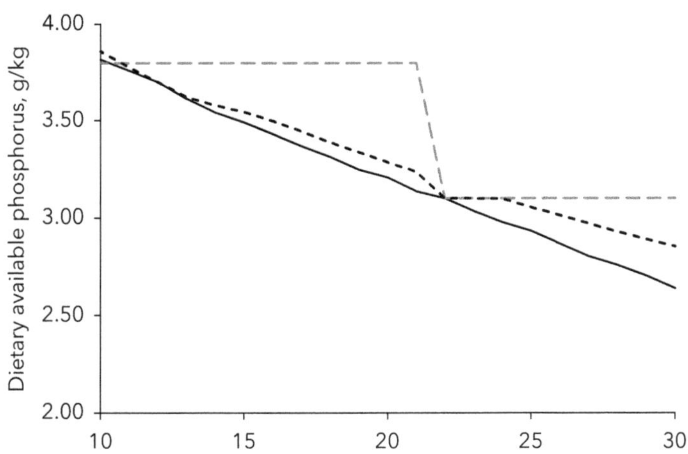

Figure 6 Precision feeding vs. conventional phase feeding in grower – finisher broiler (from 10 to 30 day of age). Black full line, P requirement (based on theorical growth); black dotted line, precision feeding diet; gay dotted line, grower-finisher strategy (2 phases).

In broilers, because of the size of the flocks, precision feeding strategies can only be implemented at the flock level rather than at the individual level. Méda et al. (2019, 2022) recently proposed a modeling approach for the implementation of precision feeding strategies combining several models designed to (1) evaluate the average growth performance 'of the next day' of the flock (using body weight data collected the previous days in the farm with automatic scales) and (2) the lysine, energy, and P requirements associated with this 'expected performance.' Based on these growth expectations and nutrient requirements, an optimal diet can be obtained by mixing two different diets and distributing them to the birds. Using this approach, Pampouille et al. (2022) tested a precision feeding strategy in fast-growing broilers and compared it to a three-phase strategy between 10 and 31 days of age. Final body weight and cumulated feed intake were similar in both strategies, but the feed conversion ratio was significantly increased in the precision feeding strategy (1.35 vs. 1.32). However, despite this higher feed conversion, the precision feeding strategy substantially reduced both feed cost and P excretion by 8% and 13%, respectively, confirming the potential of the approach to improve the sustainability of the broiler sector.

5.2 Potential for adaptation and genetic determinism

Optimizing the efficiency of P utilization is crucial in reducing dietary supply while maintaining bone health and growth performance. In this regard, we can question whether this efficiency can be improved in the long term not only by stimulating physiological processes but also by selecting more efficient animals. If possible, co-adapting the animal and the feed may be a promising way to reduce P supply and excretion. The concentrations of minerals can be reduced in the diets of efficient animals without detrimental effects on performance or health. Conversely, the genotype used may be adapted to dietary constraints, such as the proportion of PP in the diet.

5.2.1 Compensatory mechanisms

In poultry, P and Ca metabolism are subjected to homeostatic mechanisms. Deficiencies or imbalances in P and Ca induce physiological changes, including increases in their absorption, through common regulatory pathways (Proszkowiec-Weglarz et al., 2013). As for Ca, when birds are fed a P-deficient diet or when P requirements are high, the blood P concentration drops. This leads to the activation of 1,25 $(OH)_2$ cholecalciferol synthesis in the kidney, which in turn leads to increased P absorption in the small intestine and reabsorption in the kidney. Animals can therefore compensate, to some extent, for the reduced mineral availability.

Studies have evaluated the long-term effects of early P and Ca restrictions on growth performance and bone mineralization (Yan et al., 2005; Rousseau et al., 2016; Valable et al., 2020). Broilers early fed a diet moderately deficient in P and Ca demonstrated the ability to partially adapt to the deficiency in the long term. This has been illustrated by increased ileal absorption of P and Ca, increased ileal PP disappearance, compensatory growth, and compensatory improvement in bone parameters. The level to which animals can compensate depends on the P and Ca concentrations of the recovery feed (Valable et al., 2018).

Currently, the need to reduce P inputs and outputs has led to the formulation of feeds that are ever closer to requirements, leaving little safety margin. In a way, the compensatory phenomenon can be seen as a safety margin in the case of manufacturing hazards or shifts in individual requirements within a batch resulting from the variability between birds. While it may be possible to consider these compensatory phenomena in the mathematical models described earlier, the underlying mechanisms of compensation and the exact conditions of recovery need to better understand. There is also a need to determine the precise origin of animal variability to adequately include it in models.

5.2.2 Genetic determinism

Previous studies have shown that P efficiency and bone traits are variable between animals and can thus be heritable. Understanding the genetic basis of P efficiency or bone mineralization thus represents a valuable approach to controlling bone quality traits as well as dietary P supply and output. More specifically, this information can be used in selection schemes to improve these traits.

Low-to-moderate heritability has been reported for P retention and plasma concentrations in chickens (Zhang et al., 2003; Wang et al., 2006; Ankra-Badu et al., 2010; de Verdal et al., 2011; Mignon-Grasteau et al., 2016) and quail (Beck et al., 2014; Vollmar et al., 2021). Regarding bone traits, higher heritability estimates have generally been observed, i.e. medium for morphological traits (Singh and Hjilani, 2008; de Verdal et al., 2013; Gonzalez-Ceron et al., 2015; Vollmar et al., 2021), and relatively high for bone ash content and breaking strength (de Verdal et al., 2013; Mignon-Grasteau et al., 2016; Vollmar et al., 2021).

Additionally, genomic regions that control bone traits have been identified in quantitative trait loci mapping studies. More than 100 loci on different chromosomes have been reported in chickens, hens, and quail (NRSP8: National Research Support Program: animalgenome.org; Johnsson et al., 2015; Faveri et al., 2019; Vollmar et al., 2021). These findings support evidence that multiple genetic loci control bone and mineral metabolism. Additionally, some of the effects on bone traits depend on gender, indicating the importance of considering the interaction between QTL and sex (Faveri et al., 2019).

Some genes have been identified as functional candidates (Mignon-Grasteau et al., 2016; Faveri et al., 2019; Vollmar et al., 2021). They are involved in many biological processes, such as skeletal system development and bone formation, mineral transmembrane transporter activity, cell differentiation, growth and morphogenesis, and cytoskeletal structure and function. In future functional studies, these candidate genes can be investigated, opening new perspectives in terms of nutrigenomic and nutrigenetic studies or in selection schemes.

5.3 Balance between calcium and phosphorus in an extended production cycle in laying hens

5.3.1 Interplay between calcium and phosphorus metabolism in laying hens

On a daily scale, laying hens require more than 2 g of Ca for eggshell formation, a process that lasts 20 h (Gautron et al., 2021), partly during the night period when the hen is not eating. This specific metabolism in regard to egg production drives the P and Ca requirements of the laying hen, with different kinetics of utilization for these two minerals. Hens draw up to 40% of the required Ca from their medullary bone (Dacke et al., 1993). Since Ca is complexed with P as a hydroxyapatite crystal with a Ca:P ratio around 2.2, osteolysis releases about 0.36 g P together with 0.79 g Ca during the dark period, when the P requirement is low (about 10 mg, Suttle, 2010). Excess P in the circulation leads bones to secrete the phosphaturic hormone, FGF23 (Fibroblast Growth Factor – 23), which inhibits renal P reabsorption and exerts negative feedback on Ca (and P) balance by lowering 1,25-OH_2-D_3 levels (Gloux et al., 2020). According to the metabolic model of Kebreab et al. (2009), urinary P excretion in laying hen is around 0.25 g/day and most of it results from osteolysis, depending on the time of lay during the day. During the non-eggshell production period, both dietary Ca and P are required to restore bone reserves. Finally, the P requirements of the laying hen can be defined as the sum of the requirement to produce the different egg compartments, around 0.12 g/day (Suttle, 2010), the P requirement for body maintenance, around 0.02 g/day (Kebreab et al., 2009), and the requirement for medullary bone due to Ca mobilization during the night, up to 0.36 g/day (Nys, 2017). However, the P requirements for laying hens have not been precisely defined. A recent review provided evidence that the dietary P supply to laying hens was in excess (Rodehutscord et al., 2023) and proposed reducing it to 2.2 g nPP/kg of feed or even less with the supply of PhytM.

Because osteoclasts resorbing medullary bone are not specific, part of the surrounding cortical bone is resorbed during the night. Unlike medullary bone, cortical bone is not reconstituted during the non-eggshell production period. Consequently, aging is characterized by an erosion of cortical bone, leading to osteoporosis (Whitehead and Fleming, 2000). Additionally, aging is also marked by alterations of P/Ca homeostasis (Gloux et al., 2020), including lower circulating concentrations of active vitamin D and increased expression of FGF23 together with the expression of candidate genes marking increased bone turnover in medullary bone. Thus, improving the dietary Ca:P balance in laying hens is a prerequisite for reducing bone mobilization and P urinary excretion in old hens. The objective of lengthening the careers of laying hens from 350 to 500 eggs (75–100 weeks of age) amplifies this problem.

5.3.2 Benefits of modeling approaches in laying hens

In laying hens, as previously mentioned, the main challenge is the management of the Ca supply to ensure egg quality and the welfare of birds. Guyot et al. (2024) recently proposed a novel modeling approach to evaluate the effect of P and Ca feeding strategies (such as dietary Ca and P, distribution time, and particle size of limestone) on Ca and P flows at three different scales: (1) one hen during 24 h; (2) one hen over its productive career; and (3) a complete flock during a full production cycle. This approach combines both mechanistic approaches for the computation of flows in the hen (absorption, bone deposition/mobilization, eggshell formation) using the modeling approach developed by Kebreab et al. (2009) and empirical ones to represent the evolution of laying parameters over time for a given animal (i.e. laying rate and laying time) and the 'composition' of a flock, considering three animal profiles (i.e. 'good', 'average,' and 'poor' layers). However, up to day, this model cannot be directly used in real-time control of feed composition in laying hens, as several improvements are still required, including a more robust prediction of P and Ca digestibility, which is crucial to limiting bone mobilization.

Hervo (2023) recently proposed the first version of a model describing the digestive utilization of P and Ca in laying hens, considering the main dietary factors (dietary concentrations of P, Ca, and PhytM, as well as the limestone particle size) and the eggshell calcification cycle. It showed that the quantity of soluble Ca diffusing from the gizzard to the small intestine during the night varies from 0.04 g when hens are fed with fine particle limestone (FLP) to 0.49 g when they are fed with a coarse particle limestone (CLP). The use of CLP allows for the diffusion of soluble Ca in the small intestine for an extended period during the night (Rao and Roland, 1989; Rao et al., 1992). Using these data in the model proposed by Kebreab et al. (2009), we can calculate the P balance in laying hens fed 100% FLP or a mix of 25% FLP and 75% CLP as is classically used by the industry. The results, shown in Fig. 7, indicate that incorporation of CLP reduces bone mobilization by 72% and consequently P urinary excretion by 46%. This model, in association with metabolic models, represents an interesting tool to define P requirements and decrease P excretion in laying hens by an approach similar to that developed in broilers (see Section 3).

5.3.3 Promising feeding strategies

Unlike CLP, FLP represents a directly available source of Ca for animals. In fact, a mixture of fine and large particles should be fed to the hen to synchronize the Ca intake with the requirements (Mongin and Sauveur, 1974). During the first part of the day, before eggshell calcification, the hen restores her bone reserves (Hurwitz, 1964) and requires a rapidly available Ca source. During the last part of the day, the hen anticipates the high Ca requirement for the eggshell, hence the need for a more slowly available source. In this view, Molnar et al. (2018) tested different Ca intakes between the morning and afternoon to hens in production. These authors showed that an intake of 50% FP in the morning and 50% LP in the afternoon was optimal for hens. Other authors also proposed that sequential feeding strategies with higher concentrations of P and lower concentrations of Ca in the morning and conversely lower concentrations of P with higher concentrations of Ca in the afternoon would reduce P loss

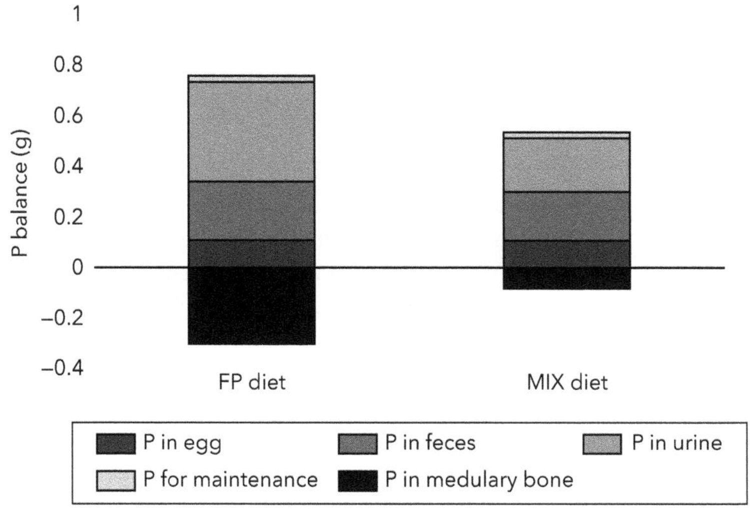

Figure 7 Phosphorus balance of hens fed with fine limestone particles (FLP diet) or with a mix of 25% FLP and 75% coarse limestone particles (CLP; MIX diet). In this simulation, FLP were 0.2 mm and CLP were 2.4 mm. Light period was 16 h (from T_{0h} to T_{16h}), and laying was considered at T_{2h}.

while improving eggshell quality and bone strength (Sauveur, 1992). These sequential feeding strategies can be tested using a modeling approach before an *in vivo* experiment.

Decreasing the P supply would reduce FGF23 expression and avoid its expected negative effect on P and Ca homeostasis. Although FGF23 cannot be directly manipulated, vitamin D3 metabolites are useful dietary supplements that can be supplied as native vitamin D3, 25-OH-vitamin D3, or 1,25-OH2-vitamin D3. Several studies have shown a beneficial effect of 25-OH-vitamin D3 on vitamin D status and egg and bone quality parameters (Swiatkiewicz et al., 2017). Some studies have considered incorporating 1,25-OH2-vitamin D3 into feed. This product does not appear to have any particular benefit on egg quality when added to a diet already providing vitamin D3, but it may have a beneficial effect on bone strength (Frost et al., 1990).

5.4 Digestible calcium system

Although digestible P is recognized as a relevant system, total Ca is still used in formulations, nutrient recommendations and in phytase matrixes (Angel et al., 2023). As previously stated, there are strong interactions between P and Ca at intestinal and metabolic level. Ultimately, the development of animals depends on the amounts of Ca and P that can be digested, absorbed, and potentially used for deposition in body tissues as well as for metabolic needs. The proportion of Ca and P that is retained is directly and quickly impacted by digestible Ca to digestible P ratios. In order to fully implement a system that allows us to feed closer to P requirements and reduce P excretions, we need to understand the digestibility of Ca in ingredients and factors that affect these, as well as estimating the requirements of digestible Ca for the different broiler growth phases which is ongoing (see Section 3).

In poultry diets, limestone accounts for 30 to 95 % of Ca supply. In recent years, research has shown that Ca digestibility in broilers varies dramatically depending on the Ca source, limestone solubility, particle size and geographic origin, as well as dietary phytate and phytase concentration in the diet (Anwar et al., 2016; Angel, 2019; Kim et al., 2019; Li et al., 2021). Recently, equations have been proposed to predict the SID of Ca in limestones based on laboratory analyses (limestone solubility, geometric mean diameter, Ca and Mg concentration; Angel, 2023). Moreover, many studies have contributed to the determination of digestible Ca requirements for broilers (David et al., 2021, 2022, 2023; Walk et al., 2021, 2022; Angel et al., 2022).

All of this work therefore contributes to move towards a system based on digestible Ca. This system will allow for providing adequate quantities of Ca at the tissue level to optimize growth and bone mineralization. This in turn will result in lower supply of limestone in the diet, better dephosphorylation of PA by phytase and thus less potential for chelation's of the phytate molecule with mineral cations, all contributing to better digestibility of Ca, P, Zn, and Cu, among others (Angel et al, 2023). Ultimately, the need for P from inorganic sources will be reduced, resulting in a lower P in the diet and lower P excretion. It is worth nothing that lowering dietary Ca and specially Ca limestone will be reduced water consumption and incidence of pododermatitis (Venter et al., 2023) and digestibility of proteins will be improved (Grynspan and Cheryan, 1989).

6 Conclusion

This review showed that it is still possible to optimize P utilization in poultry to mitigate P impact on sustainability and improve bird performance. This relies first on a precise and robust estimation of Ca and P values of feedstuffs considering the main modulating factors other than a unique value per ingredient. The poultry P system based on digestible P is more advanced than digestible Ca system, but active research is ongoing. Exogenous phytase is still a key element to fine-tune with precise and robust matrices. Secondly, precise requirements have to be available to nutritionists using factorial multi-criteria approach and again active research is ongoing in both broilers and laying hens. Bird adaptation and genetics are also key elements allowing improving P utilization. In conclusion, there is still a place for improvement toward a parsimonious use of phosphate, a non-renewable resource necessary for all the food system.

7 Where to look for further information

- Chowdhury, R.B., Moore, G.A., Weatherley, A.J. and Arora, M. 2017. Key sustainability challenges for the global phosphorus resource, their implications for global food security, and options for mitigation. *J. Clean. Prod.* 140, 945–963.
- Gerber, P.J., Uwizeye, A., Schulte, R.P.O, Opio, C.I. and de Boer, I.J.M. 2014. Nutrient use efficiency: a valuable approach to benchmark the sustainability of nutrient use in global livestock production? *Curr. Opin. Env. Sust.* 9–10, 122–130.
- Nesme, T., Withers, P.J.A. 2016. Sustainable strategies towards a phosphorus circular economy. *Nutr. Cycl. Agroecosyst.* 104, 259–264.
- Modelling nutrient digestion and utilisation in farm animals. 2010. Wageningen, Brill - Wageningen Academic. Editors Sauvant, D., van Milgen, J., Faverdin, P. and Friggens, N.
- Poultry and pig nutrition. 2019. *Challenges of the 21stcentury.* Wageningen: Brill - Wageningen Academic. Editors Hendriks, W., Verstegen M. and Babinszky L.

- Sturkie's Avian Physiology. 2021. Seventh edition. Cambridge, Massachusetts : Academic Press. Editors: Scanes, C. and Dridi, S.
- https://www.unep.org/news-and-stories/story/what-phosphorus-and-why-are-concerns-mounting-about-its-environmental-impact.

8 References

Angel, R. 2019. Update on ingredient calcium digestibility: impact of presence and source of phytate and source of calcium. *Proceedings of the Australian poultry science symposium*, Sidney, 51–56.

Angel, R., Li, W. and Plumstead, P. W. 2022. Towards digestible calcium requirements and a digestible calcium system. *26th World's Poultry Congress*, Paris, 324–34.

Angel, R., Tamim, N. M., Applegate, T. J. et al. 2002. Phytic acid chemistry: Influence on phytin phosphorus availability and phytase efficacy. *J. Appl. Poult. Res.* 11(4), 471–480.

Angel, R., Wenter, K. and Li, W. 2023. Sustainanility as related to optimal mineral nutrition. *23rd European Symposium on Poultry Nutrition*, Rimini, 55–57.

Ankra-Badu, G. A., Pesti, G. M. and Aggrey, S. E. 2010. Genetic interrelationships among phosphorus, nitrogen, calcium, and energy bioavailability in a growing chicken population. *Poult. Sci.* 89(11), 2351–2355.

Anwar, M. N., Ravindran, V., Morel, P. C. H. et al. 2016. Effect of limestone particle size and calcium to non-phytate phosphorus ratio on true ileal calcium digestibility of limestone for broiler chickens. *Br. Poult. Sci.* 57(5), 707–713.

Applegate, T. J., Angel, R., and Classen, H. L. 2003. Effect of dietary calcium, 25-hydroxycholecalciferol, or bird strain on small intestinal phytase activity in broiler chickens. *Poult. Sci.* 82(7), 1140–1148.

Aviagen. 2022. Ross broiler: nutrition specifications. Aviagen Group, Huntsville. https://aviagen.com/assets/Tech_Center/Ross_Broiler/Ross-BroilerNutritionSpecifications2022-EN.pdf

Babatunde, O. O., Osho, S. O., Park, C. S. and Adeola, O. 2020. Additivity of apparent and standardized ileal digestibility of phosphorus in mixed diets containing corn and soybean meal fed to broiler chickens. *Poult. Sci.* 99(12), 6907–6913.

Beck, P., Rodehutscord, M., Bennewitz, J. and Bessei, W. 2014. A pilot study on the genetic variation of phosphorus utilization in young Japanese quail (*Coturnix japonica*). *Poult. Sci.* 93(8), 1916–1921.

Besançon, P. and Guéguen, L. 1969. Les principales voies du métabolisme calcique chez le porc en croissance.*Ann. Biol. Anim. Biochim. Biophys*. 9(4), 537–553.

Coon, C. Leske, K. Seo, S. et al. (2002. The availability of calcium and phosphorus in feedstuffs. In J. M. McNab and K. M. Borrman (eds.), *Poultry Feedstuffs Supply, Composition and Nutritive Value*, pp. 151–179. New York, NY: CABI Pub.

Cordell, D., Drangert, J.-O. and White, S. 2009. The story of phosphorus: global food security and food for thought. *Glob. Environ. Change* 19(2), 292–305.

Crenshaw, T. D. 2001. Calcium, phosphorus, vitamin D, and vitamin K in swine nutrition. In A. J. Lewis and L. L. Southern (eds.), *Swine Nutrition*, pp. 187–212. Boca Raton, FL: CRC Press,.

CVB - Table Booklet Feeding of Poultry. 2018. Feeding standards, feeding advices and nutritional values of feed ingredients for poultry, CVB report nr.61. Pub. CVB, Wageningen: Wageningen Livestock Research, 23 p.

Dacke, C. G., Arkle, S., Cook, D. J. et al. 1993. Medullary bone and avian calcium regulation. *J. Exp. Biol.* 184(1), 63–88.
David, L. S., Abdollahi, M. R., Bedford, M. R. and Ravindran, V. 2021. Requirement of digestible calcium at different dietary concentrations of digestible phosphorus for broiler chickens. 1. Broiler starters (d 1 to 10 post-hatch). *Poult. Sci.* 100(11), 101439.
David, L. S., Abdollahi, M. R., Bedford, M. R. and Ravindran, V. 2022. Requirement of digestible calcium at different dietary concentrations of digestible phosphorus for broiler chickens. 2. Broiler growers (d 11 to 24 post-hatch). *Poult. Sci.* 101(11), 102135.
David, L. S., Abdollahi, M. R., Bedford, M. R. and Ravindran, V. 2023. Requirement of digestible calcium at different dietary concentrations of digestible phosphorus for broiler chickens 3. Broiler finishers (d 25 to 35 post-hatch). *Poult. Sci.* 102(4), 102492.
de Verdal, H., Narcy, A., Bastianelli, D. et al. 2011. Improving the efficiency of feed utilization in poultry by selection. 1. Genetic parameters of anatomy of the gastro-intestinal tract and digestive efficiency. *BMC Genet.* 12, 59.
de Verdal, H., Narcy, A., Bastianelli, D. et al. 2013. Genetic variability of metabolic characteristics in chickens selected for their ability to digest wheat. *J. Anim. Sci.* 91(6), 2605–2615.
Dilger, R. N. and Adeola, O. 2006. Estimation of true phosphorus digestibility and endogenous phosphorus loss in growing chicks fed conventional and low-phytate soybean meals. *Poult. Sci.* 85(4), 661–668.
Doeschl-Wilson, A. B., Knap, P. W. and Kinghorm, B. P. 2006. Evaluating animal genotypes through model inversion. In R. Gous, T. Morris, and C. Fisher (eds.), *Mechanistic Modelling in Pig and Poultry Production*, pp. 163–187. Trowbridge, UK: CABI.
Dourmad, J.-Y., Garcia-Launay, F., Meda, et al. 2019. Nutrition and environmental sustainability. In W. H. Hendriks, M. W. A. Verstegen, and L. Babinsky (eds.), *Poultry and Pig Nutrition*, pp. 315–339. Wageningen: Wageningen Academic Publishers.
Engelen, A. J., van der Heeft, F. C. Randsdorp, P. H. G. and Smit, E. L. C. 1994. Simple and rapid determination of phytase activity. *J. AOAC Inter.* 77(3), 760–764.
Faveri, J. C., Pinto, L. F. B., de Camargo, G. M. F. et al. 2019. Quantitative trait loci for morphometric and mineral composition traits of the tibia bone in a broiler × layer cross. *Animal* 13(8), 1563–1569.
Fernández, J. A. 1995. Calcium and phosphorus metabolism in growing pigs. III. A model resolution. *Livest. Prod. Sci.* 41(3), 255–261.
Fisher, C. 2015. An overview of poultry models. In N. K. Sakomura, R. M. Gous, I. Kyriazakis, and L. Hauschild (eds.), *Nutritional Modelling for Pigs and Poultry*, pp. 1–21. Wallingford, UK: CAB International.
Frost, T. J., Roland Sr., D. A. and Untawale, G. G. 1990. Influence of vitamin D_3, 1α-hydroxyvitamin D_3, and 1,25-dihydroxyvitamin D_3 on eggshell quality, tibia strength, and various production parameters in commercial laying hens. *Poult. Sci.* 69(11), 2008–2016.
Gaillard, C., Brossard, L. and Dourmad, J.-Y. 2020. Improvement of feed and nutrient efficiency in pig production through precision feeding. *Anim. Feed Sci. Technol.* 268, 114611.
Gautron, J., Stapane, L., Le Roy, N. et al. 2021. Avian eggshell biomineralization: an update on its structure, mineralogy and protein tool kit. *BMC Mol. Biol.* 22, 11.
Gloux, A., Le Roy, N., Ezagal, J. et al. 2020. Possible roles of parathyroid hormone, $1,25(OH)_2D_3$, and fibroblast growth factor 23 on genes controlling calcium

metabolism across different tissues of the laying hen. *Domest. Anim. Endocrinol.* 72, 106407.

Gonçalves, C. A., Sakomura, N. K., Sarcinelli, M. F. et al. 2020. In vivo assessment of body composition and growth potential of modern broiler using dual-energy X-ray absorptiometry. *Anim. Prod. Sci.* 60(16), 1959–1968.

Gonzalez-Ceron, F. Rekaya, R. and Aggrey, S. E. 2015. Genetic analysis of bone quality traits and growth in a random mating broiler population. *Poult. Sci.* 94(5), 883–889.

Greiner, R. and Konietzny, U. 2006. Phytase for food application. *Food Technol. Biotechnol.* 44(2), 125–140.

Grynspan, F. and Cheryan M. 1989. Phytate-calcium interaction with soy protein. *J. Am. Oil Chem. Soc.* 66(1), 93–97.

Guyot, Y., Pampouille, E., Burlot, T. et al. 2024. OCalM: Un modèle pour étudier l'effet de stratégies alimentaires sur les flux de minéraux chez la poule pondeuse. *15èmes Journées de la Recherche Avicole et Palmipèdes à Foie Gras*, Tours (in press).

Hamdi, M., Solà-Oriol, D., Franco-Rosseló, R. et al. 2017. Comparison of how different feed phosphates affect performance, bone mineralization and phosphorus retention in broilers. *Span. J. Agric. Res.*, 5(3), e0605.

Heaney, R. P. 2004. Phosphorus nutrition and the treatment of osteoporosis. *Mayo Clin. Proc.* 79(1), 91–97.

Hervo, F. 2023. *Optimisation de l'utilisation digestive du calcium et du phosphore chez la poule pondeuse. Vers un nouveau système d'évaluation des besoins par une approche de modélisation*. doctoral thesis, Laval University & NRAE, University of Tours. https://corpus.ulaval.ca/entities/publication/227d9c12-f788-470f-b265-a7b360a1900a.

Hervo, F., Létourneau-Montminy, M.-P. and Même, N. et al. 2023. Effect of phytase and limestone particle size on mineral digestibility, performance, eggshell quality and bone mineralization in laying hens. *Poult. Sci.* 102(5), 102613.

Hurwitz, A. 1964. Bone composition and Ca^{45} retention in fowl as influenced by egg formation. *Am. J. Physiol.* 206(1), 198–204.

Johnsson, M., Jonsson, K. B., Andersson, L. et al. 2015. Genetic regulation of bone metabolism in the chicken: similarities and differences to mammalian systems. *PLoS Genet.* 11(5), e1005250.

Kebreab, E., France, J., Kwakkel, R. P. et al. 2009. Development and evaluation of a dynamic model of calcium and phosphorus flows in layers. *Poult. Sci.* 88(3), 680–689.

Khaksar, V., Méda, B. and Narcy, A. 2017. Working group report: modeling available phosphorus requirements of broilers. In *Working Group N°2: Nutrition of the European Federation of Branches of WPSA*, 35 p.

Khaksar, V., Narcy, A., Maamer, J. et al. 2022. Vers une meilleure digestibilité de phosphore chez le poulet de chair. *14èmes Journées de la Recherche Avicole et Palmipèdes à Foie Gras*, Tours, 175–179.

Kim, S. W., Li, W., Angel, R. and Plumstead, P. W. 2019. Modification of a limestone solubility method and potential to correlate with *in vivo* limestone calcium digestibility. *Poult. Sci.* 98(12), 6837–6848.

Kim, S. W., Li, W., Angel, R. and Proszkowiec-Welgraz, M. 2018. Effects of limestone particle size and dietary Ca concentration on apparent P and Ca digestibility in the presence or absence of phytase. *Poult. Sci.* 97(12), 4306–4314.

Lautrou, M., Pomar, C., Dourmad, J. Y. et al. 2020. Phosphorus and calcium requirements for bone mineralisation of growing pigs predicted by mechanistic modelling. *Animal* 14(S2), 313–322.

Létourneau-Montminy, M.-P., Narcy, A., Dourmad, J.-Y. et al. (2015. Modeling the metabolic fate of dietary phosphorus and calcium and the dynamics of body ash content in growing pigs. *J. Anim. Sci.* 93(3), 1200–1217.

Li, W., Angel, R., Plumstead, P. W. and Enting, H. 2021. Effects of limestone particle size, phytate, calcium source, and phytase on standardized ileal calcium and phosphorus digestibility in broilers. *Poult. Sci.* 100(2), 900–909.

Liu, S. B., Liao, X. D., Lu, L. et al. (2017. Dietary non-phytate phosphorus requirement of broilers fed a conventional corn-soybean meal diet from 1 to 21 d of age. *Poult. Sci.* 96(1), 151–159.

Lott, J. N. A., Ockenden, I., Raboy, V. and Batten, G. D. 2000. Phytic acid and phosphorus in crop seeds and fruits: a global estimate. *Seed Sci. Res.* 10(1), 11–33.

Lutrell, B. M. 1993. The biological relevance of the binding of calcium ions by inositol phosphates. *J. Biol. Chem.* 268(3), 1521–1524.

Manangi, M. K. and Coon, C. N. 2007. The effect of calcium carbonate particle size and solubility on the utilization of phosphorus from phytase for broilers. *Int. J. Poult. Sci.* 6(2), 85–90.

Manangi, M. K., Maharjan, P. and Coon, C. N. 2018. Calcium particle size effects on plasma, excreta, and urinary Ca and P changes in broiler breeder hens. *Poult. Sci.* 97(8), 2798–2806.

Matuszewski, A., Łukasiewicz, M. and Niemiec, J. 2020. Calcium and phosphorus and their nanoparticle forms in poultry nutrition. *Worlds Poult. Sci. J.* 76(2), 328–345.

McCuaig, L. W., Davies, M. I. and Motzok, I. 1972. Intestinal alkaline phosphatase and phytase of chicks: effect of dietary magnesium, calcium, phosphorus and thyroactive casein. *Poult. Sci.* 51(2), 526–530.

Méda, B., Pampouille, E., Talineau, M. et al. 2022. From measuring average body weight of the flock to precision feeding in broilers: a modelling approach to adjust daily feed composition. *26th World's Poultry Congress*, Paris, 7–11.

Méda, B., Talineau, M., Chéret, C. et al. 2019. Vers une alimentation de précision chez le poulet de chair : Modélisation des besoins en énergie, lysine et phosphore à partir des performances de croissance. *13èmes Journées de la Recherche Avicole et Palmipèdes à Foie Gras*, Tours, 8–12.

Mignon-Grasteau, S., Chantry-Darmon, C., Boscher, M. Y. et al. 2016. Genetic determinism of bone and mineral metabolism in meat-type chickens: A QTL mapping study. *Bone Reports* 5, 43–50.

Molnar, A., Maertens, L., Ampe, B. et al. (2018. Effect of different split-feeding treatments on performance, egg quality, and bone quality of individually housed aged laying hens. *Poult. Sci.* 97(1), 88–101.

Mongin, P. and Sauveur, B. 1974. Voluntary food and calcium intake by the laying hen. *Br. Poult. Sci.* 15(4), 349–359.

Mutucumarana, R. K., Ravindran, V., Ravindran, G. and Cowieson, A. J. 2014. Measurement of true ileal digestibility of phosphorus in some feed ingredients for broiler chickens. *J. Anim. Sci.* 92(12), 5520–5529.

Narcy, A., Khaksar, V., Jlali, M. et al. 2023. Towards a better prediction of phosphorus digestibility in broilers. *23rd European Symposium on Poultry Nutrition*, Rimini, Italy.

Narcy, A., Rousseau, X., Même, N., Magnin, M. and Nys, Y. 2015. Impact of dietary phosphorus and calcium concentrations on their deposition in soft and bone tissues in chicks. *11èmes Journées de la Recherche Avicole et Palmipèdes à Foie Gras*, Tours, 545–549.

Nolan, K. B. and Duffin, P. A. 1987. Effects of phytate on mineral bioavailability. *In vitro* studies on Mg^{2+}, Ca^{2+}, Fe^{3+}, Cu^{2+}, and Zn^{2+} (also Cd^{2+}) solubilities in the presence of phytate. *J. Sci. Food Agric.* 40(1), 79–85.

NRC (National Research Council). 1994. *Nutrient requirements of poultry. Ninth revised edition.* Washington, DC: National Academy Press.

NRSP8: National Research Support Program, Animal QTL db. https://www.animalgenome.org/cgi-bin/QTLdb/GG/lstref?trait_ID=2286&traitnm=Production-Other; Accessed September 23, 1998.

Nys, Y. 2017. Laying hen nutrition: optimising hen performance and heath, bone and eggshell quality. In J. Roberts (ed.), *Achieving sustainable production of eggs, Volume 2: animal welfare and sustainability (1st edition).* Cambridge: Burleigh Dodds Science Publishing Limited.

Pampouille, E., Guyot, Y., Souchet, J. et al. 2022. Experimental validation of a precision feeding strategy to reduce feeding cost and nutrient excretion in broilers. *26th World's Poultry Congress*, Paris, 460.

Plumstead, P. W., Leytem, A. B., Maguire, R. O. et al. 2008. Interaction of calciu and phytate in broiler diets. 1. Effects on apparent prececal digestibility and retention of phosphorus. *Poult. Sci.* 87(3), 449–458.

Proszkowiec-Weglarz, M. and Angel, R. 2013. Calcium and phosphorus metabolism in broilers: effect of homeostatic mechanism on calcium and phosphorus digestibility. *J. Appl. Poult. Res.* 22(3), 609–627.

Rao, K. S. and Roland Sr., D. A. 1989. Influence of dietary calcium level and particle size of calcium source on *in vivo* calcium solubilization by commercial Leghorns. *Poult. Sci.* 68(11), 1499–1505.

Rao, K. S., Roland Sr., D. A., Adams, J. L. and Durboraw, W. M. 1992. Improved retention in the gizzard of commercial Leghorn hens. *J. Appl. Poult. Res.* 1(1), 6–10.

Ravindran, V., Kornegay, E. T., Potter, L. M. et al. 1995. An evaluation of various response criteria in assessing biological availability of phosphorus for broilers. *Poult. Sci.* 74(11), 1820–1830.

Reis, M. P., Couture, C., Sakomura, N. K. et al. 2023a. A first model of the fate of dietary calcium and phosphorus in broiler chickens. *Animal* (in press). https://doi.org/10.1016/j.animal.2023.100896.

Reis, M. P., Gous, R. M. Hauschild, L. and Sakomura, N. K. 2023b. Evaluation of a mechanistic model that estimates feed intake, growth and body composition, nutrient requirements, and optimum economic response of broilers. *Animal*, 101016. https://doi.org/10.1016/j.animal.2023.101016.

Reis, M. P., Sakomura, N. K., Teixeira, I. A. et al. 2018. Partitioning the efficiency of utilization of amino acids in growing broilers: multiple linear regression and multivariate approaches. *PloS One* 13(12). https://doi.org/10.1371/journal.pone.0208488.

Rizwanuddin, S., Kumar, V., Naik, B. et al. 2023. Microbial phytase: their sources, production, and role in the enhancement of nutritional aspects of food and feed additives. *J. Agri. Food Res.* 12, 100559.

Rodehutscord, M. and Shastak, Y. 2013. Phosphorus evaluation assays: scientific approaches and practical implications. *24th Australian Poultry Science Symposium*, Sidney, 3–9.

Rodehutscord, M., Sommerfeld, V. C., Angel, R. and Korver, D. R. 2023. Minimum phosphorus requirements for laying hen feed formulations. *Poult. Sci.* 102, 102344.

Rostagno, H. S., Albino, L. F. T., Hannas, M. I. et al. 2017. *Brazilian tables for poultry and swine.* Viçosa, Brazil: UFV.

Rousseau, X., Létourneau-Montminy, M.-P., Même, N. et al. 2012. Phosphorus utilization in finishing broiler chickens: effects of dietary calcium and microbial phytase. *Poult. Sci.* 91(11), 2829–2837.

Rousseau, X., Valable, A.-S., Létourneau-Montminy, M.-P. et al. 2016. Adaptive response of broilers to dietary phosphorus and calcium restrictions. *Poult. Sci.* 95(12), 2849–2860.

Sauvant, D., Schmidely, P., Daudin, J. J. and St-Pierre, N. R. 2008. Meta-analyses of experimental data in animal nutrition. *Animal* 2(8), 1203–1214.

Sauvant, D., Perez, J. M. and Tran, G. 2004. *Tables de composition et de valeur nutritive des matières premières destinées aux animaux d'élevage : porcs, volailles, bovines, ovins, caprins, lapins, chevaux, poisons. 2ème édition revue et corrigée*, INRA Editions, Paris, France.

Sauveur, B. 1992. Adaptation des apports alimentaires aux variations journalières des besoins en calcium et phosphore de la poule. *INRA Prod. Anim.* 5(1), 19–28.

Scott, M. L., Hull, S. J. and Mullenhoff, P. A. 1971. The calcium requirements of laying hens and effects of dietary oyster shell upon egg shell quality. *Poult. Sci.* 50(4), 1055–1063.

Selle, P. H., Cowieson, A. J. and Ravindran, V. 2009. Consequences of calcium interactions with phytate and phytase for poultry and pigs. *Livest. Sci.* 124(1–3), 126–141.

Selle, P. H. and Ravindran, V. 2007. Microbial phytase in poultry nutrition. *Anim. Feed Sci. Technol.* 135(1–2), 1–41.

Shastak, Y. and Rodehutscord, M. 2013. Determination and estimation of phosphorus availability in growing poultry and their historical development. *World's Poult. Sci. J.* 69(3), 569–586.

Shastak, Y., Zeller, E., Witzig, M. and Schollenberger, M. 2014. Effects of the composition of the basal diet on the evaluation of mineral phosphorus sources and interactions with phytate hydrolysis in broilers. *Poult. Sci.* 93(10), 2548–2559.

Singh, C. B. and Hjilani, M. 2008. Inheritance of six-week body weight, shank length, keel bone length and breast angle in broiler chickens. *Pantnagar J. Res.* 6, 279–280.

Stilborn, H. L., Moran Jr., E. T., Gous, R. M. and Harrison, M. D. 1997. Effect of age on feather amino acid content in two broiler strain crosses and sexes. *J. Appl. Poult. Res.* 6(2), 205–209.

Suttle, N. 2010. *Mineral Nutrition of Livestock 4th edition*. Wallingford, UK: CABI.

Swiatkiewicz, S., Arczewska-Wlosek, A., Bederska-Lojewska, D. and Jozefiak, D. 2017. Efficacy of dietary vitamin D and its metabolites in poultry – review and implications if the recent studies. *Wolrds Poult. Sci. J.* 73(1), 57–68.

Tamim, N. M., Angel, R. and Christman, M. 2004. Influence of dietary calcium and phytase on phytate phosphorus hydrolysis in broiler chickens. *Poult. Sci.* 83(8), 1358–1367.

Valable, A.-S., Létourneau-Montminy, M.-P., Klein, S. et al. 2020. Early life conditioning strategies to reduce dietary phosphorus in broilers; underlying mechanisms. *J. Nutr. Sci.* 9, e28.

Valable, A.-S., Narcy, A., Duclos, M. J. et al. 2018. Effects of dietary calcium and phosphorus deficiency and subsequent recovery on broiler chicken growth performance and bone characteristics. *Animal* 12(8), 1555–1563.

van der Klis, J. D., Verstegen, M. W. A. and De Witt, W. 1990. Absorption of minerals and retention time of dry matter in the gastrointestinal tract of broilers. *Poult. Sci.* 69(12), 2185–2194.

van der Klis, J. D. and Blok, M. C. 1997. *Definitief systeem opneembaar fosfor pluimvee*. Centraal Veevoederbureau.

Vargas, L., Sakomura, N. K., Leme, B. B. et al. 2020. A description of the potential growth and body composition of two commercial broiler strains. *Br. Poult. Sci.* 61(3), 266–273.

Venter, K., Angel, R. and Plumstead, P. W. 2023. Effects of varying dietary calcium levels on broiler performance, litter quality, and pododermatitis. *23rd European Symposium on Poultry Nutrition*, Rimini, 163.

Vollmar, S., Haas, V., Schmid, M. et al. 2021. Mapping genes for phosphorus utilization and correlated traits using a 4k SNP linkage map in Japanese quail (*Coturnix japonica*). *Anim. Genet.* 52(1), 90–98.

Walk, C. L., Jenn, P., Sorbora, J. O. B. et al. 2022. Research note: formulating broiler diets using digestible calcium significantly improved growth performance but reduced apparent ileal digestibility of calcium and phosphorus. *Poult. Sci.* 101(10), 102069.

Walk, C. L., Romero, L. F. and Cowieson, A. J. 2021. Towards a digestible calcium system for broiler chicken nutrition: a review and recommendations for the future. *Anim. Feed Sci. Technol.* 276, 114930.

Wang, K., Fan, M., Tong, H. et al. 2006. Study of biochemistry index of RuGao chicken plasma at 2–12 weeks of age. *Anim. Biotechnol. Bull.* 10, 151–156.

Whitehead, C. C. and Fleming, R. H. 2000. Osteoporosis in cage layers. *Poult. Sci.* 79(7), 1033–1041.

Wise, A. 1983. Dietary factors determining the biological activity of phytates. *Nutr. Abstr. Rev. Clin. Nutr.* 53, 791–806.

WPSA (Worlds's Poultry Science Association, working group 2 -nutrition- of the European federation of branches) (2013. Determination of phosphorus availability in poultry. World's *Poult. Sci. J.* 69(3), 687–698.

Yan, F., Angel, R., Aschwell, C. M. et al. 2005. Evaluation of the broiler's ability to adapt to an early moderate deficiency of phosphorus and calcium. *Poult. Sci.* 84(8), 1232–1241.

Zhang, W., Aggrey, S. E., Pesti, G. M. et al. 2003. Genetics of phytate phosphorus bioavailability: heritability and genetic correlations with growth and feed utilization traits in a randombred chicken population. *Poult Sci* 82(7), 1075–1079.

Zuidhof, M. J. 2020. Precision livestock feeding: matching nutrient supply with nutrient requirements of individual animals. *J. Appl. Poult. Res.*, 29(1), 11–14.

Chapter 13

The use of probiotics and prebiotics to improve gut function and immunity in poultry: unlocking the secrets of avian wellness

Guillermo Tellez-Isaias[1], University of Arkansas, USA; Daniel Hernandez-Patlan and Bruno Solis-Cruz, Universidad Nacional Autonoma de Mexico (UNAM), Mexico; Saeed El-Ashram, Kafrelsheikh University, Egypt; and Ebtsam Al-Olayan, King Saud University, Saudi Arabia

1 Introduction

Animal welfare is crucial in modern agriculture. Chicken growers need birds with good immunological and gastrointestinal health. The gut is responsible for water and feed absorption and nutrient digestion. Nutrient digestion and absorption affect chicken growth, development, and immunity. The gut also acts as a primary barrier to potentially harmful bacteria. While the term 'gut health' is unclear, scientists agree that it refers to the ability of the gastrointestinal

1 Corresponding author: gtellez@uark.edu.

http://dx.doi.org/10.19103/AS.2024.0143.18

tract (GIT) to maintain homeostasis and withstand both infection and non-infectious stresses.

The complexity and importance of the GIT is reflected in the fact that the mammalian gut microbiota contains a significantly greater amount of genetic material than that of somatic cells in the host (Zhu et al., 2010; Fathima et al., 2022). The enteric/intestinal nervous system is called animals' 'second brain' because it has more neurons than the peripheral nervous system (Schneider et al., 2019). Up to 80% of the body's immune cells are located in gut-associated lymphoid tissue (GALT). Over 50 gut hormones and bioactive peptides have been identified, showing that the gut is the body's largest endocrine organ with endocrinological, neuroendocrine, autocrine, and paracrine functions (Gribble and Reimann, 2017). Interestingly, the gut microbiome controls enteroendocrine (EEC)-produced serotonin, dopamine, oxytocin, and endorphins (Mayer et al., 2014). The neuroendocrine network links the central nervous system (CNS), enteric nervous system (ENS), intestinal microbiota, and GALT affects the fragile intestinal epithelial barrier (Megur et al., 2021). This barrier, a single layer of enterocytes with strong intercellular connections, regulates non-self-antigen tolerance and immunization (Maslowski and Mackay, 2010). Dysbiosis of the GIT microbiota causes intestinal damage (Weiss and Hennet, 2017). This chapter discusses the use of probiotics and prebiotics to improve various aspects of gut function and immunity in poultry.

2 The role of gut microbiota in poultry health

As noted previously, trillions of microorganisms residing in the GIT as part of the microbiome play a crucial role in the health and overall well-being of poultry (Hu et al., 2023). The connections between the host and its gut microbiome are complex and dynamic, influencing various aspects of poultry physiology, metabolism, immune function, and disease resistance. Some of the most important roles include the following:

1. Nutrient digestion and absorption: The gut microbiota of poultry is involved in the breakdown and fermentation of complex dietary components that the host cannot digest on its own, such as dietary fiber, non-starch polysaccharides, and some proteins (He et al., 2023). These microorganisms produce enzymes that aid in the degradation of these complex substrates, leading to the production of short-chain fatty acids (SCFAs) and other metabolites (Beldowska et al., 2023). Short-chain fatty acids serve as an important energy source for the host, promote intestinal health, and influence the development of the intestinal epithelium (Proszkowiec-Weglarz et al., 2023; Liu et al., 2023a).

2 Immune development and function: the gut microbiota plays a vital role in the development and maturation of the immune system in poultry (Liu et al., 2023b). It helps educate the host's immune system, ensuring appropriate immune responses to pathogens and preventing excessive immune activation. The presence of specific gut microbial communities early in life is critical for the development of immune tolerance, preventing immune-mediated disorders, and shaping the host's immune response throughout its life (Cheng et al., 2023).
3 Pathogen exclusion and competitiveness: a healthy gut microbiota acts as a barrier against colonization and infection by pathogens. Beneficial bacteria in the gut can outcompete pathogenic microorganisms for nutrients and space, thereby preventing their establishment and reducing the risk of infections (Lee et al., 2023a). These beneficial bacteria also produce antimicrobial compounds and stimulate the production of antimicrobial peptides by the host, further enhancing the defense mechanisms against pathogens (Lee et al., 2023b).
4 Nutrient utilization and growth performance: the composition and activity of the gut microbiota have a significant impact on the efficiency of nutrient utilization and growth performance in poultry. A balanced and diverse gut microbiota is associated with improved nutrient absorption, better feed conversion, and enhanced growth rates. Conversely, dysbiosis or an imbalance in the gut microbiota can lead to nutrient malabsorption, reduced growth performance, and increased susceptibility to various diseases (Brugaletta et al., 2023).
5 Stress response and behavior: there is accumulating evidence that the gastrointestinal microbiota can influence the stress response and behavior of poultry (Ye et al., 2023). The gut–brain axis, a bidirectional communication pathway between the intestine and the central nervous system, enables the gut microbiota to influence neural function, the release of stress hormones, and behavioral responses. Imbalances in the composition of the gastrointestinal microbiota have been associated with altered stress responses, anxiety-like behaviors, and compromised welfare in poultry (Fu et al., 2023).

The gut microbiome of poultry thus plays a multifaceted role in maintaining host health and well-being. Understanding the intricate interactions between the host and its gut microbiome can help in the development of strategies to optimize poultry production, enhance disease resistance, improve nutrient utilization, and promote animal welfare in the poultry industry.

3 The use of probiotics to improve gut function and immunity in poultry

As noted, the GIT serves a crucial role in the digestion, absorption, and overall health of poultry. Essential for optimal growth, feed utilization, and disease resistance is the maintenance of a healthy intestine. In recent years, there has been a growing interest in using probiotics to improve the digestive health of poultry. Probiotics are live microorganisms that impart health benefits to the host when administered in adequate quantities (Isolauri et al., 2004). They can include bacteria, yeasts, or other beneficial microorganisms. In poultry, commonly used probiotics include *Lactobacillus*, *Bifidobacterium*, *Bacillus*, and *Saccharomyces* species. Probiotics can be administered orally or through feed supplementation, allowing for easy integration into poultry farming practices (Al-Otaibi et al., 2023).

Probiotics exert their beneficial effects through various mechanisms. They compete with pathogenic bacteria for nutrients and adhesion sites in the gut, thereby preventing their colonization. Probiotics also produce antimicrobial substances, such as organic acids, bacteriocins, and hydrogen peroxide, which inhibit the growth of harmful bacteria. Furthermore, probiotics enhance the intestinal barrier function, modulate the immune response, stimulate mucus production, and promote gut health (Abreu et al., 2023).

Choosing the right probiotic strains is vital for achieving desired outcomes. Probiotics should be derived from the host species or closely related species to ensure compatibility and effective colonization. The strains should be safe, non-pathogenic, and capable of surviving and functioning in the avian gut. Important characteristics to consider include acid and bile tolerance, adherence to intestinal epithelial cells, and the ability to produce beneficial metabolites (Yaqoob et al., 2022).

The administration of probiotics has been associated with several benefits for poultry gut health. Probiotics can improve nutrient digestion and absorption, leading to enhanced growth performance and feed efficiency. They can also alleviate the negative effects of stress on gut integrity and function. Probiotics have shown efficacy in preventing and controlling intestinal infections, reducing the need for antibiotics. Additionally, probiotics have been found to modulate the immune response, reducing inflammation, and enhancing disease resistance (Ahmad et al., 2022).

Probiotics can be administered to poultry through various routes, including oral gavage, drinking water, feed incorporation, and spray application. The choice of administration method depends on factors such as age, management practices, and the desired effect. The viability and stability of probiotics during storage and feed processing should also be considered to ensure their effectiveness (Khalid et al., 2022).

Numerous field studies have demonstrated the positive effects of probiotics on poultry health and immunity. For instance, supplementation with specific probiotic strains has been shown to reduce the incidence of enteric diseases, improve feed conversion efficiency, and increase weight gain in broilers (Gul and Alsayeqh 2022). Similarly, laying hens supplemented with probiotics have exhibited enhanced egg production and improved egg quality. These findings highlight the potential of probiotics as an effective tool for promoting poultry health and productivity (Darboe, 2022). Some of the benefits of the use of probiotics in commercial poultry include the following:

1 Enhanced gut health: probiotics are live bacteria that, when consumed in sufficient quantities, provide health advantages to the host. Prebiotics, on the other hand, are food elements that encourage the growth and activity of good gut bacteria. Both probiotics and prebiotics contribute to maintaining a healthy gut microbiota in poultry, which is crucial for optimal digestion, nutrient absorption, and overall health. Improved gut health leads to better feed conversion, reduced mortality rates, and improved performance, all of which have a positive economic impact on poultry production (Shehata et al., 2022).
2 Disease prevention: probiotics have been shown to have antimicrobial properties and can inhibit the growth of harmful bacteria in the gut. By colonizing the gut with beneficial bacteria and creating a competitive environment, probiotics help prevent the colonization of pathogens such as *Salmonella* and *Campylobacter*. This reduces the incidence of poultry diseases, decreases the need for antibiotic treatments, and lowers the associated costs of disease management (Shini et al., 2021).
3 Improved immune function: probiotics have immunomodulatory effects, meaning they can positively influence the immune system of poultry. They enhance the production of antibodies, stimulate the activity of immune cells, and promote the production of anti-inflammatory compounds. By boosting the immune response, probiotics and prebiotics help poultry to better withstand disease challenges and stressors, leading to reduced morbidity and mortality rates. This translates into cost savings and improved productivity in the poultry industry (Park and Seo, 2023).
4 Anti-inflammatory properties: inhibition of NF-κB pathway: nuclear factor kappa B (NF-κB) is a key regulator of inflammation. Some probiotics have been found to inhibit the activation of NF-κB, thereby reducing the production of inflammatory mediators (Wlazlak et al., 2023).
5 Better feed utilization: probiotics improve the efficiency of nutrient utilization in poultry. They enhance the breakdown of complex feed components, such as fibers, into simpler compounds that can be readily absorbed by the bird's digestive system. This improves feed efficiency

and reduces feed costs. Additionally, probiotics and prebiotics can help alleviate digestive disorders, such as diarrhea and enteritis, thereby reducing the need for medications and veterinary interventions (Alagawany et al., 2023).

6 Reduced environmental impact: the use of probiotics in poultry production can have positive environmental implications. Improved gut health and nutrient absorption lead to reduced excretion of undigested nutrients, including nitrogen and phosphorus, in poultry waste. This lowers the environmental pollution associated with poultry farming, such as eutrophication of water bodies. Additionally, the decreased reliance on antibiotics and antimicrobials contributes to reducing the risk of antimicrobial resistance development, which is a global concern (Chen et al., 2023).
7 Reduction of stress: stress is a major immunosuppressive factor in poultry production. Factors such as transportation, high stocking density, and environmental changes can lead to stress, compromising immune function. Probiotics have been shown to mitigate the negative effects of stress on poultry by reducing stress-induced corticosterone levels and improving gut integrity. By alleviating stress and supporting immune function, probiotics contribute to the overall health and well-being of poultry (Mangan et al., 2023).

While probiotics offer numerous advantages, their practical application in poultry nutrition faces some challenges. One significant challenge is strain selection. Not all probiotic strains exhibit the same efficacy or compatibility with specific poultry species. Therefore, careful consideration and scientific research are essential in selecting strains with proven effectiveness and safety (Li et al., 2023).

While probiotics offer promising potential for improving poultry gut health, several challenges need to be addressed. The variability in individual bird responses to probiotics and the need for strain-specific selection poses challenges for practical applications (Abd El-Hack et al., 2020). Ensuring probiotic survival and functionality during storage, feed processing, and passage through the GIT is another area that requires further research (Krysiak et al., 2021). Additionally, the optimal dosage, timing, and duration of probiotic administration need to be determined for different production systems and bird ages.

The variability in individual bird responses to probiotics, along with the need for strain-specific selection, presents significant challenges for practical applications in poultry farming and avian research. This phenomenon arises due to several interrelated factors:

1 Host-specific microbiota: birds have unique gut microbiomes that are influenced by various factors including genetics, diet, environment, and

early-life experiences. This means that a probiotic strain that benefits one bird species or even one individual within a species may not have the same effect on another. A specific *Lactobacillus* strain might enhance gut health in broiler chickens, but it may not yield the same results in ducks or turkeys due to their distinct microbial communities.

2 Dietary variations: the composition of a bird's diet can significantly influence its gut microbiota. Birds with different dietary ingredients may have different microbial communities, which can affect how they interact with probiotics. A probiotic that is effective in birds raised on a standard commercial diet might not produce the same benefits in free-range or organic poultry, which may have a more diverse and dynamic microbiome due to foraging. The same applies to the source of the grain used for energy (corn, wheat, rye, barley, or sorghum) which have different levels of non-starch polysaccharides.
3 Genetic diversity: birds of the same species can exhibit substantial genetic variation. This genetic diversity can influence the composition of their gut microbiota and, consequently, their response to probiotics. Within a flock of broiler chickens, individual birds may have genetic differences that result in varying responses to a particular probiotic strain. Some birds may show improved growth rates and disease resistance, while others may not experience the same benefits.
4 Environmental factors: factors such as temperature, humidity, and housing conditions can influence the gut microbiota of birds. These environmental variables can alter the conditions under which probiotics interact with the host. Birds raised in different climates or housing systems may have distinct gut microbiota compositions. Therefore, a probiotic strain that is effective in one environment may not produce the same results in another.
5 Interactions with existing microbiota: the existing gut microbiota of a bird can significantly impact how probiotics are absorbed and utilized. Some strains may be more compatible with the existing microbial community, while others may face competition or antagonism. If a bird already has a well-established population of beneficial bacteria, introducing a probiotic strain that interacts negatively with these bacteria could lead to suboptimal results.

Due to these complexities, selecting the right probiotic strain for a specific bird population or individual is crucial. This requires detailed knowledge of the host's biology, genetics, diet, and environment, as well as a deep understanding of the probiotic strains under consideration. Additionally, conducting controlled trials in the target population is essential to evaluate the efficacy of a chosen probiotic. The variability in individual bird responses to probiotics and the need for strain-specific selection are major hurdles in realizing the full potential of probiotics in avian farming and research. Overcoming these challenges necessitates a multidisciplinary approach that combines microbiology, genetics, nutrition, and veterinary science.

In vitro-culture systems play a crucial role in the selection of potential probiotic candidates for poultry. These systems aim to mimic the gastrointestinal environment to evaluate the survivability, adhesion properties, and functionality of candidate strains. However, despite their significance, there are several inherent difficulties associated with these *in vitro* models, as well as notable gaps when transitioning from *in vitro* to *in vivo* applications. These include:

1 Simplicity and over-simplification: the simplicity *in vitro* models are designed to replicate specific aspects of the GIT, but they often oversimplify the complex interactions between microbes, host cells, and various components of the gut environment. This can lead to a skewed representation of a probiotic's true potential.
2 Static *versus* dynamic environments: many *in vitro* models are static, meaning they do not account for the dynamic nature of the gut, where peristaltic movements, pH variations, and mucosal flow play vital roles in shaping microbial interactions. This can lead to inaccurate predictions of how probiotics will behave *in vivo*.
3 Limited nutrient availability: *in vitro* models may not adequately represent the competitive nature of the gut environment, where nutrients are limited, and different microbial species vie for resources. This can lead to an overestimation of a probiotic's potential impact on the gut.
4 Adhesion and colonization: assessing probiotic adhesion to epithelial cells *in vitro* might not accurately reflect the true adhesive capabilities of the strain in the dynamic and mucus-rich environment of the gut. This can lead to the selection of strains that are not optimally suited for *in vivo* colonization.
5 Host–pathogen interactions: some *in vitro* models do not incorporate pathogenic bacteria that probiotics are meant to compete against. This can limit the ability to accurately evaluate a probiotic's potential to inhibit harmful pathogens *in vivo*.
6 Host immunomodulation: *in vitro* models often lack the presence of immune cells and their interactions with probiotics. This is a critical aspect as probiotics can modulate immune responses, which is a key factor in their beneficial effects.
7 Long-term sability: *in vitro* studies are typically conducted over a relatively short period. This does not account for potential changes in

the probiotic's behavior or the gut microbiota composition over time, which is crucial for long-term benefits.

8 Strain variability: probiotic strains can have significantly different effects on the host. An *in vitro* model might not capture these nuances, potentially leading to the selection of a strain that is less effective *in vivo*.
9 Ethical and cost considerations: some advanced *in vitro* models can be resource intensive and may still require animal-derived components. This raises ethical concerns and may limit scalability for widespread use in research.

While *in vitro*-culture systems are invaluable tools for initial probiotic screening, they have inherent limitations that may not fully reflect the complex interactions that occur *in vivo*. To bridge this gap, a multi-faceted approach that combines *in vitro* assessments with more sophisticated *ex vivo* and *in vivo* models is necessary to better select and understand probiotic candidates for poultry applications. This integrated approach will yield more reliable results and improve the overall effectiveness of probiotic interventions in poultry production.

Another challenge lies in maintaining probiotic viability during feed processing and storage. Probiotics are sensitive to heat, moisture, and acidic conditions. Strategies like microencapsulation and proper feed formulation need to be employed to ensure sufficient viability and efficacy of probiotics until they reach the poultry gut (Vaseji et al., 2022).

4 The use of prebiotics to improve gut function and immunity in poultry

Prebiotics are indigestible dietary components, primarily consisting of non-starch polysaccharides, oligosaccharides, and fiber, which selectively stimulate the growth and activity of beneficial bacteria in the gut (Roberfroid, 2007). They act as substrates for the beneficial bacteria, particularly lactic acid bacteria and bifidobacteria, promoting their growth and metabolic activity. Prebiotics selectively stimulate the growth of beneficial bacteria in the gut, leading to a shift in the microbial composition toward a more favorable and diverse community. This modulation of the gut microbiota contributes to improved gut health and performance (Gu and Deng, 2023).

The article 'Roles of Prebiotics in Intestinal Ecosystem of Broilers' by Teng and Kim (2018) is perhaps the most comprehensive review of the beneficial effects that prebiotics have on poultry intestinal health as they discuss various prebiotic formulations and their specific modes of action in broilers, particularly because in recent years, prebiotics have been considered as potential alternatives to antibiotics.

Mechanisms by which prebiotics modulate the ecosystem of the gut include alternation of the intestinal microbiota, improvement of the epithelium, and stimulation of the immune system. It is suggested that the administration of prebiotics not only influences these aspects but also regulates the interaction between the host and the intestinal microbiota expansively, by modulating microbial community which in turn regulates the production of cytokines and antibodies, improving gut development and the overall broiler health. As brilliantly is described by Teng and Kim (2018), understanding the cross-talk between prebiotics and the intestinal ecosystem is crucial for preventing pathogen invasion and improving the health and productivity of broilers

Different prebiotic formulations may have varying degrees of efficacy in these modes of action, and their effectiveness can also depend on factors such as dosage, duration of use, and the specific strain of broilers being raised. It's essential to choose the right prebiotic formulation and tailor its usage to the specific needs and conditions of the broiler farm (Teng and Kim, 2018).

Several prebiotics have been studied and applied in poultry nutrition. Some commonly used prebiotics include the following:

- Inulin: Derived from chicory root, inulin is a soluble dietary fiber that exhibits prebiotic effects by selectively promoting the growth of beneficial bacteria (Bucław, 2016).
- Fructo oligosaccharides (FOSs): FOS are short-chain carbohydrates that occur naturally in various plant materials, such as fruits and vegetables. They serve as a substrate for beneficial bacteria in the gut (Jahan et al., 2022).
- Mannan oligosaccharides (MOSs): MOS are derived from the cell walls of yeast and are known to bind to pathogens, preventing their attachment to the gut epithelium and facilitating their removal from the digestive tract (Chacher et al., 2017).

Prebiotics also have several mechanisms of action that contribute to improved gut function in poultry. These include the following:

1 Modulation of gut microbiota: the gut microbiota plays a crucial role in the development and function of the avian immune system. Prebiotics act as a substrate for beneficial bacteria, leading to their proliferation and colonization in the gut (Yaqoob et al., 2021). This modulation of gut microbiota helps maintain a healthy microbial balance, which is essential for optimal immune function. By promoting the growth of beneficial bacteria, prebiotics can enhance the production of SCFAs, which provide an energy source for gut epithelial cells and contribute to a favorable gut environment. Prebiotics provide nourishment to

beneficial gut bacteria, leading to their proliferation. This helps maintain a healthy gut microbiota balance and supports optimal digestive function in poultry (Ricke et al., 2020).

2 Enhanced nutrient utilization: the fermentation of prebiotics produces metabolites that enhance the absorption of nutrients, such as minerals and vitamins, in the gut. This results in improved nutrient utilization, better feed efficiency, reduced wastage, and overall performance of poultry (Ricke, 2021).

3 Reduced pathogen colonization: prebiotics influence the development and function of the immune system in the gut, promoting the production of antimicrobial peptides and enhancing the immune response against pathogens. Certain prebiotics can inhibit the adhesion of pathogenic bacteria to the intestinal lining, preventing their colonization and subsequent negative impacts on health. In addition, beneficial bacteria stimulated by prebiotics compete with harmful pathogens for nutrients and attachment sites in the gut, reducing their colonization and pathogenicity.

4 Immune modulation: prebiotics can also positively influence the immune system in poultry, promoting a balanced and efficient immune response against pathogens. Prebiotics have been shown to stimulate the immune system in poultry. They can enhance the production of antibodies, particularly immunoglobulin A (IgA), which plays a crucial role in mucosal immunity. Prebiotics also promote the production of antimicrobial peptides, such as defensins, which have antimicrobial properties and help combat pathogens. Additionally, prebiotics can stimulate the activity of natural killer (NK) cells and macrophages, enhancing the innate immune response (Kim et al., 2019).

5 Increased production of short-chain fatty acids (SCFAs): prebiotics are fermented by beneficial bacteria in the gut, leading to the production of SCFAs such as acetate, propionate, and butyrate. Short-chain fatty acids serve as an energy source for the gut epithelial cells, promoting their integrity and function (Wu et al., 2022b).

6 Improved gut morphology: prebiotics have been shown to enhance the development and integrity of the gut epithelium in poultry. This leads to increased villus height, crypt depth, and surface area for nutrient absorption, improving nutrient digestion and absorption (Soumeh et al., 2021).

7 Enhanced gut barrier function: prebiotics strengthen the gut barrier by promoting the production of mucus and enhancing tight junction integrity between gut epithelial cells. This reduces the translocation of pathogens across the gut barrier, preventing gut inflammation and improving overall gut health (Ali et al., 2022).

8 Anti-inflammatory effects: chronic inflammation in the GIT can negatively impact poultry health and performance. Prebiotics have demonstrated anti-inflammatory effects by reducing the production of pro-inflammatory cytokines and modulating the expression of genes related to inflammation. By promoting a balanced immune response and reducing inflammation, prebiotics can improve gut health and overall immunity in poultry (Wu et al., 2022a).
9 Reduction in antibiotic dependency: the use of prebiotics in poultry nutrition has the potential to reduce reliance on antibiotics, which is an important aspect in the context of antimicrobial resistance. By promoting a healthy gut microbiota, prebiotics can help to maintain gut health and minimize the need for antibiotics as growth promoters or disease control agents (Reuben et al., 2021).

The incorporation of prebiotics into poultry diets has gained attention as a practical approach to enhance immunity and improve production performance. Prebiotics can be added to feed in the form of additives or included in functional feed ingredients. The dosage, duration, and timing of prebiotic supplementation should be carefully considered to achieve optimal results. Furthermore, the selection of appropriate prebiotics and their combination with other additives, such as probiotics, can have synergistic effects on immune enhancement.

5 The use of symbiotics to improve gut function and immunity in poultry

Many researchers and poultry producers combine probiotics and prebiotics to achieve a synergistic effect, often referred to as synbiotics. This approach combines the growth-promoting benefits of prebiotics with the beneficial effects of probiotics, leading to improved gut health, enhanced disease resistance, and better performance in poultry (Ganguly, 2013).

It's important to note that while probiotics and prebiotics can be valuable tools to reduce antibiotic use in poultry, their effectiveness may vary depending on factors such as the specific strains used, dosage, and management practices. Further research is still ongoing to optimize their application and determine the most suitable combinations for different poultry production systems (Ajuwon, 2016).

Probiotics and prebiotics offer several economic benefits in poultry production. They improve gut health, prevent diseases, enhance immune function, optimize feed utilization, and reduce environmental impact. By promoting the overall well-being and performance of poultry, these additives contribute to increased productivity, cost savings, and sustainability in the poultry industry (Dhama et al., 2008).

6 Antioxidant properties of probiotics and prebiotics

Both probiotics and prebiotics have been studied for their potential antioxidant properties and their ability to promote overall health and well-being. Some strains of probiotics have been shown to possess antioxidant properties, primarily through their ability to scavenge free radicals and reduce oxidative stress in the body. Free radicals can cause damage to cells and tissues, leading to various diseases and aging (Mounir et al., 2022).

Probiotics have antioxidant properties in a variety of ways. For starters, they increase the synthesis of antioxidant enzymes that help neutralize free radicals, such as superoxide dismutase (SOD), catalase, and glutathione peroxidase. Second, by altering the gut microbiota and lowering inflammation, probiotics can directly reduce the generation of reactive oxygen species (ROS). Thirdly, certain strains of probiotics produce metabolites, such as SCFAs and bacteriocins, which have antioxidant properties and can protect against oxidative stress (Gurram et al., 2022; Wang et al., 2018).

Prebiotics, on the other hand, are non-digestible fibers that serve as food for beneficial bacteria in the gut. They promote the growth and activity of probiotics, thereby supporting a healthy gut microbiota. While prebiotics themselves do not possess direct antioxidant activity, they indirectly contribute to antioxidant defense by improving the balance of gut bacteria (Mounir et al., 2022). By promoting the growth of beneficial bacteria, prebiotics help maintain a healthy gut microbiota composition, which is essential for proper digestion, absorption of nutrients, and immune function. A dysbiotic gut microbiota, characterized by an imbalance of harmful bacteria, can contribute to oxidative stress and inflammation. By modulating the gut microbiota, prebiotics can help reduce oxidative stress and support antioxidant defenses (Ricke, 2018).

Furthermore, prebiotics have been shown to enhance the production of SCFAs in the gut. Short-chain fatty acids, such as butyrate, acetate, and propionate, have been linked to numerous health benefits, including antioxidant effects. They can scavenge free radicals, reduce inflammation, and improve the integrity of the gut barrier, thereby protecting against oxidative damage (Alloui et al., 2014).

Overall, probiotics and prebiotics have the potential to exert antioxidant effects by reducing oxidative stress, enhancing the activity of antioxidant enzymes, modulating the gut microbiota, and promoting the production of beneficial metabolites. However, more research is needed to fully understand the specific mechanisms involved and the clinical implications of these antioxidant properties.

7 Gut barrier properties of probiotics and prebiotics

The gut barrier refers to the lining of the GIT that acts as a protective barrier between the gut contents and the bloodstream. It plays a crucial role in maintaining gut health by selectively allowing the absorption of nutrients while preventing the entry of harmful substances and pathogens. Probiotics and prebiotics can both contribute to improving the gut barrier function, although they do so through different mechanisms, including:

1 Enhancing mucus production: probiotics can stimulate the production of mucus, a gel-like substance that forms a physical barrier on the inner lining of the gut. Mucus helps to prevent the direct contact of harmful substances, such as toxins and pathogens, with the intestinal cells (Abd El-Hack et al., 2020).
2 Tight junction regulation: tight junctions are specialized protein structures that connect adjacent cells in the intestinal lining, forming a tight seal between them. Probiotics can regulate the expression and integrity of these tight junction proteins, promoting a stronger barrier and reducing the permeability of the gut (Mohsin et al., 2022).
3 Modulating immune responses: probiotics interact with immune cells in the GALT, influencing the immune response and promoting a balanced and appropriate reaction to pathogens. This modulation can help reduce inflammation and maintain the integrity of the gut barrier. Prebiotics, on the other hand, are non-digestible dietary fibers that serve as a food source for beneficial bacteria in the gut. While prebiotics do not directly improve the gut barrier, they indirectly support its function through their effects on the gut microbiota (Ahmad et al., 2022).
4 Promoting the growth of beneficial bacteria: prebiotics selectively nourish beneficial bacteria, such as *Bifidobacteria* and *Lactobacilli*, which are known to contribute to a healthy gut environment. These beneficial bacteria produce SCFAs as metabolic byproducts, which help strengthen the gut barrier (Lin et al., 2022).
5 Modulating the gut microbiota composition: prebiotics can shift the balance of the gut microbiota toward a more diverse and beneficial microbial community. A diverse gut microbiota is associated with a stronger gut barrier and improved overall gut health (Khomayezi and Adewole, 2022).

In summary, probiotics improve the gut barrier by enhancing mucus production, regulating tight junction proteins, and modulating immune responses. Prebiotics indirectly support the gut barrier by promoting the growth of beneficial bacteria and modulating the gut microbiota composition. Combining probiotics and prebiotics, known as synbiotics, can provide a synergistic effect in enhancing gut barrier function and overall gut health.

8 Conclusion

This chapter delves into the fascinating world of avian gut health and highlights the crucial role of probiotics and prebiotics in enhancing overall poultry well-being. Throughout this chapter, we have explored the intricate mechanisms underlying the gut microbiota and its impact on poultry health, productivity, and immunity. From the initial exploration of the avian gut ecosystem to the in-depth analysis of the various probiotic and prebiotic strains, this chapter has provided a comprehensive understanding of how these beneficial microorganisms can positively influence gut function and strengthen the immune system of poultry. The research presented within these pages has showcased significant advancements in the field, shedding light on the potential applications of probiotics and prebiotics as natural alternatives to antibiotics in poultry production.

Moreover, the chapter serves as a valuable resource for poultry farmers, veterinarians, and researchers seeking to optimize the health and performance of their flocks. By incorporating the knowledge and recommendations presented here, industry professionals can implement effective strategies to promote avian gut health and bolster the birds' immune responses, leading to improved growth rates, disease resistance, and overall welfare.

Nevertheless, the field of probiotics and prebiotics in poultry is ever evolving, and further research is necessary to deepen our understanding of their precise mechanisms of action and their full potential in avian wellness. Future investigations should focus on optimizing dosage, delivery methods, and specific strain selection to maximize the beneficial effects of probiotics and prebiotics in different poultry species and production systems. The aim of this chapter was to provide a comprehensive overview of the current state of knowledge and research in this vital area. By continuing to explore and harness the potential of probiotics and prebiotics, we can unlock the secrets of avian wellness, revolutionize poultry production practices, and contribute to the sustainable and healthy future of the poultry industry. It is also important to recognize that there is still, several limitations associated with current experimental methods for delivering probiotics and quantifying their attachment and proliferation in broilers. These limitations include the following:

1 Variability in gut microbiota composition: boilers have a diverse gut microbiota, which can vary based on factors like age, diet, and environmental conditions. This natural variability can make it challenging to accurately measure the attachment and proliferation of specific probiotic strains.
2 Complexity of gut environment: the avian GIT is a complex ecosystem with various niches, pH levels, and microbial populations. This

complexity can influence the survival and establishment of probiotic bacteria, making it difficult to isolate their specific effects.

3 Survival during passage through GIT: probiotic bacteria must survive the harsh conditions of the stomach, including low pH levels and digestive enzymes. Some strains may have higher resistance to these conditions than others, impacting their ability to proliferate in the gut.
4 Competition with indigenous microbiota: the indigenous microbial communities in the broiler gut may already occupy niches and compete with introduced probiotics for resources and attachment sites. This competition can limit the colonization and proliferation of probiotic strains.
5 Inconsistent dosage and application: administering a consistent dosage of probiotics to broilers can be challenging, especially in large-scale commercial settings. Variations in dosage and application methods can lead to inconsistent results in terms of attachment and proliferation.
6 Sampling techniques: obtaining representative samples from different segments of the GIT for analysis can be technically challenging. Variations in sampling techniques can introduce bias and affect the accuracy of quantifying probiotic attachment and proliferation.
7 Molecular detection method: molecular methods used to quantify probiotics often rely on DNA-based techniques, such as qPCR (quantitative polymerase chain reaction). However, these methods may not distinguish between live and dead cells, potentially overestimating the actual proliferation of probiotics.
8 Limited understanding of strain-specific interactions: different probiotic strains may have varying abilities to attach and proliferate in the broiler gut due to strain-specific characteristics. Understanding these interactions requires in-depth genetic and physiological studies, which may not always be feasible.

Addressing these limitations requires a multidisciplinary approach, incorporating advanced molecular techniques, controlled experimental designs, and a deep understanding of both probiotic strains and the broiler gut environment. Additionally, conducting research under standardized conditions that closely mimic commercial poultry production settings can help bridge the gap between experimental findings and real-world applications.

9 Where to look for further information

To further explore the topic of using probiotics and prebiotics to enhance gut function and immunity in poultry, you can consider the following sources of information:

1 *Academic Journals and Research Papers*: Conduct a literature search on academic databases such as PubMed, ScienceDirect, and Google Scholar. Look for research articles, reviews, and studies related to probiotics, prebiotics, gut health, and immunity in poultry. Some relevant journals include *Poultry Science*, *Journal of Applied Poultry Research*, *Veterinary Immunology and Immunopathology*, and *Avian Pathology*.
2 Books and reference materials: explore books that focus on poultry nutrition, health, and welfare. Look for titles written by experts in the field, such as renowned veterinarians or animal scientists. Check out university libraries, online bookstores, and academic publishers for books that cover the specific topics of probiotics, prebiotics, and avian wellness.
3 Industry reports and publications: keep an eye out for reports or publications released by reputable organizations, associations, or companies related to the poultry industry. These sources often provide insights into the latest trends, advancements, and best practices in poultry nutrition and health. Examples include publications from the World Poultry Science Association (WPSA) and poultry-focused magazines or newsletters.
4 Conferences and symposia: stay informed about scientific conferences, symposia, or seminars dedicated to poultry research and production. Attend these events or search for proceedings and abstracts online to access cutting-edge information and presentations related to avian wellness, gut function, and immunity. Often, leading researchers and industry experts share their findings and perspectives in such forums.
5 Websites and online resources: Explore websites of renowned research institutions, universities, and government agencies that specialize in poultry science and animal health. These websites often provide access to research articles, publications, and other resources related to probiotics, prebiotics, and avian wellness. Examples include the websites of the U.S. Department of Agriculture (USDA), the Food and Agriculture Organization (FAO), and universities with strong agricultural or veterinary programs.
6 Expert interviews and case studies: look for interviews or case studies featuring experts in poultry science, nutrition, or veterinary medicine. These interviews could be found in industry publications, podcasts, or websites focused on poultry production. Expert opinions and real-world examples can provide valuable insights into the practical application of probiotics, prebiotics, and their impact on gut health and immunity in poultry.

Remember to critically evaluate the sources you find, consider the credibility of the authors or organizations, and cross-reference information to ensure accuracy and reliability. The following are some of our publications that our laboratory has published during the last 22 years related to the topic of this chapter book.

Electronic open access books

- Tellez-Isaias, G. ed. 2022. *The Global Antimicrobial Resistance Epidemic - Innovative Approaches and Cutting-Edge Solutions.* London: IntechOpen. https://doi.org/10.5772/intechopen.98171.
- Tellez-Isaias, G., Latorre, J. D. and Martínez-Aguilar, Y. eds. 2022. *Broiler Industry*. London: IntechOpen. https://doi.org/10.5772/intechopen.98143.
- Tellez, G. and Latorre, J. D. eds. 2017. *Alternatives to Antimicrobial Growth Promoters and Their Impact in Gut Microbiota, Health and Disease.* Lausanne: Frontiers Media. https://doi.org/10.3389/978-2-88945-375-7.
- Tellez-Isaias, G. and Latorre, J. D. eds. 2022. *Alternatives to Antimicrobial Growth Promoters and Their Impact in Gut Microbiota, Health and Disease: Volume II*. Lausanne: Frontiers Media SA. https://doi.org/10.3389/978-2-88974-750-4.
- Hernandez-Patlan, D., Solis-Cruz, B. and Tellez-Isaias, G. eds. 2023. *Technological Strategies to Improve Animal Health and Production.* Lausanne: Frontiers Media SA. https://doi.org/10.3389/978-2-8325-2645-3.

Journals

- Tellez, G., Higgins, S. E., Donoghue, A. M. and Hargis, B. M. 2006. Digestive physiology and the role of microorganisms. *Journal of Applied Poultry Research* 15(1), pp.136–144.
- Tellez-Isaias, G., Vuong, C. -N., Graham, B. D., Selby, C. M., Graham, L. E., Senas-Cuesta, R., Barros, T. L., Beer, L. C., Coles, M. E., Forga, A. J. and Ruff, J. 2021. Developing probiotics, prebiotics, and organic acids to control Salmonella spp. in commercial turkeys at the University of Arkansas, USA. *German Journal of Veterinary Research* 3, pp.7–12.
- Shehata, A. A., Attia, Y., Khafaga, A. F., Farooq, M. Z., El-Seedi, H. R., Eisenreich, W. and Tellez-Isaias, G. 2022. Restoring healthy gut microbiome in poultry using alternative feed additives with particular attention to phytogenic substances: challenges and prospects. *German Journal of Veterinary Research* 2, pp.32–42.
- Tellez-Isaias, G., Eisenreich, W., Shehata, A. A. 2022. Nutraceuticals to mitigate the secret killers in animals. *Veterinary Research*. 9, 435. https://doi.org/10.3390/vetsci9080435.

- Shehata, A. A., Yalçın, S., Latorre-Cárdenas, J. D., Basiouni, S., Attia, Y. A., El-Wahab, A. A., Visscher, C., El-Sweedi, H., Huber, C., Hafez, H. M., Eisenreich, W. and Tellez, G. 2022. Probiotics, prebiotics, and phytogenic substances for optimizing gut health in poultry. *Microorganisms* 10, 395. https://doi.org/10.3390/microorganisms10020395.
- Hernandez-Patlan, D., Solis-Cruz, B., Latorre, J. D., Merino-Guzman, R., Morales Rodríguez, M., Ausland, C., Hernandez-Velasco, X., Ortiz Holguin, O., Delgado, R., Hargis, B. M., Singh, P. and Tellez-Isaias, G. 2022. Whole-genome sequence and interaction analysis in the production of six enzymes from the three bacillus strains present in a commercial direct-fed microbial (NorumTM) using a bliss independence test. *Frontiers in Veterinary Science* 9, 784387. https://doi.org/10.3389/fvets.2022.784387.
- Tellez-Isaias, G. and Latorre, J. D. 2022. Editorial: alternatives to antimicrobial growth promoters and their impact in gut microbiota, health and disease: Volume II. *Frontiers in Veterinary Science* 9, 857583. https://doi.org/10.3389/fvets.2022.857583.
- Tellez-Isaias, G., Eisenreich, W., Petrone-Garcia, V. M., Hernandez-Velasco, X., Castellanos-Huerta, I., Tellez Jr. G., Latorre, J. D., Bottje, W. G., Senas-Cuesta, R., Coles, M. E., Hargis, B. M., El-Ashram, S., Graham, B. D. and Shehata, A. A. 2023. Effects of chronic stress and intestinal inflammation on commercial poultry health and performance: a review. *German Journal of Veterinary Research* 3(1), 38–57. https://doi.org/10.51585/gjvr.2023.1.0051.

10 Acknowledgments

Research for this chapter was supported in part by funds provided by USDA-NIFA, Sustainable Agriculture Systems, Grant no. 2019-69012-29905. Title of project: Empowering U.S. Broiler Production for Transformation and Sustainability USDA-NIFA (Sustainable Agriculture Systems): no. 2019-69012-29905. In preparing the chapter, the authors also benefitted from the Supporting Researchers Project (RSP2023R111), King Saud University, Riyadh, Saudi Arabia.

11 References

Abd El-Hack, M. E., El-Saadony, M. T., Shafi, M. E., Qattan, S. Y., Batiha, G. E., Khafaga, A. F., Abdel-Moneim, A. M. E. and Alagawany, M. 2020. Probiotics in poultry feed: a comprehensive review. *Journal of Animal Physiology and Animal Nutrition* 104(6), pp.1835–1850.

Abreu, R., Semedo-Lemsaddek, T., Cunha, E., Tavares, L. and Oliveira, M. 2023. Antimicrobial -drug resistance in poultry production: current status and innovative strategies for bacterial control. *Microorganisms* 11(4), p.953.

Ahmad, R., Yu, Y. H., Hsiao, F. S. H., Su, C. H., Liu, H. C., Tobin, I., Zhang, G. and Cheng, Y. H. 2022. Influence of heat stress on poultry growth performance, intestinal inflammation, and immune function and potential mitigation by probiotics. *Animals* 12(17), p.2297.

Ajuwon, K. M. 2016. Toward a better understanding of mechanisms of probiotics and prebiotics action in poultry species. *Journal of Applied Poultry Research* 25(2), pp.277–283.

Alagawany, M., Bassiony, S. S., El-Kholy, M. S., El-Naggar, K., El-Metwally, A. E. and Al-Sagheer, A. A. 2023. Comparison of the effects of probiotic-based formulations on growth, feed utilization, blood constituents, cecal fermentation, and duodenal morphology of rabbits reared under hot environmental conditions. *Annals of Animal Science*. https://doi.org/10.2478/aoas-2023-0004.

Ali, Q., Ma, S., La, S., Guo, Z., Liu, B., Gao, Z., Farooq, U., Wang, Z., Zhu, X., Cui, Y. and Li, D. 2022. Microbial short-chain fatty acids: a bridge between dietary fibers and poultry gut health–a review. *Animal Bioscience* 35(10), p.1461.

Alloui, M. N., Agabou, A. and Alloui, N. 2014. Application of herbs and phytogenic feed additives in poultry production: a review. *Global Journal of Animal Scientific Research* 2(3), pp.234–243.

Al-Otaibi, A. M., Abd El-Hack, M. E., Dmour, S. M., Alsowayeh, N., Khafaga, A. F., Ashour, E. A., Nour-Eldeen, M. A. and Świątkiewicz, S. 2023. A narrative review on the beneficial impacts of probiotics on poultry: an updated knowledge. *Annals of Animal Science*. https://doi.org/10.2478/aoas-2023-0001.

Beldowska, A., Barszcz, M. and Dunislawska, A. 2023. State of the art in research on the gut-liver and gut-brain axis in poultry. *Journal of Animal Science and Biotechnology* 14(1), p.37.

Brugaletta, G., Zampiga, M., Laghi, L., Indio, V., Oliveri, C., De Cesare, A. and Sirri, F. 2023. Feeding broiler chickens with arginine above recommended levels: effects on growth performance, metabolism, and intestinal microbiota. *Journal of Animal Science and Biotechnology* 14(1), p.33.

Bucław, M. 2016. The use of inulin in poultry feeding: a review. *Journal of Animal Physiology and Animal Nutrition* 100(6), pp.1015–1022.

Chacher, M. F. A., Kamran, Z., Ahsan, U., Ahmad, S., Koutoulis, K. C., Qutab Ud Din, H. G. and Cengiz, Ö. 2017. Use of mannan oligosaccharide in broiler diets: an overview of underlying mechanisms. *World's Poultry Science Journal* 73(4), pp.831–844.

Chen, X., Zhou, X., Li, S., Zhang, H. and Liu, Z. 2023. Effects of tea residues-fermented feed on production performance, egg quality, antioxidant capacity, caecal microbiota, and ammonia emissions of laying hens. *Frontiers in Veterinary Science* 10, p.1195074.

Cheng, J., Lei, H., Xie, C., Chen, J., Yi, X., Zhao, F., Yuan, Y., Chen, P., He, J., Luo, C. and Shu, D. 2023. B lymphocyte development in the bursa of fabricius of young broilers is influenced by the gut microbiota. *Microbiology Spectrum* 11(2), p.e04799-22.

Darboe, A. K. 2022. Review on the use of probiotics in poultry production (Layers and broilers) as feed additives. *International Journal of Veterinary Sciences and Animal Husbandry*. https://doi.org/10.22271/veterinary.2022.v7.i5a.442.

Dhama, K., Mahendran, M., Tomar, S. and Chauhan, R. S. 2008. Beneficial effects of probiotics and prebiotics in livestock and poultry: the current perspectives. *Intas Polivet* 9(1), pp.1–12.

Fathima, S., et al. 2022. Gastrointestinal microbiota and their manipulation for improved growth and performance in chickens. *Foods* 11, p.1401.

Fu, Y., Hu, J. and Cheng, H. W. 2023. Research note: probiotic, bacillus subtilis, alleviates neuroinflammation in the hippocampus via the gut microbiota-brain axis in heat-stressed chickens. *Poultry Science* 102(6), p.102635.

Ganguly, S. 2013. Supplementation of prebiotics, probiotics and acids on immunity in poultry feed: a brief review. *World's Poultry Science Journal* 69(3), pp.639–648.

Gribble, F. M. and Reimann-, F. 2017. Signalling in the gut endocrine axis. *Physiology & Behavior* 176, pp.183–188.

Gu, W. H and Deng, F-L. 2023. Research progress on application of prebiotics in poultry production. *Feed Research*, *46*(8): 104–108.

Gul, S. T. and Alsayeqh, A. F. 2022. Probiotics as an alternative approach to antibiotics for safe poultry meat production. *Pakistan Veterinary Journal* 42(3), pp.285–291

Gurram, S., Chinni Preetam, V., Vijaya Lakshmi, K., Raju, M. V. L. N., Venkateswarlu, M. and Bora, S. 2022. Synergistic effect of probiotic, chicory root powder and coriander seed powder on growth performance, antioxidant activity and gut health of broiler chickens. *Plos One* 17(6), p.e0270231.

He, Y., Li, J., Wang, F., Na, W. and Tan, Z. 2023. Dynamic changes in the gut microbiota and metabolites during the growth of hainan wenchang chickens. *Animals* 13(3), p.348.

Hu, Y., Liu, D., Jin, X., Feng, Y. and Guo, Y. 2023. Synthetic microbiome for a sustainable poultry industry. *The Innovation* 4(1), p.100357.

Isolauri, E., Salminen, S. and Ouwehand, A. C. 2004. Probiotics. *Best Practice & Research Clinical Gastroenterology* 18(2), pp.299–313.

Jahan, A. A., González Ortiz, G., Moss, A. F., Bhuiyan, M. M. and Morgan, N. K. 2022. Role of supplemental oligosaccharides in poultry diets. *World's Poultry Science Journal* 78(3), pp.615–639.

Khalid, A., Khalid, F., Mahreen, N., Hussain, S. M., Shahzad, M. M., Khan, S. and Wang, Z. 2022. Effect of spore-forming probiotics on the poultry production: a review. *Food Science of Animal Resources* 42(6), p.968.

Khomayezi, R. and Adewole, D. 2022. Probiotics, prebiotics, and synbiotics: an overview of their delivery routes and effects on growth and health of broiler chickens. *World's Poultry Science Journal* 78(1), pp.57–81.

Kim, S. A., Jang, M. J., Kim, S. Y., Yang, Y., Pavlidis, H. O. and Ricke, S. C. 2019. Potential for prebiotics as feed additives to limit foodborne campylobacter establishment in the poultry gastrointestinal tract. *Frontiers in Microbiology* 10, p.91.

Krysiak, K., Konkol, D. and Korczyński, M. 2021. Overview of the use of probiotics in poultry production. *Animals* 11(6), p.1620.

Lee, M. D., Pedroso, A. A. and Maurer, J. J. 2023a. Bacterial composition of a competitive exclusion product and its correlation with product efficacy at reducing Salmonella in poultry. *Frontiers in Physiology* 13, p.2737.

Lee, M. D., Pedroso, A. A., Lumpkins, B., Cho, Y. and Maurer, J. J. 2023b. Pioneer colonizers: bacteria that alter the chicken intestinal morphology and development of the microbiota. *Frontiers in Physiology* 14, p.1139321.

Li, C., Wang, S., Chen, S., Wang, X., Deng, X., Liu, G., Chang, W., Beckers, Y. and Cai, H. 2023. Screening and characterization of pediococcus acidilactici LC-9-1 toward selection as a potential probiotic for poultry with antibacterial and antioxidative properties. *Antioxidants* 12(2), p.215.

Lin, Y., Teng, P. Y. and Olukosi, O. A. 2022. The effects of xylo-oligosaccharides on regulating growth performance, nutrient utilization, gene expression of tight junctions, nutrient transporters, and cecal short chain fatty acids profile in Eimeria-challenged broiler chickens. *Poultry Science* 101(11), p.102125.

Liu, T., Ruan, S., Mo, Q., Zhao, M., Wang, J., Ye, Z., Chen, L. and Feng, F. 2023a. Evaluation of dynamic effects of dietary medium-chain monoglycerides on performance, intestinal development and gut microbiota of broilers in large-scale production. *Animal Nutrition*. https://doi.org/10.1016/j.aninu.2023.05.003.

Liu, Y., Feng, Y., Yang, X., Lv, Z., Li, P., Zhang, M., Wei, F., Jin, X., Hu, Y., Guo, Y. and Liu, D. 2023b. Mining chicken ileal microbiota for immunomodulatory microorganisms. *The ISME Journal* 17(5), pp.758–774.

Mangan, M. and Siwek, M. 2023. Strategies to combat heat stress in poultry production: a review. *Animal Physiol. and Animal Nutri.* 108(3): 576–595.

Maslowski, K. M. and Mackay, C. R. 2010. Diet, gut microbiota and immune responses. *Nature Immunology* 12, pp.5–9.

Mayer, E. A., Knight, R., Mazmanian, S. K., Cryan, J. F. and Tillisch, K. 2014. Gut microbes and the brain: paradigm shift in neuroscience. *The Journal of Neuroscience* 34, pp.15490–15496.

Megur, A., Baltriukienė, D., Bukelskienė, V. and Burokas, A. 2021. The Microbiota-Gut-Brain Axis and Alzheimer's Disease: Neuroinflammation Is to Blame? *Nutrients* 13, p.37.

Mohsin, M., Zhang, Z. and Yin, G. 2022. Effect of probiotics on the performance and intestinal health of broiler chickens infected with Eimeria tenella. *Vaccines* 10(1), p.97.

Mounir, M., Ibijbijen, A., Farih, K., Rabetafika, H. N. and Razafindralambo, H. L. 2022. Synbiotics and their antioxidant properties, mechanisms, and benefits on human and animal health: a narrative review. *Biomolecules* 12(10), p.1443.

Park, S. O. and Seo, K. H. 2023. Digital livestock systems and probiotic mixtures can improve the growth performance of swine by enhancing immune function, cecal bacteria, short-chain fatty acid, and nutrient digestibility. *Frontiers in Veterinary Science* 10, p.1126064.

Proszkowiec-Weglarz, M., Oakley, B., Ellestad, L. E. and Javed, S. 2023. Avian microbiome: from embryonic development to adulthood. *Frontiers in Physiology* 14, p.1211911.

Reuben, R. C., Sarkar, S. L., Roy, P. C., Anwar, A., Hossain, M. A. and Jahid, I. K. 2021. Prebiotics, probiotics and postbiotics for sustainable poultry production. *World's Poultry Science Journal* 77(4), pp.825–882.

Ricke, S. C. 2018. Focus: nutrition and food science: impact of prebiotics on poultry production and food safety. *The Yale Journal of Biology and Medicine* 91(2), p.151.

Ricke, S. C. 2021. Prebiotics and alternative poultry production. *Poultry Science* 100(7), p.101174.

Ricke, S. C., Lee, S. I., Kim, S. A., Park, S. H. and Shi, Z. 2020. Prebiotics and the poultry gastrointestinal tract microbiome. *Poultry Science* 99(2), pp.670–677.

Roberfroid, M. 2007. Prebiotics: the concept revisited. *The Journal of Nutrition* 137(3), pp.830S–837S.

Schneider, S., Wright, C. M. and Heuckeroth, R. O. 2019. Unexpected roles for the second brain: enteric nervous system as master regulator of bowel function. *Annual Review of Physiology* 81, pp.235–259.

Shehata, A. A., Yalçın, S., Latorre, J. D., Basiouni, S., Attia, Y. A., Abd El-Wahab, A., Visscher, C., El-Seedi, H. R., Huber, C., Hafez, H. M. and Eisenreich, W. 2022. Probiotics, prebiotics, and phytogenic substances for optimizing gut health in poultry. *Microorganisms* 10(2), p.395.

Shini, S. and Bryden, W. L. 2021. Probiotics and gut health: linking gut homeostasis and poultry productivity. *Animal Production Science* 62(12), pp.1090–1112.

Soumeh, E. A., Cedeno, A. D. R. C., Niknafs, S., Bromfield, J. and Hoffman, L. C. 2021. The efficiency of probiotics administrated via different routes and doses in enhancing production performance, meat quality, gut morphology, and microbial profile of broiler chickens. *Animals* 11(12), p.3607.

Teng, P-Y. and Kim, W. K. 2018. Review: roles of prebiotics in intestinal ecosystem of broilers. *Front. Vet Sci.* 5: 245. https://doi.org/10.3389/fvets.2018.00245.

Vaseji, N., Mojgani, N. and Moteshafi, H. 2022. The mechanism of action, selection criteria and application of probiotics in poultry nutrition. *Applied Animal Science Research Journal* 11(42), pp.37–54.

Wang, Y., Dong, Z., Song, D., Zhou, H., Wang, W., Miao, H., Wang, L. and Li, A. 2018. Effects of microencapsulated probiotics and prebiotics on growth performance, antioxidative abilities, immune functions, and caecal microflora in broiler chickens. *Food and Agricultural Immunology* 29(1), pp.859–869.

Weiss, G. A. and Hennet, T. 2017. Mechanisms and consequences of intestinal dysbiosis. *Cellular and Molecular Life Sciences* 74, pp.2959–2977.

Wlazlak, S., Pietrzak, E., Biesek, J. and Dunislawska, A. 2023. Modulation of the immune system of chickens a key factor in maintaining poultry production-a review. *Poultry Science*, p.102785. https://doi.org/10.1016/j.psj.2023.102785.

Wu, M., Zuo, S., Maiorano, G., Kosobucki, P. and Stadnicka, K. 2022a. How to employ metabolomic analysis to research on functions of prebiotics and probiotics in poultry gut health? *Frontiers in Microbiology* 13, p.1040434.

Wu, Y., Yang, F., Jiang, W., Hu, A., Xiong, Z., Yang, S., Cao, P., Cao, Z., Xiong, Z. and Cao, H. 2022b. Effects of compound probiotics on intestinal barrier function and caecum microbiota composition of broilers. *Avian Pathology* 51(5), pp.465–475.

Yaqoob, M. U., Abd El-Hack, M. E., Hassan, F., El-Saadony, M. T., Khafaga, A. F., Batiha, G. E., Yehia, N., Elnesr, S. S., Alagawany, M., El-Tarabily, K. A. and Wang, M. 2021. The potential mechanistic insights and future implications for the effect of prebiotics on poultry performance, gut microbiome, and intestinal morphology. *Poultry Science* 100(7), p.101143.

Yaqoob, M. U., Wang, G. and Wang, M. 2022. An updated review on probiotics as an alternative of antibiotics in poultry–a review. *Animal Bioscience* 35(8), p.1109.

Ye, J., Yang, H., Hu, W., Tang, K., Liu, A. and Bi, S. 2023. Changed cecal microbiota involved in growth depression of broiler chickens induced by immune stress. *Poultry Science* 102(5), p.102598.

Zhu, B., Wang, X. and Li, L. 2010. Human gut microbiome: the second genome of human body. *Protein & Cell* 1, pp.718–725.

Part 4

Novel feed sources

Chapter 14

Emerging protein sources for poultry feed

Archibold G. Bakare, Fiji National University, Fiji Islands; Taiye Olugbemi, Ahmadu Bello University, Nigeria; Mohammed M. Ari, Nasarawa State University, Nigeria; and Paul A. Iji, Fiji National University, Fiji Islands and University of New England, Australia

1 Introduction

Commercial poultry diets contain from 15% to 24% total protein, depending on the age of birds, level of production and end product. Most of the protein is supplied from major plant and animal sources. Although the range of ingredients that can supply this protein is wide, the global poultry industry tends to depend on a limited number of feed ingredients due to supply, ingredient quality and safety issues.

The supply of key protein sources around the world is variable and this has a major impact on the costs of diets and production. The number one vegetable protein source is soybean, with supply coming mostly from North and South America. Other key vegetable sources such as peanut, canola, cotton seed and sunflower are produced across many more regions than soybean, but they tend to have more constraints in supply and quality. The sources of animal protein ingredients and their quality are even more variable. However,

http://dx.doi.org/10.19103/AS.2024.0143.19

such products, including meat and bone meal, and fishmeal remain the most important ingredients that can complement vegetable protein sources in most commercial diets. Globally, there are emerging sources of protein for the poultry industry. Many ingredients have always been available but have been under-explored or not considered as protein sources for poultry. The objective of this chapter is to examine some of these emerging ingredients for potential use locally or for the development into global protein sources for the poultry industry.

2 Challenges in using conventional protein sources in poultry diets

Conventional protein sources are of both plant and animal origins. Plant sources include soybean, peanut, cottonseed, sunflower seed, rapeseed and linseed, while animal sources include fishmeal, meat and bone meal. Soybean meal and fishmeal, however, are the most used conventional protein sources used in poultry feeds. Despite the many positive attributes of these conventional protein sources in poultry diets, their use is not without challenges associated with nutritional limitations, animal and human health concerns, availability and cost as well as environmental and sustainability issues (Sajid et al., 2023). Some sources are known to contain anti-nutritional factors that can reduce nutrient absorption and utilization by poultry, leading to poor growth and development. There is concern that the use of animal by-products in poultry feed can have an adverse impact on animal and human health, leading to food safety concerns and potential regulatory issues (Jedrejek et al., 2016; an overview e.g. of recent developments in EU policy and regulation in this area can be found at: https://www.efsa.europa.eu/en/topics/animal-by-products). Allergens contained in some conventional protein sources can cause allergic reactions in poultry and may lead to decreased performance or health of birds.

The availability of conventional protein sources varies according to location and season, which can lead to disruptions and fluctuations in supply, resulting in price fluctuations as well as availability. Some sources may not be available in certain countries or regions, leading to increased costs and decreased profitability. Growing demand for these ingredients has resulted in higher feed costs. The use of conventional protein sources for poultry feed may also not be sustainable in the long term, as it can lead to over-exploitation of natural resources (e.g. deforestation to grow protein crops) and environmental degradation (Fraanje and Garnett, 2020). The production of soybeans, for instance, also requires considerable resources and energy. Fishmeal can also have a significant impact on the marine ecosystem.

3 Potential alternative sources of protein: animal sources

As the world seeks sustainable protein sources, alternative protein sources like insect meal, earthworms, rumen digesta, tankage and blood meal present viable options in non-ruminant diets, especially poultry feeds. These alternatives contribute to the evolution of modern animal nutrition given their nutritional profiles, which, with proper processing, formulation and inclusion strategies, would be beneficial for the growth and well-being of poultry.

3.1 Insect meal

Insect meals are obtained by processing various insect species, such as mealworms, black soldier fly larvae and crickets. The nutrient diversity of insect meals effectively makes them suitable as a replacement for conventional protein sources like fish and meat meals. Insect meal protein content ranges from 39% to 65% (Moyo and Moyo, 2022). Amino acid profiles vary among insect species but often include essential amino acids. They also contain vitamins, minerals and useful fats (Kourimska and Adamkova, 2016; Spranghers et al., 2017). Sustainable sourcing of insects can be guaranteed since they can be reared on organic waste and reducing environmental impact. The high protein content of insect meals makes it highly suitable for poultry. With proper amino acid supplementation, insect meal inclusion in poultry diets has resulted in positive effects on growth rate, feed efficiency and egg quality (Moyo and Moyo, 2022). The insect species used for making insect meal are found in abundance in tropical climates and are reared by some small-scale farmers under natural environmental conditions, making them a cheap source of protein for smallholder farmers. In developing countries, where environmental conditions are not conducive for rearing insects, modern facilities with controlled environments are used, which may affect cost (Gasco et al., 2023).

Insects can be offered to animals as live larvae or as processed feed. The most noticeable improvement in growth performance is found when live larvae are provided at the highest frequency of about four times per day (Ipema et al., 2020). Insects are usually processed using methods such as freezing, heat treatment or exposure to carbon dioxide. Preservation techniques, like drying or freeze-drying, are then employed to prevent spoilage. Dried insects are ground into a fine powder using mechanical processes like grinding or milling or pelleted to improve feed handling, reduce dust and enhance feed conversion efficiency.

3.2 Earthworms

Earthworms play an important role in the ecosystem where they improve nutrient availability to plants, soil structure, drainage and productivity. Harvesting of earthworms should always take account of their importance to soil quality control. Earthworms are also proving to be important in the feed industry where they provide valuable nutrients for animals. Earthworm meal contains a substantial amount of protein, ranging from 58% to 71%, depending on factors such as species, diet and processing methods (Bahadori et al., 2021; Khan et al., 2016). The protein in earthworm meals is considered of good quality due to its balanced amino acid profile, including indispensable amino acids such as lysine, methionine and tryptophan, which are crucial for animal growth, tissue development and protein synthesis. Earthworm meals are also a rich source of minerals, vitamins and beneficial fatty acids like omega-3 and omega-6 fatty acids (Kumlu et al., 2021). The consumption by earthworms of organic matter rich in fibre results in increased dietary fibre content in earthworm meal. This supports gut health and digestive processes in animals. Increased growth and higher-quality meat can be observed in broilers-fed diets containing up to 30% earthworm meal (Bahadori et al., 2021). However, a major drawback of the use of earthworm meal is low dry matter content of between 15% and 20% (Cayot et al., 2009; Pérez-Corría et al., 2019).

3.3 Rumen digesta

Rumen digesta is a byproduct of the content of the rumen, containing microbial protein and undigested feed particles. It is typically a post-slaughter waste product. The contents of the first compartment of a ruminant animal's stomach are rich in nutrients due to the fermentation processes that occur (Ari et al., 2006). While it is not typically used as a direct feed ingredient in non-ruminant diets, understanding its nutritional profile can provide insights into its potential as an alternative protein source. Rumen digesta contains microbial protein produced during the fermentation of ingested feed (Cherdthong, 2020). The protein content and the amino acid profile of rumen digesta are influenced by microbial activity, the composition of the original feed and the stage of digestion. The microbes involved in the fermentation process include bacteria, protozoa and fungi. These microbes can contribute to the nutrient profile, providing protein and other beneficial compounds. Other benefits are the production of volatile fatty acids (VFAs) as the end products of microbial fermentation in the rumen (Elfaki and Abdelatti, 2016). They include acetate, propionate and butyrate, which are a source of energy for the animal. Rumen biodigesta also contains vitamins (B vitamins) and minerals such as potassium, phosphorus and magnesium, which are released from feed materials during fermentation and are important for animal health. Rumen digesta can be collected after slaughter, dried and ground into a meal.

3.4 Tankage

Tankage is the residue from tanks used in rendering of animal carcasses and is rich in protein, minerals and amino acids (Cromwell, 2006). Animal tissues are cooked, dried and ground to create tankage, providing a concentrated source of protein and essential nutrients, typically containing around 50–70% protein. The high protein content makes it a valuable ingredient for promoting growth and development in animals, including poultry (Khan, 2018). Tankage provides a diverse amino acid profile, including essential amino acids required by poultry. It is rich in lipids and minerals such as calcium, phosphorus and trace minerals. Calcium supports bone health while phosphorus is essential for energy metabolism and overall growth. Tankage may contain collagen and gelatin derived from connective tissues and have potential health benefits for animals. The strong odour and flavour of tankage can act as a palatability enhancer in animal diets, encouraging consumption and utilization by poultry. Animal tissues, such as bones and offal, are rendered through cooking and drying. The rendered material is finely ground to create tankage. Proper rendering processes ensure the removal of unwanted materials to meet feed industry standards.

3.5 Blood by-products

Most of the blood used to make blood by-products is sourced from abattoirs during the slaughter of livestock. The composition of blood is comparable across species, consisting of 40% blood cells and 60% plasma (Kuan et al., 2018). Blood meal is mainly processed from fresh whole blood, whereas blood plasma meal is from blood plasma. For blood meal, fresh clean blood can be dried using different methods like spray-drying, flash drying and drum drying (Hardy and Kaushik, 2021). Blood plasma, on the other hand, is created by treating blood with an anticoagulant to stop the clotting process (Makara et al., 2016). Plasma and blood cells are separated from one another. The resulting solution of plasma is converted into a powder using spray drying. Both blood and blood plasma meals are rich in protein and essential amino acids (Table 1). However, nutrients in blood plasma meal are highly digestible compared to nutrients in blood meal. The dietary inclusion of spray-dried porcine plasma in broiler chickens challenged with *Salmonella sofia* improved performance, reduced the negative physiological effects of the pathogen and enhanced immune response (Beski et al., 2015). Spray-dried porcine plasma can be seen to improve overall health and well-being in poultry.

4 Potential alternative sources of protein: plant sources

The use of alternative protein sources for all classes of poultry has become more necessary because, like other non-ruminant animals, poultry requires

Table 1 Nutrient composition of blood by-products

	Blood meal[1]	Plasma meal[2]
Nutrients (g/kg)		
Dry matter	935.7	920
Crude protein	904	780
Crude fat	3.7	3
Ash	245	100
Metabolisable energy, poultry (MJ/kg)	–	15.99
Amino acid (total) (g/kg)		
Arginine	39.4	47
Lysine	86.0	68
Methionine	15.5	7
Cysteine	11.8	28
Methionine + cysteine	–	35
Tryptophan	40.7	14
Glycine	40.4	30
Histidine	54.0	28
Leucine	109.5	78
Isoleucine	11.6	29
Phenylalanine	64.1	4.6
Threonine	40.7	48
Valine	83.3	9
Minerals (g/kg)		
Calcium	2.6	1.5
Sodium	–	22
Phosphorus (available)	–	13
Phosphorus (total)	16	13
Chloride	–	11
Magnesium	–	0.3
Iron (mg/kg)	–	90

[1]Liu et al., 1989.
[2]Beski et al. 2015.

high-quality protein sources in their diets for optimal growth and development. Several oilseeds and leaf meals have been identified as potential replacements for conventional oilseeds as protein meals in poultry feed given their nutritional, environmental and cost benefits. Leaf meals, seaweed, algae and yeast also offer emerging alternative feed resources for poultry. Their utilization enhances nutrient diversity, health and sustainability in poultry feeding. As the industry evolves, these resources contribute to efficient and innovative feed ingredients for poultry nutrition.

4.1 Leaf meals

Leaf meals are derived from various plants and can provide protein, fibre and micronutrients. Leaves are dried and ground to produce leaf meals for poultry (Bamdad et al., 2019; Bakare et al., 2021). The following are some examples.

Moringa: Native to north India, moringa can be grown in tropical and subtropical regions of Asia and Africa. Moringa leaves are rich in protein, essential amino acids, vitamins (A, C, B), minerals (calcium, iron) and antioxidants (flavonoids, polyphenols) (Yadav et al., 2024). Leaves provide high-quality protein, contributing to growth and development in poultry (Musa-Azara et al., 2013). Moringa has a high vitamin A content and supports vision, immunity and overall health in poultry. Calcium and iron enhance bone strength and blood health, respectively, in birds, and antioxidants support immune function and protect cells from oxidative stress.

Baobab: Baobab (*Adansonia digitata*) is a widely distributed tree found in Africa. Young leaves of baobab trees can be a potential source of protein (Chadare et al., 2008). They are also rich in essential amino acids, vitamins (A, C, B), minerals (calcium, potassium) and dietary fibre. The high vitamin C content enhances immune function and reduces stress in birds. The available calcium and potassium in baobab leaf meal support bone health, muscle function and electrolyte balance. Dietary fibre in leaf meals promotes gut health, aiding in digestion and nutrient absorption.

4.2 Oilseeds

Oilseed meals are mainly used as primary sources of protein. As noted earlier, there is a need to come up with other seed meals to replace soyabean meal. Some of these oil seed meals with the potential to replace soyabean meal are the Locust Bean seed meal, Baobab seed meal, African Mesquite seed meal and Roselle seed meal.

4.3 Locust bean seed meal

Parkia filicoidea is a leguminous tree found in certain regions of Africa. The meal is obtained by processing the Locust bean seeds, resulting in a protein-rich material. It contains essential amino acids, fibre, vitamins (especially B vitamins), minerals (calcium, phosphorus) and other nutrients.

The protein content of *Parkia filicoidea* seed meal ranges from 25% to 35% (Hassan and Umar, 2005). The amino acid profile of this oilseed includes essential amino acids, with lysine and methionine potentially being limiting (Ari and Ayanwale, 2012). The meal also contains dietary fibre, which can contribute to gut health and digestion. The opportunities for *P. filicoidea* utilization include its wide availability as trees are well-suited to certain African climates, contributing to local sustainability. The high protein content of the seed meal makes it suitable as an alternative protein source, and utilizing *P. filicoidea* seed meal can create economic opportunities in regions where the tree is abundant thus, its agroforestry potential for local economies. The meal contains compounds like tannins and trypsin inhibitors, which can affect nutrient absorption. Processing methods are employed to mitigate these factors.

4.4 Baobab seed meal

Baobab is known for its nutritious fruits and seeds. It is a drought-resistant plant requiring minimal inputs which is not consumed by humans so can be a sustainable source. Baobab seed meal has a protein content ranging from 20% to 30%, making it a suitable potential protein source for monogastric animals (Fatima et al., 2022). It contains a well-balanced amino acid profile, although lysine and methionine may be limiting, requiring supplementation. The baobab seed meal is obtained by removing the seed kernels from the fruit and processing them into a protein-rich meal. The meal is also rich in fibre, vitamins (B-complex and vitamin C), minerals (calcium, potassium and magnesium) and antioxidants. Baobab seed meal can be included in poultry diets as a partial replacement for traditional protein sources with resultant improvement in egg production, weight gain and meat quality (Nxele, 2016). Baobab seeds may contain compounds like tannins and phytates that could limit nutrient absorption. Therefore, processing techniques are required to improve oilseed meal.

4.5 African mesquite seed meal

Prosopis africana, a tree native to Africa, offers potential as an alternative protein source in non-ruminant animal nutrition. Seed meal contains essential amino acids, fibre, vitamins (vitamin B_6), minerals (calcium, potassium) and antioxidants. The protein content ranges from 20% to 28%, and the amino acid profile includes both essential and non-essential amino acids (Akande and Alabi, 2021). *Prosopis africana* trees are drought-resistant and can thrive in arid regions. The utilization of *Prosopis africana* seed meal can contribute to income generation in regions where the tree grows. Amino acid imbalances, especially lysine and methionine are the major setbacks. Similarly, the seed meal may contain compounds like tannins that affect nutrient absorption. Processing methods can mitigate these factors (Yusuf et al., 2013).

4.6 Roselle seed meal

The utilization of Roselle seed meal (*Hibiscus sabdariffa*) as an alternative protein source in the nutrition of non-ruminant animals, especially poultry, is gaining acceptability. The protein content of Roselle seed meal can be close to 40% (Fagbenro et al., 2004). The amino acid profile is well-balanced, although certain amino acids may require supplementation. The meal contains dietary fibre, promoting gut health and digestion. The meal contains essential amino acids, fibre, vitamins (B vitamins and vitamin C), minerals (iron and calcium) and antioxidants. The antinutritional factors associated with Roselle seed meal might contain compounds like tannins that affect nutrient absorption. There are processing methods that can mitigate these factors. Roselle seed meal can be included in poultry diets as a partial protein source with positive effects on growth performance and egg production.

4.7 Algae

Algae are microorganisms rich in protein, omega-3 fatty acids and pigments. The main algae currently used in animal nutrition are *Chlorella vulgaris* (green, single-celled microalgae), *Spirulina platensis* (filamentous and multicellular blue-green algae), *Ulva lactuca* (flat green algae) and *Sargassum* sp (large brown seaweed like algae). The chemical composition and nutritional quality vary with species of algae (Table 2). Algae are harvested, dried, and processed into meal form. Algae offer essential nutrients and bioactive compounds, promoting poultry health and performance. According to Zheng et al. (2012), egg productivity increased from 55.4% in the control group to 59% in 80-week-old Hy-Line layers supplemented with diets containing 2% *Chlorella vulgaris*. The egg yolk colour of 35–43-week-old White Leghorn hens improved after supplementing of diets with 4% or 8% *Sargassum* spp (Carrillo et al., 2012). The addition of algae to broiler diets can also modify the quality of meat.

Table 2 Nutritional composition of algae species

	Chlorella vulgaris	*Spirulina platensis*	*Ulva lactuca*	*Sargassum* sp.
Nutrient composition				
Protein (%)	57	60	19	14
Lipids (%)	20	6	0.3	6
Carbohydrates (%)	23	34	80.7	80
Minerals (mg/kg)				
Na	134.6	189.7	351.7	389.3
K	4.99	132.6	209	244.3
Ca	59.3	88.3	180.7	176
Fe	25.9	10.3	34.5	32.2
Zn	1.2	2.5	1.78	5.8
Cu	0.6	0.32	1.83	1.62
Se	0.7	0.37	1.6	49.8
Mn	0.21	0.38	4.8	3.3

Source: Adapted from: Coudert et al. (2020).

5 Improving the nutritional value of alternative protein sources for poultry: chemical, biological and physical treatments

Feed for poultry can be made from a variety of feed ingredients. However, the nutritional value of these feed ingredients varies. Ingredients have different physicochemical properties, which make some of the nutrients unavailable for use by the birds. Some of the properties of ingredients that prevent utilization of the nutrients are plant fibres, plant secondary metabolites and waxes in plants and insects (Meremikwu et al., 2013; Nascimento et al., 2020, 2021). Plant secondary compounds and waxes prevent the use of several nutrients and minerals (Soetan et al., 2009). In addition, some plant secondary compounds cause toxicity to animals (Tadele, 2015). Over the past 20 years, much work has been done to improve the nutritional value of alternative feed ingredients. The nutritional value of the ingredients can be improved by using methods that fall into three categories: chemical, biological and physical treatment.

5.1 Chemical treatment

Chemical treatments use substances that can alter molecular structures through chemical interactions. The treatment processes involve the use of alkalis such as sodium hydroxide, ammonium hydroxide, and sodium bicarbonate or reducing agents include the use of sodium metabisulfite, metabisulfite, *tris* (2-carboxyethyl-phosphine [TCEP]) (Guerrero-Beltran et al., 2009; Avilés-Gaxiola et al., 2018). Most proteins in plants are prevented from being utilised by animals because of trypsin inhibitors (TIs). TIs inhibit enzyme activity by blocking the active site of the enzymes (Cristina et al., 2019). The N- or C-terminus and the exposed loop of protease inhibitors are frequently considered critical structural features for inhibiting enzyme activity. Acids, bases and reducing agents help by cleaving the disulphide bond and making proteins available for use by poultry. A study by Avilés-Gaxiola et al. (2018) found that the temperature required to deactivate TI in soybeans is lowered from 100 to 90.5°C and 73.9°C, respectively, when cooked in the presence of 1% NaOH or NH_4OH. Likewise, soyabeans in 0.1M $Na_2S_2O_5$ aqueous solution boiled for 2 h at 65°C inactivated trypsin inhibitor by more than 94%. Most leguminous tree species produce seeds that can be used as feed for animals (Mehari et al., 2019). Examples of some of these leguminous trees are *Laecana leucociphala*, *Calliandra*, and *Acacia* spp. The seeds are rich in crude protein and can also be processed the same way as soya beans to make proteins available for use by poultry.

Insect meal is now a popular protein ingredient in some countries. The exoskeleton of insects contains a cuticle consisting of alternate layers of protein and chitin impregnated with calcium carbonate and pigments and interspersed with polyphenols (Muthukrishnan et al., 2020). Sclerotin is the form of protein found in the exoskeleton of insects. It is formed by cross-linking classes of protein molecules. Acids, bases and reducing agents also help break down the complexes and make protein available for use by poultry (Pellis et al., 2022).

Apart from acids and bases, polyethylene glycol (PEG) can also be used to make proteins available for use by animals. Polyethylene glycol is a synthetic polymer poorly digested and absorbed in the intestines of animals (Grosell and Genz, 2006). It has a high affinity with tannins and can bind with tannins to form a tannin–PEG complex (Makkar et al., 1995). A study by Ngwa et al. (2002) reported the use of PEG to improve crude protein digestibility in sheep. Similarly, Xie et al. (2021) found that adding PEG to diets containing high tannins improved CP digestibility. Van Niekerk et al. (2020) found the use of PEG to inactivate tannins and hence improved the utilization of protein and the overall performance of broilers. Polyethylene glycol may, therefore, be used for novel ingredients high in tanniferous compounds when formulating diets for poultry.

5.2 Biological treatment

Different biological methods are used to inactivate trypsin inhibitors and make protein available for animal use. Examples of some of the methods are germination and fermentation (Nkhata et al., 2018). All plants naturally go through the biological process of germination. In this process, the seed emerges from dormancy, where some quantitative and qualitative changes occur. These changes depend on plant species, plant variety and germination conditions. During germination, seed enzymatic systems are activated, protease activity increases, protein nitrogen decreases and peptides, polypeptides and non-protein amino acids increase (Bamdad et al., 2009). The production of proteases is believed to inactivate trypsin inhibitors and make proteins available for use by the animal. A study by Sangronis and Machado (2007) found that germination increased *in vitro* protein digestibility by 2% (white beans), 3% (black beans) and 4%.

Fermentation is another biological method used to preserve feed for livestock. There are so many physicochemical changes that occur during the fermentation process. Proteolysis is a process where proteins are broken down into individual amino acids (Fijałkowska et al., 2015). Animals can then easily assimilate the amino acids. Anti-nutritional factors have also been reported to decrease because of anaerobic fermentation (Kumar et al., 2022). The fermentation process can be improved by additives such as probiotics, urea and molasses to mention a few (Kholif et al., 2022). Enzymes released by the probiotics facilitate the breaking down of fibres and proteins. A plethora of studies have used this method to ensile forage for ruminants (Keady et al., 2013; Evans, 2018; Campbell et al., 2020). Lately, this method has also been used to improve forage to be used by non-ruminant animals. Cassava leaves, a potential alternative protein source, can be processed using this method (Bhavna et al., 2023).

5.3 Physical/mechanical treatment

Physical and/or mechanical treatment methods are used to process feed for animal feeding. The methods have an advantage in that they destroy pathogenic microorganisms, increase the palatability of feed, make the storage easy and safe, increase nutrient content and availability, and change the particle size or density of the feed. Dry heat and hydrothermal treatment are classified under physical/mechanical treatment. The methods reduce anti-nutritional factors in plants, including tannins and trypsin inhibitors, which prevent protein utilization (Ojo, 2022). According to Kadam and Smithard (1987), 80% and 90% of the inhibitors in soybeans are inactivated when boiled for 14 min and for 30 min, respectively. The same methods can be used for treating non-conventional protein sources.

Heat is used to dehydrate the insects, before grinding them into powder and incorporating into feed. Heat is well known for altering the structure of proteins, thereby exposing sites for enzyme action to facilitate protein digestion by animals. Many of these treatments are not practical to use in commercial farming settings if electric power from grid lines is used, as they require large machines for processing. As a result, these treatments are economically unprofitable for large-scale commercial farms as the benefits may be too low or even negative.

6 Environmental sustainability of using alternative protein sources feed for poultry

Feed production has negative and positive effects on the environment, both direct and indirect. For instance, cultivating feed crops adds to the eutrophication and acidification of the environment, primarily due to nitrate leaching into the water and NH_3 emissions into the atmosphere. Chemicals like herbicides, pesticides and fertilisers used during the cultivation of crops all contribute to the pollution of the environment (Baweja et al., 2020). Insect meals, on the other hand, can be reared easily at an economically and environmentally sustainable cost. For example, black soldier fly larvae and maggots can be reared on substrates such as manure and food waste (Makkar et al., 2014). Black soldier fly larvae and maggots can convert organic waste into valuable biomass (van Huis et al., 2013), solving several environmental problems associated with manure and other organic wastes.

Poultry manure management practices can contribute to reducing environmental pollution. The amount of nitrogen excreted depends on the birds' ability to convert feed into energy and the diet's protein content. Reducing the feed-related impacts of poultry production on the environment can be done in two ways. First, improving the feed efficiency, i.e. reducing the amount of feed needed for a particular body weight gain or egg production, would reduce the emissions from feed production and manure management (Oxenboll et al., 2011). Second, selecting feed ingredients with lower environmental impacts during the production stage or more balanced nutrient content should be possible, reducing the excretion of nutrients such as nitrogen and phosphorus (Leinonen and Kyriazakis, 2016). In an ideal situation, diets that fulfil both criteria would be expected to produce a maximal reduction of the environmental impacts. Studies have shown insect meals to be highly digestible. Insect meals (housefly larvae) have up to 98.5% CP digestibility; hence, lower nitrogen losses, and lower eutrophication levels (Hwangbo et al., 2009).

7 Consumer perceptions of feeding animals with alternative feed sources

There has been a growing interest in the consumption of organic food (Frewer *et al.*, 2003). When it comes to meat consumption, consumers would like to know how the animals were reared and how they were fed. Most people are now leaning towards organic feed, which is not linked to some of the chronic ailments in the human population. Feeding chickens leaf meals will likely make poultry meat more acceptable by consumers. The meat is likely to be rejected if there are changes to sensory characteristics of the meat. A plethora of studies have investigated the use of different novel ingredients on the sensory characteristics of meat. Ncube et al. (2018) fed acacia leaves to broiler chickens with no adverse effects on of the organoleptic properties of meat. Similarly, meat quality attributes were not affected by feeding broiler chickens cassava leaves (Bakare et al., 2021). Feeding grasshopper meal had no discernible effects on the sensory panel results for breast and thigh meats (Khan, 2018). There is, therefore, a potential for using alternative plant and insect protein sources.

Human entomophagy, also known as eating insects, is a widespread and culturally ingrained practice of many populations in developing countries in Central and South America, Africa and Asia. Some examples of insects that form a delicacy in these countries include crickets, cicadas, grasshoppers, ants, mealworms (larvae of the darkling beetle) and certain types of caterpillars like mopane worms (Bernard and Womeni, 2017). In countries where insects have long served as traditional foods, they can also be used as feed for livestock when in abundance. For example, in some countries experiencing locust outbreaks, farmers take advantage of this by harvesting the locusts and feeding them to animals (Cerritos and Cano-Santana, 2008). Consumption of products from animals fed insect meals is highly acceptable in these countries. For example, in the Philippines, free-range chickens fed on grasshoppers were found to taste better compared to those fed on feed with conventional protein sources (Khusro et al., 2012). This is not the case in Western societies where entomophagy is not embedded in culinary traditions (Ghosh et al., 2018). Consumption of insects in some of the Western countries is regarded as unacceptable by most of the human populace (Wendin and Nyberg, 2021).

8 Conclusion

Emerging protein sources will play a strong role in poultry feeding in the coming decades. This is mainly due to the result of recurring constraints in the supply of conventional sources due to climate change or political upheavals in major areas of production. It is likely that many of the emerging sources will be produced in commercial quantities in line with the general growth of agriculture in many developing countries. Such a change will have only minimal impact on the prices of conventional sources, which will continue to be used by major producers around the world.

9 References

Akande, K. E. and Alabi, O. J. 2021. The utilization of African mesquite (Prosopis africana) as potential feedstuff for monogastric animals: a review. *Nigerian Journal of Animal Science* 23(1), pp.168–172.

Ari, M. M. and Ayanwale, B. A. 2012. Nutrient retention and serum profile of broilers fed fermented African Locust beans (Parkia filicoide). *Asian Journal of Agricultural Research* 6(3), pp.129–136.

Ari, M. M., Maikaffi, Y. M., Barde, R. E., Aya, V. E. and Ogah, D. M. 2006. Effect of biostatic agents on the short-term preservation of wet rumen content. *PAT* 2(1).

Avilés-Gaxiola, S., Chuck-Hernández, C. and Serna Saldívar, S. O. 2018. Inactivation methods of trypsin inhibitor in legumes: a review. *Journal of Food Science* 83(1), pp.17–29.

Bahadori, Z., Esmaylzadeh, L. and Torshizi, M. A. K. 2021. The effect of earthworm meal on growth performance, immunity, intestinal microbiota, carcass characteristics and meat quality of broiler chickens. *Livestock Science* 202:74–81.

Bakare, A. G., Cawaki, P., Ledua, I., Bautista-Jimenez, V., Kour, G., Sharma, A. C. and Tamani, E. 2021. Quality evaluation of breast meat from chickens fed cassava leaf meal-based diets. *Animal Production Science* 61(6), pp.613–619.

Bamdad, F., Dokhani, S. and Keramat, J. 2009. Functional assessment and subunit constitution of Lentil (lens culinaris) proteins during Germination. *International Journal of Agriculture and Biology* 11(6), pp.690–694.

Baweja, P., Kumar, S. and Kumar, G. 2020. Fertilizers and pesticides: -their impact on soil health and environment. In Giri, B. and Varma, A. (eds). Soil health, Springer Nature, Geneva, Switzerland (pp.265–285). Cham: Springer.

Bernard, T. and Womeni, H. M. 2017. Entomophagy: insects as food. *Insect Physiology and Ecology* 2017, pp.233–249.

Beski, S. S. M., Swick, R. A. and Iji, P. A. 2015. The effect of the concentration and feeding duration of spray-dried plasma protein on growth performance, digestive enzyme activities, nutrient digestibility and intestinal mucosal development of broiler chickens. *Animal Production Science* 56(11), pp.1820–1827.

Bhavna, A., Zindove, T. J., Iji, P. A. and Bakare, A. G. 2023. Growth performance, carcass characteristics and meat sensory evaluation of broiler chickens fed diets with fermented cassava leaves. *Animal Bioscience*. https://doi.org/10.5713/ab.23.0362.

Campbell, M., Ortuño, J., Ford, L., Davies, D. R., Koidis, A., Walsh, P. J. and Theodoridou, K. 2020. The effect of ensiling on the nutritional composition and fermentation characteristics of brown seaweeds as a ruminant feed ingredient. *Animals* 10(6), p.1019.

Carrillo, S., Bahena, A., Casas, M., Carranco, M. E. and Calvo, C. C. 2012. The alga Sargassum spp. as alternative to reduce egg cholesterol content. *Cuban Journal of Agricultural Science* 46(2), pp.1–6.

Cayot, N., Cayot, P., Bou-Maroun, E., Laboure, H., Abad-Romero, B., Pernin, K., Seller-Alvarez, N., Hernández, A. V., Marquez, E. and Medina, A. L. 2009. Physico-chemical characterisation of a non-conventional food protein source from earthworms and sensory impact in arepas. *International Journal of Food Science & Technology* 44(11), pp.2303–2313.

Cerritos, R. and Cano-Santana, Z. 2008. Harvesting grasshoppers *Sphenarium purpurascens* in Mexico for human consumption: a comparison with insecticidal control for managing pest outbreaks. *Crop Protection* 27(3–5), pp.473–480.

Chadare, F. J., Linnemann, A. R., Hounhouigan, J. D., Nout, M. J. R. and Van Boekel, M. A. J. S. 2008. Baobab food products: a review on their composition and nutritional value. *Critical Reviews in Food Science and Nutrition* 49(3), pp.254–274.

Cherdthong, A. 2020. Potential use of rumen digesta as ruminant diet–a review. *Tropical Animal Health and Production* 52(1), pp.1–6.

Coudert, E., Baéza, E. and Berri, C. 2020. Use of algae in poultry production: a review. *World's Poultry Science Journal* 76(4), pp.767–786.

Cristina Oliveira de Lima, V., Piuvezam, G., Leal Lima Maciel, B. and Heloneida de Araújo Morais, A. 2019. Trypsin inhibitors: promising candidate satietogenic proteins as complementary treatment for obesity and metabolic disorders? *Journal of Enzyme Inhibition and Medicinal Chemistry* 34(1), pp.405–419.

Cromwell, G. L. 2006. Rendered products in swine nutrition. In D. L. Meeker (ed.), *Essential Rendering*, p.141. Alexandria, VA: Renderers Association.

Elfaki, M. O. and Abdelatti, K. A. 2016. Rumen content as animal feed: a review. *Journal of Veterinary Medicine and Animal Production* 7(2), pp.80–88.

Evans, B. 2018. The role ensiled forage has on methane production in the rumen. *Animal Husbandry, Dairy and Veterinary Science* 2(4), pp.1–4..

Fagbenro, O. A., Akande, T. T., Fapohunda, O. O. and Akegbejo-Samsons, Y. 2004. Comparative assessment of roselle (Hibiscus sabdariffa var. sabdariffa) seed meal and kenaf (Hibiscus sabdariffa var. altissima) seed meal as replacement for soybean meal in practical diets for fingerlings of Nile tilapia Oreochromis niloticus. In *6th International Symposium on Tilapia in Aquaculture*, pp. 277–287. Manila, Philippines: Philippine International Convention Center Roxas Boulevard.

Fatima, G. A., Hamdon, A. A. and Mekki, D. M. 2022. Impacts of feeding graded levels of baobab seed meal on growth performance and feed utilization of broiler chicks. *Acta Sci. Pharma.* 3(10): 2–7.

Fijałkowska, M., Pysera, B., Lipiński, K. and Strusińska, D. 2015. Changes of nitrogen compounds during ensiling of high protein herbages–a review. *Annals of Animal Science* 15(2), pp.289–305.

Fraanje, W. and Garnett, T. 2020. *Soy: food, feed and land use change* (Foodsource: Building Blocks). Oxford, UK: Food Climate Research Network, University of Oxford.

Frewer, L., Scholderer, J. and Lambert, N. 2003. Consumer acceptance of functional foods: issues for the future. *British Food Journal* 105(10), pp.714–731.

Gasco, L., Renna, M., Bellezza Oddon, S., Rezaei Far, A., Naser El Deen, S. and Veldkamp, T. 2023. Insect meals in a circular economy and applications in monogastric diets. *Animal Frontiers* 13(4), pp.81–90.

Ghosh, S., Jung, C. and Meyer-Rochow, V. B. 2018. What governs selection and acceptance of edible insect species? In A. Halloran et al. (eds.), *Edible insects in sustainable food systems*, pp.331–351. Cham: Springer Publication.

Grosell, M. and Genz, J. 2006. Ouabain-sensitive bicarbonate secretion and acid absorption by the marine teleost fish intestine play a role in osmoregulation. *American Journal of Physiology-Regulatory, Integrative and Comparative Physiology* 291(4), pp.R1145–R1156.

Guerrero-Beltrán, J. A., Estrada-Girón, Y., Swanson, B. G. and Barbosa-Cánovas, G. V. 2009. Pressure and temperature combination for inactivation of soymilk trypsin inhibitors. *Food Chemistry* 116(3), pp.676–679.

Hardy, R. W. and Kaushik, S. J. eds. 2021. *Fish nutrition*. San Diego, CA: Academic Press.

Hassan, L. G. and Umar, K. J. 2005. Protein and amino acids composition of African locust bean (Parkia biglobosa). *Tropical and Subtropical Agroecosystems* 5(1), pp.45–50.

Hwangbo, J., Hong, E. C., Jang, A., Kang, H. K., Oh, J. S., Kim, B. W. and Park, B. S. 2009. Utilization of house fly-maggots, a feed supplement in the production of broiler chickens. *Journal of Environmental Biology* 30(4), pp.609–614.

Ipema, A. F., Gerrits, W. J., Bokkers, E. A., Kemp, B. and Bolhuis, J. E. 2020. Provisioning of live black soldier fly larvae (Hermetia illucens) benefits broiler activity and leg health in a frequency-and dose-dependent manner. *Applied Animal Behaviour Science* 230, p.105082.

Jedrejek, D. et al. 2016. Animal by-products for feed: characteristics, European regulatory framework, and potential impacts on human and animal health and the environment. *Journal of Animal and Feed Sciences* 25, pp.189–202.

Kadam, S. S. and Smithard, R. R. 1987. Effects of heat treatments on trypsin inhibitor and hemagglutinating activities in winged bean. *Plant Foods for Human Nutrition* 37, pp.151–159.

Keady, T., Hanrahan, S., Marley, C. and Scollan, N. D. 2013. Production and utilization of ensiled forages by beef cattle, dairy cows, pregnant ewes and finishing lambs-A review. *Agricultural and Food Science* 22(1), pp.70–92.

Khan, S., Naz, S., Sultan, A., Alhidary, I. A., Abdelrahman, M. M., Khan, R. U., Khan, N. A., Khan, M. A. and Ahmad, S. 2016. Worm meal: a potential source of alternative protein in poultry feed. *World's Poultry Science Journal* 72(1), pp.93–102.

Khan, S. H. 2018. Recent advances in role of insects as alternative protein source in poultry nutrition. *Journal of Applied Animal Research* 46(1), pp.1144–1157.

Kholif, A. E., Hamdon, H. A., Gouda, G. A., Kassab, A. Y., Morsy, T. A. and Patra, A. K. 2022. Feeding date-palm leaves ensiled with fibrolytic enzymes or multi-species probiotics to Farafra ewes: Intake, digestibility, ruminal fermentation, blood chemistry, milk production and milk fatty acid profile. *Animals* 12(9), p.1107.

Khusro, M., Andrew, N. R. and Nicholas, A. 2012. Insects as poultry feed: a scoping study for poultry production systems in Australia. *World's Poultry Science Journal* 68(3), pp.435–446.

Kouřimská, L. and Adámková, A. 2016. Nutritional and sensory quality of edible insects. *NFS Journal* 4, pp.22–26.

Kuan, D. H., Wu, C. C., Su, W. Y. and Huang, N. T. 2018. A microfluidic device for simultaneous extraction of plasma, red blood cells, and on-chip white blood cell trapping. *Scientific Reports*, 8(1), p.15345.

Kumar, Y., Basu, S., Goswami, D., Devi, M., Shivhare, U. S. and Vishwakarma, R. K. 2022. Anti-nutritional compounds in pulses: implications and alleviation methods. *Legume Science* 4(2), p.e111.

Kumlu, M., Beksari, A., Yilmaz, H. A., Sariipek, M., Kinay, E., Turchini, G. and Eroldogan, O. T. 2021. n-3 LC-PUFA enrichment protocol for red earthworm, Eisenia fetida: a cheap and sustainable method. *Turkish Journal of Fisheries and Aquatic Sciences*. https://doi.org/10.4194/1303-2712-v21_7_03.

Leinonen, I. and Kyriazakis, I. 2016. How can we improve the environmental sustainability of poultry production? *Proceedings of the Nutrition Society* 75(3), pp.265–273.

Liu, J. K., Waibel, P. E. and Noll, S. L. 1989. Nutritional evaluation of blood meal and feather meal for turkeys. *Poultry Science* 68(11), pp.1513–1518.

Makara, A., Kowalski, Z., Fela, K. and Generowicz, A. 2016. Utilization of animal blood plasma as example of using cleaner technologies methodology. *Technical Transactions/ Czasopismo Techniczne* 11. https://doi.org/10.4467/2353737XCT.16.197.5946.

Makkar, H. P., Blümmel, M. and Becker, K. 1995. In vitro effects of and interactions between tannins and saponins and fate of tannins in the rumen. *Journal of the Science of Food and Agriculture* 69(4), pp.481–493.

Makkar, H. P., Tran, G., Heuzé, V. and Ankers, P. 2014. State-of-the-art on use of insects as animal feed. *Animal Feed Science and Technology* 197, pp.1–33.

Mehari, R., Shumuye, B., Kinfe, M. and Adugna, G. 2019. Performance of egg-laying hens fed Acacia saligna seed meal. Livestock Research for Rural Development 31: Article 27.

Meremikwu, V. N., Ibekwe, H. A. and Essien, A. 2013. Improving broiler performance in the tropics using quantitative nutrition. *World's Poultry Science Journal* 69(3), pp.633–638.

Moyo, S. and Moyo, B. 2022. Potential utilization of insect meal as livestock feed. In *Animal feed science and nutrition-production, health and environment.* IntechOpen.

Musa-Azara, S. I., Jibrin, M., Ari, M. M., Hassan, D. I. and Ogah, D. M. 2014. Effects of Moringa oleifera Linn seed administration on sperm production rate and gonadal sperm reserve in rabbits. *British Biotechnology Journal* 4(7), p.801.

Musa-Azara, S. I., Ogah, D. M., Yakubu, A., Ari, M. M. and Hassan, D. I. 2013. Effects of Hibiscus calyx extracts on the blood chemistry of broiler chickens. *Egyptian Poultry Science* 33(1), pp.309–312.

Muthukrishnan, S., Mun, S., Noh, M. Y. Geisbrecht, E. R. and Arakane, Y. 2020. Insect cuticular chitin contributes to form and function. *Current Pharmaceutical Design* 26(29), pp.3530–3545.

Nascimento, M. Q. D., Gous, R. M., Reis, M. P., Viana, G. S., Nogueira, B. R. F. and Sakomura, N. K. 2021. Gut capacity of broiler breeder hens. *British Poultry Science* 62(5), pp.710–716.

Nascimento, M. Q. D., Gous, R. M., Reis, M. D. P., Fernandes, J. B. K. and Sakomura, N. 2020. Prediction of maximum scaled feed intake in broiler chickens based on physical properties of bulky feeds. *British Poultry Science* 61(6), pp.676–683.

Ncube, S., Halimani, T. E., Chikosi, E. V. I. and Saidi, P. T. 2018. Effect of Acacia angustissima leaf meal on performance, yield of carcass components and meat quality of broilers. *South African Journal of Animal Science* 48(2), pp.271–283.

Ngwa, A. T., Nsahlai, I. V. and Iji, P. A. 2002. Effect of supplementing veld hay with a dry meal or silage from pods of Acacia sieberiana with or without wheat bran on voluntary intake, digestibility, excretion of purine derivatives, nitrogen utilization, and weight gain in South African Merino sheep. *Livestock Production Science* 77(2–3), pp.253–264.

Nkhata, S. G., Ayua, E., Kamau, E. H. and Shingiro, J. B. 2018. Fermentation and Germination improve nutritional value of cereals and legumes through activation of endogenous enzymes. *Food Science & Nutrition* 6(8), pp.2446–2458.

Nxele, S. K. 2016. *The Potential Use of Baobab (Adansonia Digitata) Seedcake as a Commercial Diet Replacement and in Feed Formulation for Broiler Chickens* (Doctoral dissertation). University of Fort Hare.

Ojo, M. A. 2022. Tannins in foods: nutritional implications and processing effects of hydrothermal techniques on legume seeds. *Prev. Nutr. Food Sci.* 27(1): 14–19.

Øverland, M., Mydland, L. M. and Skrede, A. 2017. Marine macroalgae as a source of protein and bioactive compounds in feed for non-ruminant animals. *Journal of the Science of Food and Agriculture* 99(1). https://doi.org/10.1002/jsfa.9143.

Oxenboll, K. M., Pontoppidan, K. and Fru-Nji, F. 2011. Use of a protease in poultry feed offers promising environmental benefits. *International Journal of Poultry Science* 10(11), pp.842–848.

Pellis, A., Guebitz, G. M. and Nyanhongo, G. S. 2022. Chitosan: sources, processing and modification techniques. *Gels*,8(7), p.393.

Pereira, A. G., Fraga-Corral, M., Garcia-Oliveira, P., Otero, P., Soria-Lopez, A., Cassani, L., Cao, H., Xiao, J., Prieto, M. A. and Simal-Gandara, J. 2022. Single-cell proteins obtained by circular economy intended as a feed ingredient in aquaculture. *Foods* 11(18), p.2831.

Pérez-Corría, K., Botello-León, A., Mauro-Félix, A. K., Rivera-Pineda, F., Viana, M. T., Cuello-Pérez, M., Botello-Rodríguez, A. and Martínez-Aguilar, Y. 2019. Chemical composition of earthworm (Eisenia foetida) co-dried with vegetable meals as an animal feed. *Ciencia y Agricultura* 16(2), pp.79–92.

Sajid, Q. U. A., Asghar, M. U., Tariq, H., Wilk, M. and Płatek, A. 2023. Insect meal as an alternative to protein concentrates in poultry nutrition with future perspectives (An updated review). *Agriculture* 13(6), p.1239.

Sangronis, E. and Machado, C. J. 2007. Influence of Germination on the nutritional quality of Phaseolus vulgaris and Cajanus cajan. *LWT-Food Science and Technology* 40(1), pp.116–120.

Soetan, K. O. and Oyewole, O. E. 2009. The need for adequate processing to reduce the anti-nutritional factors in plants used as human foods and animal feeds: a review. *African Journal of food Science* 3(9), pp.223–232.

Spranghers, T., Ottoboni, M., Klootwijk, C., Ovyn, A., Deboosere, S., De Meulenaer, B., Michiels, J., Eeckhout, M., De Clercq, P. and De Smet, S. 2017. Nutritional composition of black soldier fly (Hermetia illucens) prepupae reared on different organic waste substrates. *Journal of the Science of Food and Agriculture* 97(8), pp.2594–2600.

Tadele, Y. 2015. Important anti-nutritional substances and inherent toxicants of feeds. *Food Science and Quality Management* 36, pp.40–47.

Van Huis, A., Van Itterbeeck, J., Klunder, H., Mertens, E., Halloran, A., Muir, G. and Vantomme, P. 2013. *Edible insects: future prospects for food and feed security* (No. 171). Rome: Food and Agriculture Organization of the United Nations.

Van Niekerk, R. F., Mnisi, C. M. and Mlambo, V. 2020. Polyethylene glycol inactivates red grape pomace condensed tannins for broiler chickens. *British Poultry Science* 61(5), pp.566–573.

Wendin, K. M. and Nyberg, M. E. 2021. Factors influencing consumer perception and acceptability of insect-based foods. *Current Opinion in Food Science* 40, pp.67–71.

Xie, B. et al. 2021. Adding polyethylene glycol to steer rations containing sorghum tannins increases crude protein digestibility. *Animal Nutrition* 7(3): 779–786.

Yadav, L. P., Gangadhara, K., Apparao, V. V., Singh, A. K., Rane, J., Kaushik, P., Sekhawat, N., Malhotra, S. K., Rai, A. K., Yadav, S. L. and Berwal, M. K. 2024. Nutritional, antioxidants and protein profiling of leaves of Moringa oleifera germplasm. *South African Journal of Botany* 165, pp.443–454.

Yusuf, N. D., Ogah, D. M., Hassan, D. I., Musa, M. M., Ari, M. M. and Doma, U. D. 2013. Carcass Evaluation of Broilers fed on Decorticated Fermented Prosopis africana G. seed meal. *Macedonian Journal of Animal Science* 3(1), pp.45–48.

Zheng, L. et al. 2012. The dietary effects of fermented *Chlorella vulgaris* on production performance, liver lipids and intestinal microflora in laying hens. *Asian-Austral. J. Anim. Sci.* 25(2): 261–266.

Chapter 15

The use of phytogenics in optimizing gut function and health in poultry

Konstantinos C. Mountzouris and Ioannis Brouklogiannis, Agricultural University of Athens, Greece

1 Introduction

This chapter begins by reviewing the main types of phytogenic compounds (terpenoids, flavonoids, and alkaloids), as well as general issues relating to phytogenic variability and bioavailability. The following sections review relevant aspects of poultry gut function, particularly critical signaling pathways and gene components that are directly linked to optimal gut function and health. The chapter will then discuss the beneficial effects of phytogenics in the maintenance and protection of gut function and health, focusing particularly on gut barrier integrity, cytoprotection, and inflammatory response.

2 Phytogenic compounds: terpenoids, flavonoids, and alkaloids

Phytogenics (also referred to as phytochemicals, phytobiotics, botanicals, or plant extracts) are plant-derived bioactive components that include the utilized parts of aromatic plants, herbs, spices, fruits, and/or their respective extracts such as essential oils and oleoresins (Hippenstiel et al., 2011; Diaz-Sanchez et al., 2015; Shehata et al., 2022). According to the current European Union regulation

http://dx.doi.org/10.19103/AS.2024.0143.20

(Regulation (EC) No 1831/2003), the majority of phytogenics are categorized as flavoring feed additives, while only a limited number of products have been classified so far as zootechnical feed additives. Phytogenics can be divided into various classes based on their botanical origin, extraction process, biological properties, and chemical structures. Based on chemical structure, some of the main groups are terpenoids, flavonoids, and alkaloids (Diaz-Sanchez et al., 2015; Panche et al., 2016; Valenzuela-Grijalva et al., 2017; Debnath et al., 2018).

2.1 Terpenoids

Terpenoids (or terpenes) consist of five carbon building blocks known as isoprene units (C5). Depending on the number of isoprene units in their structure, terpenoids are sub-classified into monoterpenoids (C10), sesquiterpenoids (C15), diterpenoids (C20), triterpenoids (C25), and tetraterpenoids (C30) (Brenes and Roura, 2010; Diaz-Sanchez et al., 2015; Valenzuela-Grijalva et al., 2017). Terpenes are the largest and most diverse group among phytogenic compounds with multiple properties, including anti-inflammatory, antimicrobial, antioxidant, metabolic-modulating effects, and beneficial effects on gut function and the overall performance of poultry (Gholami-Ahangaran et al., 2022; Basiouni et al., 2023). Examples of important terpenes include thymol, carvacrol, limonene, and menthol, which are among the most frequently studied compounds in poultry nutrition (Paraskeuas and Mountzouris, 2019; Mountzouris et al., 2020; Wang et al., 2022a; Gholami-Ahangaran et al., 2022).

2.2 Flavonoids

Flavonoids are an important class of phytogenic compounds, characterized by water-soluble, low molecular-weight phenolic compounds (Panche et al., 2016). They include flavones (e.g. apigenin and luteolin), flavonols (e.g. quercetin, kaempferol, myricetin, and fisetin), flavanones (e.g. hesperetin and naringenin), isoflavones (e.g. genistein), anthocyanins, and chalcones. Various phytogenic substances, including flavonoids, are associated with an array of health-promoting effects. In addition to their anti-inflammatory and antioxidant properties, they also have antimutagenic and anticarcinogenic actions (Diaz-Sanchez et al., 2015; Panche et al., 2016; Valenzuela-Grijalva et al., 2017; Shehata et al., 2022).

2.3 Alkaloids

Alkaloids represent one of the largest groups of nitrogen-containing compounds, biosynthesized from amino acid precursors or occasionally from purine bases. They play a crucial role in plant defense against herbivores and other threats. As feed additives, alkaloids possess multiple biological activities that potentially promote poultry health and performance. More specifically, isoquinoline alkaloids (IQs), including quaternary-benzo(c)phenanthridine alkaloids (e.g. chelerythrine and sanguinarine) and protopine alkaloids (e.g. protopine and allocryptopine) sourced from plant materials, e.g. *Macleaya cordata*, have been reported to enhance growth performance (Khadem et al., 2014; Kikusato et al., 2021) and beneficially modulate intestinal barrier function (Song et al., 2023), intestinal inflammation (Khadem et al., 2014; Kikusato et al., 2021; Song et al., 2023), gut microbiota composition (Song et al., 2023), and antioxidant capacity of broiler chickens (Kikusato et al., 2021).

3 Phytogenic variability and bioavailability

As naturally occurring plant components, phytogenics are subject to large variability in their chemical composition and concentration due to an array of factors such as the geographical origin of plants, climatic conditions, cultural techniques, harvesting time, state of plant maturity, and extraction/processing techniques. In addition, storage conditions and duration amplify naturally occurring compositional variation (Brenes and Roura, 2010; Valenzuela-Grijalva et al., 2017; Stevanović et al., 2018; Basiouni et al., 2023). Furthermore, possible synergistic or antagonistic effects of phytogenic compounds may also strongly affect overall phytogenic efficacy with respect to poultry performance and health (Stevanović et al., 2018; Basiouni et al., 2023). Despite significant advancements in phytogenic characterization and quantification techniques, there is still a need for further development with respect to standardization, protection during storage, feed processing, and targeted release in the gut so as to guarantee the benefits and reproducibility of phytogenic applications in poultry and other livestock (Hippenstiel et al., 2011; Diaz-Sanchez et al., 2015; Valenzuela-Grijalva et al., 2017; Stevanović et al., 2018; Shehata et al., 2022).

Despite the large number of studies of phytogenics in poultry, there is limited information on their absorption and metabolism kinetics in poultry. Absorption in the proximal gut has been reported in the cases of carvacrol and thymol (Michiels et al., 2008; Brenes and Roura, 2010; Oceľová et al., 2019), while absorption in the large intestine has been reported in the case of quercetin (Aura et al., 2022; Suzuki and Hara, 2011). Limited evidence on the fate of absorbed phytogenics in the proximal gut suggests that phytogenic components are metabolized and/or eliminated by the kidneys without significant deposition in the body (Lee et al., 2004; Michiels et al., 2008). Thymol partitioning in the blood and duodenal wall was shown to be dose–response related (Oceľová et al., 2019). A phytogenic dose–response nutrigenomic study profiling the chicken gut has shown that the duodenum and the ceca were the most responsive intestinal sites to modulation by phytogenic inclusion, followed by the jejunum and then by the ileum (Mountzouris et al., 2020). In summary, phytogenic-specific absorption and metabolism kinetics have to account for primary intestinal sites of phytogenic absorption, enterohepatic circulation, and the fact that, in birds, urinary excreted phytogenic components and/or metabolites may get re-introduced in the ceca via the known reflux of uric substances in birds (Sacranie et al., 2012; Mountzouris et al., 2020).

4 Gut function and health

Gut health is a complicated area linked with zootechnical performance responses (Yegani and Korver, 2008; Kogut et al., 2017; Ducatelle et al., 2023). Gut health can be defined as the ability of the gut to perform physiological functions and maintain homeostasis, thereby supporting its ability to withstand stressors (Kogut et al., 2017). Gut health can be seen as a binary state in terms of the presence or absence of disease. However, the occurrence of a 'pre-disease state', as an intermediate state in which organisms, organs, and tissues are extremely vulnerable to homeostasis disruption, adds to the complexity of the topic (Zeineldin et al., 2019; Koizumi et al., 2019).

Overall, homeostasis maintenance depends on an array of essential functions that occur in the gut, such as nutrient absorption, physical and immune barrier function to pathogens, and the cross-talk between gut microbiota and the host (Pan and Yu, 2013; Shini and Bryden, 2022; Wang et al., 2022b; Ducatelle et al., 2023). Some of the key homeostasis components that compose and regulate gut function and health are (Vomund et al., 2017; Lingappan, 2018; Paraskeuas and Mountzouris, 2019; Mountzouris et al., 2020):

- gut barrier integrity;
- adaptive cytoprotection; and
- management of inflammation.

The following sections review critical signaling pathways and gene components that are directly linked to optimal gut function and health. The chapter will then discuss the beneficial effects of phytogenics in the maintenance and protection of gut function and health.

4.1 Gut barrier integrity

As the gut is a very dynamic environment, continuous interactions take place between diet, the birds' immune system, the gut microbiota, and gut integrity (Yegani and Korver, 2008; Paraskeuas and Mountzouris, 2019; Shini and Bryden, 2022; Ducatelle et al., 2023). Gut integrity, in particular, is determined by the junctional complexes between enterocytes (Fig. 1). The epithelial cells are

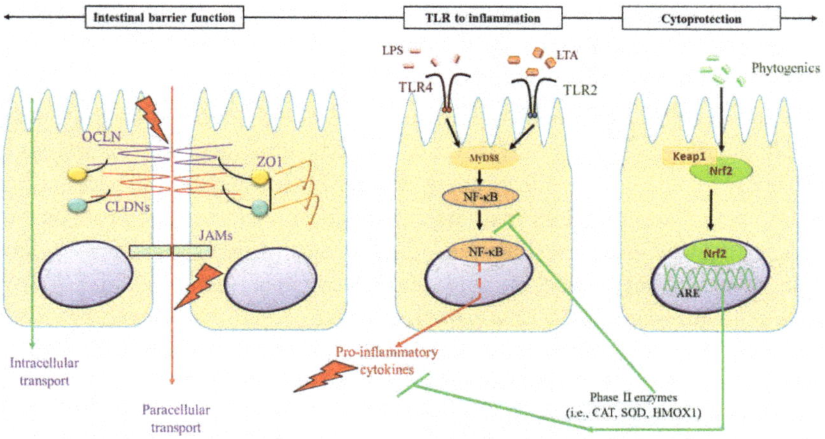

Figure 1 Schematic representation of the main key pathways/components that regulate gut function and health. Enterocytes are connected by tight junction proteins that compose a dynamic complex consisting of integral membrane proteins OCLN, CLDNs, and peripheral membrane proteins such as ZO1, that act as a barrier and regulate paracellular permeability. Recognition of LTA and LPS by TLR2 and TLR4, respectively, stimulates the recruitment of intracellular adaptors like MyD88 (MyD88-dependent pathway). Recruitment of MyD88 causes the nuclear translocation of the transcription factor NF-κB, that regulates the transcription of mainly pro-inflammatory cytokines, that may disrupt the physiological TJ structure and composition. Potential inducers such as phytogenics cause the disruption of Nrf2 with Keap1 and allow Nrf2 to translocate into the nucleus, where it binds to the ARE, leading to the transcription of cytoprotective genes known as phase II enzymes. These antioxidant enzymes (e.g. CAT, SOD, and HMOX1) block the activation of NF-κB and subsequent pro-inflammatory cytokines. OCLN, occludin; CLDNs, claudins; ZO1, zonula occluden 1; JAMs, junctional adhesion molecules; LTA, lipoteichoic acid; LPS, lipopolysaccharide; TLR2, toll-like receptor 2; TLR4, toll-like receptor 4; MyD88, myeloid differentiation primary response 88; NFκB, nuclear factor kappa B; Nrf2, nuclear factor erythroid-2-like2; Keap1, kelch-like ECH-associated protein 1; ARE, antioxidant response element; CAT, catalase; SOD1, superoxide dismutase; HMOX1, heme oxygenase 1.

tightly bound together by intercellular junctions (i.e. tight junctions, adherens junctions and desmosomes) that regulate the passage of ions, nutrients, and water through the paracellular pathway (Kogut et al., 2017; Awad et al., 2017; Cheng et al., 2019; Shini and Bryden, 2022). In particular, tight junction proteins compose a dynamic complex consisting of transmembrane proteins such as claudins (e.g. CLDN1, CLDN2, and CLDN3), occludin (OCLN), junctional adhesion molecule-A (JAM-A), and intracellular proteins, such as Zonula occludens (i.e. ZO1 and ZO2), that act as a barrier and regulate paracellular permeability (Suzuki, 2013; Awad et al., 2017).

4.2 Cytoprotection

In large-scale intensive poultry farming, birds face various stressor challenges, including nutritional, biological, and environmental (e.g. xenobiotics, pathogens, and heat), that may cause excessive production of reactive oxygen species (ROS) (Mountzouris et al., 2020; Basiouni et al., 2023; Du et al., 2023). ROS function in cells as signaling molecules and play a crucial role in maintaining cellular homeostasis. However, when ROS overwhelm the cellular antioxidant defense system, a dysregulation of the intracellular redox status can occur (Ray et al., 2012; Mountzouris et al., 2022). Subsequently, stressors may cause direct or indirect ROS-mediated damage of nucleic acids, proteins, and lipids, negatively impacting birds' productive and reproductive performance, health, and product quality (Ray et al., 2012; Sahin et al., 2013; Lauridsen, 2019; Brouklogiannis et al., 2023; Zhang et al., 2023a).

Birds may counterbalance oxidative stress via direct and indirect mechanisms. The first involves the immediate engagement of dietary nonenzymatic antioxidant compounds such as vitamins (e.g. vitamins C and E), phytochemicals, trace minerals (e.g. copper, zinc, and selenium), carotenoids, and cofactors such as folic and uric acids (Mountzouris et al., 2022). The indirect mechanism protects against oxidative stress by inducing gene expression of cytoprotective enzymes with detoxifying, antioxidant, and anti-inflammatory functions. Enhancing cellular antioxidant capacity by upregulating cytoprotective genes is critical in cellular protection against oxidative damage (Köhle and Bock, 2006; Mountzouris et al., 2022; Basiouni et al., 2023). More specifically, the signaling pathway referred to as nuclear factor-erythroid-derived 2-like 2 (Nrf2), from the transcription factor that activates it, mediates the inducible cellular defense against oxidation and overall cytoprotection (Fig. 1). The Nrf2 is a critical transcription factor, widely regarded as the master regulator of cellular antioxidant response and inflammation, which regulates the expression of antioxidant and cytoprotective genes known as phase II enzymes (Köhle and Bock, 2006; Ray et al., 2012; Stefanson and Bakovic, 2014; Vomund et al., 2017).

4.3 Toll-like receptor signaling to inflammation

In addition to its role in nutrient absorption and as a critical barrier to pathogens, the avian gut is the most complex component of the immune system. It is densely populated by a vast variety of microorganisms, with extensive interactions of this intestinal microbiome with the poultry host and diet (Pan and Yu, 2013; Kogut et al., 2018; Chen et al., 2021a). Immune homeostatic balance is therefore critical for gut health and is under threat during infections and stressor challenges. However, to counteract these threats and facilitate host homeostasis, the innate immune system has developed a coordinated adaptable strategy (i.e. inflammation), which allows for an immediate response in case of an arising threat (Kogut et al., 2018; Jiang et al., 2019; Shini and Bryden, 2022). In this sense, innate immune receptors on immune and intestinal cells recognize and detect infection and danger-signaling molecules. More specifically, these cells express pattern recognition receptors (PRRs), including toll-like receptors (TLRs) and Nod-like receptors (NLRs), located on their surface, which have a crucial role in the restoration of homeostasis (Yegani and Korver, 2008; Kogut, 2013; Shini and Bryden, 2022; Ducatelle et al., 2023).

In particular, TLRs are a group of PRRs that are triggered by the presence of microbial structures, microbe-associated molecular patterns, and excess nutrients and metabolites (Zhang and Ghosh, 2001; Kogut, 2013; Paraskeuas and Mountzouris, 2019). Subsequently, TLR signaling results in the activation of the transcription factor nuclear factor-kappa B (NF-κB), which controls the expression of several pro- (i.e. IL-1β, IL-6 and TNF-α) and anti- (i.e., TGF-β and IL-10) inflammatory cytokines, as well as factors (i.e. COX2 and iNOS) related to the innate immune system and inflammatory response (Zhang and Ghosh, 2001; Lingappan, 2018; Paraskeuas and Mountzouris, 2019; Chen et al., 2021b).

All of the pathways mentioned above are essential for gut health. The following section reviews relevant research on how phytogenics affect these components in the poultry intestine.

5 Effects of phytogenics on gut function and health: gut barrier integrity, cytoprotection, and inflammatory response

During the last two decades, dietary phytogenics have gained increasing research and commercial interest as feed additives in poultry nutrition. Ongoing research highlights the beneficial potential of dietary phytogenics in enhancing poultry gut health. Table 1 presents selected examples of dietary phytogenics and their beneficial role in modulating gut function (gut barrier integrity, cytoprotection, and inflammation) markers.

5.1 Phytogenic effects on gut barrier integrity

Phytogenics as feed additives have the potential to improve gut barrier function directly and indirectly. The first involves the promotion of TJ protein assembly that may result in more effective protection against toxic feed-derived molecules or endogenously produced toxic metabolites. The indirect mechanism refers to the modulation of signaling pathways (e.g. TLR and NF-κB) in which inflammatory cytokines are produced (Paraskeuas and Mountzouris, 2019). In laying hens, different types of dietary-supplemented phytochemicals have been shown to increase the expression of critical TJ proteins (Feng et al., 2021; Tian et al., 2022; Wang et al., 2022a; Anagnostopoulos et al., 2023). The positive modulatory influence of phytogenics on TJ expression was also

Table 1 Dietary phytogenic effects on gut barrier, cytoprotection, and inflammation-related genes in poultry

			Gut function			
Phytogenic	Species	Tissue/ Mucosa	Barrier integrity	Cytoprotection	Inflammatory response	References
Oregano Essential Oil (EO)	Laying hens	Ileal mucosa	ZO-1 ↑	–	TLR4, IL-1β, IFN-γ ↓	Feng et al. (2021)
Carvacrol and cinnamaldehyde EO Thymus and cinnamaldehyde EO	Laying hens	Duodenal mucosa	OCLN ↑	–	–	Wang et al. (2022a)
Montmorillonite and essential oils (eugenol and cinnamaldehyde)	Laying hens	Ileal mucosa	–	–	NF-Kb, MyD88, IL-1β ↓	Chen et al. (2020)
Premix with ginger, lemon balm, oregano and thyme	Laying hens	Duodenum	–	Nrf2, CAT, SOD, GPX7, GSR, PRDX1, HO-1 ↑	–	Anagnostopoulos et al. (2023)
Fermented Chinese medicinal	Laying hens	Jejunum	OCLN, ZO-1 ↑	–	–	Tian et al. (2022)
Blend with carvacrol, anethol, and limonene	Broilers	Ileum mucosa Cecal mucosa	ZO-2↑ ZO-2, CLDN5, OCLN↑	–	TLR2↓ TLR2↓	Paraskeuas and Mountzouris (2019)
Encapsulated carvacrol, thymol, D-carvone, methyl salicylate and L-menthol	Broilers	Duodenum Ceca	–	CAT, SOD, NQO1, GSR, PRDX1, HO-1 ↑ GST ↑	TLR2, MyD88, IL8↓ TLR3, MyD88↓	Griela and Mountzouris (2023)

(*Continued*)

Table 1 (Continued)

Phytogenic	Species	Tissue/ Mucosa	Gut function: Barrier integrity	Gut function: Cytoprotection	Gut function: Inflammatory response	References
Blend of extract powders of curcuma roots, chamomile flowers, licorice roots, and olive leaf.	Broilers	Jejunal mucosa	OCLN, JAM-2 ↑	–	–	Farahat et al. (2021)
Encapsulated thymol and carvacrol	Broilers	Jejunal mucosa	OCLN, ZO-1 ↑	Nrf2, GSH-Px ↑	NF-Kb, IL-1β, TNF-α ↓	Li et al. (2022)
Camellia oleifera extract	Broilers	Ileal mucosa	OCLN ↑	–	TNF-α ↓	Dong et al. (2016)
Resveratrol	Broilers	Ileum	ZO-1, OCLDN, CLDN1 ↑	–	TLR4, NF-Kb, IL1-β, IL6, TNF-α ↑	Zhang et al. (2023b)
Dendrobium officinale leaf powder	Broilers	Duodenum Jejunum Ileum	ZO-2,OCLN, CLDN2, CLDN3 ↑ OCLN, CLDN2 ↑ ZO-1, ZO-2, CLDN2	SOD1, CAT ↑ SOD1, SOD2, CAT, Nrf2 ↑ SOD1, SOD2, CAT ↑	–	Zhao et al. (2023)

observed in broilers (Zhang and Ghosh, 2001; Dong et al., 2016; Paraskeuas and Mountzouris, 2019; Farahat et al., 2021; Li et al., 2022; Zhao et al., 2023).

5.2 Phytogenic effects on cytoprotection

Dietary phytogenics may serve as an indirect source of natural antioxidant compounds in poultry nutrition (Seymour et al., 2013; Lee et al., 2019). Accumulating evidence indicates that most phytochemicals can target an array of signaling pathways and molecules involved in cellular antioxidant and inflammatory responses. Moreover, these effects are strongly associated with the modulation of Nrf2 and NF-κB pathways (Sahin et al., 2013; Seymour et al., 2013; Ahmed et al., 2017; Mountzouris et al., 2022). Recent studies have demonstrated the cytoprotective potential of phytogenics via beneficial changes in Nrf2-pathway genes (Li et al., 2022; Zhao et al., 2023; Anagnostopoulos et al., 2023; Griela and Mountzouris, 2023).

5.3 Phytogenic effects on inflammatory response

In addition to their roles in antioxidant defense, phytogenics exert inflammation-modulating effects. Phytogenics have resulted in downregulated expression levels of the TLR- and NF-κB-related genes in the poultry intestine. The upregulation of the Nrf2 pathway could inhibit NF-κB activation, resulting in increased protection against oxidative damage and inflammation (Sahin et al., 2013; Mountzouris et al., 2020). In addition, phytogenics, via the downregulation of TLR signaling, reduce NF-κB activation and the subsequent transcription of inflammatory response (Kawai and Akira, 2007; Mountzouris et al., 2020).

6 Conclusion

Current research highlights the negative effects of various stressors (e.g. mycotoxins, heat stress, and pathogens) suppressing gut barrier function (Zhang et al., 2020; Chen et al., 2021a; Tao et al., 2023) and cytoprotection (Song et al., 2018; Yang et al., 2022; Guo et al., 2023), and exacerbating inflammatory responses (Song et al., 2019; Yang et al., 2022; Sun et al., 2023; Feng et al., 2023).

Phytogenics represent valuable nutritional tools for the prevention and management of the effects of stressors by beneficially enhancing gut integrity and adaptive cytoprotection components, as well as the management of inflammatory responses summarized in Table 2. In summary, it can be shown (Table 1) that dietary phytogenic supplementation beneficially modulates the expression of the intestinal gut barrier and cytoprotective genes of poultry and mitigates inflammatory response by regulating TLR signaling and NF-κB pathway. Carefully researched and validated phytogenic components could be

Table 2 Summary of phytogenic and stressors effects on critical components of gut homeostasis

Case	Gut barrier integrity	Cytoprotection	Inflammatory response
Phytogenics	↑	↑	↓
Stressors (e.g. mycotoxins, heat stress, and pathogens	↓	↓	↑
Phytogenics with stressors	↑	↑	↓

part of overall disease prevention strategies and sustainable poultry production operations, contributing to the increasing global demand for animal protein and global food security.

7 References

Ahmed, S. M. U., Luo, L., Namani, A., Wang, X. J. and Tang, X. 2017. Nrf2 signaling pathway: pivotal roles in inflammation. *Biochim. Biophys. Acta Mol. Basis Dis.* 1863(2):585–597.

Anagnostopoulos, E. C., Brouklogiannis, I. P., Griela, E., Paraskeuas, V. V. and Mountzouris, K. C. 2023. Phytogenic effects on layer production performance and cytoprotective response in the duodenum. *Animals* 13(2):294.

Aura, A. M., O'Leary, K. L., Williamson, G., Ojala, M., Bailey, M., Puuoponen-Pimia, R., Nuutila, A. M., Oksman-Caldentey, K. M. and Poutanen, K. 2022. Quercetin derivatives are deconjugated and converted into hydroxyphenylacetic acids but not methylated by human fecal flora in vitro. *J. Agric. Food Chem.* 50:1725e30.

Awad, W. A., Hess, C. and Hess, M. 2017. Enteric pathogens and their toxin-induced disruption of the intestinal barrier through alteration of tight junctions in chickens. *Toxins* 9(2):60.

Basiouni, S., Tellez-Isaias, G., Latorre, J. D., Graham, B. D., Petrone-Garcia, V. M., El-Seedi, H. R., Yalçın, S., El-Wahab, A. A., Visscher, C., May-Simera, H. L., Huber, C., Eisenreich, W. and Shehata, A. A. 2023. Anti-inflammatory and antioxidative phytogenic substances against secret killers in poultry: current status and prospects. *Vet. Sci.* 10(1):55.

Brenes, A. and Roura, E. 2010. Essential oils in poultry nutrition: main effects and modes of action. *Anim. Feed Sci. Technol.* 158(1–2):1–14.

Brouklogiannis, I. P., Anagnostopoulos, E. C., Griela, E., Paraskeuas, V. V. and Mountzouris, K. C. 2023. Dietary phytogenic inclusion level affects production performance and expression of ovarian cytoprotective genes in laying hens. *Poult. Sci.* 102(4):102508.

Chen, F., Zhang, H., Zhao, N., Yang, X., Du, E., Huang, S., Guo, W., Zhang, W. and Wei, J. 2021a. Effect of chlorogenic acid on intestinal inflammation, antioxidant status, and microbial community of young hens challenged with acute heat stress. *Anim. Sci. J.* 92(1):e1369.

Chen, Y., Lin, J., Zhao, Y., Ma, X. and Yi, H. 2021b. Toll-like receptor 3 (TLR3) regulation mechanisms and roles in antiviral innate immune responses. *J. Zhejiang Univ. Sci. B* 22(8):609–632.

Chen, J., Geng, X. and Ali, A. S. 2020. Montmorillonite combined organic acid or essential oil complex improves the intestinal health status of laying hens. doi: 10.21203/rs.3.rs-52591/v1.

Cheng, Y. F., Chen, Y. P., Chen, R., Su, Y., Zhang, R. Q., He, Q. F., Wang, K., Wen, C. and Zhou, Y. M. 2019. Dietary mannan oligosaccharide ameliorates cyclic heat stress-induced damages on intestinal oxidative status and barrier integrity of broilers. *Poult. Sci.* 98(10):4767–4776.

Commission Regulation (EC) No 1831/2003 of the European Parliament and of the Council of 22 September 2003 on Additives for Use in Animal Nutrition, Official Journal of the European Union L 268/29.

Debnath, B., Singh, W. S., Das, M., Goswami, S., Singh, M. K., Maiti, D. and Manna, K. 2018. Role of plant alkaloids on human health: a review of biological activities. *Mater. Today Chem.* 9:56–72.

Diaz-Sanchez, S., D'Souza, D., Biswas, D. and Hanning, I. 2015. Botanical alternatives to antibiotics for use in organic poultry production. *Poult. Sci.* 94(6):1419–1430.

Dong, Z. L., Wang, Y. W., Song, D., Hou, Y. J., Wang, W. W., Qi, W. T., Yun, T. T. and Li, A. K. 2016. The effects of dietary supplementation of pre-microencapsulated *Enterococcus fecalis* and the extract of *Camellia oleifera* seed on growth performance, intestinal morphology, and intestinal mucosal immune functions in broiler chickens. *Anim. Feed Sci. Technol.* 212:42–51.

Du, M., Cheng, Y., Chen, Y., Wang, S., Zhao, H., Wen, C. and Zhou, Y. 2023. Dietary supplementation with Synbiotics improves growth performance, antioxidant status, immune function, and intestinal barrier function in broilers subjected to cyclic heat stress. *Environ. Sci. Pollut. Res.* 30(7):18026–18038.

Ducatelle, R., Goossens, E., Eeckhaut, V. and Van Immerseel, F. 2023. Poultry gut health and beyond. *Anim. Nutr.* 13:240–248.

Farahat, M., Ibrahim, D., Kishawy, A. T. Y., Abdallah, H. M., Hernandez-Santana, A. and Attia, G. 2021. Effect of cereal type and plant extract addition on the growth performance, intestinal morphology, caecal microflora, and gut barriers gene expression of broiler chickens. *Animal* 15(3):10056.

Feng, J., Lu, M., Wang, J., Zhang, H., Qiu, K., Qi, G. and Wu, S. 2021. Dietary oregano essential oil supplementation improves intestinal functions and alters gut microbiota in late-phase laying hens. *J. Anim. Sci. Biotechnol.* 12:1–15.

Feng, J., Li, Z., Ma, H., Yue, Y., Hao, K., Li, J., Xiang, Y. and Min, Y. 2023. Quercetin alleviates intestinal inflammation and improves intestinal functions via modulating gut microbiota composition in LPS-challenged laying hens. *Poult. Sci.* 102(3):102433.

Gholami-Ahangaran, M., Ahmadi-Dastgerdi, A., Azizi, S., Basiratpour, A., Zokaei, M. and Derakhshan, M. 2022. Thymol and carvacrol supplementation in poultry health and performance. *Vet. Med. Sci.* 8(1):267–288.

Griela, E. and Mountzouris, K. C. 2023. Nutrigenomic profiling of reduced specification diets and phytogenic inclusion effects on critical toll-like receptor signaling, mitogen-activated protein kinase-apoptosis, and PI3K-Akt-mTOR gene components along the broiler gut. *Poult. Sci.* 102(6):102675.

Guo, S., Li, K., Zhang, P., Jin, X., Xu, Y., Xing, Y., Yan, S. and Shi, B. 2023. Effects of *Artemisia ordosica* aqueous extract on intestinal inflammation and antioxidant-related gene expression in lipopolysaccharide-challenged broilers. *J. Anim. Feed Sci.* 33(1):47–55.

Hippenstiel, F., Abdel-Wareth, A. A. A., Kehraus, S. and S¨udekum, K.-H. 2011. Effects of selected herbs and essential oils, and their active components on feed intake and performance of broilers: a review. *Archiv. Gefl¨ugelk* 75:226–234.

Jiang, J., Qi, L., Lv, Z., Jin, S., Wei, X. and Shi, F. 2019. Dietary stevioside supplementation alleviates lipopolysaccharide-induced intestinal mucosal damage through anti-inflammatory and antioxidant effects in broiler chickens. *Antioxidants* 8(12):575.

Kawai, T. and Akira, S. 2007. Signaling to NF-κB by toll-like receptors. *Trends Mol. Med.* 13(11):460–469.

Khadem, A., Soler, L., Everaert, N. and Niewold, T. A. 2014. Growth promotion in broilers by both oxytetracycline and *Macleaya cordata* extract is based on their anti-inflammatory properties. *Br. J. Nutr.* 112(7):1110–1118.

Kikusato, M., Xue, G., Pastor, A., Niewold, T. A. and Toyomizu, M. 2021. Effects of plant-derived isoquinoline alkaloids on growth performance and intestinal function of broiler chickens under heat stress. *Poult. Sci.* 100(2):957–963.

Kogut, M. H. 2013. The gut microbiota and host innate immunity: regulators of host metabolism and metabolic diseases in poultry? *J. Appl. Poult. Res.* 22(3):637–646.

Kogut, M. H., Genovese, K. J., Swaggerty, C. L., He, H. and Broom, L. 2018. Inflammatory phenotypes in the intestine of poultry: not all inflammation is created equal. *Poult. Sci.* 97(7):2339–2346.

Kogut, M. H., Yin XiaoNan, Y. X., Yuan JianMin, Y. J. and Broom, L. 2017. Gut health in poultry. *CAB Rev.* 12:1–7.

Köhle, C. and Bock, K. W. 2006. Activation of coupled Ah receptor and Nrf2 gene batteries by dietary phytochemicals in relation to chemoprevention. *Biochem. Pharmacol.* 72(7):795–805.

Koizumi, K., Oku, M., Hayashi, S., Inujima, A., Shibahara, N., Chen, L., Igarashi, Y., Tobe, K., Saito, S., Kadowaki, M. and Aihara, K. 2019. Identifying pre-disease signals before metabolic syndrome in mice by dynamical network biomarkers. *Sci. Rep.* 9(1):8767.

Lauridsen, C. 2019. From oxidative stress to inflammation: redox balance and immune system. *Poult. Sci.* 98(10):4240–4246.

Lee, K. W., Everts, H. and Beynen, A. C. 2004. Essential oils in broiler nutrition. *Int. J. Poult. Sci.* 3:738.e52.

Lee, M. T., Lin, W. C. and Lee, T. T. 2019. Potential crosstalk of oxidative stress and immune response in poultry through phytochemicals: a review. *Asian-Australas. J. Anim. Sci.* 32(3):309–319.

Li, Z., Jin, X., Wu, Q., Long, L., Li, Y., Zhang, Q., Liu, A., Chen, X., Geng, Z. and Zhang, C. 2022. Effects of encapsulated thymol and carvacrol mixture on growth performance, antioxidant capacity, immune function and intestinal health of broilers. *Ital. J. Anim. Sci.* 21(1):1651–1659.

Lingappan, K. 2018. NF-κB in oxidative stress. *Curr. Opin. Toxicol.* 7:81–86.

Michiels, J., Missotten, J., Dierick, N., Fremaut, D., Maene, P. and Smet, S. D. 2008. In vitro degradation and in vivo passage kinetics of carvacrol, thymol, eugenol and trans-cinnamaldehyde along the gastrointestinal tract of piglets. *J. Sci. Food Agric.* 88(13):2371–2381.

Mountzouris, K. C., Paraskeuas, V. V. and Fegeros, K. 2020. Priming of intestinal cytoprotective genes and antioxidant capacity by dietary phytogenic inclusion in broilers. *Anim. Nutr.* 6(3):305–312.

Mountzouris, K. C., Paraskeuas, V. V. and Griela, E. 2022. Adaptive poultry gut capacity to resist oxidative stress. In: Kogut, M. H. and Zhang, G. (Eds) *Gut Microbiota, Immunity, and Health in Production Animals*. Springer Nature, Switzerland AG, pp. 243–262.

Oceľová, V., Chizzola, R., Battelli, G., Pisarcikova, J., Faix, S., Gai, F. and Placha, I. 2019. Thymol in the intestinal tract of broiler chickens after sustained administration of thyme essential oil in feed. *J. Anim. Physiol. Anim. Nutr. (Berl)* 103(1):204–209.

Pan, D. and Yu, Z. 2013. Intestinal microbiome of poultry and its interaction with host and diet. *Gut Microbes* 5:108–119.

Panche, A. N., Diwan, A. D. and Chandra, S. R. 2016. Flavonoids: an overview. *J. Nutr. Sci.* 5:e47.

Paraskeuas, V. V. and Mountzouris, K. C. 2019. Modulation of broiler gut microbiota and gene expression of toll-like receptors and tight junction proteins by diet type and inclusion of phytogenics. *Poult. Sci.* 98(5):2220–2230.

Ray, P. D., Huang, B. W. and Tsuji, Y. 2012. Reactive oxygen species (ROS) homeostasis and redox regulation in cellular signaling. *Cell. Signal.* 24(5):981–990.

Sacranie, A., Svihus, B., Denstadli, V., Moen, B., Iji, P. A. and Choct, M. 2012. The effect of insoluble fiber and intermittent feeding on gizzard development, gut motility, and performance of broiler chickens. *Poult. Sci.* 91(3):693–700.

Sahin, K., Orhan, C., Smith, M. O. and Sahin, N. 2013. Molecular targets of dietary phytochemicals for the alleviation of heat stress in poultry. *Worlds Poult. Sci. J.* 69(1):113–124.

Seymour, E. M., Bennink, M. R. and Bolling, S. F. 2013. Diet-relevant phytochemical intake affects the cardiac AhR and nrf2 transcriptome and reduces heart failure in hypertensive rats. *J. Nutr. Biochem.* 24(9):1580–1586.

Shehata, A. A., Attia, Y., Khafaga, A. F., Farooq, M. Z., El-Seedi, H. R., Eisenreich, W. and Tellez-Isaias, G. 2022. Restoring healthy gut microbiome in poultry using alternative feed additives with particular attention to phytogenic substances: challenges and prospects. *Ger. J. Vet. Res.* 2(3):32–42.

Shini, S. and Bryden, W. L. 2022. Probiotics and gut health: linking gut homeostasis and poultry productivity. *Anim. Prod. Sci.* 62(12):1090–1112.

Song, B., He, J., Pan, X., Kong, L., Xiao, C., Keerqin, C. and Song, Z. 2023. Dietary Macleaya cordata extract supplementation improves the growth performance and gut health of broiler chickens with necrotic enteritis. *J. Anim. Sci. Biotechnol.* 14(1):113.

Song, J., Lei, X., Luo, J., Everaert, N., Zhao, G., Wen, J. and Yang, Y. 2019. The effect of epigallocatechin-3-gallate on small intestinal morphology, antioxidant capacity and anti-inflammatory effect in heat-stressed broilers. *J. Anim. Physiol. Anim. Nutr.* 103(4):1030–1038.

Song, Z. H., Cheng, K., Zheng, X. C., Ahmad, H., Zhang, L. L. and Wang, T. 2018. Effects of dietary supplementation with enzymatically treated Artemisia annua on growth performance, intestinal morphology, digestive enzyme activities, immunity, and antioxidant capacity of heat-stressed broilers. *Poult. Sci.* 97(2):430–437.

Stefanson, A. L. and Bakovic, M. 2014. Dietary regulation of Keap1/Nrf2/ARE pathway: focus on plant-derived compounds and trace minerals. *Nutrients* 6(9):3777–3801.

Stevanović, Z. D., Bošnjak-Neumüller, J., Pajić-Lijaković, I., Raj, J. and Vasiljević, M. 2018. Essential oils as feed additives–future perspectives. *Molecules* 23(7):1717. https://doi.org/10.3390/molecules23071717

Sun, H., Zheng, X., Yang, B., Yan, M., Wang, H., Yang, S., Shi, D., Guo, S. and Liu, C. 2023. Effect of Wu Zhi San supplementation in LPS-induced intestinal inflammation and barrier damage in broilers. *Front. Vet. Sci.* 10:1234769.

Suzuki, T. 2013. Regulation of intestinal epithelial permeability by tight junctions. *Cell. Mol. Life Sci.* 70(4):631–659.

Suzuki, T. and Hara, H. 2011. Role of flavonoids in intestinal tight junction regulation. *J. Nutr. Biochem.* 22(5):401–408.

Tao, W., Zhu, W., Nabi, F., Li, Z. and Liu, J. 2023. *Penthorum chinense* Pursh compound flavonoids supplementation alleviates aflatoxin b1-induced liver injury via modulation of intestinal barrier and gut microbiota in broiler. *Ecotoxicol. Environ. Saf.* 255:114805.

Tian, Y., Li, G., Zhang, S., Zeng, T., Chen, L., Tao, Z. and Lu, L. 2022. Dietary supplementation with fermented plant product modulates production performance, egg quality, intestinal mucosal barrier, and cecal microbiota in laying hens. *Front. Microbiol.* 13:955115.

Valenzuela-Grijalva, N. V., Pinelli-Saavedra, A., Muhlia-Almazan, A., Domínguez-Díaz, D. and González-Ríos, H. 2017. Dietary inclusion effects of phytochemicals as growth promoters in animal production. *J. Anim. Sci. Technol.* 59:1–17.

Vomund, S., Schäfer, A., Parnham, M. J., Brüne, B. and Von Knethen, A. 2017. Nrf2, the master regulator of anti-oxidative responses. *Int. J. Mol. Sci.* 18(12):2772.

Wang, Y., Wang, Y., Su, C., Wang, L., Lv, X., Cui, G., Ji, L., Huang, Y., Zhang, H. and Chen, W. 2022a. Dietary cinnamaldehyde with carvacrol or thymol improves the egg quality and intestinal health independent of gut microbiota in post-peak laying hens. *Front. Vet. Sci.* 9:994089.

Wang, Z., Shao, D., Wu, S., Song, Z. and Shi, S. 2022b. Heat stress-induced intestinal barrier damage and dimethylglycine alleviates via improving the metabolism function of microbiota gut brain axis. *Ecotoxicol. Environ. Saf.* 244:114053.

Yang, H., Wang, Y., Yu, C., Jiao, Y., Zhang, R., Jin, S. and Feng, X. 2022. Dietary resveratrol alleviates AFB1-induced ileum damage in ducks via the Nrf2 and NF-κB/NLRP3 signaling pathways and CYP1A1/2 expressions. *Agriculture* 12(1):54.

Yegani, M. and Korver, D. R. 2008. Factors affecting intestinal health in poultry. *Poult. Sci.* 87(10):2052–2063.

Zeineldin, M., Lowe, J. and Aldridge, B. 2019. Contribution of the mucosal microbiota to bovine respiratory health. *Trends Microbiol.* 27(9):753–770.

Zhang, G. and Ghosh, S. 2001. Toll-like receptor-mediated NF-κB activation: a phylogenetically conserved paradigm in innate immunity. *J. Clin. Invest.* 107(1):13–19.

Zhang, J., Fang, Y., Fu, Y., Jalukar, S., Ma, J., Liu, Y., Guo, Y., Ma, Q., Ji, C. and Zhao, L. 2023a. Yeast polysaccharide mitigated oxidative injury in broilers induced by mixed mycotoxins via regulating intestinal mucosal oxidative stress and hepatic metabolic enzymes. *Poult. Sci.* 102(9):102862.

Zhang, L. Z., Gong, J. G., Li, J. H., Hao, Y. S., Xu, H. J., Liu, Y. C. and Feng, Z. H. 2023b. Dietary resveratrol supplementation on growth performance, immune function and intestinal barrier function in broilers challenged with lipopolysaccharide. *Poult. Sci.* 102(10):102968.

Zhang, X., Zhao, Q., Ci, X., Chen, S., Xie, Z., Li, H., Zhang, H., Chen, F. and Xie, Q. 2020. Evaluation of the efficacy of chlorogenic acid in reducing small intestine injury, oxidative stress, and inflammation in chickens challenged with Clostridium perfringens type A. *Poult. Sci.* 99(12):6606–6618.

Zhao, W., Chen, Y., Tian, Y., Wang, Y., Du, J., Ye, X., Lu, L. and Sun, C. 2023. Dietary supplementation with Dendrobium officinale leaves improves growth, antioxidant status, immune function, and gut health in broilers. *Front. Microbiol.* 14:1255894.

Chapter 16

Using macroalgae/seaweed and microalgae in poultry feed

Elisabeth Baéza and Cécile Berri, INRAE Nouzilly, France; and Edouard Coudert, Roullier, France

1 Introduction

Today the concept of natural, organic, and sustainable farming has become more and more important for consumers, especially in developed countries. The farming industry has thus grown interested in using natural forms of vitamins and minerals instead of synthetically produced ones. Considering these dynamics, the possibility of using micro- and macroalgae as a new source of nutrients and health additives in animal feed formulations has been evaluated. Algae are a very heterogeneous group of plants that can be found in different sizes, shapes, and colors. Thus, the taxonomy linked to this group is complex and controversial in the scientific community. It includes green (Chlorophyceae), brown (Phaeophyceae), and red (Rhodophyceae) algae. Species, habitat, growing conditions (e.g. temperature, light exposure), and collection methods play a key role in the highly variable composition of each type of algae. Algae, as living species, are found in freshwater as well as in salt water and are separated into two main groups: the large macroalgae, also called seaweed, that are often found in the littoral zone, and the small microalgae that are found both on the coast and in the ocean (El Gamal, 2010). Cultivated algae, both microalgae and macroalgae, have long been incorporated into fish feed, in both hatcheries and farms, and microalgae are considered a reliable source of nutrients for aquaculture. However, algae are

http://dx.doi.org/10.19103/AS.2024.0143.21

less widely integrated in terrestrial animal nutrition although there is currently momentum to include both micro- and macroalgae in their feed. The main algae currently used in animal nutrition are presented in Table 1.

Both macro- and microalgae contain various active molecules, such as polyphenols, polysaccharides, lipids, peptides and proteins, minerals, polyphenols, and pigments (Harnedy and Fitzgerald, 2011; Michalak and Chojnacka, 2015; Michalak and Mahrose, 2020; Kalia and Lei, 2022). Most of the molecules they contain are well characterized and some are known for their antioxidant, antibacterial, or anti-inflammatory properties (Shalaby, 2011; Park, 2015). Even when incorporated at a low level in diets of both aqueous and terrestrial livestock species, algae have been demonstrated to improve animal immune response, lipid metabolism (Nakagawa, 1997; Norambuena et al., 2015),

Table 1 Main algae currently used in animal nutrition

Name	Type	Picture
Chlorella vulgaris	Green microalga	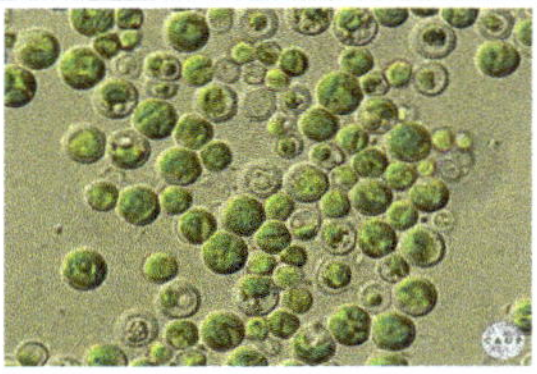
Spirulina platensis	Green microalga	
Ulva lactuca[a]	Green seaweed	
Sargassum sp[a]	Brown seaweed	

[a]Photo credits to ©Olmix.

and gut function (Michiels et al., 2012). They also act as antiviral and antibacterial agents (Pulz and Gross, 2004; Becker, 2007; Gouveia et al., 2010; Kulshreshtha et al., 2020). They have been studied as alternative sources of macro- and micronutrients such as proteins, lipids, minerals (Table 2), amino acids, and n-3 fatty acids (FA), or as a source of specific biologically active phytochemicals. In addition to their high protein content, seaweeds are also rich in polysaccharides, polyphenols, and pigments (Korczyński et al., 2015). The detailed average chemical composition of the main algae used in animal feed is available in the review published by Costa et al. (2020). So far, few studies have examined the use of algae as a major source of macronutrients because of the large amount of biomass needed. They are used 'rather' as supplements, enhancing standard feed formulations to the benefit of animal health or product quality.

This chapter aims to present the main characteristics of algae, as used in poultry nutrition and the way they are produced, with a special focus on compounds beneficial for animal health and performance, and final product quality. Both laying hen and broiler chicken production will be considered.

2 What interest may algae present for poultry nutrition?

Algal research represents a new field of interest for animal nutrition and health, because both microalgae and macroalgae contain several compounds that

Table 2 Nutritional and chemical composition[a] of the four main algae used in animal nutrition

	Chlorella vulgaris	*Spirulina platensis*	*Ulva lactuca*	*Sargassum* sp.
Proximate composition[b]				
Protein (%)	57	60	19	14
Lipids (%)	20	6	0.3	6
Carbohydrates (%)	23	34	80.7	80
Minerals (mg/kg)				
Na	134.6	189.7	351.7	389.3
K	4.99	132.6	209	244.3
Ca	59.3	88.3	180.7	176
Fe	25.9	10.3	34.5	32.2
Zn	1.2	2.5	1.78	5.8
Cu	0.6	0.32	1.83	1.62
Se	0.7	0.37	1.6	49.8
Mn	0.21	0.38	4.8	3.3

[b]Dry matter composition.

[a]**Data issued from Tokusoglu and Ünal (2003); Volkman and Brown (2005); Debbarma et** al. (2016); Rasyid (2017); Praiboon et al. (2018).

could be beneficial for animal production (Fig. 1). They are present or produced all around the globe, and the number of publications devoted to or related to this subject has increased regularly over the past three decades as represented on Fig. 2 and highlighted by Garrido-Cardenas et al. (2018) and Rumin et al. (2020).

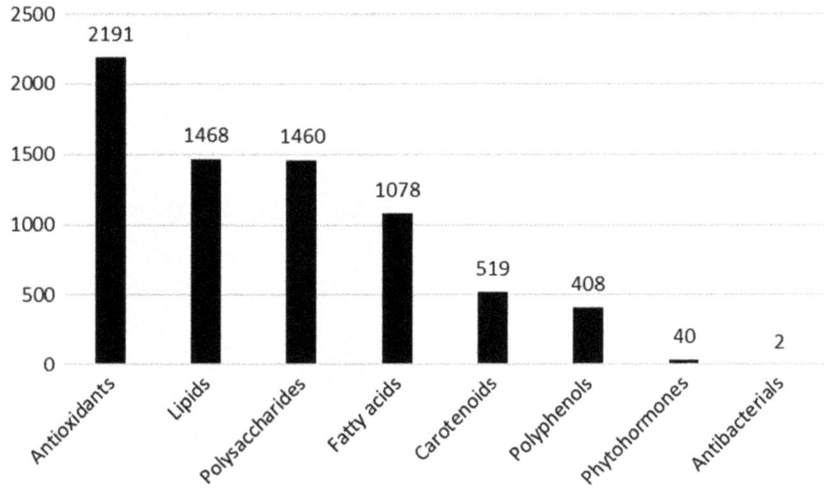

Figure 1 Number of articles concerning the different compounds extracted from algae and published in English language between 1991 and 2022 according to the database 'Web of Science'.

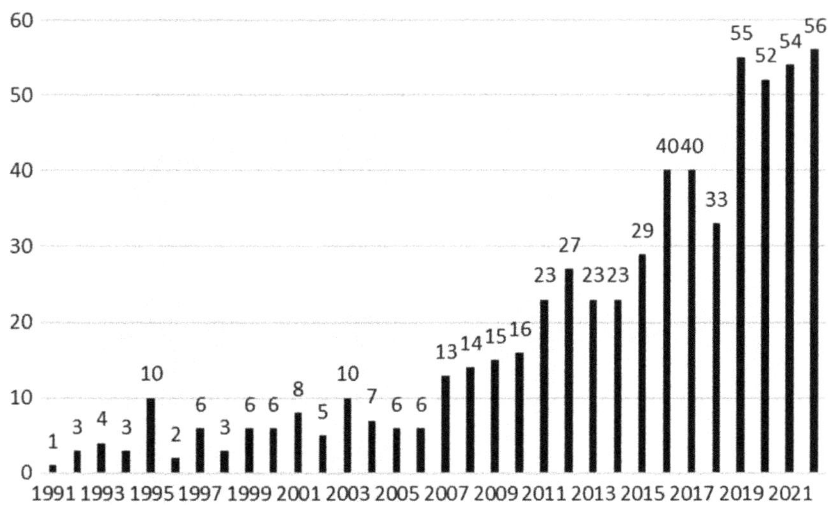

Figure 2 Number of articles concerning algae and animal nutrition and published in English language between 1991 and 2022 according to the database 'Web of Science'.

2.1 Chemical and nutritional value of algae

Chemical composition and nutritional quality are specific to algal sources. Differences exist according to algae type, size (micro or macro), and classification (brown, red, or green). Seasonality is also an important factor of variation in algal nutritional quality and chemical composition (Kumar et al., 2015; Stengel et al., 2011). Changes in culture conditions also affect microalgae composition or production yield. For instance, a lower nitrogen level in *Chlorella* culture medium leads to higher lipid accumulation (up to 85%) whereas mixotrophic culture of *Spirulina* increases its productivity fivefold (Kovač Blagojević et al., 2013). Table 2 presents the nutritional values and chemical composition of several algae commonly used in animal nutrition. Algae are added as a supplementary component of diets. Diets are enriched with algae directly as raw or processed, the latter to concentrate the potential beneficial effects of specific extracts.

2.2 Extraction methods

To be able to produce and use specific algal compounds, several extraction methods have been developed. They differ depending on the primary nature of seaweeds, the amount of extract needed and the potential applications for the products obtained. The most common extraction method is based on solvent extraction with water or ethanol for example because of its low cost and relatively high efficiency (Chojnacka, 2012; Michalak and Chojnacka, 2015). However, the use of organic solvents is being reconsidered due to their flammability, toxicology and hazardous nature for people and the environment (Tobiszewski et al., 2017). Furthermore, these methods usually have a heating step that can cause thermo-degradation of the compounds of interest which is likely to deteriorate their beneficial activity and produce harmful molecules not suitable for food or feed applications in humans, animals, or both (Starmans and Nijhuis, 1996). For all these reasons, more environmentally, friendly and safe new methods are being developed. This is the case for microwave-assisted extraction that is based on the use of enzymes able to hydrolyze the seaweed cell wall to improve the transfer of active substances to the liquid phase. Microwaves induce vibrations of water molecules that lead to an increase of pressure inside the cell, hence the cell wall breaks down, releasing intracellular content into the medium, which facilitates the extraction of target compounds (Hahn et al., 2012; Michalak and Chojnacka, 2014). This method is used to extract polysaccharides from *Enteromorpha prolifera* (Lin et al., 2015), lipids from *Chlorella vulgaris* (Liang et al., 2013), pigments from *Dunaliella tertiolecta* (Pasquet et al., 2011), phenols from *Caulerpa racemosa* (Li et al., 2012), and carotenoids from *D. salina* (Macías-Sánchez et al., 2009).

Supercritical fluid extraction is another method currently used for algal extract preparation. It consists of increasing fluid temperature and pressure above their critical points so the extracted fluid density is lowered. This technique is efficient for isolating substances of medium molecular weight or with low polarity (Skala et al., 2002; Kadam et al., 2013). Supercritical fluid extraction uses mainly CO_2 because of its low cost, availability, and high affinity to volatile compounds (Hosikian et al., 2010; Crampon et al., 2011). This method is mainly used to extract algae compounds intended for plant care, as biostimulants or fertilizers, like polyphenols from *C. vulgaris* (Wang et al., 2010), vitamin E from *Spirulina platensis* (Mendiola et al., 2007), chlorophyll from *Monoraphidium* sp. (Fujii, 2012), and carotenoids from *C. vulgaris* (Mendes et al., 1995).

Pressurized liquid extraction has been proposed for algae extraction. This is still in development but has already shown promising results because it is faster, more selective, and requires less biomaterial compared to traditional extraction methods (Alvarez-Rivera et al., 2020). It uses a combination of both high temperature and pressure and includes accelerated 'green' solvent extraction (Jaime et al., 2005; Ibañez et al., 2012). Antioxidants from *S. platensis* (Herrero et al., 2005), phenols from both *Ulva intestinalis* and *Fucus spiralis* (Tierney et al., 2014), and fatty acids from *Himanthalia elongata* (Santoyo et al., 2011) can be extracted using this technique.

Both new and traditional methods are currently used to extract biological molecules from both macro- and microalgae. Novel technologies contribute to the development of new 'green' and sustainable methodologies to explore the vast domain of target compounds from seaweeds. However, the choice of extraction method used ultimately depends on the algal biomass available, chemical compounds to target and their final use (plant, animal, or human care). The different processing of algae biomass including bead milling, pasteurization, freeze-drying, high-pressure homogenization, pulse electric field, ultrasound, microwave, chemical and enzymatic treatment will disrupt the cell membranes allowing the release of intracellular nutrients for the animals and increasing the digestibility of algae compounds (Annamalai et al., 2021).

It is necessary also, to take into account the stability of algae extracts during the animal feed processing. For example, Martinez-Delgado et al. (2020) examined the effect of extrusion process at high temperature (80–100°C) on the stability of astaxanthin and antioxidant capacity provided by microalgae incorporated in fish feed. Their study showed an increase of the antioxidant value from 23 to 32% compared to feeds without microalgae and the retention of astaxanthin in extruded feeds was 85–92% even after 5 months of storage in dark at 4°C.

2.3 Benefits of algae for animals

The mode of production and the quality of final animal products (eggs and meat) are key challenges for the sustainability of animal husbandry. Several questions related to animal welfare, product quality, and environmental impacts must be addressed to improve consumer acceptability and confidence in livestock production. From this perspective, animal feed companies are looking for natural solutions likely to improve the health of farm animals and the quality of their products (Nocella et al., 2009; Napolitano et al., 2013). Among the possible natural solutions, algae are a valuable source of functional additives, especially because the EU considers them as potential feed material (EU regulation No 575/2011 of 16 June 2011). The main interest in algae is due to their richness in several compounds. Essential n-3 polyunsaturated fatty acids (PUFA) are valuable for the production of high nutritional quality products (Shapira et al., 2008). Polyphenols and pigments that have antioxidative activities can prevent or cure diseases or improve product shelf life (Scalbert et al., 2005). Essential amino acids and proteins may help to control intestinal microflora, and some polysaccharides have antimicrobial activity (Kuorwel et al., 2011). In addition, some pigments are used to improve product color (Shields and Lupatsch, 2012). Algae can be used in different forms, as raw materials fresh, frozen, or dried; oil obtained by extraction; or extracts used as a supplement in formulated diets. The properties of algal extracts have been less studied than those of other forms, but there is currently a growing interest in them, mainly because they contain biologically active compounds likely to improve animal health and product quality (Korczyński et al., 2015). Haberecht et al. (2018) reviewed potential beneficial effects or antimicrobial properties of macroalgae for poultry health and performance. Moreover, most of these algae molecules are not authorized for direct use in human food, which increases their interest in animal nutrition.

3 Laying hen production

3.1 Egg productivity

For a long time, the nutrition of laying hens focused on sustaining egg production with the objective to maintain or improve it in parallel with other traits related to egg quality and animal health. Dietary supplementation in 80-week-old Hy-Line Brown layers with 2% of *C. vulgaris* increased egg productivity, from 55.4% in the control to 59% in the supplemented group (Zheng et al., 2011). Supplementing diets of 22–36-week-old Lohmann Brown hens with 0.5% of the same microalgae also increased egg weight (60.4 g vs. 58.9 g for the control group), but only at the beginning (first to third month) and end (sixth to eighth month) of the laying period. However, this supplementation had no effect on hen health or the protein content of eggs (Halle et al., 2009). A sulfated polysaccharide extract from the green seaweed Ulva was added to diets of 61-weeks-old laying hens at 0.5%, 0.8%, and 1% for 8 weeks (Li et al., 2018). These supplementations improved the laying rate (82–83% vs. 76% in the control group at 68 weeks of age), total egg weight (1.27–1.35 kg vs. 1.23 kg in the control group at 68 weeks of age) and decreased FCR (2.08–1.93 vs. 2.13 in the control group at 68 weeks of age; Li et al., 2018). The red microalga *Porphyridium* sp. has been tested in 30-week-old White Leghorn hens' diets at 5% or 10% for 10 days. These treatments had no effect on body weight, egg weight, and laying rate but feed consumption decreased (10% compared to control; Ginzberg et al., 2000), likely because *Porphyrium* sp. comprises 70% polysaccharides which can act as a fiber source (Dvir et al., 2000). Despite no difference in feed intake, supplementing diets in laying hens with 0.5% of the brown seaweed *Undaria pinnatifida* induced a higher laying rate (73.8% vs. 65.2%) and egg weight (65.4 g vs. 64.1 g) compared to the control group, with no effect on egg quality (eggshell color, strength, thickness, egg yolk color; Choi et al., 2018). These authors hypothesized that this positive effect was linked to the high oligosaccharide content in this brown seaweed (Heo et al., 2005). Finally, Carrillo et al. (2012) tested the effects of dietary supplementation with *Sargassum* spp. at 2%, 4%, and 8% in laying hens. Unfortunately, this resulted in reduced laying rate, inversely proportional to *Sargassum* spp. incorporation rate (85.9% vs. 91.3% for 8% supplementation).

3.2 Egg quality

Several studies have shown that algae can affect the quality of eggs. For instance, supplementing *ad libitum*-fed laying hens with 10% *Entermorpha* sp. in the diet for 8 weeks induced a lighter yolk color compared to the control group (Carrillo et al., 2008). In contrast, 4% or 8% dietary supplementation with *Sargassum* spp. in 35- to 43-week-old White Leghorn hens resulted in a more intense yolk color (Carrillo et al., 2012). These diverging results could be a consequence of the different types of carotenoids contained in these two seaweeds, *Enteromorpha* sp. containing mainly lutein, neoxanthin, and cryptoxanthin, which are particularly sensitive to sunlight and being stored in a dry condition, compared to the carotenoids contained in *Sargassum* spp. (mainly lutein, zeaxanthin, and fucoxanthin). Other studies have shown a positive effect with algae supplementation whereby yolk color has become more yellow. This was the case for dietary supplementation with *S. platensis* (at 1.5%, 2%, and 3%) in 63- to 67-week-old Hy-line W36 hens (Zahroojian and Morajev, 2013), with *C. vulgaris* in 80 week-old Hy-Line Brown layers (Zheng et al., 2011), and with *Porphyridium* sp. (at 5% and 10%) in 30-week-old White Leghorn hens (Ginzberg et al., 2000). In addition to the effect on yolk color, supplementing 42-week-old

Highland Brown hens with 3% of the green alga *Enteromorpha prolifera* in the diet increased eggshell thickness and calcium content, compared to the control group (Wang et al., 2013a). Finally, Englmaierová et al. (2013) showed that 1.25% *C. vulgaris* dietary supplementation for 25- to 39-week-old ISA Brown hens induced an increase in egg weight (62.3 g vs. 61.1 g for control hens), shell weight (6.1 g vs. 5.9 g for control hens), and yolk color, which was more intense for the supplemented group (increased redness and yellowness).

3.3 Egg composition

Some algae are rich in fatty acids that are able to influence egg lipid composition. Several studies have reported that alga supplementation can significantly reduce egg cholesterol content (Table 3). Cholesterol content reduction varied between 19% and 23% depending on the study, the alga source, and the dietary supplementation level. The algae tested were *S. platensis* (0.3%; Sujatha and Narahari, 2011), *Sargassum* sp. (8%; Carrillo et al., 2012), *E. prolifera* (3%; Wang et al., 2013a), and *Sargassum dentifebium* (6%; Al-Harthi and El-Deek, 2012). The effect of algae on cholesterol content in eggs is likely due to the effect of sterols, fatty acids, and polysaccharides on cholesterol synthesis in the liver of hens, either by direct interference with formative pathways or by competition between diet and alga cholesterol in the small intestine absorption sites (Nishide et al., 2003). A 0.5% dietary supplementation of the brown seaweed *Undaria pinnatifida* (Choi et al., 2018) induced higher albumin in blood as also observed in the case of dried brown seaweed (*Sargassum* spp.) supplementation (Rizk, 2017). Because albumin serves as a steroid and fatty acid transporter in the blood, increasing its amount with brown seaweeds could help laying hens to control blood cholesterol.

Alga supplementation induces significant changes in the FA composition of eggs. In the different studies reported by Fraeye et al. (2012), the n-3 long-chain (LC) FA content was 3.8- to 7.0-fold higher in eggs from hens fed microalgae at various incorporation rates compared to a control group. Thereby, dietary supplementation (2.9%) with *Schizochytrium* sp. in 21-week-old Line 477 hens affected both yolk color (redder) and n-3 FA content, which was increased by a factor of 2.9 compared to controls, particularly for LC FA. However, such supplementation negatively affected egg taste, by introducing a fishy flavor (Baéza et al., 2015b). However, studies testing incorporation rates between 2% and 4.8% of microalgae in diets found no effect on the sensorial quality of eggs (Fraeye et al., 2012). A recent study compared the sensorial quality of table eggs from hens fed diets supplemented with graded levels of docosahexaenoic acid (DHA: 1.25 mg/g feed, 2.50 mg/g feed, and 5.00 mg/g feed) derived from microalgae oil or fish oil, for 10 weeks (Feng et al., 2020). The overall efficiency of DHA incorporation into eggs was similar for the two

Table 3 Summary of algae used in hen nutrition

Reference	Seaweed	Type	Incorporation rate in diet	Animals	Effects
Zahroojian and Morajev (2013)	*Spirulina platensis*	Microalgae	1.5%, 2.0%, 2.5%	Hy-Line W36 hens, 63–67 weeks	Increased yolk color
Sujatha and Narahari (2011)	*Spirulina platensis*	Microalgae	0.30%	White Leghorn hens, 53 weeks	Decreased total cholesterol content in eggs
Halle et al. (2009)	*Chlorella vulgaris*	Microalgae	0.25%, 0.50%, 0.75%	Lohmann Brown hens, 22–36 weeks	Increased laying performance, number and egg quality, shell thickness, and yolk color
Janczyk et al. (2009)	*Chlorella vulgaris*	Microalgae	0.50%	Lohmann Brown hens, 23 weeks	Increased egg quality and productivity
Janczyk et al. (2009)	*Chlorella vulgaris*	Microalgae	0.50%	Lohmann Brown hens, 23 weeks	Increased universal bacteria diversity in intestinal tract
Englmaierová et al. (2013)	*Chlorella vulgaris*	Microalgae	1.25%	ISA Brown hens, 25–39 weeks	Increased egg weight, shell quality, yolk color, lutein, and zeaxanthin concentrations
Kotrbáček et al. (2015)	*Chlorella vulgaris*	Microalgae	1.0%	Laying hens, 56–63 weeks	Increased yolk carotenoids, (lutein, beta-carotene, and zeaxanthin)
Zheng et al. (2011)	*Chlorella vulgaris*	Microalgae	2.0%	Hy-Line Brown hens, 80 weeks	Increased laying rate and yolk color
Baéza et al. (2015b)	*Schizochytrium* sp.	Microalgae	2.9%	Line 477 hens, 21 weeks	Increased egg yolk redness and n-3 FA[a] content
Ginzberg et al. (2000)	*Porphyrium* sp.	Macroalgae	5.0%, 10.0%	White Leghorn hens, 30 weeks	Decreased daily feed consumption, increases yolk darkness, and modify yolk FA composition

(*Continued*)

Table 3 (*Continued*)

Reference	Seaweed	Type	Incorporation rate in diet	Animals	Effects
Carrillo et al. (2008)	*Enteromorpha* sp.	Macroalgae	10%	White Leghorn hens, 35–43 weeks	Increased yolk lightness
Carrillo et al. (2008)	*Macrocystis pyrifera*	Macroalgae	10%	White Leghorn hens, 35–43 weeks	Increased egg lipid and n-3 FA contents
Wang et al. (2013a)	*Enteromorpha prolifera*	Macroalgae	3.0%	Highland Brown laying hens, 42 weeks	Increased shell thickness and egg quality, decreased egg cholesterol content, improved immune system
Kulshreshtha et al. (2014)	*Chondrus crispus* and *Sarcodiotheca gaudichaudii*	Macroalgae	2.0%	Lohmann Brown classic laying hens, 67 weeks	Increased villus height, ceca weight, probiotic bacteria concentration in intestinal tract, and concentration of short-chain FA in eggs
Al-Harthi and El-Deek (2012)	*Sargassum dentifebium*	Macroalgae	6.0%	Hy-Line laying hens, 23–42 weeks	Decreased yolk cholesterol, triglycerides and n-6 FA contents
Carrillo et al. (2012)	*Sargassum* spp.	Macroalgae	4.0, 8.0%	White Leghorn hens, 35–43 weeks	Decreased egg cholesterol, improved yolk colour, and reduces egg production
Choi et al. (2018)	*Undaria pinnatifida*	Macroalgae	0.5%	Hy-Line Brown hens, 70 weeks	Increased egg production and egg weight, increased albumin blood content
Rizk (2017)	*Sargassum* spp.	Macroalgae	0.2%	Sinai hens, 52–64 weeks	Increased albumen and FA contents in eggs

[a]FA, fatty acid.

sources but in comparison with fish oil, dietary microalgae oil supplementation resulted in higher scores for egg flavor, and overall acceptability, both of which declining linearly in response to DHA supplementation. Supplementing diets for White Leghorn hens with 8% of the red alga *Porphyridium* sp. increased the linoleic and arachidonic acid content in eggs by 29% and 24%, respectively (Ginzberg et al., 2000). Similarly, 10% *Macrocystis pyrifera* supplementation in 35- to 43-week-old White Leghorn hens resulted in both increased lipid and n-3 FA content (Carrillo et al., 2008), whereas 6% *Sargassum dentifebium* dietary supplementation in 23- to 42-week-old Hy-Line hens decreased both triglycerides and n-6 FA percentages of egg yolk (Al-Harthi and El-Deek, 2012). The high level of docosahexaenoic acid (DHA) in *Aurantiochytrium limacinum* microalgae (172 mg/g) enabled to increase DHA level in egg yolk from 257 to 476 mg/100 g yolk by supplementing diet for ISA Brown laying hens with 1% of this microalgae without any detrimental effect on laying performance of hens (Moran et al., 2019).

The high level of carotenoids in microalgae can reduce the susceptibility of eggs to oxidation, improving storage ability. Finally, supplementation with algae can alter or increase egg carotenoid content and affect yolk color, which is an important consumer requirement in certain regions. Providing dietary 1% or 1.25% of *C. vulgaris* in either 56- to 63-week-old and 25- to 39-week-old ISA brown laying hens, significantly increased lutein, zeaxanthin, and beta-carotene in egg yolk (Englmaierová et al., 2013; Kotrbáček et al., 2015).

3.4 Hen health

Algae are considered a potential prebiotic, and hence, could be used to improve gastrointestinal health and the ability to absorb minerals and nutrients (Sako et al., 1999), limit the overuse of antibiotics, and reduce adverse effects of antibiotics on animal microbiota and the risk of antibiotic resistance. Both *Chondrus crispus* and *Sarcodiotheca gaudichaudii* fed to 67-week-old Lohmann Brown classic hens at a 2% or 4% dietary inclusion induced significant changes in the gastrointestinal tract by increasing villus height (+31%), caecum weight (+75%), and concentration of the beneficial bacteria *Bifidobacterium longum*, *Lactobacillus acidophilus*, and *Streptococcus salivarius* (Kulshreshtha et al., 2014, 2017). It also increased the short-chain fatty acid (SCFA) concentration in the gut microbiome that may help to limit colon diseases and preserve the integrity of the intestinal mucosa and reduced the pathogenic microbes (*Clostridium perfringens*) in the ileal fillings. Another positive impact of algae on gut health and microbiota is to decrease inflammation caused by pathogens compared to birds fed non-supplemented diets (Baurhoo et al., 2007). The microbial profile in the crop and caecum of 23-week-old Lohmann Brown laying hens fed 5% *C. vulgaris* in the diet was determined using 16S rRNA sequencing technology, and compared to that of non-supplemented hens (Janczyk et al., 2009). The microalga *C. vulgaris* strongly affected the caecum and crop bacterial population, especially *Lactobacillus* spp., enhancing their diversity. This study also showed that hen productivity and health, feed digestibility, and nutrient availability were directly correlated to changes in the microbial community in the crop and caecum, which were induced by alga consumption. Specifically, increasing the diversity of the universal bacterial community in the intestinal tract resulted in improved animal welfare and productivity. Supplementing 42-week-old Highland Brown hens with 3% *E. prolifera* in the diet increased the intestinal populations of beneficial *Lactobacillus* spp. and reduced *Escherichia coli* in the feces compared to an unsupplemented group (Wang et al., 2013a). Hens fed with the diet supplemented with 0.2% of brown seaweed had significantly higher serum total antioxidant capacity and concentration of glutathione peroxidase enzyme whereas supplementing the hen diet with 0.2% of red seaweed significantly increased the activity of serum superoxide dismutase (Rizk, 2017).

From the available literature, algae seem to be a promising ingredient to improve egg quality and egg productivity of layers (Table 3). Even though there is already evidence that algae can be beneficial in the digestive tract of laying hens, it is important to evaluate their potential effect as a source of calcium. Indeed, seaweeds show a high level of calcium that is essential for laying hen production because it sustains bone health and eggshell quality, particularly in the current context, which promotes the lengthening of productive laying period.

4 Utilization for broiler chickens

Seaweeds have been used as feed additives for ruminants and pigs for a long time. Many benefits have been found in livestock species, providing protein and energy, and improving general and digestive health. An increasing number of studies have aimed to evaluate if seaweeds can be efficiently used to improve meat quality traits and animal immune status in chickens.

4.1 Growth performance

The use of microalgae as a new source of protein in poultry feed developed in the early 1980s (Lipstein et al., 1980; Becker, 1986), although macroalgae (seaweeds) was more recent. Ventura et al. (1994) tested the possibility of complementing broiler and breeder diet with the green seaweed *Ulva rigida*. They concluded that incorporating 10% or more in diet was not suitable because it significantly decreased feed intake and growth rate in both adult cockerels and 1–21-day-old broilers. This finding pointed out that algae incorporation rate is a crucial parameter to consider for getting the most benefit in the animal and products. Moreover, the recommendation for algae incorporation rate in poultry diet is much higher than that recommended for fishmeal not exceeding 5%. A 4% dietary supplementation of the green seaweed *E. prolifera* increased feed intake and body weight gain in 35-day-old Arbor Acre broilers, and enhanced nutrient availability and apparent metabolizable energy (Wang et al., 2013b). In another study, Baéza et al. (2015a) reported that 2% dietary supplementation with *Schizochytrium* sp. in Ross 308 broiler chickens improved their feed conversion ratio over the total rearing period, due to decreased feed consumption for an equivalent growth rate. However, Yan and Kim (2013) did not observe any effect of inclusion of this microalga at 10 g/kg and 20 g/kg on growth performance of Ross 708 chickens. In contrast, Ribeiro et al. (2013) demonstrated that the dietary supply of this microalga at 74 g/kg between 21 and 35 days of age had a positive effect on growth (+22%) and improved the gain to feed ratio (+15%) in comparison with control broilers fed a diet containing 56.5 g/kg soybean oil. Ribeiro et al. (2014) confirmed this beneficial effect of microalgae on the growth performance of chickens in a further study.

In the same way, dietary supplementation with 1% of the brown alga *Ascophyllum nodosum* increased the body weight of 42-day-old Ross broilers by 15% (Evans and Critchley, 2014). Such a positive effect on body weight was also reported with 1% dietary supplementation of *S. platensis* in 1–35-day-old Cobb 500 broilers (Ross et al., 1994). More precisely, increasing the rate of *S. platensis* supplementation (0.2%, 0.4%, and 0.8%) in growing diets of Cobb 500 broilers (7-28 days) induced a proportional increase in body weight and decrease in feed conversion ratio (FCR; Jamil et al., 2015). At 28 days, body weight was 10% higher and FCR 10% lower (1.48 vs. 1.65 for the control group) when the incorporation rate in diet was 0.8%. Incorporation of 0.14% microalga *C. vulgaris* in the diet fed to 1–42-day-old Ross 308 broilers also led to a reduction in FCR from 1.60 to 1.54 (Rezvani et al., 2012). When supplemented at a rate of 1%, *C. vulgaris* induced a 4.6% body weight increase in 28-day-old Ross broilers (Kang et al., 2013). Dried *Sargassum wightii* powder at 1%, 2%, 3%, and 4% improved dietary palatability resulting in higher feed intake and enhanced digestibility and intestinal absorption, leading to increased body weight, and decreased FCR of broilers as compared to controls (Kumar, 2018).

4.2 Meat production and quality

Adding algae to broiler diets can alter carcass and meat quality by changing their proximate composition or color. Dietary supplementation with 4% of *E. prolifera* in 1–35-day-old broilers decreased the abdominal and breast Pectoralis major muscle fat content (Wang et al., 2013b). Incorporating 3% of *Ulva lactuca* to replace 3% of corn in a broiler diet from 12 to 33 days of age had no effect on feed intake and body weight, but increased dressing percentage and breast meat yield and reduced the proportion of abdominal fat (Abudabos et al., 2013). The greater breast-to-fat ratio observed in the treated group could be a consequence of the hypocholesterolemic role of this seaweed due to its high soluble fiber and PUFA content (Carvalho et al., 2009). A 5% dietary supplementation of *S. platensis* changed the FA composition of thigh muscles in 1–42-day-old broiler chickens by increasing arachidonic, docosahexaenoic-, and total PUFA, and decreasing arachidic, eicosenoic, and total monounsaturated fatty acids (MUFA; Bonos et al., 2016). Supplementing Ross 308 chicken diets with 2% *Schizochytrium* spp. increased the breast meat n-3 FA content by almost fourfold compared to control birds, mostly due to the high linoleic and n-3 LC FA content of this microalga (Baéza et al., 2015a). Such supplementation increased the ultimate pH of breast meat and decreased its lightness. It also strongly increased its oxidation susceptibility, whereby the thio-barbituric acid reactive substances (TBARS) value increased almost fourfold confirming the observations of Ribeiro et al. (2013, 2014), but without impacting any sensory characteristics of breast meat. Only thigh meat from broilers supplemented with microalgae had an abnormal flavor after cooking (Baéza et al., 2015a). Supplementation of diets for 21–42-day-old Ross 308 broilers at a rate of 1% or 2% for 168 days resulted in 47 and 97 mg DHA/100 g of breast meat and 97 and 156 mg DHA/100 g thigh meat, respectively, without any detrimental effect on growing performance (Keegan et al., 2020).

Due to their richness in carotenoids, some algae can affect poultry meat color. This was seen for *S. platensis,* which is rich in zeaxanthin, that increased both the yellowness and redness of breast (Pectoralis major) and thigh (Sartorius) muscles of growing 1–21-day-old Cobb 500 broilers when supplemented in the diet at a level of 0.4% (Toyomizu et al., 2001). More recently, Martins et al. (2021) reviewed the main effects of different microalgae used in diets on the nutritional and sensorial properties of poultry meat showing their interest to enrich meat with n-3 LC FA and carotenoid pigments.

4.3 Broiler health

C. vulgaris has been used to improve broiler health (Kang et al., 2013). Its incorporation at a rate of 1% in the diet of 1–28-day-old Ross broilers stimulated some innate immunity traits, such as white blood cells, lymphocytes, and plasma immunoglobulin IgA, IgM, and IgG that were increased by 7%, 6%, and 25.5%, respectively. Similarly, supplementing 1–35-day-old Ross broilers with 0.5 % *C. vulgaris* in diet increased blood IgG and IgM content by 46% and 107%, respectively (An et al., 2016). Significantly higher concentrations of IgA and IgM and lower IgG in the serum of broilers 1–35 day old fed diets supplemented at 0.5% with fermented seaweeds (*Undaria pinnatifida* or *Hizikia fusiformis*) than in the control group suggested the effectiveness of the feed additive used (Choi et al., 2014).

As leg weakness is a wide problem in poultry farming, algae could be an interesting alternative source of calcium (see Table 2). Nowadays, dietary calcium is mainly provided by limestone that decreases phosphorus digestibility, resulting in skeletal frailty. Red calcified seaweeds constitute an alternative organic and more digestible source of calcium for broilers that can contribute to a decrease in the use of inorganic forms, and consequently improved phosphorus digestibility, resulting in better bone development and leg strength, and reduced lameness (Bradbury et al., 2012).

Beyond its effects on animal basal health status, algae supplementation can help fight negative effects due to heat or bacterial challenges. Supplementing diets for heat-challenged Cobb 500 chickens (36°C, 6 h/day between 38 and 44 days of age) with 2% *S. platensis* increased the activity of both superoxide dismutase and glutathione peroxidase, resulting in improved antioxidant defense against free radical damage (Mirzaie et al., 2018). A basal diet supplementation with 100 mg/kg and 200 mg/kg of a fucoxanthin extract from brown seaweed *Undaria pinnatifida* increased catalase and superoxide dismutase activities and glutathione levels, and decreased malondialdehyde levels in the liver, breast, and drumstick tissues (Gumus et al., 2018). The fucoxanthins could then be used to regulate the antioxidant metabolism of broilers. The performance and health status of Arbor Acres broilers, challenged with *Salmonella enteritidis* between 12 and 42 days, were improved by dietary supplementation with 0.2% of brown seaweed extract, rich in oligosaccharides (Yan et al., 2011). It muted the negative effects of *Salmonella* infection by upregulating mRNA expression of interferon-γ and interleukin-10 in the cecum and increasing blood IgA production. This resulted in decreased *S. enteritidis* colonization and improvement of the intestinal barrier. It was shown that the brown alga *Undaria pinnatifida* (provided at 2% in the diet) reduced the effects of acute phase response following *Salmonella typhimurium* infection in Ross broilers (Koh et al., 2005). The acute phase response is an early defense system activated by infection and involving an innate immune response to prevent further infection and start the healing process (Cray et al., 2009). After 7 days of age, chicks fed diets supplemented with or without *U. pinnatifida,* were infected with *Salmonella* spp. and their acute phase response was activated, leading to changes in plasma protein concentrations. Brown alga supplementation did not affect chicken performance, but increased protein retention by decreasing protein breakdown and lowering tissue nitrogen balance.

For broiler chicken production, algae are likely to be valuable ingredients to improve growth, FCR, meat quality, and basal health (Table 4). However, when included at too high concentration in the diet, algae may have a deleterious effect on feed consumption, body weight gain and meat quality due to their composition and digestibility.

5 Conclusion and future trends

The available literature is quite abundant, and generally shows the beneficial effect of algae supplementation on many parameters related to animal performance, basal health, and product quality. This review does not particularly reveal the functional properties that are linked to the color of algae: green, red, or brown. Many varieties of algae exist across the world, and their chemical composition and digestibility vary widely and are likely to be influenced by how they are obtained. For this reason, acquiring knowledge regarding composition is important to evaluate any potential benefit for poultry. According to published results, a dietary incorporation rate of 2% for microalgae or a range comprised between 1% and 5% for macroalgae is suitable for both laying hens and broiler chickens, even though these ranges greatly depend on the type of algae and extract used and the expected benefits. Because raw algae are quite expensive, poultry producers do not consider them for mass utilization as a feed material. For this reason, there is a growing interest in producing alga extracts allowing the acquisition and concentration of valuable compounds, such as polyphenols and polysaccharides, targeting specific effects on animals' health and performance. However, it is still necessary to increase the levels of knowledge on the properties of algae to assess the impact they may have on the different production parameters, whether related to health and animal welfare, or the quality of products.

Algae represent a large and new source of nutrients with other health benefits as supplements in animal feed formulations. 'Algae' is a generic term that groups brown, green, and red types of both macro- and microalgae. These marine plants may play a key role in the future for poultry production, as they constitute a new and valuable nutrient source thanks to their nutritional composition and richness in polyphenols, polysaccharides, and fatty and amino acids. Many studies have evaluated the advantages and inconvenience of using micro- and macroalgae in poultry nutrition and their ability to improve animal health and, thus, welfare. This chapter describes the main nutritional characteristics of algae and the current knowledge on their effects in poultry production, considering impacts on animal health, growth performance, and product quality (eggs and meat). The increase in laying rate and egg weight can reach +4.0 to 8.6 percentage points and +1.3 to 1.5 g, respectively. The increase in body weight of broilers and decrease in feed conversion ratio can vary from 5% to 22% and from 4% to 15%, respectively. According to the

Table 4 Summary of algae used in broiler nutrition

Reference	Seaweed	Algae	Incorporation rate in diet	Animals	Effects
Toyomizu et al. (2001)	*Spirulina platensis*	Microalgae	0.40%	Cobb 500, 21 days	Increased yellowness and redness of pectoralis major and sartorius muscles
Mirzaie et al. (2018)	*Spirulina platensis*	Microalgae	2.0%	Cobb 500, 38-44 days	Increased SOD[a] and Gpx[b] blood levels in case of heat challenge
Jamil et al. (2015)	*Spirulina platensis*	Microalgae	0.20%, 0.40%, 0.80%	Cobb 500, 7–28 days	Increased body weight and decreases FCR[c]
Ross et al. (1994)	*Spirulina platensis*	Microalgae	1.0%	Cobb 500, 35 days	Increased body weight
Bonos et al. (2016)	*Spirulina platensis*	Microalgae	5.0%	Broilers, 42 days	Increased polyunsaturated FA[d] content in thigh muscle
Rezvani et al. (2012)	*Chlorella vulgaris*	Microalgae	0.07%, 0.14%	Ross 308, 42 days	Decreased FCR
Kang et al. (2013)	*Chlorella vulgaris*	Microalgae	1.0%	Ross chickens, 28 days	Increased body weight and stimulated immune system (number of white cells and lymphocytes)
An et al. (2016)	*Chlorella vulgaris*	Microalgae	0.50%	Ross chickens, 35 days	Increased IgG and IgM blood levels
Yan and Kim (2013)	*Schizochytrium* sp.	Microalgae	0.1–0.2%	Ross 308 chickens	Increased n-3 FA content in breast muscle
Ribeiro et al. (2013, 2014)	*Schizochytrium* sp.	Microalgae	3.7-7.4%	Ross 308 chickens	Increased n-3 FA content in breast muscle
Baéza et al. (2015a)	*Schizochytrium* sp.	Microalgae	2.0%	Ross 308 chickens	Decreased FCR and increased n-3 fatty acid content in breast muscle
Wang et al. (2013b)	*Enteromorpha prolifera*	Macroalgae	3.0%, 4.0%	Arbor Acres, 35 days	Increased daily weight gain and decreased FCR, abdominal fat and fat content in pectoralis major muscle

(*Continued*)

Table 4 (*Continued*)

Reference	Seaweed	Algae	Incorporation rate in diet	Animals	Effects
Abudabos et al. (2013)	*Ulva lactuca*	Macroalgae	3.0%	Ross chickens, 12–33 days	Increased dressing percentage and breast meat yield; decreased abdominal fat percentage
Evans and Critchley (2014)	*Ascophyllum nodosum*	Macroalgae	1.0%	Ross chickens, 42 days	Improved growth performance
El-Deek et al. (2011)	*Sargassum* spp.	Macroalgae	2.0, 4.0%	Broilers, 18–39 d	Decreased cholesterol and increased n-3 FA content in blood
Koh et al. (2005)	*Untaria pinnatifida*	Macroalgae	2.0%	Ross chickens, 7 dāys	Decreased negative impact of acute phase response to *Salmonella* strain; increased protein retention in thigh muscle
Yan et al. (2011)	Mix of brown seaweeds (oligosaccharide-rich extract)	Macroalgae	0.2%	Arbor Acres, 12-42 days	Muted negative effects of *Salmonella enteritidis* on growth performance, improved intestinal barrier and defense
Ventura et al. (1994)	*Ulva lactuca*	Macroalgae	10%, 20%, 30%	Adult cockerels and chicks (3 weeks)	Decreased body weight and feed intake
Bradbury et al. (2012)	Red calcified seaweed	Macroalgae	0.77% Ca concentration	Cobb 500, 1–40 days	Improved bone health and leg strength

[a]SOD, superoxide dismutase.
[b]Gpx, gluthation peroxidase.
[c]FCR, feed conversion ratio.
[d]FA: fatty acid.

literature, it seems that a dietary incorporation rate of 2% for microalgae or a range between 1% and 5% for macroalgae is suitable for both laying hens and broiler chickens, even though these ranges greatly depend on the type of algae used and the expected benefits for poultry production.

Even if the market for macro- and microalgae use in feed has a great potential, the actual contribution of those products to animal nutrition is quite limited. First, the global seaweed industry largely relies on harvesting seaweed as natural resource. Over-harvesting due to increases in seaweed demand could negatively affect the environment and the sustainability of supply. The second limit is due to the small scale of production and to the high production costs, especially in comparison with traditional commodities, and the competition with the use for human food, food supplements, and nutraceuticals (Araujo et al., 2021). Moreover, in Europe, algae production remains limited by a series of technological, regulatory, and market-related barriers. On the other hand, the efficiency of macro- and microalgae in poultry feed could also be improved by optimizing their processing combined for example with fermentation or associate with other natural bioactive compounds or clay (Kulshreshtha et al., 2020). Regarding safety concerns, seaweeds (cultivated or directly harvested) for food and feed are always tested to measure levels for trace elements (As, Cd, Pb, Sn, and Hg) in order to meet national and international regulation and safety standards. Other safety hazards for seaweeds may include anti-nutritional factors, radioactive isotopes, ammonium, dioxins and pesticides to guarantee contaminant-free materials (Stengel et al., 2011). Given current knowledge from the literature, data from the Rapid Alert System for Food and Feed and results from a stakeholder survey, Banach et al. (2020) identified 22 food safety hazards in the European seaweed chain among which, arsenic, cadmium, iodine, and *Salmonella* were identified as major hazards.

6 Where to look for further information

Further reading

- Seaweed and microalgae as alternative sources of protein. Edited by Prof. Xin Gen Lei, Cornell University, USA, 2021, Burleigh Dodds Science Publishing, London, UK, p. 1-344.
- Report of the European Commission on Algae production in Europe: status challenges and future developments. Workshop on 27/02/2019. https://knowledge4policy.ec.europa.eu/sites/default/files/jrc116858_report_workshop_kcb_algae_workshop.pdf (accessed 10 May 2023).

Conferences

- Annual conferences organized by the European Algae Biomass Association.
- https://eaba-association.org (accessed 10 May 2023).

Major international research projects

- EU project HORIZON 2020: Microalgae protein ingredients for the food and feed of the future. https://cordis.europa.eu/project/id/862980 (accessed 10 May 2023).
- Project from the Cornell University Agricultural Experiment Station: Developing a new, healthy, value-added meat by feeding chicks with defatted marine microalgae from biofuel production. https://cals.cornell.edu/agricultural-experiment-station/research-impacts/ researchers-use-algae-based-animal-feed-produce-healthier-meat (accessed 10 May 2023).
- FranceAgriMer project SAGA (P3A-PSA-SAGA n°2017-1808): Sans Antibiotique Grâce aux Algues (Olmix group, INRAE). https://www.franceagrimer.fr/content/download/54605/document/P3A%20-%20fiche%20laur%C3%A9at-SAGA.pdf (accessed 10 May 2023).
- Major seaweed companies supplying poultry feed and additives.
- The major seaweed companies supplying poultry feed and additives are presented in the review published by Kulshreshtha et al. (2020).

7 References

Abudabos, A. M., Okab, A. B., Aljumaah, R. S., Samara, E. M., Abdoun, K. A. and Al-Haidary, A. A. (2013). Nutritional value of green seaweed (*Ulva Lactuca*) for broiler chickens. *Ital. J. Anim. Sci.* 12(2): e28.

Al-Harthi, M. and El-Deek, A. (2012). Nutrient profiles of brown marine algae (*Sargassum dentifebium*) as affected by different processing methods for chickens. *J. Food Agricult. Environ.* 10: 475–480.

Alvarez-Rivera, G., Bueno, M., Ballesteros-Vivas, D., Mendiola, J. A. and Ibañez, E. (2019). Pressurized liquid extraction. In: Poole, C. F. (Eds.), *Liquid-Phase Extraction*. Elsevier, Amsterdam, Netherlands, pp. 375–398.

An, B. K., Kim, K. E., Jeon, J. Y. and Lee, K. W. (2016). Effect of dried *Chlorella vulgaris* and Chlorella growth factor on growth performance, meat qualities and humoral immune responses in broiler chickens. *SpringerPlus* 5(1): 718.

Annamalai, S. N., Das, P., Thaher, M. I. A., Quadir, M. A., Khan, S., Mahata, C. and Al Jabri, H. (2021). Nutrients and energy digestibility of microalgal biomass for fish feed applications. *Sustainability* 13(23): 19 p.

Araujo, R., Vazquez Calderon, F., Sanchez Lopez, J., Costa Azevedo, I. C., Brhun, A., Fluch, S., Garcia Tasende, M., Ghaderiardakani, F., Ilmjärv, T., Laurans, M., Mac Monagail, M., Mangini, S., Peteiro, C., Rebours, C., Stefansson, T. and Ullmann, J. (2021). Current status of the algae production industry in Europe: an emerging sector of the blue bioeconomy. *Front. Mar. Sci.* 7: 626389, 24 p.

Baéza, E., Chartrin, P., Lessire, M., Méteau, K., Chesneau, G., Guillevic, M. and Mourot, J. (2015a). Is it possible to increase n-3 fatty acid content of meat without affecting its technological and/or sensory quality and the growing performance of chickens? *Br. Poult. Sci.* 56(5): 543–550.

Baéza, E., Chartrin, P., Lessire, M., Méteau, K., Chesneau, G., Guillevic, M. and Mourot, J. (2015b). Is it possible to increase the n-3 fatty acid content of eggs without affecting their technological and/or sensorial quality and the laying performance of hens? *Br. Poult. Sci.* 56(6): 748–754.

Banach, J. L., Hoek-van den Hil, E. F. and van der Fels-Klerx, H. J. (2020). Food safety hazards in the European seaweed chain. *Comp. Rev. Food Sci. Food Saf.* 19(2): 332–364.

Baurhoo, B., Letellier, A., Zhao, X. and Ruiz-Feria, C. A. (2007). Cecal populations of lactobacilli and bifidobacteria and *Escherichia coli* populations after in vivo *Escherichia coli* challenge in birds fed diets with purified lignin or mannanoligosaccharides. *Poult. Sci.* 86(12): 2509–2516.

Becker, E. W. (1986). Nutritional properties of microalgae: potentials and constraints. In: Richmond, A. (Ed.), *CRC Handbook of Microalgal Mass Culture*. CRC Press, Boca Raton, FL, pp. 339–420.

Becker, E. W. (2007). Microalgae in human and animal nutrition. In: Richmond, A. (Eds.), *Handbook of Microalgal Culture*. Blackwell Publishing Ltd, Oxford, UK, pp. 312–351.

Bonos, E., Kasapidou, E., Kargopoulos, A., Karampampas, A., Christaki, E., Florou-Paneri, P. and Nikolakakis, I. (2016). Spirulina as a functional ingredient in broiler chicken diets. *S. Afr. J. Anim. Sci.* 46(1): 94.

Bradbury, E. J., Wilkinson, S. J., Cronin, G. M., Walk, C. L. and Cowieson, A. J. (2012). The effect of marine calcium source on broiler leg integrity. In: Proceedings of the 23rd Annual Australian Poultry Science Symposium, Sydney, New South Wales, Australia, 19–22 February 2012, pp. 85–88.

Carrillo, S., López, E., Casas, M. M., Avila, E., Castillo, R. M., Carranco, M. E., Calvo, C. and Pérez-Gil, F. (2008). Potential use of seaweeds in the laying hen ration to improve the quality of n-3 fatty acid enriched eggs. In: 19th International Seaweed Symposium, Springer Netherlands, Dordrecht, pp. 721–728.

Carrillo, S., Bahena, A., Casas, M. and Perez-Gil, F. (2012). The alga *Sargassum* spp. as alternative to reduce egg cholesterol content. *Cuban J. Agricult. Sci.* 46: 181–186.

Carvalho, A. F. U., Portela, M. C. C., Sousa, M. B., Martins, F. S., Rocha, F. C., Farias, D. F. and Feitosa, J. P. A. (2009). Physiological and physico-chemical characterization of dietary fibre from the green seaweed Ulva fasciata Delile. *Braz. J. Biol.* 69(3): 969–977.

Choi, Y. J., Lee, S. R. and Oh, J. W. (2014). Effects of dietary fermented seaweed and seaweed fusiforme on growth performance, carcass parameters and immunoglobulin concentration in broiler chicks. *Asian Australas. J. Anim. Sci.* 27(6): 862–870.

Choi, Y. J., Lee, E. C., Na, Y. and Lee, S. R. (2018). Effects of dietary supplementation with fermented and non-fermented brown algae by-products on laying performance, egg quality, and blood profile in laying hens. *Asian Austral. J. Anim. Sci.* 31: 1654–1659.

Chojnacka, K. (2012). Biologically active compounds in seaweed extracts – the prospects for the application. *The Open Conference Proceedings Journal* 3: 20–28.

Costa, M., Cardoso, C., Afonso, C., Bandarra, N. M. and Prates, J. A. M. (2020). Current knowledge and future perspectives of the use of seaweeds for livestock production and meat quality: a systematic review. *J. Anim. Physiol. Anim. Nutr.* 105: 1075–1102.

Crampon, C., Boutin, O. and Badens, E. (2011). Supercritical carbon dioxide extraction of molecules of interest from microalgae and seaweeds. *Ind. Eng. Chem. Res.* 50(15): 8941–8953.

Cray, C., Zaias, J. and Altman, N. H. (2009). Acute phase response in animals: a review. *Comp. Med.* 59(6): 517–526.

Debbarma, J., Madhusudana Rao, B., Murthy, L. N., Mathew, S., Venkateshwarlu, G. and Ravishankar, C. N. (2016). Nutritional profiling of the edible seaweeds *Gracilaria edulis*, *Ulva lactuca* and *Sargassum* sp. *Indian J. Fisher.* 63(3): 81–87.

Dvir, I., Chayoth, R., Sod-Moriah, U., Shany, S., Nyska, A., Stark, A. H., Madar, Z. and Arad, S. M. (2000). Soluble polysaccharide and biomass of red microalga *Porphyridium* sp. alter intestinal morphology and reduce serum cholesterol in rats. *Br. J. Nutr.* 84(4): 469–476.

El-Deek, A., Al-Harthi, M., Abdalla, A. and Elbanoby, M. M. (2011). The use of brown algae meal in finisher broiler diets. *Egypt. Poult. Sci.* 31(IV): 767–781.

El Gamal, A. A. (2010). Biological importance of marine algae. *Saudi Pharm. J.* 18(1): 1–25.

Englmaierová, M., Skřivan, M. and Bubancová, I. (2013). A comparison of lutein, spray-dried Chlorella, and synthetic carotenoids effects on yolk colour, oxidative stability, and reproductive performance of laying hens. *Czech J. Anim. Sci.* 58(9): 412–419.

Evans, F. D. and Critchley, A. T. (2014). Seaweeds for animal production use. *J. Appl. Phycol.* 26(2): 891–899.

Feng, J., Long, S., Zhang, H. J., Wu, S. G., Qi, G. H. and Wang, J. (2020). Comparative effects of dietary microalgae oil and fish oil on fatty acid composition and sensory quality of table eggs. *Poult. Sci.* 99(3): 1734–1743.

Fraeye, I., Bruneel, C., Lemahieu, C., Buyse, J., Muylaert, K. and Foubert, I. (2012). Dietary enrichment of eggs with omega-3 fatty acids: a review. *Food Res. Int.* 48(2): 961–969.

Fujii, K. (2012). Process integration of supercritical carbon dioxide extraction and acid treatment for astaxanthin extraction from a vegetative microalga. *Food Bioprod. Process.* 90(4): 762–766.

Garrido-Cardenas, J. A., Manzano-Agugliaro, F., Acien-Fernandez, F. G. and Molina-Grima, E. (2018). Microalgae research worldwide. *Algal Res.* 35: 50–60.

Ginzberg, A., Cohen, M., Sod-Moriah, U. A., Shany, S., Rosenshtrauch, A. and Arad, S. (2000). Chickens fed with biomass of the red microalga *Porphyridium* sp. have reduced blood cholesterol level and modified fatty acid composition in egg yolk. *J. Appl. Phycol.* 12(3/5): 325–330.

Gouveia, L., Marques, A. E., Sousa, J. M., Moura, P. and Bandarra, N. M. (2010). Microalgae – source of natural bioactive molecules as functional ingredients. *Food Sci. Technol. Bull. Funct. Foods* 7(2): 21–37.

Gumus, R., Urcar Gelen, S. U., Koseoglu, S., Ozkanlar, S., Ceylan, Z. G. and Imik, H. (2018). The effects of fucoxanthin dietary inclusion on the growth performance, antioxidant metabolism and meat quality of broilers. *Braz. J. Poult. Sci.* 20(3): 487–496.

Haberecht, S., Wilkinson, S., Roberts, J., Wu, S. and Swick, R. (2018). Unlocking the potential health and growth benefits of macroscopic algae for poultry. *World's Poult. Sci. J.* 74(1): 5–20.

Hahn, T., Lang, S., Ulber, R. and Muffler, K. (2012). Novel procedures for the extraction of fucoidan from brown algae. *Process Biochem.* 47(12): 1691–1698.

Halle, I., Janczyk, P. and Souffrant, W. (2009). Effect of microalgae *Chlorella vulgaris* on laying hen performance. *Archiv. Zootech.* 12: 5–13.

Harnedy, P. A. and Fitzgerald, R. J. (2011). Bioactive proteins, peptides, and amino acids from macroalgae. *J. Phycol.* 47(2): 218–232.

Heo, S. J., Park, E. J., Lee, K. W. and Jeon, Y. J. (2005). Antioxidant activities of enzymatic extracts from brown seaweeds. *Bioresour. Technol.* 96(14): 1613–1623.

Herrero, M., Simó, C., Ibáñez, E. and Cifuentes, A. (2005). Capillary electrophoresis-mass spectrometry of Spirulina platensis proteins obtained by pressurized liquid extraction. *Electrophoresis* 26(21): 4215–4224.

Hosikian, A., Lim, S., Halim, R. and Danquah, M. K. (2010). Chlorophyll extraction from microalgae: a review on the process engineering aspects. *Int. J. Chem. Eng.*: 391632: 11 p.

Ibañez, E., Herrero, M., Mendiola, J. A. and Castro-Puyana, M. (2012). Extraction and characterization of bioactive compounds with health benefits from marine resources: macro and micro algae, cyanobacteria, and invertebrates. In: *Marine Bioactive Compounds: Sources, Characterization and Applications*. Springer US, pp. 55–98.

Jaime, L., Mendiola, J. A., Herrero, M., Soler-Rivas, C., Santoyo, S., Señorans, F. J., Cifuentes, A. and Ibáñez, E. (2005). Separation and characterization of antioxidants from *Spirulina platensis* microalga combining pressurized liquid extraction, TLC, and HPLC-DAD. *J. Sep. Sci.* 28(16): 2111–2119.

Jamil, A., Akanda, M., Rahman, M., Hossain, M. and Islam, M. (2015). Prebiotic competence of spirulina on the production performance of broiler chickens. *J. Adv. Vet. Anim. Res.* 2(3): 304–309.

Janczyk, P., Halle, B. and Souffrant, W. B. (2009). Microbial community composition of the crop and ceca contents of laying hens fed diets supplemented with *Chlorella vulgaris*. *Poult. Sci.* 88(11): 2324–2332.

Kadam, S. U., Tiwari, B. K. and O'donnell, C. P. (2013). Application of novel extraction technologies for bioactives from marine algae. *J. Agric. Food Chem.* 61(20): 4667–4675.

Kalia, S. and Lei, X. G. (2022). Dietary microalgae on poultry meat and eggs: explained versus unexplained effects. *Curr. Opin. Biotechnol.* 75: 102689, 5 p.

Kang, H. K., Salim, H. M., Akter, N., Kim, D. W., Kim, J. H., Bang, H. T., Kim, M. J., Na, J. C., Hwangbo, J., Choi, H. C. and Suh, O. S. (2013). Effect of various forms of dietary Chlorella supplementation on growth performance, immune characteristics, and intestinal microflora population of broiler chickens. *J. Appl. Poult. Res.* 22(1): 100–108.

Keegan, J. D., Fusconi, G., Morlacchini, M. and Moran, C. A. (2020). Comparing docosahexaenoic acid supplementation strategies in terms of broiler tissue enrichment productivity, and cost. *J. Appl. Poult. Res.* 29(3): 636–652.

Koh, T., Im, J., Park, I., Lee, H. and Choi, Y. (2005). Effect of dietary brown seaweed levels on the protein and energy metabolism in broiler chicks activated acute phase response. *J. Anim. Sci. Technol.* 47(3): 379–390.

Korczyński, M., Witkowska, Z., Opaliński, S., Świniarska, M. and Dobrzański, Z. (2015). Algae extract as a potential feed additive. In: Kim, S.-K. and Chojnacka, K. (Eds.), *Marine Algae Extracts*. Wiley-VCH Verlag GmbH & Co. KGaA, Weinheim, pp. 603–626.

Kotrbáček, V., Doubek, J. and Doucha, J. (2015). The chlorococcalean alga Chlorella in animal nutrition: a review. *J. Appl. Phycol.* 27(6): 2173–2180.

Kovač Blagojević, D., Simeunović, J., Babić, O. and Milovanović, I. (2013). Algae in food and feed. *Food Feed Res.* 40: 21–31.

Kulshreshtha, G., Rathgeber, B., Stratton, G., Thomas, N., Evans, F., Critchley, A., Hafting, J. and Prithiviraj, B. (2014). Feed supplementation with red seaweeds, *Chondrus crispus* and *Sarcodiotheca gaudichaudii*, affects performance, egg quality, and gut microbiota of layer hens. *Poult. Sci.* 93(12): 2991–3001.

Kulshreshtha, G., Rathgeber, B., MacIsaac, J., Boulianne, M., Brigitte, L., Stratton, G., Thomas, N. A., Critchley, A. T., Hafting, J. and Prithiviraj, B. (2017). Feed supplementation with red seaweeds, *Chondrus crispus* and *Sarcodiotheca gaudichaudii*, reduce *Salmonella enteritidis* in laying hens. *Front. Microbiol.* 8(567): 12 p.

Kulshreshtha, G., Hincke, M. T., Prithiviraj, B. F. and Critchley, A. (2020). A review on the varied uses of macroalgae as dietary supplements in selected poultry with special reference to laying hen and broiler chickens. *J. Mar. Sci. Eng.* 8(536): 28 p.

Kumar, S., Sahoo, D. and Levine, I. (2015). Assessment of nutritional value in a brown seaweed *Sargassum wightii* and their seasonal variations. *Algal Res.* 9: 117–125.

Kumar, A. K. (2018). Effect of *Sargassum wightii* on growth, carcass and serum qualities of broiler chickens. *Vet. Sci. Res.* 3(156): 10 p.

Kuorwel, K. K., Cran, M. J., Sonneveld, K., Miltz, J. and Bigger, S. W. (2011). Antimicrobial activity of biodegradable polysaccharide and protein-based films containing active agents. *J. Food Sci.* 76(3): R90–R102.

Li, Z., Wang, B., Zhang, Q., Qu, Y., Xu, H. and Li, G. (2012). Preparation and antioxidant property of extract and semipurified fractions of *Caulerpa racemosa*. *J. Appl. Phycol.* 24(6): 1527–1536.

Li, Q., Luo, J., Wang, C., Tai, W., Wang, H., Zhang, X., Liu, K., Jia, Y., Lyv, X., Wang, L. and He, H. (2018). Ulvan extracted from green seaweeds as new natural additives in diets for laying hens. *J. Appl. Phycol.* 30(3): 2017–2027.

Liang, K., Zhang, Q., Gu, M. and Cong, W. (2013). Effect of phosphorus on lipid accumulation in freshwater microalga *Chlorella* sp. *J. Appl. Phycol.* 25(1): 311–318.

Lin, W., Wang, W., Liao, D., Chen, D., Zhu, P., Cai, G. and Kiyoshi, A. (2015). Polysaccharides from *Enteromorpha prolifera* improve glucose metabolism in diabetic rats. *J. Diabetes Res.*: 675201, 12 p. doi: 10.1155/2015/675201.

Lipstein, B., Hurwitz, S. and Bornstein, S. (1980). The nutritional value of algae for poultry. Dried chlorella in layer diets. *Br. Poult. Sci.* 21(1): 23–27.

Macías-Sánchez, M. D., Mantell, C., Rodríguez, M., Martínez De La Ossa, E., Lubián, L. M. and Montero, O. (2009). Comparison of supercritical fluid and ultrasound-assisted extraction of carotenoids and chlorophyll a from *Dunaliella salina*. *Talanta* 77(3): 948–952.

Martinez-Delgado, A. A., Khandual, S., Morales-Hernandez, N., Martinez-Bustos, F., Vélez-Medina, J. J. and Nolasco-Soria, H. (2020). Fish feed formulation with microalgae *H. Pluvialis* and *A. Platensis*: effect of extrusion process on stability of astaxanthin and antioxidant capacity. *J. Food Nutr. Sci.* 7(1): 1–8.

Martins, C. F., Ribeiro, D. M., Costa, M., Coelho, D., Alfaia, C. M., Lordelo, M., Almeida, A. M., Freire, J. P. B. and Prates, J. A. M. (2021). Using microalgae as a sustainable feed resource to enhance quality and nutritional value of pork and poultry meat. *Foods* 10(12): 12 p.

Mendes, R. L., Coelho, J. P., Fernandes, H. L., Marrucho, I. J., Cabral, J. M. S., Novais, J. M. and Palavra, A. F. (1995). Applications of supercritical CO_2 extraction to microalgae and plants. *J. Chem. Technol. Biotechnol.* 62(1): 53–59.

Mendiola, J. A., Jaime, L., Santoyo, S., Reglero, G., Cifuentes, A., Ibañez, E. and Señoráns, F. J. (2007). Screening of functional compounds in supercritical fluid extracts from *Spirulina platensis*. *Food Chem.* 102(4): 1357–1367.

Michalak, I. and Chojnacka, K. (2014). Algal extracts: technology and advances. *Eng. Life Sci.* 14(6): 581–591.

Michalak, I. and Chojnacka, K. (2015). Algae as production systems of bioactive compounds. *Eng. Life Sci.* 15(2): 160–176.

Michalak, I. and Mahrose, K. (2020). Seaweeds, intact and processed, as a valuable component of poultry feeds. *J. Mar. Sci. Eng.* 8(620): 28 p.

Michiels, J., Skrivanova, E., Missotten, J., Ovyn, A., Mrazek, J., De Smet, S. and Dierick, N. (2012). Intact brown seaweed (*Ascophyllum nodosum*) in diets of weaned piglets: effects on performance, gut bacteria and morphology and plasma oxidative status. *J. Anim. Physiol. Anim. Nutr.* 96(6): 1101–1111.

Mirzaie, S., Zirak-Khattab, F., Hosseini, S. A. and Donyaei-Darian, H. (2018). Effects of dietary Spirulina on antioxidant status, lipid profile, immune response and performance characteristics of broiler chickens reared under high ambient temperature. *Asian-Australas. J. Anim. Sci.* 31(4): 556–563.

Moran, C. A., Morlacchini, M., Keegan, J. D. and Fusconi, G. (2019). Increasing the omega-3 content of hen's eggs through dietary supplementation with *Aurantiochytrium limacinum* microalgae: effect of inclusion rate on the temporal pattern of docosahexaenoic acid enrichment, efficiency of transfer, and egg characteristics. *J. Appl. Poult. Res.* 28(2): 329–338.

Nakagawa, H. (1997). Effect of dietary algae on improvement of lipid metabolism in fish. *Biomed. Pharmacother.* 51(8): 345–348.

Napolitano, F., Serrapica, M. and Braghieri, A. (2013). Contrasting attitudes towards animal welfare issues within the food chain. *Animals (Basel)* 3(2): 551–557.

Nishide, E., Uchida, N., Chapman, A., Anderson, R. J., Vreeland, V. and Davison, I. (2003). Effects of Ulva powder on the ingestion and excretion of cholesterol in rats. In: Chapman, A. R. O., Anderson, R. J., Vreeland, V. J. and Davison, I. R. (Eds), Proceedings of the 17th International Seaweed Symposium. Oxford University Press, Oxford, pp. 165–168.

Nocella, G., Hubbard, L. and Scarpa, R. (2009). Farm animal welfare, consumer willingness to pay, and trust: results of a cross-national survey. *Appl. Econ. Perspect. Policy* 32: 275–297.

Norambuena, F., Hermon, K., Skrzypczyk, V., Emery, J. A., Sharon, Y., Beard, A. and Turchini, G. M. (2015). Algae in fish feed: performances and fatty acid metabolism in juvenile Atlantic salmon. *PLoS ONE* 10(4): e0124042, 17 p.

Park, J. K. (2015). Algal polysaccharides: properties and applications. *Biochem. Anal. Biochem.* 4(176): 2 p.

Pasquet, V., Morisset, P., Ihammouine, S., Chepied, A., Aumailley, L., Berard, J. B., Serive, B., Kaas, R., Lanneluc, I., Thiery, V., Lafferriere, M., Piot, J. M., Patrice, T., Cadoret, J. P. and Picot, L. (2011). Antiproliferative activity of violaxanthin isolated from bioguided fractionation of *Dunaliella tertiolecta* extracts. *Mar. Drugs* 9(5): 819–831.

Praiboon, J., Palakas, S., Noiraksa, T. and Miyashita, K. (2018). Seasonal variation in nutritional composition and anti-proliferative activity of brown seaweed, *Sargassum oligocystum*. *J. Appl. Phycol.* 30(1): 101–111.

Pulz, O. and Gross, W. (2004). Valuable products from biotechnology of microalgae. *Appl. Microbiol. Biotechnol.* 65(6): 635–648.

Rasyid, A. (2017). Evaluation of nutritional composition of the dried seaweed *Ulva lactuca* from Pameungpeuk Waters, Indonesia. *Trop. Life Sci. Res.* 28(2): 119–125.

Rezvani, M., Zaghari, M. and Moravej, H. (2012). A survey on *Chlorella vulgaris* effect's on performance and cellular immunity in broilers. *Int. J. Agricult. Sci. Res.* 3: 9–15.

Ribeiro, T., Lordelo, M. M., Alves, S. P., Bessa, R. J. B., Costa, P., Lemos, J. P. C., Ferreira, L. M. A., Fontes, C. M. G. A. and Prates, J. A. M. (2013). Direct supplementation of diet is the most efficient way of enriching broiler meat with n-3 long-chain polyunsaturated fatty acids. *Br. Poult. Sci.* 54(6): 753–765.

Ribeiro, T., Lordelo, M. M., Costa, P., Alves, S. P., Benevides, W. S., Bessa, R. J. B., Lemos, J. P. C., Pinto, R. M. A., Ferreira, L. M. A., Fontes, C. M. G. A. and Prates, J. A. M. (2014). Effect of reduced dietary protein and supplementation with a docosahexaenoic acid product on broiler performance and meat quality. *Br. Poult. Sci.* 55(6): 752–765.

Rizk, Y. (2017). Effect on dietary green tea and dried seaweed on productive and physiological performance of laying hens during late phase of production. *Egypt. Poult. Sci. J.* 37: 685–706.

Ross, E., Puapong, D. P., Cepeda, F. P. and Patterson, P. H. (1994). Comparison of freeze-dried and extruded *Spirulina platensis* as yolk pigmenting agents. *Poult. Sci.* 73(8): 1282–1289.

Rumin, J., Nicolau, E., Gonçalves de Oliveira Junior, R., Fuentes-Grünewald, C., Flynn, K. J. and Picot, L. (2020). A bibliometric analysis of microalgae research in the world, Europe and the European Atlantic area. *Mar. Drugs* 18(79): 28 p.

Sako, T., Matsumoto, K. and Tanaka, R. (1999). Recent progress on research and applications of non-digestible galacto-oligosaccharides. *Int. Dairy J.* 9(1): 69–80.

Santoyo, S., Plaza, M., Jaime, L., Ibañez, E., Reglero, G. and Señorans, J. (2011). Pressurized liquids as an alternative green process to extract antiviral agents from the edible seaweed *Himanthalia elongata*. *J. Appl. Phycol.* 23(5): 909–917.

Scalbert, A., Johnson, I. T. and Saltmarsh, M. (2005). Polyphenols: antioxidants and beyond. *Am. J. Clin. Nutr.* 81(1): 215S–217S.

Shalaby, E. A. (2011). Algae as promising organisms for environment and health. *Plant Signal. Behav.* 6(9): 1338–1350.

Shapira, N., Weill, P. and Loewenbach, R. (2008). Egg fortification with n-3 polyunsaturated fatty acids (PUFA): nutritional benefits versus high n-6 PUFA western diets, and consumer acceptance. *Isr. Med. Assoc. J.* 10(4): 262–265.

Shields, R. and Lupatsch, I. (2012). Algae for aquaculture and animal feeds. In: Posten, C. and Walter, C. (Eds), *Microalgal Biotechnology: Integration and Economy*. De Gruyter, Berlin, Germany, pp. 23–37.

Skala, D., Zizovic, I. and Gavrancic, S. (2002). Supercritical fluid extraction: application in the food industry. *Hemijska Ind.* 56(5): 179–190.

Starmans, D. A. J. and Nijhuis, H. H. (1996). Extraction of secondary metabolites from plant material: a review. *Trends Food Sci. Technol.* 7(6): 191–197.

Stengel, D. B., Connan, S. and Popper, Z. A. (2011). Alga chemodiversity and bioactivity: sources of natural variability and implications for commercial application. *Biotechnol. Adv.* 29(5): 483–501.

Sujatha, T. and Narahari, D. (2011). Effect of designer diets on egg yolk composition of 'White Leghorn' hens. *J. Food Sci. Technol.* 48(4): 494–497.

Tierney, M. S., Soler-Vila, A., Rai, D. K., Croft, A. K., Brunton, N. P. and Smyth, T. J. (2014). UPLC-MS profiling of low molecular weight phlorotannin polymers in *Ascophyllum nodosum*, *Pelvetia canaliculata* and *Fucus spiralis*. *Metabolomics* 10(3): 524–535.

Tobiszewski, M., Namieśnik, J. and Pena-Pereira, F. (2017). Environmental risk-based ranking of solvents using the combination of a multimedia model and multi-criteria decision analysis. *Green Chem.* 19(4): 1034–1042.

Tokusoglu, Ö. and Ünal, M. K. (2003). Biomass nutrient profiles of three microalgae: *Spirulina platensis*, *Chlorella vulgaris*, and *Isochrisis galbana*. J. Food Sci. 68(4): 1144–1148.

Toyomizu, M., Sato, K., Taroda, H., Kato, T. and Akiba, Y. (2001). Effects of dietary *Spirulina* on meat colour in muscle of broiler chickens. *Br. Poult. Sci.* 42(2): 197–202.

Ventura, M. R., Castañon, J. I. R. and Mc Nab, J. M. (1994). Nutritional value of seaweed (*Ulva rigida*) for poultry. *Anim. Feed Sci. Technol.* 49(1–2): 87–92.

Volkman, J. and Brown, M. (2005). Nutritional value of microalgae and applications. *Algal Cult. Analogues Blooms Appl.* 1: 407–457.

Wang, L., Min, M., Li, Y., Chen, P., Chen, Y., Liu, Y., Wang, Y. and Ruan, R. (2010). Cultivation of green algae *Chlorella* sp. in different wastewaters from municipal wastewater treatment plant. *Appl. Biochem. Biotechnol.* 162(4): 1174–1186.

Wang, S., Jia, Y., Wang, L., Zhu, F. and Lin, Y. (2013a). Enteromorpha prolifera supplemental level: effects on laying performance, egg quality, immune function and microflora in feces of laying hens. *Chin. J. Anim. Nutr.* 25(6): 1346–1352.

Wang, S., Shi, X., Zhou, C. and Lin, Y. (2013b). Enteromorpha prolifera: effects on performance, carcass quality and small intestinal digestive enzyme activities of broilers. *Chin. J. Anim. Nutr.* 25(6): 1332–1337.

Yan, G. L., Guo, Y. M., Yuan, J. M., Liu, D. and Zhang, B. K. (2011). Sodium alginate oligosaccharides from brown algae inhibit *Salmonella enteritidis* colonization in broiler chickens. *Poult. Sci.* 90(7): 1441–1448.

Yan, L. and Kim, I. H. (2013). Effects of dietary ω-3 fatty acid-enriched microalgae supplementation on growth performance, blood profiles, meat quality and fatty acid composition of meat in broilers. *J. Appl. Anim. Res.* 41(4): 392–397.

Zahroojian, N. and Morajev, H. M. S. (2013). Effects of dietary marine algae (*Spirulina platensis*) on egg quality and production performance of laying hens. *J. Agricult. Sci. Technol.* 15: 1353–1360.

Zheng, L., Oh, S. T., Jeon, J. Y., Moon, B. H., Kwon, H. S., Lim, S. U., An, B. K. and Kang, C. W. (2011). The dietary effects of fermented *Chlorella vulgaris* (CBT) on production performance, liver lipids and intestinal microflora in laying hens. *Asian Austral. J. Anim. Sci.* 25: 261–266.

Chapter 17

Advances in synthetic biology to improve poultry production[1]

Emile Fierro-Morel, Annika Kirsten Madler, Xin Gen Lei, Cornell University, USA

1 Introduction

Synthetic biology involves re-engineering organisms for targeted applications and can be characterized as an interdisciplinary field comprising the concepts and tools needed for organism modification (Hanczyc, 2020). A branch of synthetic biology is biotechnology, which may be defined as the manipulation, genetic or otherwise, of certain natural processes to provide new, functional products such as proteins, hormones, and antibiotics (National Academies of Sciences and Medicine, 2018). From the beginning of 1982, with the first recombinant expression of insulin in *Escherichia coli*, to now, where there are over 170 approved recombinant protein drugs (i.e. substances combining genetic material from different sources) which have been approved by the FDA, synthetic biology has had a growing impact (Baeshen et al., 2014). Over the past decade, in particular, the field has evolved significantly and made major advances.

One of the most groundbreaking discoveries has been the characterization of the Cas9 enzyme used in the technique, CRISPR. This has allowed

[1] All authors contributed equally

http://dx.doi.org/10.19103/AS.2024.0143.23

implementation of powerful gene editing capabilities in a number of organisms. It is, to date, the simplest, most versatile, and most precise method of genetic manipulation. Founded by 2020 Nobel prize winning French scientists, Jennifer Doudna and Emmanuelle Charpentiere, CRISPR is an acronym for 'Clustered Regularly Interspaced Short Palindromic Repeats.' The system includes sgRNA that guides a Cas9 enzyme to a specific site allowing for sight-directed DNA cuts to occur to the double stranded DNA and, therefore, site-directed mutagenesis (Li et al., 2023). Although this is one of the most impactful discoveries of the past decade, there are various other genetic tools used for protein design.

Two strategies used to modify proteins of interest are as follows:

- Rational design; and
- Directed evolution.

Each has its advantages and limitations. Rational design is a process in which site-specific mutations are conducted, based on prior knowledge of the protein of interest (Hua et al., 2018). Since this strategy typically requires prior knowledge of the structure, it can be a bottleneck to create meaningful mutations.

On the other hand, directed evolution typically performs random mutations on the protein sequence, which means that prior knowledge of the protein structure or mechanism is typically not required. Once the mutations are complete, assays are done with the desired outcome in mind, and the mutation that achieves the desired effect is then used (Zhang et al., 2019). This strategy can be time-consuming and costly due to high material usage since high-throughput screening of desired protein traits is required.

Alphafold is a recent development from DeepMind which has made a fundamental change to synthetic biology. The current version-Alphafold2 (Jumper et al., 2021) is available free through Google's CoLab and allows anyone to input their peptide sequences. This then enables the program to predict protein structures, usually with very high accuracy comparable with experimentally solved structures. Smaller proteins are generally predicted accurately, however, this tends to decrease as the protein gets larger, or when the prediction contains several domains. Experimental techniques, such as cryo-EM and especially X-ray crystallography, can struggle to create the right environment and sufficient protein yield to conduct the structure analysis. Improvements to structure prediction technology are constantly being made, and the path toward even more accurate structure prediction is becoming ever clearer. As one can imagine, this technology allows rapid structure information of desired mutations. With tools such as Alphafold2, rational design approaches for enzyme improvements have become increasingly easier and less costly than directed evolution approaches.

Cell-free expression systems are popular amongst biotech companies due to their high-throughput screening of proteins. This is particularly useful in protein engineering, making it possible to collect proteins in a matter of hours as opposed to days to weeks with traditional molecular cloning and expression systems. Briefly, these systems use a master mix containing all the compounds required for protein translation. The components are mixed with phospholipids and emulsified together to simulate a cell structure which allows for safe protein translation and folding. These translation components can be harvested from several cell types of prokaryotic and eukaryotic origin. The highest protein yields have been reported from the *E. coli* cell-free platform (Gregorio et al., 2019), with concentrations over 2 g/l in batch mode reactions (Garamella et al., 2016). For eukaryotic-based platforms (using wheat germ components), protein concentrations up to 1.6 g/l can be produced in batch mode reactions (Harbers, 2014). However, several grams of protein per liter can be collected from both systems if conducting repeat batches by re-feeding the system with the required compounds for maintaining reactions, for example amino acids. Protein engineers thus have a variety of systems to use, depending on whether their protein is of eukaryotic or prokaryotic origin.

The use of enzymes as tools has also become increasingly popular due to new technologies for functional improvement and resulting applications. Research strategies mentioned earlier can be used to improve certain traits of enzymes such as thermostability, solubility, activity, and substrate affinity, depending on the desired end application. The global industrial enzyme market, which includes feed enzymes, is expected to reach $7 billion in 2023 with a 4.9% growth rate (Saldarriaga-Hernández et al., 2020). Other estimates predict a higher growth rate of 6.3% up to 2026, with a higher range of up to $8.7 billion in market value. Over 20% of globally produced enzymes are lignocellulolytic (Toushik et al., 2017), while proteases comprise about 60% of the market (Naveed et al., 2021). Of the top ten enzyme companies, seven are headquartered in the European Union, with the remainder in the United States, India, and Japan. Currently, Asian countries are the highest customers for enzyme products (Großmann, 2023). This is predominantly in the form of research and development, as companies and laboratories are adapting to newer techniques, as well as developing everyday consumer products.

2 Syntheses of proteins: bacterial and fungal producers

The uses of bacteria, yeast, and filamentous fungi have been instrumental in producing recombinant proteins for uses in poultry feed. These expression systems have significant advantages over mammalian cell systems, for example high enzyme production and low-cost media. On the other hand, mammalian cell cultures are generally more appropriate for human-grade pharmaceuticals as their protein production most closely resembles ours and eliminates the issue of contaminating secondary metabolites.

The most widely used organism for recombinant expression is *E. coli* (Katz et al., 2018). It is perhaps the easiest and most time-efficient expression system to determine whether the recombinant protein is active. However, not every protein expressed by *E.coli* will demonstrate proper post-translational modifications such as glycosylation. Additionally, disulfide bond formation can be more difficult as the *E. coli* cytoplasm is naturally a reducing environment. Glycosylation is especially important for functional expression of eukaryotic genes, as this type of post-translational modification helps stabilize and increase recognition of proteins (Francis and Page, 2010). Despite *E. coli* having trouble secreting proteins extracellularly, advancements have been made to improve secretion (Kleiner-Grote et al., 2018). However, this may not always be the case, depending on the protein of interest. The protein can potentially form inclusion bodies, in which the recombinant protein must be denatured and refolded. All of these factors are on a case-by-case basis, so expressing novel proteins may be simple or more complicated when using *E. coli*.

Other bacterial expression systems are commonly used, such as *Bacillus* spp., particularly *B. subtilis* and *B. licheniformis* (Cai et al., 2019). These systems are particularly used for the expression of proteases, as they are well equipped to secrete these in their natural state. It is important to note that native proteases secreted by the bacteria can potentially interact with the protein of interest. Two native proteases (subtilisin and neutral protease) account for up to 98% of all native secreted protease activity (He et al., 1991, Demain and Vaishnav, 2009). The protein may degrade prior to harvest. Fortunately, research has been done to design protease deficient strains (Zhao et al., 2019).

Fungal systems include both yeasts and filamentous fungi. Commonly used yeasts include *Saccharomyces cerevisiae* and *Pichia pastoris*, also known as *Komagataella pastoris*. These are widely used hosts in the enzyme industry due to their rapid growth, secretory abilities, and relatively clean media when extracting the protein of interest. Unlike *Bacillus*, these yeasts secrete very little native enzyme into the culture, which makes the overall purification process significantly easier. Another benefit of fungal systems is their ability to glycosylate and conduct post-translational modifications of eukaryotic genes (Byrne et al., 2005).

Filamentous fungi are also attractive hosts for producing large amounts of commercially relevant enzymes such as cellulases, amylases, and phytases (Wösten, 2019). Some of the more common hosts are *Aspergillus* species and *Trichoderma reesei*. *Aspergillus* hosts are generally favored for the expression of proteases (Boer and Peralta, 2000), while *Trichoderma* is commonly used

Table 1 Microbial expression system types and their respective advantages and disadvantages

Expression system	Pros	Cons
Bacteria	Simple transformation, inexpensive, rapid growth	Prokaryotic PTM, inclusion bodies, misfolded product
Yeast	Simple transformation, rapid growth	UPR, hyperglycosylation, more expensive media
Filamentous fungus	Complex glycosylation, huge expression of native enzymes, SSF	UPR, complicated transformation, mycotoxins, mitotic instability

Source: Adapted from: Shanmugaraj, B., Bulaon, C. J. I. and Phoolcharoen, W. (2020). "Plant molecular farming: A viable platform for recombinant biopharmaceutical production." *Plants* 9(7): 842.

for lignocellulolytic enzyme production (Druzhinina and Kubicek, 2017). However, both hosts can serve as viable platforms for recombinant expression depending on the gene. Generally, expression of heterologous genes in filamentous fungi is most successful when genes are of fungal origin (Lubertozzi and Keasling, 2009, Nevalainen and Peterson, 2014). When it comes to non-fungal genes, protein yield is commonly not as high, and more susceptible to unfolded protein response (UPR). UPR can be triggered in fungi when the host detects misfolded proteins or incorrect glycosylation. Furthermore, some *Aspergillus* species may naturally secrete mycotoxins, although this is typically not the case in industrial settings with controlled environments without survival competition. Other species such as *Trichoderma reesei* do not secrete any mycotoxins, providing an even safer expression platform (Frisvad et al., 2018). The expression hosts mentioned here are some of the most commonly used in industry and academic research settings. However, research is continuously exploring new hosts of prokaryotic and eukaryotic origin as well as to improve yields of successful expression. Table 1 describes different microbial expression systems, with their advantages and disadvantages.

3 Syntheses of proteins: plant producers

Although microbes are a robust production platform for synthetic biology, plants have various advantageous properties that also make them ideal candidates for recombinant protein expression in the poultry industry. They offer a more long-term, cost-effective platform than most microbes (Schillberg et al., 2019). Plants also inherently conserve many of the metabolic pathways and allow for compartmentalization of various toxic intermediates that could diminish protein function (Barnum et al., 2021). Originally, plants underwent genetic modification for resistance to certain bacterial infections and other similar resistances. However, recently, this technology has grown from prevention to enhancement of these strains, including the overexpression or novel expression of nutraceutical components. Using synthetic biology, researchers have been able to increase fat and fatty acid content, modify the starch components, increase trace elements, increase bioavailability, decrease environmental pollutant output, decrease undesirable constituents in feedstuff, and more to enhance feed (Barnum et al., 2021).

There are several differences in using plants compared with other expression systems. The major goal of plant production platforms is to utilize inherent biosynthetic pathways to express compounds in plant systems that are lower cost, easier to use, etc. There are two primary expression pathways that are utilized within a plant species (Barnum et al., 2021):

- Transient expression; and
- Direct expression.

Transient expression refers to temporary expression of genes that are most frequently from an expression cassette, such as a plasmid (Sainsbury and Lomonossoff, 2014). *Agrobacterium tumefaciens,* a rod-shaped, Gram-negative soil bacterium responsible for crown gall disease in plants, is typically used for transient expression in plant species. This allows for expression in 3–6 days of the genes of interest; furthermore, it is also possible through the system to co-insert genes in the plasmid allowing for the creation of fewer constructs. Once the gene is in the bacterium, the species, *Nicotiana benthamiana,* a close relative of the common tobacco plant, is the primary expression platform used. This plant naturally diminishes its pathogen defense, in order to accelerate its production cycle. This characteristic makes it highly susceptible to pathogens, such as the *Agrobacterium tumefaciens,* and an ideal target for transient expression (Barnum et al., 2021). Reed and Osbourne have made significant advances in optimization of this method allowing for larger-scale production (Reed et al., 2017). Screening is still generally difficult and lengthy in these lines without the use and development of smaller-scale production platforms first, such as protoplasts of the plant species of interest. This is, however, a technique that involves utilizing fragile cells as a platform and requires greater optimization before being able to utilize these quickly and efficiently (Barnum et al., 2021).

The other expression system commonly used is a stable plant line with direct expression. Although expression directly into stable plant lines is a lengthy process involving various steps until the researcher is able to confirm whether the gene was properly inserted, the procedure is well characterized. These steps include transgene delivery, generation of calli, the growth of the plant, and then confirmation of gene insertion. This also comes with various other limitations, such as gene size, transformation efficiency, and regeneration

Table 2 Examples of promoters in plants. Constitutive promoters always allow for moderate levels of gene expression. Tissue-specific only expresses in certain organs of the plant. Inducible promoters can be forced to express genes under certain conditions[1]

Promoter	Source	Activity	Category
CaMV35S	Cauliflower mosaic virus	Almost always active	Constitutive
SCR, SRK	*Brassica rapa*	Pollen and stigma specific	Tissue specific
Exo70C2	*Arabidopsis*	Pollen and root specific	Tissue specific
NtHSP3A	Tobacco	Stress inducible	Inducible promoter

Source: Adapted from: Low, L.-Y. (2018). *Transgenic Plants: Gene Constructs, Vector and Transformation Method*. sine loco, IntechOpen.

time. There have been various recent advancements that have focused on optimizing these methods with significant improvements though. A series of studies have begun to improve the generation time, the size of gene insert, and even utilized a Transgene Stacking I1[2] method to insert a cassette encoding an entire pathway. Although this method takes considerably longer than transient expression, it does allow for larger-scale production in an open field generating overall higher yields (Barnum et al., 2021).

Along with gene insertion, there is the ability to undergo site-specific modification with targeted genes that could circumvent tricky approval processes seen with whole genome engineering. For example, in 2016, Ma and collaborators were able to utilize TALEN site-directed mutagenesis, a targeted mutation system similar to CRISPR but with higher binding specificity and fewer sequence restraints (Ma et al., 2016), to modify soybeans to reduce undesirable linoleic acid and, simultaneously, increase oleic acid through a targeted pathway disruption. As CRISPR-Cas9 and its derivatives continue to gain popularity, these modifications will allow access to other pathways within plants enhancing their use (Barnum et al., 2021).

Appropriate vectors and gene constructs must be carefully chosen to successfully integrate novel genes into plants and to produce desired proteins, similar to those from microbial and fungal systems. The gene construct always includes a promoter region, gene coding region, and a terminator/stop region. The promoters are plant-specific and must be chosen with care. A few baseline examples are outlined in Table 2. It is also vital to ensure that a reporter gene is encoded in the construct: a gene joined to the promoter sequence that allows for a measurement of promoter activity, as to allow the protein to be quantified in the study in the future (Low, 2018) (Clark et al., 2019). There exist various vectors that can be chosen to deliver the gene construct of choice. In plants,

2 Transgene Stacking 11 is a method utilized to synthesize multiple genes for complex metabolic pathways that requires specialized *E. coli* strains, reagents, and molecular biology tools (Zhu and Liu, 2021).

the Ti plasmid is typically used; however, there have been developments in the utilization of a binary vector that encodes for an origin of replication for *E. coli* and *A. tumefaciens*, that allows for a lower plasmid size and, therefore, higher copy number (Low, 2018). This is a rapidly changing field of study with optimizations constantly arising; however, more specifics can be found from Low (2018).

Utilizing these techniques, researchers have been able to produce pharmaceutical compounds, enzymes, and nutraceuticals within plant platforms. One of the most common additives in animal feed is biofortified plants that are created and expressed using synthetic biology. Traditionally, cross-breeding of plants has been the primary means of introduction of more favorable varieties. However, with the increased popularity of CRISPR-Cas9, there have been more site-directed mutations within plants. Utilizing an understanding of the nutrient production pathway, researchers have been able to target different points in that pathway and knock-out-specific aspects, e.g. to increase the production of an intermediate compound or decrease degradation of the target nutrient. This must be done carefully, however, given that the greater the change, the greater harm it can do to the proper development of the plant (Barnum et al., 2021). For example, in tomatoes, researchers attempted to overproduce a compound, γ-aminobutyric acid, which led to developmental issues that did not allow the study to continue. It is therefore vital to have a complete understanding of the pathway that is being manipulated and the unforeseen effects it could have (Li et al., 2018). As plants have very complex pathways that have the ability to create various products, this is an exciting prospect for the future.

The primary benefits for plant production platforms are the low-cost scalability, ability to grow in non-axenic environments, and the fact that they are autotrophic, making them a potentially more sustainable option. They are ideal candidates for introduction into feed, as plants are already a staple, and increasing scalability would be easy to integrate into the feed system. This does, however, generally come with a high-generation time that accompanies lengthy and expensive approval processes. Although approval processes are generally long for all recombinant systems, the high-generation time in plants leads to very time-consuming experiments and will require further optimization in the future. This is, nevertheless, a promising prospect for the future of poultry feed additives.

4 Syntheses of feed supplements: nutrient supplements

Utilizing the myriad of synthetic biology techniques discussed above, many agricultural innovators have begun to also focus on overexpression and supplementation of key nutrients that could benefit the poultry industry and its consumers. Recombinant supplementation ranges from direct nutrient supplementation to supplementation of the production platform itself, i.e. yeast, bacteria, etc., to improve the health of the chickens and consumers alike.

The direct supplementation of the recombinant protein with its manufacturing platforms is a novel method that saves time and cost, as it removes the need for a purification method to recover the recombinant protein of interest. This would allow for induction conditions to be met within a bioreactor and direct supplementation into the feed with a simple removal of possible harmful impurities. A recent study has proven that supplementation of yeast culture has beneficial effects on growth performance, antioxidant utilization, and the immune function of broiler chickens. With the expression of the various enzymes, antibiotics, and other beneficial components and the yeast, this could be an effective nutritional supplement (Wang et al., 2021) (Kiarie et al., 2019). More details of this method are explained below.

Various purified recombinant products from different organisms have been explored as nutrient supplementation platforms within the poultry industry. The recombinant protein that is most frequently supplemented into animal feed is phytase that breaks down phytic acid to aid in phosphorous, as well as other micromineral release within the gut. Although not a novel enzyme, recombinant use is new and an excellent example of the successful use of these systems. As of now, supplementation of phytase expressed from *Aspergillus niger, Aspergillus oryzae, Trichoderma longibrachiatum,* and *Pichia pastoris* has been approved to be added into poultry feed (Chase McGrath, 2021; Lei and Porres, 2003). There are further innovations that can be made. In *P. pastoris,* there have been improvements in increasing secretory capabilities through the use of various techniques, including a novel bidirectional promoter technique, co-expression of folding and secretion chaperones, and non-methanol induction strategies (Vogl et al., 2018; Navone et al., 2021).

Another significant benefit of recombinant production of enzymes, such as phytase, is that these platforms allow for direct modification to the amino acid sequence in an attempt to improve activity. For example, a study was performed utilizing site-specific mutations to increase thermostability of an AppA2 phytase, by introducing a side-chain-side-chain hydrogen bond that stabilized the loop region (Kim et al., 2008). N-glycosylation sites on phytase allow for stronger pepsin resistance, i.e. increased resistance to gastric digestion (Yang et al., 2012). Along with *P. pastoris,* there are over a dozen other approved microorganisms that have the capabilities of expression and modification of recombinant enzymes that can be used as production platforms for various new and improved proteins and enzymes.

Along with protein production platforms for nutraceutical uses, there are also enhancements that can be made in the pathways that exist within microbes/plants to increase other, non-protein components. Astaxanthin is an example of a nutrient that can be produced in a recombinant system to create a cost-effective alternative to the natural source. It is a valuable pigment that has antioxidant, carotenoid, and other biological activities. Currently, non-recombinant astaxanthin is purified directly from microalgae, made into a powder, and sold at a higher market price (Casella et al., 2020). It is, however, also able to be cultivated from various yeast compounds and great strides have been made, specifically in the realm of bacterial astaxanthin production (Kanwugu et al., 2021). Astaxanthin is one of many examples of cheaper, recombinant nutrient expression for feed purposes within the poultry industry. There are also examples of overexpression of vitamin A, vitamin D, docosahexaenoic acid (DHA), and more.

As the industry continues to push toward more sustainable options in the feed industry, various other components of microalgae have been gaining popularity, due to the reduced land requirement, CO_2 sequestration capabilities, and ability to grow in various types of environments. There are also over 50 000 different species (Kalia and Lei 2022). By modifying the microalgae strain to overproduce what is already naturally found, it is an easy, cost-effective, sustainable platform for feed supplementation. It may even quickly become a staple of animal feed, as the competition for land becomes an even greater issue in the agricultural industry.

Microalgae can be modified to express and secrete various agriculturally relevant compounds, including n-3 fatty acids, and polyphenols. Along with their high protein content (14–61%) and complete amino acid profile, these are ideal candidates for whole mass supplementation into feed. In one study salmon were fed microalgae that naturally overexpressed eicosapentaenoic acid (EPA) and DHA containing higher amounts of antioxidants, anti-inflammatory and immune-related amino acid transport genes (Johnson, 2021; Kalia and Lei, 2022). These findings led to enriched microalgae being fed to poultry, resulting in increased DHA, polyunsaturated fatty acids (PUFA), EPA, astaxanthin amounts found in the egg yolk and chicken meat (Kalia and Lei, 2022). A recent study even found that feeding microalgae allowed for a triple enrichment of DHA, 25-OH D3, and AST in egg yolks that was significant enough to be relevant for human health (Ou et al., 2023). Clearly, there are various other potential types of supplementation using this approach.

Nutrient supplementation through recombinant systems such as these could be vital for the future of the poultry industry. Small supplements, such as astaxanthin, can be very expensive. These methods allow for inexpensive supplementation that could yield higher-quality meat and eggs from poultry. This would benefit the consumer and the animal alike.

5 Syntheses of feed supplements: antimicrobial peptides and prebiotics

In 2020, it was estimated that global antibiotic usage was over 99 000 tons and is expected to increase to over 107 000 tons by 2030 (Mulchandani et al., 2023). Despite this global increase, the use of antibiotics in poultry feed has decreased. In the United States, for example antibiotic usage in broiler hatcheries decreased from 90% to <1% from 2013 to 2022. Turkey hatchery antibiotic usage decreased from 97% to 40% over the same period (U.S. Poultry and Egg Association, 2022). The FDA also considers ionophores as antibiotics, and they are the second most used class of antibiotics in the Unites States (Wong, 2019). Adoption of slower-growing breeds may cause a decrease in overall antibiotic usage as these are generally healthier than fast-growing breeds. Although ionophores are generally safe to use, they must be handled properly to prevent overdosing in feeds. Additionally, there is concern about ionophores ending up in the environment, as up to 80% of ionophores consumed are excreted in poultry (Munaretto et al., 2016). Alternative antimicrobials may prove to be safer than ionophores.

Antimicrobial peptides (AMPs) are short amino acid sequences using various mechanisms to kill microbes, whether it be membrane disruption or prevention of some essential function. As AMPs are broad spectrum, their likelihood of microbial resistance is less likely than using antibiotics (Nguyen et al., 2011). AMPs can be ubiquitously found in nearly all organisms, but methods for extracting and concentrating them from tissues can be time consuming, costly and with low purity. The production of recombinant AMPs with expression hosts provides an attractive alternative to traditional extraction. They also have the potential to diminish or replace the use of antibiotics in livestock production.

The most important step to make this viable is scaling up to produce sufficient AMP quantities for mass production and supplementation in feeds. Some examples of AMPs produced in high quantities include melittin, a component of honeybee venom, with a yield of 159 mg/l in *B. subtilis* (Ji et al., 2017). In the case of plectasin, a fungal defensin in *P. pastoris* a total yield of 831 mg/ml of protein has been reported (Cao et al., 2015). Many other examples in both prokaryotic and eukaryotic systems have recorded high protein yields with functional activity (Deo et al., 2022).

A number of studies have demonstrated the antimicrobial activity of different AMPs, but there is less literature on direct AMP usage in poultry diet supplementation (Choi et al., 2013, Forkus et al., 2017, Tai et al., 2020). Recently though, a patent for a live microbe supplementation for poultry diets has been accepted. The microbes in question were lactic acid bacteria (*Lactococcus, Lactobacillus* sp.) genetically engineered to overexpress enterocin A to produce an AMP effective against *Clostridia* and *Enterococcus*. Additionally, it was found that less than 1% of engineered microbes were detected in the environment (Kaznessis et al., 2020; Kaznessis and Kruziki, 2022). There is an increasing amount of evidence relating to AMP efficacy and overall safety.

Prebiotics are an umbrella term for compounds that are not digested by the animal but are instead used by specific microorganisms in the gut, improving the overall health of the host animal (Gibson and Roberfroid, 1995). Oligosaccharides are probiotics used by bacteria and are a well-studied group of organisms that provide benefits to their host. They do so by competing with harmful bacteria through the production of lactic acid, which lowers the pH of the gut environment, which means that harmful bacteria such as *E. coli* and *Salmonella* are unable to grow. Oligosaccharides include mannan-oligosaccharides (MOS) which prevent attachment of pathogenic bacteria to mucosal lining in the intestines (Spring et al., 2015). Companies sell products consisting of hydrolyzed yeast, which naturally contains mannan-oligosaccharides from yeast cell walls. It is directly added to the feed which consequently increases feed efficiency and immune system health.

Synthesis of prebiotics can also use recombinant enzymes such as mannanases or glucansucrases using sucrose as substrate. Prebiotic saccharides can be harvested and potentially supplemented in feed. An example of production of oligosaccharides by enzymatic means is the production of isomelezitose from recombinant alpha glucosidase in *E. coli.*, converting 25% of sucrose into isomelezitose (Garcia-Gonzalez et al., 2020). Oligosaccharides can also be produced by non-enzymatic means using catalysts with typically cationic properties. These may be in a polymer form, or attached by a solid support such as silica (Ma and Zaera, 2006). For example, researchers used a recently patented oligosaccharide product as a 'precision biotic' to better manage the chicken gut microbiome. It was found that feed conversion ratio decreased and average body weight increased (Walsh et al., 2021, Bortoluzzi et al., 2023).

6 Culture of microorganisms as alternative protein sources

Previously, we have discussed the value of using microorganisms to produce recombinant proteins, whether it be supplements, AMPs or enzymes. Here, we discuss microorganisms themselves as a source of protein. Despite long use, single-cell proteins are of growing recent interest. It has been estimated that by 2050, 10–19% of protein used in animal feed industries can be of microbial origin, whilst up to 60% of grain feeds used to feed poultry can potentially be replaced with microbial proteins (Pikaar et al., 2018). Microalgae use in poultry is well studied with generally favorable results, but research on microbial protein in feed is scarce. Here, we discuss the contents of microbial proteins and their potential addition to poultry feeds.

Protein production in microorganisms is generally very efficient, with protein content of 14–61% (as previously mentioned) (Johnson, 2021), 30–70% in fungi, and up to 75–80% in some bacteria by dry biomass weight (Ravindra, 2000, Matassa et al., 2016). Microorganisms can be cultured with very little land use, low energy requirements, and generally cheap media. Much research is being done to improve production yields and decrease emissions. For example, Gassler and collaborators developed an engineered yeast strain (*P. pastoris*) capable of using CO_2 as the sole source of energy (Gassler et al., 2020). This was done by completely changing the methylotrophic pathway and replacing it with a Calvin cycle and RuBisCO pathway. The strain achieved a maximal growth rate of 0.018 h^{-1}, which is slightly lower than typical growth rates of *P. pastoris* using methanol as a carbon source. This is a good example of using biotechnology in both engineering microorganisms for feed protein and in carbon sequestration. Other potential sources of microbial protein include Knallgas bacteria, which can fix CO_2 from hydrogen through the Calvin cycle.

One issue with microbial protein, particularly of bacterial origin, is high nucleic acid content, predominantly RNA, with levels up to 15–16% of dry weight (Ravindra, 2000). Other microbial nucleic acid contents, however, are significantly lower (fungi: 9–10%, microalgae: ≈7%) (Frazier and Westhoff, 1990) (Finkel et al., 2016). High levels of nucleic acids can potentially cause visceral gout in chickens due to excess uric acid formation (Hadi and Brightwell, 2021). Urate crystals can form in several organs such as the kidneys and liver (Lakshmi Namratha et al., 2019). Additionally, there is the issue of endotoxins, which is mostly found in Gram-negative bacteria (Ravindra, 2000). In terms of amino acid content, fungi typically contain the highest levels of methionine and lysine. Methionine is generally the first limiting amino acid in conventional poultry diets, therefore making fungal proteins a good candidate for alternative protein sources in poultry diets. The main reasons for caution in the poultry industry about use of protein of microbial origin may be concern about the risk of mycotoxins from fungal protein, accumulation of contaminants such as heavy metals, and the high nucleic acid content in bacteria. Purification and quality measurements will be paramount for optimal nutrition in poultry diets.

7 Waste valorization

The idea of circular economies has become increasingly popular with the re-use and re-cycling of materials seen as key to achieving net zero. Circular principles can be applied to biotechnology practices such as enzyme production and genetic engineering. Food and agricultural wastes can be valorized by using them as substrates and alternative media for culturing of microorganisms and filamentous fungi (Sharma et al., 2022).

Solid-state fermentation (SSF) uses solid substrates to cultivate microorganisms, usually fungi since they have appropriate properties such as resistance to contamination. However, bacteria and yeast can be used with SSF with even lower water usage since they can survive on dryer substrates than filamentous fungi (Šelo et al., 2021). The substrate is suspended and homogenized followed by centrifugation. The resulting supernatant liquid containing the expressed proteins is then collected. Techniques such as SSF can provide some advantages to conventional submerged fermentation since they use waste products to culture fungi and other microorganisms, as opposed to more expensive prepared media. SSF also uses much less water than conventional fermentation, as the substrate is only kept moist to provide adequate growth (Pandey, 2003). Energy costs for culturing under SSF are also lower, as no shaking or liquid fermenter is needed.

Many examples have been published demonstrating that several types of wastes can be used as substrates for culturing production hosts. Substrates ranging from food to agricultural waste have been described. For example, xylanase is an important enzyme used in the feed industry, particularly in poultry and swine, as it helps with digestion of plant matter and decreases viscosity of digestate. SSF has been used as a platform to produce this enzyme. Rice straw, e.g. is a good candidate for xylanase production in *Aspergillus flavus* (Singh et al., 2021). Palm waste is also a good substrate for xylanase production in *Aspergillus tubingensis* (Intasit et al., 2022).

Agricultural and food waste products can provide other uses besides enzyme production. Waste products can be fermented via SSF using harmless, edible fungi such as *Pleurotus* spp. and *Ganoderma* spp. to provide a more digestible product with higher protein content to use as animal feed. White-rot fungi degrade lignin to leave cellulose which is easier to digest, particularly for ruminant livestock. Brown-rot fungi degrade cellulose and leave behind lignin with lower digestibility in livestock (Mahesh and Mohini, 2013). One problem with the use of waste products and residues is the potential concentration of contaminants. The *Trametes versicolor* laccase enzyme can then be used to capture and remove a variety of harmful compounds such as aflatoxins, BPA, and other phenolic compounds (Tišma et al., 2021).

Feather waste is a significant and increasing issue in poultry production. Feathers make up roughly 5–10% of an individual chicken's weight (Sharma et al., 2016; de Q. Souza et al., 2023). Statista recorded 20.5 million tons of poultry meat produced in the United States in 2022. Based on the individual feather weight, at least 1 million tons of feather waste produced. Feather meal is currently used as a feed component for alternative protein source as well as a fertilizer. Unfortunately, digestibility is rather low due to incomplete hydrolysis of peptide bonds and rendering requires high temperatures and mechanical stress. Keratinolytic enzymes pose as an eco-friendly strategy to upcycle feather wastes into valuable commodities such as those previously mentioned (Sypka et al., 2021). The feather is a keratin complex composed primarily of β sheets with a high cysteine content containing many disulfide bonds (Taskin et al., 2011). Much research has been done to identify microbial strains with high keratinolytic abilities and candidate enzymes for hydrolysis (Li, 2019). Despite this, feathers remain an underutilized resource. Future work should focus on safe upscaling of keratinolytic strains and increasing production of recombinant enzymes in higher volumes.

8 Conclusion

Although there is no evidence that synthetic biology poses any risks to health, this approach is still regarded with some suspicion. For example, India has yet to approve genetically modified plants in their food system, though it has approved recombinant protein expression in production of food, feed, and medicine (Macovei, 2022). There are issues to decide whether a protein derived from a pig that is expressed in a yeast and fed to as a predominant feed source to chickens would be considered Halal, or if ovalbumin expressed from a yeast source can be considered vegetarian. These are complex issues to resolve that will vary according to person, country, and religion.

Synthetic biology will increasingly be used with other innovative technologies. An example being nanoencapsulation techniques to enhance characteristics such as thermostability, pH stability, and stability during gastric digestion. Enzymes, antimicrobial peptides and probiotics have, e.g. been encapsulated by proteins, polysaccharides, and other substances (Rajendran et al., 2022). Future research should focus on recombinant and other microbial and plant-based additives in such area as safety and potential for upscaling. As the market for recombinant feed additives grows, innovation will continue to help the industry achieve a more efficient, cost-effective, and sustainable future.

9 Where to look for further information

Further reading

- A comprehensive guide to "cell-free gene expression" by Garenne and collaborators in *Nature Reviews* (Garenne et al., 2021).
- Up to date information on alternative meat space (magazine): https://www.alt-meat.net/

- Basics of bioprocess engineering microbe-derived products: Bioprocess Engineering: Basic Concepts (Prentice Hall International Series in the Physical and Chemical Engineering Sciences) 3rd Edition Matthew DeLisa.
- Recent open-acces book on plant genome engineering: "A Roadmap for Plant Gene Engineering" edited by top researchers in the field (2024).

10 References

"Alt Meat." Home | Alt Meat, www.alt-meat.net/. Accessed 2 Apr. 2024.

Baeshen, N. A., Baeshen, M. N., Sheikh, A., Bora, R. S., Ahmed, M. M. M., Ramadan, H. A., Saini, K. S. and Redwan, E. M. (2014). Cell factories for insulin production. *Microbial Cell Factories* 13(1): 1–9.

Barnum, C. R., Endelman, B. J. and Shih, P. M. (2021). Utilizing plant synthetic biology to improve human health and wellness. *Frontiers in Plant Science* 12: 691462.

Boer, C. G. and Peralta, R. M. (2000). Production of extracellular protease by *Aspergillus tamarii*. *Journal of Basic Microbiology* 40(2): 75–81.

Bortoluzzi, C., Tamburini, I. and Geremia, J. (2023). Microbiome modulation, microbiome protein metabolism index, and growth performance of broilers supplemented with a precision biotic. *Poultry Science* 102(5): 102595.

Byrne, L. J., O'Callaghan, K. J. and Tuite, M. F. (2005). Heterologous gene expression in yeast. *Therapeutic Proteins: Methods and Protocols*: 51–64.

Cai, D., Rao, Y., Zhan, Y., Wang, Q. and Chen, S. (2019). Engineering Bacillus for efficient production of heterologous protein: current progress, challenge and prospect. *Journal of Applied Microbiology* 126(6): 1632–1642.

Cao, X., Zhang, Y., Mao, R., Teng, D., Wang, X. and Wang, J. (2015). Design and recombination expression of a novel plectasin-derived peptide MP1106 and its properties against *Staphylococcus aureus*. *Applied Microbiology and Biotechnology* 99(6): 2649–2662.

Casella, P., Musmarra, D., Dimatteo, S., Chianese, S., Karatza, D., Mehariya, S. and Molino, A. (2020). Purification of astaxanthin from microalgae by using commercial activated carbon. *Chemical Engineering Transactions* 79: 295–300.

Chase McGrath, A. H. (2021). *Chinese Feed Additive Catalog Re-organized*. United States Department of Agriculture Foreign Agriculture Service. G. A. I. Network, Beijing.

Choi, S. C., Ingale, S. L., Kim, J. S., Park, Y. K., Kwon, I. K. and Chae, B. J. (2013). An antimicrobial peptide-A3: effects on growth performance, nutrient retention, intestinal and faecal microflora and intestinal morphology of broilers. *British Poultry Science* 54(6): 738–746.

Clark, D. P., Pazdernik, N. J. and McGehee, M. R. (2019). Chapter 21: Analysis of Gene Expression. Molecular Biology (3rd edn.). Clark, D. P., Pazdernik, N. J. and McGehee, M. R. *Academic Cell*: 654–690.

de Q. Souza, G. E., Burin, G. R., de Muniz, G. I. and Alves, H. J. (2023). Valorization of feather waste in Brazil: structure, methods of extraction, and applications of feather keratin. *Environmental Science and Pollution Research* 30(14): 39558–39567.

Demain, A. L. and Vaishnav, P. (2009). Production of recombinant proteins by microbes and higher organisms. *Biotechnology Advances* 27(3): 297–306.

Deo, S., Turton, K. L., Kainth, T., Kumar, A. and Wieden, H.-J. (2022). Strategies for improving antimicrobial peptide production. *Biotechnology Advances* 59: 107968.

Druzhinina, I. S. and Kubicek, C. P. (2017). Genetic engineering of *Trichoderma reesei* cellulases and their production. *Microbial Biotechnology* 10(6): 1485–1499.

Finkel, Z. V., Follows, M. J., Liefer, J. D., Brown, C. M., Benner, I. and Irwin, A. J. (2016). Phylogenetic diversity in the macromolecular composition of microalgae. *PLoS ONE* 11(5): e0155977.

Forkus, B., Ritter, S., Vlysidis, M., Geldart, K. and Kaznessis, Y. N. (2017). Antimicrobial probiotics reduce *Salmonella enterica* in turkey gastrointestinal tracts. *Scientific Reports* 7(1): 40695.

Francis, D. M. and Page, R. (2010). Strategies to optimize protein expression in *E. coli*. *Current Protocols in Protein Science* 61(1): 5.24.21–25.24.29.

Frazier, W. and Westhoff, D. (1990). *Food Microbiogy*. Tata Mc Graw Hill Publishing Company Limited. Págs, New Delhi.

Frisvad, J. C., Møller, L. L. H., Larsen, T. O., Kumar, R. and Arnau, J. (2018). Safety of the fungal workhorses of industrial biotechnology: update on the mycotoxin and secondary metabolite potential of *Aspergillus niger*, *Aspergillus oryzae*, and *Trichoderma reesei*. *Applied Microbiology and Biotechnology* 102(22): 9481–9515.

Garamella, J., Marshall, R., Rustad, M. and Noireaux, V. (2016). The all *E. coli* TX-TL toolbox 2.0: a platform for cell-free synthetic biology. *ACS Synthetic Biology* 5(4): 344–355.

Garcia-Gonzalez, M., Cervantes, F. V., Ipiales, R. P., de la Rubia, A., Plou, F. J., Fernández-Gassler, M. T., Sauer, M., Gasser, B., Egermeier, M., Troyer, C., Causon, T., Hann, S., Mattanovich, D. and Steiger, M. G. (2020). The industrial yeast Pichia pastoris is converted from a heterotroph into an autotroph capable of growth on CO_2. *Nature Biotechnology* 38(2): 210–216.

Garenne, D., Haines, M. C., Romantseva, E. F., Freemont, P., Strychalski, E. A. and Noireaux, V. (2021). Cell-free gene expression. *Nature Reviews Methods Primers* 1(1): 49.

Gibson, G. R. and Roberfroid, M. B. (1995). Dietary modulation of the human colonic microbiota: introducing the concept of prebiotics. *The Journal of Nutrition* 125(6): 1401–1412.

Gregorio, N. E., Levine, M. Z. and Oza, J. P. (2019). A user's guide to cell-free protein synthesis. *Methods and Protocols* 2(1): 24.

Großmann, K. K. (2023). Enzymatic hydrolysis of vegetable and insect proteins using technical enzyme preparations, University of Hohenheim.

Hadi, J. and Brightwell, G. (2021). Safety of alternative proteins: technological, environmental and regulatory aspects of cultured meat, plant-based meat, insect protein and single-cell protein. *Foods* 10(6): 1226.

Hanczyc, M. M. (2020). Engineering life: a review of synthetic biology. *Artificial Life* 26(2): 260–273.

Harbers, M. (2014). Wheat germ systems for cell-free protein expression. *FEBS Letters* 588(17): 2762–2773.

He, X.-S., Brückner, R. and Doi, R. H. (1991). The protease genes of Bacillus subtilis. *Research in Microbiology* 142(7–8): 797–803.

Hua, W., El Sheikha, A. F. and Xu, J. (2018). Molecular techniques for making recombinant enzymes used in food processing. *Molecular Techniques in Food Biology: Safety, Biotechnology, Authenticity and Traceability*: 95–114.

Intasit, R., Cheirsilp, B., Suyotha, W. and Boonsawang, P. (2022). Purification and characterization of a highly-stable fungal xylanase from *Aspergillus tubingensis* cultivated on palm wastes through combined solid-state and submerged fermentation. *Preparative Biochemistry and Biotechnology* 52(3): 311–317.

Ji, S., Li, W., Baloch, A. R., Wang, M., Li, H., Cao, B. and Zhang, H. (2017). Efficient biosynthesis of a cecropin A-melittin mutant in *Bacillus subtilis* WB700. *Scientific Reports* 7(1): 40587.

Johnson, Z. (2021). Marine AlGae Industrialization Consortium (MAGIC). Algae Platform review. D. B. T. O. (BETO). Duke University Press. Available at: https://www.energy.gov/sites/default/files/2021-2004/beto-2011-peer-review-2021-algae-johnson.pdf.

Jumper, J., Evans, R., Pritzel, A., Green, T., Figurnov, M., Ronneberger, O., Tunyasuvunakool, K., Bates, R., Žídek, A., Potapenko, A., Bridgland, A., Meyer, C., Kohl, S. A. A., Ballard, A. J., Cowie, A., Romera-Paredes, B., Nikolov, S., Jain, R., Adler, J., Back, T., Petersen, S., Reiman, D., Clancy, E., Zielinski, M., Steinegger, M., Pacholska, M., Berghammer, T., Bodenstein, S., Silver, D., Vinyals, O., Senior, A. W., Kavukcuoglu, K., Kohli, P. and Hassabis, D. (2021). Highly accurate protein structure prediction with AlphaFold. *Nature* 596(7873): 583–589.

Kalia, S. and Lei, X. G. (2022). Dietary microalgae on poultry meat and eggs: explained versus unexplained effects. *Current Opinion in Biotechnology* 75: 102689.

Kanwugu, O. N., Ranga Rao, A., Ravishankar, G. A., Glukhareva, T. V. and Kovaleva, E. G. (2021). Chapter 31 – Astaxanthin from bacteria as a feed supplement for animals. In:Ravishankar, G. A. and Ranga Rao, A. (Eds) *Global Perspectives on Astaxanthin*. Elsevier, Academic Press, pp. 647–667.

Kargi, M. S. L. F. and DeLisa, M. (2017). *Bioprocess Engineering: Basic Concepts*. Prentice Hall.

Katz, L., Chen, Y. Y., Gonzalez, R., Peterson, T. C., Zhao, H. and Baltz, R. H. (2018). Synthetic biology advances and applications in the biotechnology industry: a perspective. *Journal of Industrial Microbiology and Biotechnology* 45(7): 449–461.

Kaznessis, Y. J., Volzing, K. G., del Pino, B.J. and Dunny, G. (2020). *Methods for Making and Using Antimicrobial Peptides* . United States patent US9925223B2, 2018 March 27.

Kaznessis, Y. and Kruziki, K. (2022). Recombinant live therapeutic for prevention and control of necrotic enteritis associated with *Clostridia perfringens* in poultry.

Kiarie, E. G., Leung, H., Kakhki, R. A. M., Patterson, R. and Barta, J. R. (2019). Utility of feed enzymes and yeast derivatives in ameliorating deleterious effects of coccidiosis on intestinal health and function in broiler chickens. *Frontiers in Veterinary Science* 6: article 473.

Kim, M. S., Weaver, J. D. and Lei, X. G. (2008). Assembly of mutations for improving thermostability of *Escherichia coli* AppA2 phytase. *Applied Microbiology and Biotechnology* 79(5): 751–758.

Kleiner-Grote, G. R. M., Risse, J. M. and Friehs, K. (2018). Secretion of recombinant proteins from E. coli. *Engineering in Life Sciences* 18(8): 532–550.

Lakshmi Namratha, M., Kumar, Y. R. and Lakshman, M. (2019). Pathology of visceral gout in layer chicken. *International Journal of Recent Scientific Research* 10: 35546–35548.

Lei, X. G. and Porres, J. M. (2003). Phytase enzymology, applications, and biotechnology. *Biotechnology Letters* 25(21): 1787–1794.

Li, Q. (2019). Progress in microbial degradation of feather waste. *Frontiers in Microbiology* 10: 2717.

Li, R., Li, R., Li, X., Fu, D., Zhu, B., Tian, H., Luo, Y. and Zhu, H. (2018). Multiplexed CRISPR/Cas9-mediated metabolic engineering of γ-aminobutyric acid levels in *Solanum lycopersicum*. *Plant Biotechnology Journal* 16(2): 415–427.

Li, T., Yang, Y., Qi, H., Cui, W., Zhang, L., Fu, X., He, X., Liu, M., Li, P.-f. and Yu, T. (2023). CRISPR/Cas9 therapeutics: progress and prospects. *Signal Transduction and Targeted Therapy* 8(1): 36.

Low, L.-Y. (2018). *Transgenic Plants: Gene Constructs, Vector and Transformation Method*. IntechOpen, sine loco.

Lubertozzi, D. and Keasling, J. D. (2009). Developing Aspergillus as a host for heterologous expression. *Biotechnology Advances* 27(1): 53–75.

Ma, A. C., Chen, Y., Blackburn, P. R. and Ekker, S. C. (2016). TALEN-mediated mutagenesis and genome editing. *Methods in Molecular Biology* 1451: 17–30.

Ma, Z. and Zaera, F. (2006). Heterogeneous catalysis by metals. *Encyclopedia of Inorganic Chemistry*.

Ma, Z. and Zaera, F. (2014). Heterogeneous Catalysis by Metals. In *Encyclopedia of Inorganic and Bioinorganic Chemistry*, R.A. Scott (Ed.). https://doi.org/10.1002/9781119951438.eibc0079.pub2

Macovei, O. (2022). Synthetic biology- cultural and anthropological perspectives. *Postmodern Openings* 13(3): 216–233.

Mahesh, M. and Mohini, M. (2013). Biological treatment of crop residues for ruminant feeding: a review. *African Journal of Biotechnology* 12(27): 4221–4231.

Matassa, S., Boon, N., Pikaar, I. and Verstraete, W. (2016). Microbial protein: future sustainable food supply route with low environmental footprint. *Microbial Biotechnology* 9(5): 568–575.

Mulchandani, R., Wang, Y., Gilbert, M. and Van Boeckel, T. P. (2023). Global trends in antimicrobial use in food-producing animals: 2020 to 2030. *PLOS Global Public Health* 3(2): e0001305.

Munaretto, J. S., Yonkos, L. and Aga, D. S. (2016). Transformation of ionophore antimicrobials in poultry litter during pilot-scale composting. *Environmental Pollution* 212: 392–400.

National Academies of Sciences, E. and Medicine (2018). *Biotechnology in the Age of Synthetic Biology. Biodefense in the Age of Synthetic Biology*. National Academies Press.

Naveed, M., Nadeem, F., Mehmood, T., Bilal, M., Anwar, Z. and Amjad, F. (2021). Protease: a versatile and ecofriendly biocatalyst with multi-industrial applications: an updated review. *Catalysis Letters* 151(2): 307–323.

Navone, L., Vogl, T., Luangthongkam, P., Blinco, J.-A., Luna-Flores, C. H., Chen, X., von Hellens, J., Mahler, S. and Speight, R. (2021). Disulfide bond engineering of AppA phytase for increased thermostability requires co-expression of protein disulfide isomerase in pichia pastoris. *Biotechnology for Biofuels* 14(1): 80.

Nevalainen, H. and Peterson, R. (2014). Heterologous expression of proteins in Trichoderma. *Biotechnology and Biology of Trichoderma*, Elsevier: 89–102.

Nguyen, L. T., Haney, E. F. and Vogel, H. J. (2011). The expanding scope of antimicrobial peptide structures and their modes of action. *Trends in Biotechnology* 29(9): 464–472.

Ou, K. M., A., Sun, Z., Kalia, S., Sun, T. and Lei, X. (2023). A triple enrichment of three bioactive nutrients in eggs of laying hens elicited moderate interactions on related gene expression. *Precision Nutrition* 2(2): e00034.

Pandey, A. (2003). Solid-state fermentation. *Biochemical Engineering Journal* 13(2–3): 81–84.

Pikaar, I., Matassa, S., Bodirsky, B. L., Weindl, I., Humpenöder, F., Rabaey, K., Boon, N., Bruschi, M., Yuan, Z., van Zanten, H., Herrero, M., Verstraete, W. and Popp, A. (2018). Decoupling livestock from land use through industrial feed production pathways. *Environmental Science and Technology* 52(13): 7351–7359.

Rajendran, D., Ezhuthupurakkal, P. B., Lakshman, R., Gowda, N. K. S., Manimaran, A. and Rao, S. B. (2022). Application of encapsulated nano materials as feed additive in livestock and poultry: a review. *Veterinary Research Communications* 46(2): 315–328.

Ravindra, P. (2000). Value-added food: single cell protein. *Biotechnology Advances* 18(6): 459–479.

Reed, J., Stephenson, M. J., Miettinen, K., Brouwer, B., Leveau, A., Brett, P., Goss, R. J. M., Goossens, A., O'Connell, M. A. and Osbourn, A. (2017). A translational synthetic biology platform for rapid access to gram-scale quantities of novel drug-like molecules. *Metabolic Engineering* 42: 185–193.

Ricroch, A., Eriksson, D., Miladinović, D., Sweet, J., Van Laere, K. and Woźniak-Gientka, E. (2024). *A Roadmap for Plant Genome Editing* (p. 561). Springer Nature.

Sainsbury, F. and Lomonossoff, G. P. (2014). Transient expressions of synthetic biology in plants. *Current Opinion in Plant Biology* 19: 1–7.

Saldarriaga-Hernández, S., Velasco-Ayala, C., Flores, P. L.-I., de Jesús Rostro-Alanis, M., Parra-Saldivar, R., Iqbal, H. M. N. and Carrillo-Nieves, D. (2020). Biotransformation of lignocellulosic biomass into industrially relevant products with the aid of fungi-derived lignocellulolytic enzymes. *International Journal of Biological Macromolecules* 161: 1099–1116.

Schillberg, S., Raven, N., Spiegel, H., Rasche, S. and Buntru, M. (2019). Critical analysis of the commercial potential of plants for the production of recombinant proteins. *Frontiers in Plant Science* 10: 720.

Šelo, G., Planinić, M., Tišma, M., Tomas, S., Koceva Komlenić, D. and Bucić-Kojić, A. (2021). A comprehensive review on valorization of agro-food industrial residues by solid-state fermentation. *Foods* 10(5): 927.

Shanmugaraj, B., Bulaon, C. J. I. and Phoolcharoen, W. (2020). Plant molecular farming: a viable platform for recombinant biopharmaceutical production. *Plants* 9(7): 842.

Sharma, S., Gupta, A., Chik, S., Gek, K., Podde, P., Thraisingam, J. and Subramaniam, M. (2016). Extraction and characterization of keratin from chicken feather waste biomass: a study. In: Proceedings of the National Conference for Postgraduate Research (NCON-PGR 2016), Universiti Malaysia Pahang (UMP), Pekan.

Sharma, V., Tsai, M. L., Nargotra, P., Chen, C. W., Kuo, C. H., Sun, P. P. and Dong, C. D. (2022). Agro-industrial food waste as a low-cost substrate for sustainable production of industrial enzymes: a critical review. *Catalysts* 12(11): 1373.

Singer, R. (2022). Antibiotic stewardship within US poultry production, United States Poultry and Egg Association. https://www.uspoultry.org/poultry-antibiotic-use-report/.

Singh, A., Bajar, S., Devi, A. and Bishnoi, N. R. (2021). Adding value to agro-industrial waste for cellulase and xylanase production via solid-state bioconversion. *Biomass Conversion and Biorefinery* 13: 7481–7490.

Spring, P., Wenk, C., Connolly, A. and Kiers, A. (2015). A review of 733 published trials on Bio-Mos®, a mannan oligosaccharide, and Actigen®, a second generation mannose rich fraction, on farm and companion animals. *Journal of Applied Animal Nutrition* 3: e8.

Sypka, M., Jodłowska, I. and Białkowska, A. M. (2021). Keratinases as versatile enzymatic tools for sustainable development. *Biomolecules* 11(12): 1900.

Tai, H.-M., Huang, H.-N., Tsai, T.-Y., You, M.-F., Wu, H.-Y., Rajanbabu, V., Chang, H.-Y., Pan, C.-Y. and Chen, J.-Y. (2020). Dietary supplementation of recombinant antimicrobial peptide *Epinephelus lanceolatus* piscidin improves growth performance and immune response in *Gallus gallus* domesticus. *PLOS ONE* 15(3): e0230021.

Taskin, M., Sisman, T., Erdal, S. and Kurbanoglu, E. B. (2011). Use of waste chicken feathers as peptone for production of carotenoids in submerged culture of *Rhodotorula glutinis* MT-5. *European Food Research and Technology* 233(4): 657–665.

Tišma, M., Žnidaršič-Plazl, P., Šelo, G., Tolj, I., Šperanda, M., Bucić-Kojić, A. and Planinić, M. (2021). Trametes versicolor in lignocellulose-based bioeconomy: state of the art, challenges and opportunities. *Bioresource Technology* 330: 124997.

Toushik, S. H., Lee, K. T., Lee, J. S. and Kim, K. S. (2017). Functional applications of lignocellulolytic enzymes in the fruit and vegetable processing industries. *Journal of Food Science* 82(3): 585–593.

Vogl, T., Kickenweiz, T., Pitzer, J., Sturmberger, L., Weninger, A., Biggs, B. W., Köhler, E. M., Baumschlager, A., Fischer, J. E., Hyden, P., Wagner, M., Baumann, M., Borth, N., Geier, M., Ajikumar, P. K. and Glieder, A. (2018). Engineered bidirectional promoters enable rapid multi-gene co-expression optimization. *Nature Communications* 9(1): 3589.

Walsh, M. C., Jacquier, V., Schyns, G., Claypool, J., Tamburini, I., Blokker, B. and Geremia, J. M. (2021). A novel microbiome metabolic modulator improves the growth performance of broiler chickens in multiple trials and modulates targeted energy and amino acid metabolic pathways in the cecal metagenome. *Poultry Science* 100(3): 100800.

Wang, Y., Tibbetts, S. M. and McGinn, P. J. (2021). Microalgae as sources of high-quality protein for human food and protein supplements. *Foods* 10(12): 3002.

Wong, A. (2019). Unknown risk on the farm: does agricultural use of ionophores contribute to the burden of antimicrobial resistance? *mSphere* 4(5): e00433–00419.

Wösten, H. A. B. (2019). Filamentous fungi for the production of enzymes, chemicals and materials. *Current Opinion in Biotechnology* 59: 65–70.

Yang, Z., Bennett, E. P., Jorgensen, B., Drew, D. P., Arigi, E., Mandel, U., Ulvskov, P., Levery, S. B., Clausen, H. and Petersen, B. L. (2012). Toward stable genetic engineering of human O-glycosylation in plants. *Plant Physiology* 160(1): 450–463.

Zhang, Y., Geary, T. and Simpson, B. K. (2019). Genetically modified food enzymes: a review. *Current Opinion in Food Science* 25: 14–18.

Zhao, L., Ye, B., Zhang, Q., Cheng, D., Zhou, C., Cheng, S. and Yan, X. (2019). Construction of second generation protease-deficient hosts of *Bacillus subtilis* for secretion of foreign proteins. *Biotechnology and Bioengineering* 116(8): 2052–2060.

Zhu, Q. and Liu, Y. G. (2021). Transgene stacking II vector system for plant metabolic engineering and synthetic biology. *Methods in Molecular Biology* 2238: 19–35.

Chapter 18

High protein corn fermentation products for poultry derived from corn ethanol production

Peter E.V. Williams, FluidQuipTechnologies, USA

1 Introduction

Crops are a primary source of protein, and for many years, soybean meal (SBM) has been the protein of choice for livestock feed. However, the pleiotropic relationship between nitrogen and protein content has, over time, resulted in the reduction in protein content of crops such as soybean and corn (maize) as the economics of grain production have prioritized increased yield. This, in turn, has exacerbated the need for higher-concentration protein supplements for feed formulation. In addition, there has also been a movement to reduce reliance on soy products and replace them with more sustainable sources of protein, particularly vegetable proteins with the absence of anti-nutritional factors.

Decisions on how a feed is formulated are already being driven by the sustainability characteristics of individual feed ingredients. Sustainability (meeting energy and carbon emissions reduction targets) alongside demographics (e.g. the age of those in the labor market) have been identified by the International Feed Industry Federation as a key global issue facing the industry. Indexes of sustainability, such as green house gas (GHG) and land use change (LUC) metrics, are likely to become two new parameters that will be

http://dx.doi.org/10.19103/AS.2024.0143.24

components of feed specification sheets, in addition to the nutritional data that has been the norm.

To improve sustainability, processed animal proteins are being reintroduced into feed formulations, and work is ongoing to produce a range of single-cell proteins, insect protein, and microbial protein. Attention is also being paid to alternative crops high in protein. However, there are considerable challenges to growing new alternative crops on any scale. It takes several years to breed an elite variety to achieve maximum production in different geographic locations with different environmental conditions. Logistical factors such as acquiring appropriate machinery and inputs, as well as acquiring agronomic knowledge to successfully grow a new crop, all serve to favor the status quo. In addition, new crops require land and resources and potentially compete with existing crop production, contributing to GHG emissions from agriculture as well as indirectly exacerbating climate change from changes in land use (through deforestation). In contrast, plant-based co-products, secondary products generated during a manufacturing process, have a unique position in that they do not demand additional acreage and do not compete with human food consumption (Mottet et al., 2017).

2 Distillers dried grains and solubles

Corn (or maize) is the major cereal crop grown in the USA and widely cultivated globally. Traditionally considered an energy crop primarily grown for the energy produced in the form of starch and oil, corn has a limited amount of protein. In addition to its use as a food crop, corn has also become an important feedstock for the sustainable biofuel industry, with over a 100 million tons of corn annually processed into ethanol biofuel in the USA. A by-product of ethanol production is distillers dried grains and solubles (DDGS). This is the dried residue remaining after the starch fraction of corn is fermented with yeasts to produce ethanol. Following fermentation, the ethanol is removed by distillation, and the remaining fermentation residues are dried to become DDGS. The growth in ethanol production from corn has resulted in a plentiful supply of DDGS as a co-product, which has found a new market as a livestock feed. However, it is important to note that DDGS, as a by-product of the dry grind ethanol industry, was never designed as a livestock feed but emerged as a convenient means of marketing the residue from the ethanol production process. This creates a potential opportunity to further develop DDGS as a livestock feed product.

More broadly, the dry grind ethanol industry benefits from several important features as a source of animal feed. A dry grind ethanol plant has established logistics for delivery and dispatch of grain feedstock and any feed products. Individual dry grind ethanol plants benefit from economies of scale, with large plants processing in excess of one million tons of grain per annum. Ethanol plants are also integrated bioprocessing facilities, typically with integrated heat and power generation. These facilities are prime sites for the installation of processing technology to produce new products with higher nutritional value than traditional DDGS products.

3 Corn-fermented protein

There are currently four different commercial processes for the production of high-concentration corn-fermented protein (CFP) products (approximately >50% crude protein) (Fig. 1). By producing products with higher levels of protein compared to conventional DDGS, these technologies offer the potential to diversify and expand the revenue streams of the bioethanol plants, in addition to providing new protein products for the feed industry.

The four processes are an evolution of the wet grind milling process and involve the separation of protein and fiber to produce separate high-value concentrated streams of protein and fiber. Mechanically separated CFP is produced from whole stillage when portions of fiber and oil are removed, concentrating residual grain proteins and yeast by methods involving centrifugation and washing. One version of CFP processing involves the separation of all fractions post fermentation, with all components of the product exposed to fermentation: the Maximized Stillage Co-products (MSC™) process. A second version of CFP manufacture incorporates pre-fermentation grinding and separation of the whole grain. A third recovers the protein via flocculation, and a fourth employs electrostatic forces to separate fiber from protein. Key differences between these processes are identified in Fig. 1.

		Pre Fermentation separation	Fermented	Yeast	Approx crude protein %
Empyreal		Protein concentrate			75
CFP mechanically separated	SEQUENCE™				60
	GPRE NexPro BP50 A+Pro AltiPro				>48
CFP flocculant separated	ProCap Gold				52
CFP	ANDVantage ANDVantage PROTOMAX				40 50 50

	full exposure to >50 fermentation
	partial exposure to fermentation
	approx 25% spent yeast in dry matter
	undefined yeast content

Figure 1 Technologies for producing a range of commercial high-protein feed products.

The MSC™ process (Fig. 2) separates the protein from the whole stillage by mechanical centrifugation and alternate washing steps. An important element to grasp is that the dry grind ethanol process starts with a 50–70 h fermentation of ground corn. The ground grain is exposed to moisture, raised temperature, and enzymatic hydrolysis in order to facilitate the conversion of starch into ethanol. There has been little focus on the impact of the fermentation process on residual components of the grain, whether fibrous components found in whole stillage or the composition of protein components. Other than testing of phytase for example as a means of potentially increasing protein recovery, the potential value of fermentation in improving product functionality is unexplored. Xu et al. (2020) listed such potential improvements as improved protein digestibility, increased number of small peptides, increased energy digestibility, reduced fiber levels, improved nutritional value, bioavailability, weight gain, and feed conversion rate. Work is in progress to investigate whether additional functional benefits can be obtained for CFP products via the use of fermentation.

In addition, little attention has been given to the fact that, prior to fermentation, there is a significant *in situ* generation of yeast to facilitate the fermentation process, and that spent yeast at the end of the ethanol process is a valuable nutritional component of DDGS. Indeed, the spent yeast of DDGS has been calculated to be approximately 6% of DDGS dry matter (Han and Liu 2010).

During fermentation, there is a significant net generation of protein from the yeast that is grown in the fermenter to facilitate fermentation. In addition to recovering corn protein, the MSC™ process also recovers a high proportion of spent yeast, in particular cell wall components. Whole yeast cells and yeast cell

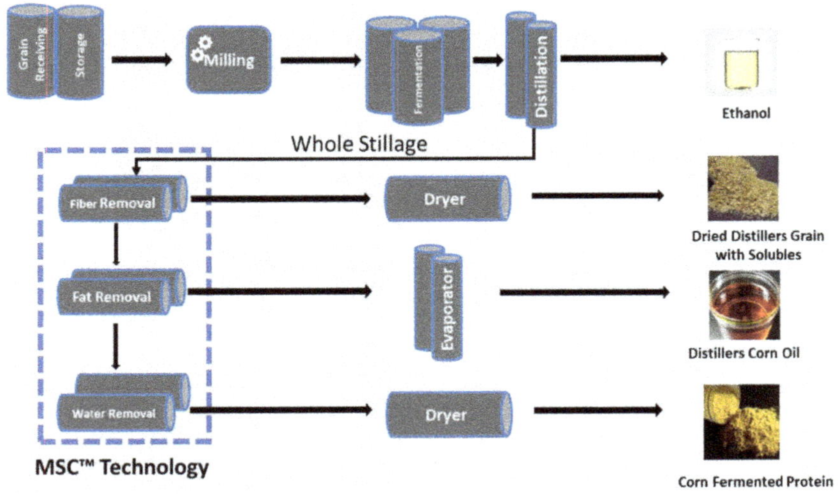

Figure 2 The maximized stillage co-products process.

components are recognized as valuable feed nutrient supplements in their own right (Shurson 2018). This produces an exceptionally high-protein CFP product (up to 60% crude protein) that has now been extensively and very successfully tested as a protein supplement for poultry.

A key challenge has been to measure the quantity of either yeast or yeast components in a dried material such as DDGS. The measurement of yeast cell wall mannose is accepted by the yeast industry as a proxy measure for yeast cell material. Yeast generated in fermentation accompanies stillage in the ethanol process and passes through the distillation columns where the yeast cells are lysed, releasing the intracellular components, including nucleotides. The components of yeast cell walls (mannan oligosaccharides and β-glucans) can be measured in dried CFP products. Using this proxy measure, the yeast cell wall material represents approximately 20–27% of the dry matter content of CFP. Approximately 9% of the protein in a 52% protein product is derived from yeast cell material. However, it is important to recognize that these spent yeast components are not equivalent to whole and viable yeast cells used as feed supplements, though the immuno-stimulatory activity of yeast cell components may be valuable functional components of CFP.

4 Challenges in producing corn-fermented protein

There are a number of challenges in producing a commercial CFP product. One potential safety issue is the use of antibiotics in the ethanol process. As a means of optimizing the efficiency of the yeast ethanol conversion step in the dry grind ethanol process, antibiotics have been used during fermentation to eliminate competitive bacterial activity. In the past, virginiamycin, erythromycin, and tylosin have been used routinely. Given that the stillage is exposed to high temperature during distillation, there is no residual antibiotic activity in the co-products, but antibiotic residues are detectable. Plants producing CFP are able to use a number of alternative products to antibiotics and, with attention to improved cleaning in place, they are able to operate without the use of antibiotics to control the fermentation process, allowing them to be compliant with No Antibiotics Ever (NAE) programs for food and feed production.

Another key aspect of product development is its environmental impact. CFP is produced in a highly competitive, sustainable manner with low GHG emission values compared to a range of different protein products (Fig. 3). Using CFP in ration formulation as a partial replacement of SBM has been reported to reduce the GHG value of feed formulation in diets for turkey poults (Burton et al., 2021).

Key to commercial sustainability is the ability to produce at scale. Up to 500000 metric tons per annum of a specific formulated feed is a norm for commercial feed production. A batch of the formulated feed of up to 100

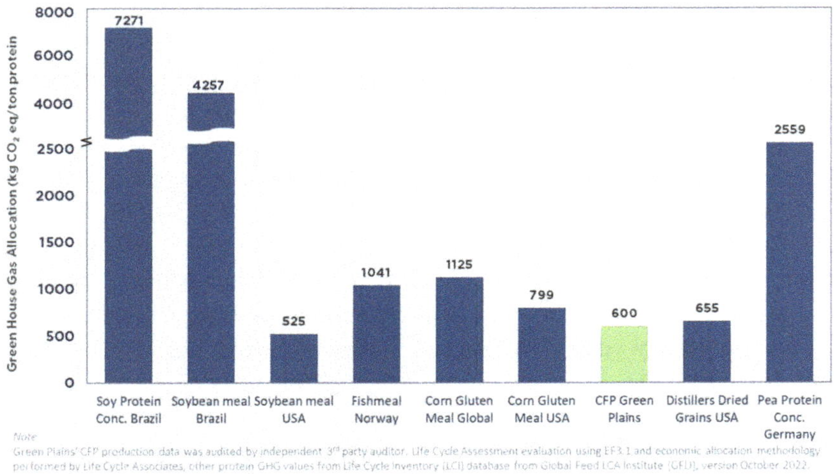

Figure 3 Greenhouse gas allocation values (kg CO_2 eq/ton protein) for a range of different protein products.

tons is made, then mixed in 10–12 ton mixer batches. Storage space for feed ingredients is limited. Generally, large-scale integrators are limited to three to four feed bins on-site, which limits the number of individual ingredients that can be stored. The availability of feed ingredients must merit bin space.

Key aspects of supply chain performance are resilience, redundancy, and reliability (the 3 R's). Changing a feed formulation, particularly for poultry, can be an expensive and time-consuming exercise. Resilience is key because feed producers will not rely on a single source of supply in case of a plant stoppage. It is essential to have one or more sites of manufacture of the product for there to be an alternative source of supply. For the same reason, there needs to be redundancy in volume production to account for increased demand. With resilience in the supply chain, there is an additional need for reliability (product consistency) across different suppliers. Producing a new novel feed protein ingredient is therefore not without significant challenge. CFP products are currently being produced from 17 different ethanol plants in the USA and Brazil with an annual production capacity close to one million tons. Production of the product in Europe is imminent.

Standard measures of product composition include amino acid digestibility using the precision-fed cecectomized rooster assay (PFR) (Parsons, 1985). The assay has been completed on nine separate occasions with samples of CFP obtained during the development of the product, from different ethanol plants as they came online over a period of approximately 6 years, in order to monitor the consistency of the product over time and from different plants. The ileal digestibilities of amino acids in CFP, as measured by the PFR technique in broilers, are presented in Table 1. Data from the full set of ten measurements of standardized digestibility is shown in Table 1 (Parsons et al., 2023), and the nutrient composition and TMEn of eight samples are presented in Table 2. The coefficient of variation of the *in vivo* mean standardized digestibility of the ten samples of CFP is 2.3%, demonstrating the excellent consistency in the nutritional *in vivo* value of CFP produced from different sites and over a period of approximately 6 years. Standardized protein digestibility values and TMEn values have been used to formulate CFP feed for broilers and turkey poults.

A number of different trials have been carried out to determine the efficacy of CFP when included as an alternative protein source in feed formulations for poultry.

5 Case study: corn-fermented protein as a feed for broilers

The goal of the study was to identify the effect of feeding CFP as a protein supplement for broilers in all phases of growth up to slaughter at 42 days of age. A total of 432 1-day old male Ross 308 birds were randomly allocated to four experimental groups:

- Control (fed on a commercial broiler concentrate for the whole of the trial);
- 5% CFP (control + 5% CFP for the whole of the trial);
- 10% CFP (control + 10% CFP for the whole of the trial); and
- 10% CFP 22+ (control + 5% CFP 22+ for the grower phase and control + for the finisher phase).

All diets were formulated to be equivalent in standardized digestible amino acid content and energy content, as well as nutrient requirements recommended for male Ross broilers. The diets were offered as crumbs in the starter phase and pellets in the grower phase. The birds had free access to feed and water and were reared in 0.8 m × 0.8 m pens, with 9 birds per pen, and pens were bedded with wood shavings. Growth performance data (live weight, live weight gain, feed intake, and feed conversion ratio) was recorded weekly (Table 3). At the age of 35 days, three birds were selected randomly from each pen, euthanized, and used to evaluate carcass yield (Table 3).

Ileal digesta was collected from four birds per pen (pooled) at D21 and D42. Ileal tissue for histology was collected from one bird/pen at D21. Villus height and crypt depth were measured, and the ratio of villus height to crypt depth was calculated. The nutritional treatments had no effect on histological measurements. Dry matter content of the excreta at day 21 and day 42 was not significantly affected by the dietary treatment. Bird growth and feed intake were

Table 1 Standardized ileal digestibility of amino acids in samples of corn-fermented protein

	Univ Illinois samples 2017–2023																Univ Geoigia		
	CFP1		CFP2		CFP3		CFP4		CFP5		CFP6		CFP7		CFP8		CFP 9	CFP 9	
Amino acid	Digest vlue	SEM[3]	Digest. value	SEM[3]	Digest. value	SEM[3]	Digest. value	SEM[3]	Digest. value	SEM[3]	Digest. value	SEM[3]	Digest. value	SEM[1]	Digest. value	SEM[5]			Mean
ALA	91.5	0.8	92.6	0.6	94.7	0.3	91.8	0.9	91.1	0.8	90.0	1.6	91.7	0.4	91.6	0.8	89.5	88.6	
ARG	93.3	1.0	96.6	0.8	98.1	0.5	93.2	0.9	93.6	0.9	91.8	2.3	93.6	0.5	92.3	0.6	91.3	91.5	
ASP	87.7	1.1	87.8	1.3	91.3	0.8	86.1	1.4	84.6	1.3	85.7	1.9	87.2	0.6	87.9	2.0	85.0	84.2	
CYS	86.6	2.2	86.3	2.2	91.8	1.1	85.7	1.9	86.5	2.1	78.6	3.7	81.2	1.5	78.9	2.6	79.8	79.5	
GLU	93.1	0.7	93.7	0.6	96.1	0.2	93.1	0.8	92.2	0.7	91.0	1.7	92.8	0.5	92.9	1.3	91.5	90.3	
GLY		—		—		—		—		—		—		—		--	68.9	52.5	
HIS	93.2	1.2	90.6	0.8	92.8	0.3	89.2	1.2	87.2	1.1	89.5	1.8	90.4	0.6	91.1	2.2	89.7	90.2	
ILE	93.8	0.8	91.9	0.7	93.2	0.2	90.5	1.0	89.9	0.9	87.3	2.0	89.3	0.4	89.3	1.0	87.6	86.3	
LEU	9*3	0.5	95.1	0.4	97.1	0.1	94.4	0.6	93.9	0.6	92.9	1.4	96.7	0.3	94.0	0.6	92.7	92.1	
LYS	83.4	1.5	83.2	1.1	90.5	0.8	81.0	2.3	80.0	1.6	85.5	2.4	88.9	0.9	89.2	2.4	80.5	90.7	
WET	91.5	0.4	93.3	0.5	96.9	0.1	92.3	0.9	92.0	0.6	92.8	1.7	93.9	0.4	93.9	0.7	91.6	90.5	
PHE	92.6	0.8	93.4	0.6	96.1	0.1	92.6	0.8	92.4	0.6	90.6	1.7	92.8	0.3	91.9	0.6	93.6	90.1	
PRO	92.4	0.9	92.4	1.1	95.4	0.2	92.0	1.0	90.9	0.7	88.5	1.9	91.1	0.7	89.7	1.0	93.8	90.2	
SER	88.5	1.7	89.6	1.9	93.0	1.0	89.1	1.3	88.2	1.1	86.5	2.8	89.2	0.9	87.7	1.1	36.6	87.9	
THR	86.1	1.5	87.0	1.6	89.9	1.2	85.7	1.6	83.5	1.1	85.5	2.6	87.1	1.0	87.2	2.3	86.2	86.1	
TRP	94.7	0.9	96.6	0.8	93.6	0.7	92.9	1.2	92.5	0.5	90.2	1.7	90.8	1.1	90.4	1.5	86.3	86.6	
TYR	93.3	0.9	96.6	0.6	96.9	0.2	93.1	1.0	92.8	0.8	90.8	1.8	92.1	0.55	91.9	1.1	91.9	91.6	
VAL	88.8	1.1	89.6	1.1	95.3	0.3	88.8	1.2	87.6	0.8	89.0	2.0	90.8	0.5	89.9	0.7	87.8	86.3	
Mean	**90.4**		**91.2**		**94.3**		**90.1**		**89.2**		**88.6**		**90.4**		**90.0**		**87.0**	**86.3**	**89.7**
															ex GLY		88.1	88.3	cv 2.3

Table 2 Composition and TMEn of corn-fermented protein (CFP) (DM basis)

	CFP1	CFP2	CFP3	CFP4	CFP5	CFP6	CFP7	CFP8
Crude protein (%)	54.8	57.7	50.7	57.8	57.7	54.3	56.5	56.3
Crude fat (%)	7.1	4.6	5.0	4.2	4.0	4.7	3.5	5.0
Ash (%)	5.9	4.7	2.3	4.3	4.1	4.6	3.2	3.1
Neutral detergent fiber (%)	27.7	29.1	39.7	35.6	32.3	24.1	36.7	37.8
Ca (%)	0.05	0.02	0.02	0.02	0.02	0.06	0.03	0.03
P (%)	1.29	0.78	0.47	0.76	0.77	1.04	0.81	0.82
Na *(%)*	0.09	0.08	0.13	0.08	0.08	0.09	0.04	0.04
Gross energy (kcal/g)	5.46	5.43	5.35	5.50	5.55	5.38	5.43	5.43
TMEn (kcal/g)	3.55	3.55	3.29	3.77	3.72	3.39	3.68	3.50
SEM of TMEn	0.08	0.10	0.04	0.04	0.17	0.05	0.21	0.11

not significantly affected by the dietary treatment. However, birds given the diet with 10% CFP had numerically higher feed intake and FCR compared to the controls. Furthermore, birds receiving 5% CFP and 10% CFP had significantly higher breast meat yield compared to the control.

A second trial (Burton et al., in press) was sponsored by a major European commercial poultry producer and was carried out to test the use of CFP in diets for commercial broilers (Burton, personal communication). The objectives of the study were to determine the effects on growth of broilers to differing levels of CFP inclusion in diets with and without an extruded canola protein product. A total of 512-day-old broilers (half male and half female) were randomly selected in a 2 × 2 factorial trial. Four diets were prepared:

- A control (commercial broiler formulation);
- A diet with 5% CFP added;
- A diet with 7.5%, CFP added; and
- A diet with 10.0% CFP added.

The CFP replaced SBM in iso-nitrogenous and iso-energetic diets. The experimental concentrate was in the form of crumb (D0-D10) and pellet (D11-D34). The birds had free access to feed and water and were reared in 0.8 m × 0.8 m pens with wood shavings as bedding. Growth performance (live weight, live weight gain, feed intake, and feed conversion ratio) was recorded every 7 days during the trial. The litter score was recorded at D31, and litter samples were collected for dry matter determination (Table 4). At D21, two birds/pen were euthanized by cervical dislocation, and ileal digesta content was collected. Both ceca from one bird per pen were collected, snap-frozen immediately, and then stored at −80°C until analysis of volatile fatty acids (VFAs) was performed.

Table 3 Effect of corn fermented protein supplementation on growth performance and carcass yields of broilers: Cumulative performance

		Treatments					
Bird weight gain (g)		Control	5% CFP	10% CFP	10% CFP 22+	SEM	*P* value
	D14	499	503	518	501	5.73	0.088
	D21	1043^{b}	1050^{b}	1081^{a}	1067^{ab}	7.12	0.003
	D28	1785^{b}	1841^{ab}	1899^{a}	1832^{ab}	23	0.014
	D35	2539	2635	2652	2606	30.7	0.064
	D42	3315	3394	3295	3304	32.7	0.144
Feed intake (g)							
	D14	604	625	624	632	7.38	0.059
	D21	1333^{b}	1354^{ab}	1365^{ab}	1377^{a}	9.16	0.013
	D28	2436	2486	2527	2518	30.9	0.172
	D35	3657^{b}	3764^{ab}	3839^{a}	3792^{ab}	44.6	0.044
	D42	4878^{b}	5042^{ab}	5151^{a}	5066^{ab}	61.8	0.028
Feed conversion ratio							
	D14	1.21^{bc}	1.24^{ab}	1.20^{c}	1.26^{a}	0.01	0.001
	D21	1.28	1.29	1.26	1.29	0.007	0.055
	D28	1.37^{a}	1.35ab	1.33b	1.38a	0.008	0.002
	D35	1.44	1.43	1.45	1.46	0.012	0.408
	D42	1.47^{b}	1.49^{b}	1.57a	1.53ab	0.02	0.005
Carcass yield (g)							
Breast		742^{c}	826^{a}	797^{ab}	771^{bc}	14.1	0.001
Drumstick		298	300	296	286	4.63	0.145
Thigh		365	362	356	345	5.93	0.114

Means within a row with similar superscript are not significantly different at $P \leq 0.05$. $n = 12$
SEM, standard error of the mean

On day 34, two birds/pen were euthanized by cervical dislocation to record carcass yield. Litter score was recorded at D31, and litter samples were collected for dry matter determination. Caecal contents were collected for VFA analysis at D21 and D42 from one bird per pen and snap-frozen immediately on dry ice before storing at −80°C. Quantification of short-chain fatty acids in mammalian feces in aqueous solution content was carried out using gas chromatography–mass spectrometry (GC-MS) with a Shimadzu 2010ultra GC-MS single quadrupole mass spectrometer equipped with a Shimadzu GC2010 and AOC20i+s autosampler (Shimadzu Corp., Kyoto, Japan). The results indicate that the presence of CFP in the diets had significant effects on the proportions of individual short-chain volatile fatty acids in the cecal contents of the birds, and there was a tendency for total VFA concentration in cecal content to be increased (Table 5). The presence of CFP in the diet was influencing and increasing cecal fermentation.

The use of CFP was further evaluated in boilers in a second commercial trial using 4000 Hubbard M99 × Cobb 500 chicks supplied from a common breeder flock. The trial was designed to evaluate the live performance of broilers fed a standard small broiler control corn–soybean meal diet with diets containing CFP at inclusion levels: 3%; 6%; 9% used to replace SBM.

Table 4 Growth performance, meat yield and litter dry matter of male and female broilers

	Control	**Extrupro 10%**	**CFP 5%**	**CFP 7.5%**	**CFP 10%**	**7.5% CFP 7.5% + Extrupro 10%**
Female birds						
D0–D34 bird weight gain (g)	2124	2234	2196	2187	2188	2149
D0–D34 feed intake (g)	3084	3117	3105	3116	3171	3189
D0–D34 FCR	1.37	1.35	1.33	1.33	1.38	1.39
Meat yield (g)	844b	890ab	888ab	917a	896ab	902ab
Litter dry matter (%)	72.3b	73.8ab	75.8ab	73.9ab	76ab	76.4a
Male birds						
D0–D34 bird weight gain (g)	2485	2574	2511	2513	2495	2464
D0–D34 feed intake (g)	3507	3551	3546	3530	3606	3467
D0–D34 FCR	1.33	1.3	1.33	1.31	1.35	1.31
Meat yield (g)	1063ab	1055ab	1019b	1065ab	1093a	1051ab
Litter dry matter (%)	69c	70.5bc	74.7ab	75.8a	71.9abc	74.6ab

Means within a row with similar superscript are not significantly different at $P \leq 0.05$. $n = 32$

All diets were balanced for protein, standardized amino acid content, and TMEn content. With $n = 8$ replicates per treatment and 120 chicks per pen, the treatments were randomized within each block. All feeds were pelleted, and birds received starter diets from 1 to 14 days, finisher diets from 15 to 25 days, and withdrawal diets from 26 to 33 days of age. All birds were weighed at 33 days of age, and average weights and feed conversion ratios were calculated for each pen. The average final weight (kg) of the control 0, 3% CFP, 6% CFP, and 9% CFP were 2.22, 2.14, 2.15, and 2.12 respectively, and FCR 1.49, 1.50, 1.49, and 1.50 respectively, with mortality in the groups 0.5, 1.4, 1.0, and 1.5. There were no significant differences in growth, feed conversion, or mortality between the four treatments. The results indicated that CFP can be used as a partial replacement for SBM in the diet of broilers.

6 Case study: corn-fermented protein as a feed for turkey poults

Turkey poults have a reputation for having a highly sensitive digestive tract, and providing an appropriate starter diet is key to optimizing performance. A comparison was made between the use of SBM and partial replacement with CFP in starter diets for turkey poults (Scholey et al., 2023). CFP was used to replace approximately 6% and 12% of the SBM in the diets, with 4% and 8% of CFP, respectively. To formulate the diets, standardized ileal digestibility and TMEn were based on evaluations made in cecetomized cockerels (Table 1). All diets were formulated to commercial specifications, with pellet integrity between 98% and 99.3% for all experimental diets. The pellet quality of the experimental diet was not influenced by the inclusion of CFP, and CFP replaced approximately 6% and 12% of the SBM in the diets (4% and 8% CFP inclusion, respectively). Litter quality and foot pad score were assessed as simple indicators of gut health. The bedding remained friable and dry throughout the study, and no extra bedding was added to pens on any treatments, suggesting that there was no adverse effect of dietary treatment on gut health. The growth performance of poults from 0 to 42 days of age across all treatments aligned well with industry performance targets (Aviagen, 2014), indicating that the diets conformed well to industry standards (Table 6).

Although the five diets were formulated to supply equivalent quantities of standardized digestible amino acids and AMEn, poults fed the CFP8% diet were significantly heavier (plus ~8.2%) and gained more weight (plus ~8.4%) than those fed the control diet based on SBM (Table 6). However, there was no significant difference in feed intake. Based on the nutrient availability measured in the poults at the termination of the trial, poults offered CFP8% had significantly higher AME, AMEn, and *N* retention over the control poults by ~26%, 27%, and 13%, respectively. Moreover, the digestibility of VAL and ILE of CFP8% poults

Table 5 Effect of including CFP in the diets on hind gut volatile fatty acid content a+B6:L22nd composition

	Formic acid	Acetic acid	Propanoic acid	Isobutyric acid	Butyric acid	Iso valeric acid	Valeric acid	Iso. hexanoic acid	Hexanoic acid	Heptanoic acid
Female										
CONTROL	129[a]	5393[b]	374[b]	46[b]	1912[b]	25.7[b]	84.6[b]	15.2[a]	19.4[b]	19.6[a]
CFP 10%	149[a]	7099[a]	502[a]	75.4[a]	2785[a]	38.1[a]	136.7[a]	16.7[a]	23.8[a]	21.5[a]
Male										
CONTROL	149[a]	4762[a]	327[b]	45.5[b]	1576[a]	25.6[a]	85[b]	17.9[a]	21[a]	23.4[a]
CFP 10%	136[a]	5425[a]	459[a]	63[a]	1636[a]	34.7[a]	108.9[a]	16.5[a]	19.2[a]	20.8[a]
SEM	9.42	509	36.8	4.24	220	3.42	6.21	0.883	0.991	1.14
P value										
Diet	0.798	0.027	0.001	<0.001	0.044	0.003	<0.001	0.884	0.249	0.76
Sex	0.62	0.032	0.228	0.172	0.002	0.634	0.042	0.179	0.169	0.187
Diet × Sex	0.092	0.314	0.946	0.17	0.076	0.636	0.032	0.117	0.004	0.061

Means within a row with similar superscript are not significantly different at $P \leq 0.05$; $n = 32$
CFP10% = 10% of the total diet was corn fermented protein

were higher than the control poults (by ~11% and ~7%, respectively). Since there was no effect on feed intake, it is reasonable to conclude, therefore, that partial replacement of SBM by an apparently equivalent supply of protein and energy from CFP actually resulted in an increased supply of protein and energy to the poults. Thus, the supply of protein and energy of CFP8% was higher compared to the controls, which might explain the improvement in poult weight gain in CFP8% poults.

Both the CFP 4% and CFP8% diets had significantly higher AME compared to the control. In addition, CFP8% had significantly higher *N* retention compared to the control (Table 7).

7 Conclusion

There are currently 17 plants producing CFP in the USA, two in South America, and one in construction in Europe, with a combined production capacity of well over one million tons of CFP per annum, providing resilience and redundancy in the supply chain. Although the dry grind ethanol process is a global source of medium-quality protein, the opportunity to turn corn into a multi-purpose energy and protein crop is a major breakthrough. The trials carried out to evaluate the use of CFP as an alternative protein supplement in diets for broilers and turkey poults demonstrate that CFP can be used in the formulation as an alternative protein.

In order to complete these trials, standardized digestibility and true metabolizable energy were measured in the target species. In all the growth experiments, the maximum inclusion rate of CFP was restricted to not exceed 10% of dietary inclusion. Since these experiments were exploratory experiments to test the use of the product in diets for poultry and CFP contains approximately 20–27% yeast cell wall material, this was a precautionary measure to not exceed the yeast cell wall dietary inclusion level of 5%. CFP can be used as a dietary replacement for SBM, and with appropriate balancing of essential amino acids, it will achieve performance at least equal to the performance of the control diets. However, there is evidence to suggest that either the content of spent yeast material or the combination of spent yeast with a fermented carbohydrate fraction may be beneficial in terms of gut health.

The trial with turkey poults is particularly interesting. Overall, the growth performance of poults across all treatments aligned well with industry standards (Aviagen, 2014), indicating that the diets conformed well to industry standards. However, in terms of improved nitrogen end energy retention, there is evidence to suggest that there may be additional benefits to the partial replacement of SBM with CFP.

CFP contains a functional component of approximately 24% (on a DM basis) spent yeast content. Several nutraceutical components of yeast are

Table 6 Effect of corn-fermented protein on body weight, feed intake, feed conversion ratio and litter dry matter of turkey poults

	Control	CFP4%	CFP8%	Econ + CFP4%*	Premium	SEM	*P* value
D0 bird weight	66.2	66.2	65.6	65.5	65.3	1.05	0.962
D42 bird weight (g)	2328b	2423ab	2518a	2357b	2430ab	52.1	0.011
D0–D42 bird weight gain (g)	2262b	2357ab	2452a	2292ab	2365b	60.2	0.009
D0–D42 feed intake (g)	3741	3850	3743	3756	3842	77.3	0.363
D0–D42 feed conversion ratio	1.66	1.64	1.61	1.64	1.63	0.028	0.797
D42 litter dry matter	74	73.7	73.9	75.5	75.1	1.15	0.752
Carbon footprint (kg CO2 eq/kg bird weight gain)	4.31a	4.18a	3.99b	3.59c	2.85d	0.066	<0.001

*CFP added in economy diet at 4%
Means within a row with similar superscript are not significantly different at $P \leq 0.05$
SEM, standard error of the mean
Control = High protein with soybean meal as primary protein source
CFP4% = 4% of the total diet was corn fermented protein used in place of high protein soybean meal
CFP8% = 8% of the total diet was corn fermented protein used in place of high protein soybean meal
Econ+CFP = 4% of the total diet was corn fermented protein with lower protein, higher fiber soybean meal, as primary protein source
Premium = 5% of the total diet was a soy protein isolate in place of high protein soybean meal
N = 10 with 5 birds per pen

Table 7 Effect of corn fermented protein on apparent metabolizable energy, apparent metabolizable energy corrected for nitrogen and nitrogen retention of turkey poults

	AME (MJ/kg)	**N retention (g/ kg of the diet)**	**AMEn (MJ/kg)**
Control	10.1b	23.3b	9.3b
CFP4%	12.4a	25.3ab	11.6a
CFP8%	12.7a	26.3a	11.8a
Econ+CFP4%	9.7b	20.0c	9.0b
Premium	12.1a	23.5b	11.3a
SEM	0.54	0.86	0.52
P value	<0.001	<0.001	<0.001

*CFP added at 4% in economy diet
[a–c] Means within the same column with no common superscript differ significantly ($P \leq 0.05$)
SEM, standard error of the mean
Control = high protein with soybean meal as primary protein source
CFP4% = 4% of the total diet was corn fermented protein replacing soybean meal
CFP8% = 8% of the total diet was corn fermented protein replacing soybean meal
Econ+CFP4% = 4% of the total diet was corn fermented protein with lower protein, higher fiber soybean meal, as primary protein source
Premium = 5% of the total diet was a soy protein isolate replacing soybean meal protein
AME = Poultry apparent metabolizable energy
AMEn = apparent metabolizable energy corrected for nitrogen

reported to have beneficial effects in poultry and may indirectly improve growth performance. β-1,3-glucans, the major component of yeast cell walls, have prebiotic effects due to their ability to bind toxins and pathogens (Vetvicka et al., 2014). Yeasts contain mannan oligosaccharides which have an overall positive effect on animal performance (Spring et al., 2015) via improving intestinal architecture, physical gut tissue turnover, changes in microbiota, or reduction in immune stimulation (Baurhoo et al., 2007). The effects on cecal VFA concentrations reported in the broiler trial are evidence that the fermented fiber component of CFP has the potential to increase hindgut fermentation although to date, there is no evidence to indicate an improvement in the histology of the gut.

Identifying new sources of high-quality vegetable protein available for use in animal feed is a priority. High-quality vegetable proteins are extensively used for food and can therefore demand a price premium. Identifying a source of high-concentration vegetable protein, such as CFP, which is not available for use in food since it is derived from a non-food industrial process, is an opportunity for the feed industry to source an alternative cost competitive product. While SBM is recognized as an excellent protein source, residual trypsin inhibitor, lipoxygenases, and antigenic proteins retain a degree of activity post-processing, which slightly but significantly reduces dietary protein digestion (Clarke and Wiseman, 2005). Thus, some of the performance improvements in the current study could be related to the reduction in anti-nutritional effects associated with partial replacement of SBM with CFP. However, it is important to acknowledge the potential limitations of working with protein fractionation co-products derived from the ethanol industry. The ethanol industry gains significant financial benefits from pursuing oil recovery. Fractionation of whole stillage significantly increases the efficacy of oil recovery, which reduces the energy content of the resultant co-product. It is essential, therefore, to refer to the declared energy level in the co-products derived from different processes. In addition, corn protein products will have significantly higher levels of isoleucine and leucine compared to an equivalent level of protein from SBM. Indeed, the ratio of leucine plus isoleucine to lysine is approximately 6× greater in corn compared to an equivalent soybean protein-based product. Acknowledging potential amino acid imbalances, which can be rectified with amino acid supplementation, and differences in energy content is essential to achieve maximum performance from corn-based ethanol co-products.

The current program covering the use of CFP in poultry is part of an extensive program on the use of protein in diets for swine, aquaculture, companion animals, and ruminants, in addition to poultry. The results reflect the beneficial effects reported in poultry. The development of CFP is an excellent example of how the introduction of tested mechanical processing technology with existing commodity feed materials can significantly improve the value of low-quality feed material and increase the flexibility and value of feed materials used by animal nutritionists in feed formulation. There have been few if any, new and significant developments in the vegetable protein space for animal feed in the past 20 years. CFP is a major development in terms of a new commercially viable high-protein feed for all classes of livestock and companion animals.

8 References

Aviagen 2014. *Ross 308 Broiler: Nutrition Specifications* [WWW Document]. http://en.aviagen.com/assets/Tech_Center/Ross_Broiler/Ross308BroilerNutritionSpecs2014-EN.pdf (accessed December 12, 2016).

Baurhoo, B., Phillip, L. and Ruiz-Feria, C. A. 2007. Effects of purified lignin and mannan oligosaccharides on intestinal integrity and microbial populations in the ceca and litter of broiler chickens. Poultry Science 86(6). 1070–1078. https://doi.org/10.1093/ps/86.6.1070.

Xu, B., Li, Z., Wang, C., Fu, J., Zhang, Y., Wang, Y. and Lu, Z. 2020. Effects of fermented feed supplementation on pig growth performance: a meta-analysis. Animal Feed Science and Technology 259. 114315. https://doi.org/10.1016/j.anifeedsci.2019.114315.

Burton, E., Scholey, D., Alkhtib, A. and Williams, P. 2021. Use of an ethanol bio-refinery product as a soy bean alternative in diets for fast-growing meat production species: a circular economy approach. *Sustainability* 13(19). 11019.

Burton, E., Scholey, D., Alkhtib, A. and Williams, P. Alternative in diets for fast-growing meat production species. *Sustainability* 13(19). 11019.

Clarke, E. and Wiseman, J. 2005. Effects of variability on trypsin inhibitor content of soy bean meals on true and apparent ileal digestibility of amino acids and pancreas size in broiler chicks. *Animal Feed Science and Technology* 121(1-2). 125-138. https://doi.org/10.1016/j.anifeedsci.2005.02.012.

Han, J. and Liu, K. 2010. Changes in composition and amino acid profile during dry grind ethanol processing from corn and estimation of yeast contribution towards DDGS proteins. *Journal of Agricultural and Food Chemistry* 58(6). 3430-3437.

Parsons, C. M. 1985. Influence of caecectomy on digestibility of amino acids by roosters fed distillers dried grains with solubles. *Journal of Agricultural Science* 104(2). 469-472.

Parsons, B. W., Utterback, P. L., Parsons, C. M. and Emmert, J. L. 2023. Standarized amino acid digestibility and true metabolizable energy for several increased protein ethanol co-products using back-end fractionation systems. *Poultry Science* 102(2). 102329.

Mottet, A., de Haan, C., Falcuucci, A., Tempio, G., Opio, C. and Gerber, P. 2017. Livestock: on our plates or eating at our table? A new analysis of the feed/food debate. *Global Food Security* 14. 1-8. https://doi.org/10.1016/j.gfs.2017.01.001.

Scholey, D., Alkhtib, A., Wiliams, P. and Burton, E. 2023. Corn fermented protein, an alternative protein to soybean meal to support growth in young turkey poults. *Journal of Applied Animal Nutrition*. https://doi.org/10.3920/JAAN2023.0002.

Shurson, G. C. 2018. Yeast and yeast derivatives in feed additives and ingredients: sources, characteristics, animal responses and quantification methods. *Animal Feed Science and Technology* 235. 60-76. https://doi.org/10.1016/j.anifeedsci.2017.11.010.

Spring, P., Wenk, C., Connolly, A. and Kiers, A. 2015. A review of 733 published trials on Bio-Moss, a mannan oligosaccharide and Actigen, a second generation mannose rich fraction, on farm and companion animals. *Journal of Applied Animal Nutrition* 3. 1-11.

Vetvicka, V. and Oliveira, C. 2014. β(1-3)(1-6)-D-glucans modulate immune status in pigs: potential importance for efficiency of commercial farming. *Annals of Translational Medicine* 2(2). 16. https://doi.org/10.3978/j.issn.2305-5839.2014.01.04.

Williams, P. E. V. 2023. High protein corn products for swine derived from corn ethanol production. In press.

Index

www.ingramcontent.com/pod-product-compliance
Ingram Content Group UK Ltd.
Pitfield, Milton Keynes, MK11 3LW, UK
UKHW021008120926
8556IPUK00028B/142

9 781801 467315